Signalling Mechanisms – from Transcription Factors to Oxidative Stress

NATO ASI Series

Advanced Science Institutes Series

A series presenting the results of activities sponsored by the NATO Science Committee, which aims at the dissemination of advanced scientific and technological knowledge, with a view to strengthening links between scientific communities.

The Series is published by an international board of publishers in conjunction with the NATO Scientific Affairs Division

A Life Sciences B Physics	Plenum Publishing Corporation London and New York
C Mathematical and Physical Sciences D Behavioural and Social Sciences E Applied Sciences	Kluwer Academic Publishers Dordrecht, Boston and London
F Computer and Systems Sciences G Ecological Sciences H Cell Biology I Global Environmental Change	Springer-Verlag Berlin Heidelberg New York London Paris Tokyo Hong Kong Barcelona Budapest

PARTNERSHIP SUB-SERIES

1. Disarmament Technologies	Kluwer Academic Publishers
2. Environment	Springer-Verlag
3. High Technology	Kluwer Academic Publishers
4. Science and Technology Policy	Kluwer Academic Publishers
5. Computer Networking	Kluwer Academic Publishers

The Partnership Sub-Series incorporates activities undertaken in collaboration with NATO's Cooperation Partners, the countries of the CIS and Central and Eastern Europe, in Priority Areas of concern to those countries.

NATO-PCO DATABASE

The electronic index to the NATO ASI Series provides full bibliographical references (with keywords and/or abstracts) to about 50000 contributions from international scientists published in all sections of the NATO ASI Series. Access to the NATO-PCO DATABASE compiled by the NATO Publication Coordination Office is possible in two ways:

- via online FILE 128 (NATO-PCO DATABASE) hosted by ESRIN, Via Galileo Galilei, I-00044 Frascati, Italy.
- via CD-ROM "NATO Science & Technology Disk" with user-friendly retrieval software in English, French and German (© WTV GmbH and DATAWARE Technologies Inc. 1992).

The CD-ROM can be ordered through any member of the Board of Publishers or through NATO-PCO, Overijse, Belgium.

Series H: Cell Biology, Vol. 92

Signalling Mechanisms – from Transcription Factors to Oxidative Stress

Edited by

Lester Packer

Department of Molecular and Cell Biology
University of California, 251 Life Science Addition
Berkeley, CA 94720, USA

Karel W. A. Wirtz

Centre for Biomembranes and Lipid Enzymology
Utrecht University, Padualaan 8
3584 CH Utrecht, The Netherlands

Springer
Published in cooperation with NATO Scientific Affairs Division

Proceedings of the NATO Advanced Study Institute on Molecular Mechanisms of Transcellular Signalling: from the Membrane to the Gene, held at the Island of Spetses, Greece, August 15–27, 1994

ISBN-13: 978-3-642-79677-7 e-ISBN-13: 978-3-642-79675-3
DOI: 10.1007/ 978-3-642-79675-3

Library of Congress Cataloging-in-Publication Data
Signalling mechanisms – from transcription factors to oxidative stress
/ edited by Lester Packer, Karel W. A. Wirtz.
P. cm. – (NATO ASI series. Series H, Cell biology ; v. 92)
Includes bibliographical references and index.

1. Cellular signal transduction. 2. Transcription factors. 3. Active oxygen.
I. Packer, Lester. II. Wirtz, Karel W. A. III Series.
QP517.C45S56 1995
574.87'6 – dc20 95-10171 CIP

Softcover reprint of the hardcover 1st edition 1995

Typesetting: Camera ready by authors/editors
SPIN 10133237 31/3136 - 5 4 3 2 1 0 - Printed on acid-free paper

PREFACE

A NATO Advanced Study Institute on "Molecular Mechanisms of Transcellular Signaling: from the Membrane to the Gene" was held on the Island of Spetsai, Greece, from August 15–27, 1994. The aim of this Institute was to bring together researchers in the field of signal transduction mechanisms, transcription factors and gene regulation with those actively involved in studies on the implications of oxygen radicals and antioxidant defence mechanisms for cell function. As diverse as these fields may be, the emergence of their interconnection during the course of the Institute was an eye-opener for students and lecturers alike.

Presentations and discussions focussed on the role of Ca^{2+}, G-proteins, protein kinase C and phospholipases in signaling mechanisms. These broad principles were extended to transcription factors and gene regulation with an emphasis on the steroid hormone receptor superfamily and NFκB. Basic principles of free radical formation and antioxidant action (vitamin E and C) were presented and discussed in connection with effects on signaling pathways.

This book present the content of the major lectures and a selection of the most relevant posters. These proceedings offer a comprehensive account of the most important topics discussed at the Institute. The book is intended to make the proceedings accessible to a large audience.

January 1995 The Editors

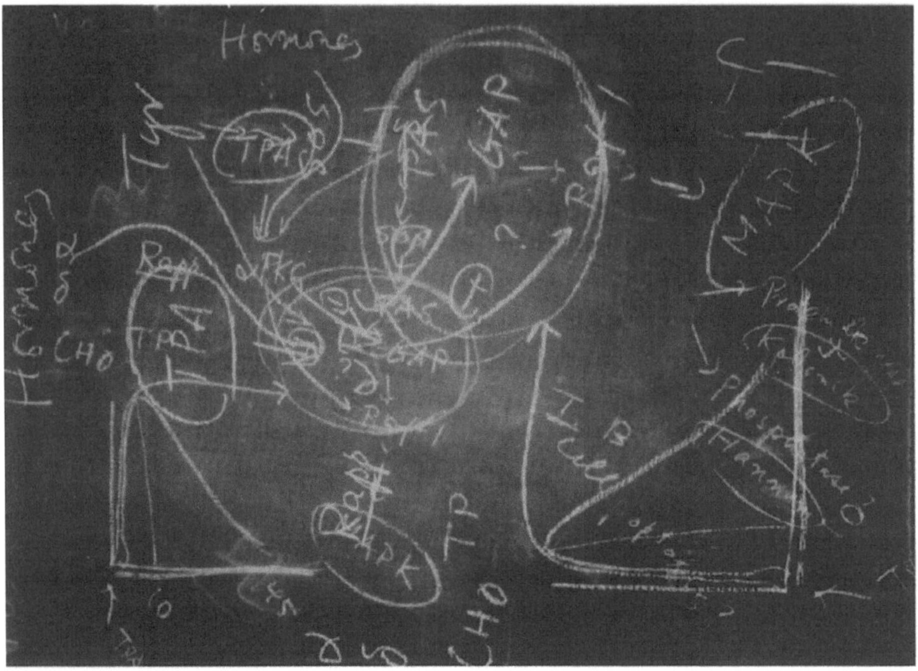

Chalk drawings on blackboard after a round-table discussion between Professor Nishizuka and the students during the 1994 ASI on "Molecular Mechanisms of Transcellular signaling: from the Membrane to the Gene", Spetses.

CONTENTS

COMPONENTS OF SIGNAL TRANSDUCTION

TRANSCRIPTION FACTORS AND GENE INDUCTION

VARIOUS FUNCTIONS OF MEMBRANE LIPIDS

ADDENDUM

The $[Ca^{2+}]_i$ Concept: from whole Cell to Microdomains

Jacopo Meldolesi and Fabio Grohovaz
Department of Pharmacology,
CNR and B. Ceccarelli Centers,
and DIBIT, S. Raffaele Scientific Institute,
Via Olgettina 58,
20132 Milano,
Italy

INTRODUCTION

The discovery that Ca ions play key roles in biology, not only as cations that contribute to the generation of currents and to the control of potential at the plasma membrane, but also in the regulation of intracellular functions and activities, stems from the pioneer experiments of Ringer in the late Century and has been uninterruptedly reinforced and expanded even since (see Pietrobon et al., 1990). The mechanisms of the second messenger function of Ca^{2+}, however, have begun to be unravelled in the early seventies. Only at that time, in fact, the basic distinction was made between the total calcium content of the cells, which is high (mmoles/liter, in the same range as the concentration of the cation in physiological extracellular fluids), and the concentration of the free cation in the cytosol ($[Ca^{2+}]_i$) which, because of its equilibrium with physiological activity regulators (proteins, enzymes etc.), is considered as the main actor in cellular control (Pietrobon et al., 1990; Carafoli, 1987; Pozzan et al., 1994). The first measurements of $[Ca^{2+}]_i$, carried out by microimpalement of large cells with Ca^{2+} specific electrodes, revealed unexpectedly low $[Ca^{2+}]_i$ levels, i.e. four orders of magnitude lower than the total cell calcium content. From those measurements the idea first emerged that cells contain molecules and structures capable of binding and segregating Ca^{2+}. The level of what remains free in the cytosol, i.e. $[Ca^{2+}]_i$, is maintained in a relatively narrow range as the result of complex and dynamic equilibria (Pietrobon et al., 1990; Carafoli, 1987; Pozzan et al., 1994). In the following years these seminal observations and ideas have been confirmed by an

NATO ASI Series, Vol. H 92
Signalling Mechanisms – from Transcription Factors
to Oxidative Stress
Edited by L. Packer, K. Wirtz

interrupted stream of findings. Here we would like to emphasize only two of them: the development and introduction into biological research of the fluorescent Ca^{2+} dyes that, beginning in 1982, have revolutioned the field since they made possible $[Ca^{2+}]_i$ measurements in all cells, no matter of their size and origin; and the identification in the cytosol of Ca^{2+} binding proteins, with their classification into two main families: the E-F hand-containing proteins (over 260 proteins nowadays!) and the annexins (Huzinken, 1994). Since both these families of proteins bind Ca^{2+} with high affinity (Kd_S in the 10^{-7} - 10^{-6} range, i.e. the $[Ca^{2+}]_i$ range revealed experimentally by the dyes in intact cells), the opinion was tacitly shared that all the Ca^{2+} -controlled events of the cytosol had to be of high affinity. As usual, diverging opinions about the existence in the cytosolic compartment of low affinity Ca^{2+} controlled phenomena were expressed also at that time. They however failed to have much impact as long as they were supported more by hypotheses than by direct experimental evidence. In addition, it was difficult to envisage the existence within the cell of cytosolic microdomains containing high $[Ca^{2+}]$ while the average $[Ca^{2+}]_i$, revealed by the dyes, was measured in the usual 10^{-7} - 10^{-6} M range. After all, the diffusion of Ca^{2+} in a solution is fast (~ 350 $\mu m^2/sec$) and no mechanisms could therefore be envisaged that could restrain the rapid dissipation of high $[Ca^{2+}]_i$ domains, if formed.

In the last few years the situation has profoundly changed, and therefore microdomains and low affinity Ca^{2+} regulated functions are now accepted. Such a progress has been due to various events, two of which we would like to mention here in some detail. The first is the recognition that Ca^{2+} diffusion in the cytosol is 10-100 fold slower than in solution (Tsien and Tsien, 1990; Albritton et al., 1992). Indeed, when Ca^{2+} ions are transferred to the cytosol they get immediately in equilibrium with the pool bound to cytosolic molecules which have on the average slow mobility. As a consequence, and contrary to what previously believed, Ca^{2+} ions move around only slowly in the cytosol, and therefore microdomains of high (or low) $[Ca^{2+}]_i$ can be maintained for quite sometime. The second point is technological and concerns the development of Ca^{2+} dyes characterized by properties appropriate for ratiometric measurements (Grynkiewicz et al., 1985). When analyzed with single cell equipments of high temporal and special resolution, these dyes provide accurate measurements throughout the entire cell, in spite of possible differences in the fluorescence intensities in various areas. Results obtained by this approach and by others developed concomitantly have finally provided the expected demonstration of the existance in the cytosol of high $[Ca^{2+}]_i$ microdomains. In the present review we will deal briefly with four different microdomain examples and will then consider the problem collectively, in the framework of the overall cell functioning.

THE NEUROMUSCULAR JUNCTION AND (MOST LIKELY) ALSO THE OTHER SYNAPSES

The neuromuscular junction has been for decades, and still remains, a model for synaptic structure and function. Classical studies carried out over twenty years ago by electrophysiology and electron microscopy had already revealed two important aspects of the system. First, release of neurotransmitter (acetylcholine, documented by the appearance of the endplate potential in appropriately impaled muscle fibers) does occur almost instantaneously (sub-msec delay) after stimulation and Ca^{2+} floading of the presynaptic compartment; second, the clear synaptic vesicles which are known to contain the neurotransmitter are distributed in the nerve terminal not at random but according to a dual criterion. They are in fact either attached (docked) at the presynaptic membrane, especially at the level of specific thickenings, the so called active zones; or distributed at some distance from the presynaptic membrane. Freeze-fracture of neuromuscular junctions revealed another important property of this synapse. The active zones, previously described at the conventional electron microscope as the sites of docking and fusion of synaptic vesicles, were shown to correspond to double rows of large intramembrane particles running perpendicular to the major axis of the nerve terminal (Fig.1; see also Meldolesi and Ceccarelli, 1981; Torri Tarelli et al., 1990). Already in the mid seventies the aligned particles were hypothesized to be themselves transmembrane Ca^{2+} channels, however direct immunocytochemical evidence was obtained only ~ 15 years later, after these channel were recognized to differ from those known since longer time, the L type, and to be in contrast N, P and recently also Q type channels (Llinas et al., 1992; Valtorta and Meldolesi, 1994).

The possibility that the $[Ca^{2+}]$ rises induced by Ca^{2+} channel activation in the cytosolic rim directly adjacent to the plasmalemma were higher than in the rest of the cytoplasm was not entirely new. Indeed, patch clamp results had already revealed that, after moderate stimulation of various cell types, the Ca^{2+}-dependent K^+ channels could get activated in the plasmalemma before (of even independently) of any measurable changes of average $[Ca^{2+}]_i$. At the neuromuscular junction, however, the issue was more complex because a function different from membrane potential, i.e. synaptic vesicle release, appeared to be coupled to the formation of the Ca^{2+} microdomain localized adjacent to the presynaptic membrane. Shortly thereafter, experimental evidence obtained in another, most favourable synapse, the giant synapse of the squid, demonstrated that indeed the microdomain builds up upon stimulation, and estimated its $[Ca^{2+}]$ well in the 10^{-5} and possibly even the 10^{-4} M range (Llinas et al., 1992). Concomitantly, biochemical studies of synaptic vesicle proteins revealed some of them (for example synaptotagmin) to be Ca^{2+} binding proteins, however of low affinity (Valtorta and Meldolesi, 1994; Brose et al., 1992). The picture that emerges from these studies is that of a

Fig.1 Freeze-fracture of the presynaptic membrane of the frog neuromuscolar junction stimulated with alfa latrotoxin. The pits visible in the figure (arrows) correspond to discharged synaptic vesicles. Notice their distribution lateral to the perpendicular double rows of large intramembrane particles that mark the localization of the active zone.

multistep release process in which the vesicles are first docked at the active zone. Under these conditions their fusion is prevented by interaction of their proteins with those of the plasmalemma. After stimulation, the high [Ca^{2+}] microdomain that is formed locally due to the opening of the active zone Ca^{2+} channels releaves this inhibition, and the vesicles can thus fuse (Valtorta and Meldolesi, 1994; Thomas et al., 1993). Interestingly, the Ca^{2+} that diffuses from the microdomain into the rest of the terminal has additional functions. Among these are the release from the cytoskeleton of the vesicles maintained at some distance from the active zones. Such a release is necessary to replace at the docking sites the vesicles already discharged; and to elicit the discharge of the second type of synaptic secretory organelles, the dense-core vesicles, which contain primarily peptides. Without the establishment of the high [Ca^{2+}] microdomain adjacent to the presynaptic membrane, however, no rapid (within hundreds of μsec!) release of the classical neurotransmitters (amines and aminoacids contained within the clear synaptic vesicles, different depending on the specificity of the various synapses) would ever occur. The rapid functioning of our nervous system depends therefore from the slow diffusion of Ca^{2+} in the cytosol!

THE PURKINJE NEURONS OF BIRDS

This story refers to Purkinje neurons because they possess peculiar molecular and structural properties and because they have been more deeply investigated with respect to other neurons and non neuronal cells. Experiments have been carried out by immunocytochemistry, focussed on the characterization of the rapidly exchanging Ca^{2+} stores, i.e. the organelles able to rapidly accumulate Ca^{2+} by pumping it from the cytosol to their lumen; to store it in direct association with Ca^{2+} binding protein(s) of adequate properties; and to release it by the activation of channel(s) localized in their limiting membrane. Stores are localized in the cell body, in the dendrites with their spines or in the axons up to the nerve terminals. The basic question was whether at all these sites the stores contain in similar portion all the essential components: pumps, buffering proteins and channels, and thus presumably operate in similar fashion in order to control the Ca^{2+} homeostasis. The results we have obtained can be summarized in two conclusions. First, we have shown that the stores, contrariwise to a widely accepted view, do not coincide with the entire endoplasmic reticulum but correspond to specialized areas of that endomembrane system; second, the areas competent in Ca^{2+} storage were found to be by no means homogeneous (see Sitia and Meldolesi, 1991; Pozzan et al., 1994).

In bird Purkinje neurons flattened, smooth cisternae, sitting one on top of the other to yield peculiar stacks, were shown to be extremely rich (500/μm^2) of the receptors/channels activated by the second messenger, inositol 1,4,5-trisphosphate (IP_3), but relatively poor of the lumenal Ca^{2+} buffering protein, calsequestrin (Fig.2; Villa et al., 1991; Volpe et al., 1991). Release from these stores is therefore expected to occur at relatively low [IP_3], to be quick, and to subside quickly due to the small quantity of Ca^{2+} segregated wihtin the lumen in association with the scarce Ca^{2+} buffering protein. Interestingly, within the dendritic spines, i.e. the sites where most impinging synapses are addressed to, the stores were only of this type, i.e. IP_3 receptors were numerous and calsequestrin almost inappreciable (Villa et al., 1991; Volpe et al., 1991). In contrast, other stores contained abundant calsequestrin in their lumen and just a few channels in their membrane (Fig.2; Volpe et al., 1991). Also these latter stores (that we have named the calciosomes) were found to be heterogeneous. Some of them did in fact express the IP_3, others the ryanodine receptor (Volpe et al., 1991). The latter is another intracellular channel activated by brisk $[Ca^{2+}]_i$ changes (calcium-induced-Ca^{2+}-release) as well as by another putative second messenger, cyclic ADP ribose. Because of their properties the calciosomes are expected to release considerable amounts of Ca^{2+} when specifically stimulated. Thus, their release appears, at least in principle, appropriate to generate local high Ca^{2+} microdomains. Some functional consequences of the heterogeneity of intracellular Ca^{2+} stores are discussed in the following section.

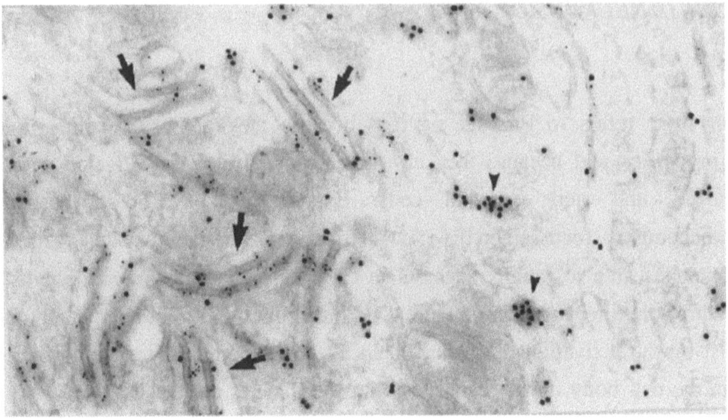

Fig. 2 Ultrathin cryosection of chicken Purkinje doubles immunolabeled for the IP_3 receptor (small gold) and the lumenal Ca^{2+} binding protein, calsequestrin (large gold). Notice that the ER cisternal stacks visible to the left of the figure (arrows) are very rich of the receptor but contain little calsequestrin, whereas the calciosomes (arrow heads) are particularly rich in the latter. The structures to the right are conventional ER cisternae, rich in neither one of the immunorevealed antigens. Courtesy of Dr. Antonello Villa.

$[Ca^{2+}]_i$ OSCILLATIONS: A HIGH $[Ca^{2+}]$ MICRODOMAIN AS THE TRIGGER

The fact that in many (but not all) cells, after moderate stimulation or even at rest, $[Ca^{2+}]_i$ does not remain constant but exhibits a rythmic spiking activity is well known. Less clear are the mechanisms that sustain this latter activity. $[Ca^{2+}]_i$ imaging experiments have recently provided new information on this matter. In individual cells of various types, the oscillations have in fact been seen to initiate from a single site, the oscillator or trigger zone, and to spread to the rest of the cell only subsequently (D'Andrea et al., 1993; Kasai et al., 1993). Indeed, not all oscillations were seen to spread. Others did in fact proceed only up to an area around the trigger zone and then got estinguished, thus explaining the irregularity of the oscillations observed in some cells. An explanation of these processes that is gaining momentum is the following. Among the rapidly exchanging Ca^{2+} stores discussed in the preceding session one or a group are particularly sensitive to IP_3. Even if the $[IP_3]$ is constant at the resting or at only moderately increased values, the probability that this store becomes activated, and thus releases its segregated Ca^{2+}, are higher than those of the other stores of the cell. Such an interpretation is not only theoretical because experimental evidence which demonstrates the heterogeneity in the store responsiveness to IP_3, with high sensitivity in the area of the trigger zone (located at the apex), has been recently shown in acinar cells of the exocrine pancreas (Kasai et al., 1993). The rationale of the trigger zone function has emerged from the observation that moderate increases of $[Ca^{2+}]_i$ (peak at 300 nM) modulate positively the responsiveness of IP_3 receptors to their ligands (Bezprozvanny et al., 1991). Discharge of the trigger, therefore, is important not much per se but because the adjacent stores become sensitized, and thus able to release their Ca^{2+}. By this process sensitization expands. Step by step, the process proceeds non-decrementally, and finally invades the entire cell. On the other hand, the activation of IP_3 receptors is transient because high $[Ca^{2+}]_i$ values (above 0.5 μM) are inhibitory of its function. After release the stores are therefore made quiet and have time to recover, i.e. to reaccumulate Ca^{2+} and get ready for the next spike.

As already mentioned not all cells express oscillations, others exhibit waves, i.e. non-decremental processes characterized by their distinct orientation. Recent results have demonstrated the mechanisms of waves to resemble those of oscillations since both these processes appear to be based on the sequential activation of IP_3 receptors (Lorenzon et al., 1994). Although apparently bizarre and initially also unexpected, the transient $[Ca^{2+}]_i$ increases have been shown to be positive for the cell inasmuch as they are energetically favourable and also less life-thretening than single persistent $[Ca^{2+}]_i$ increases. Their occurrance therefore provides the cell with new, dynamic regulatory aspects of Ca^{2+} homeostasis.

MITOCHONDRIA AS $[Ca^{2+}]_i$ SENSORS

The fourth example of $[Ca^{2+}]$ microdomain concerns mitochondria. Back to the seventies these organelles were believed to be the major Ca^{2+} stores in the cells. More recently, their accumulation process (working at the expenses of the internal membrane potential) was shown to bind Ca^{2+} with relatively low affinity (10^{-6}-10^{-5} M) and to be therefore inadequate to keep $[Ca^{2+}]_i$ in the 10^{-7} M level. Interest about mitochondria therefore decreased although it remained clear that a few matrix dehydrogenases of the Krebs cycle require Ca^{2+} to operate. In active mitochondria the matrix $[Ca^{2+}]$, $[Ca^{2+}]_m$, needs therefore to reach relatively high (μM) values (Carafoli, 1987; Pozzan et al., 1994).

The big change in the Ca^{2+} mitochondria field occurred two years ago when Rosario Rizzuto and Tullio Pozzan succeded in expressing in various cell types a construct in which the cDNA of a photoprotein, aequorin (which emits light when bound to Ca^{2+}), had been coupled with the mitochondrial targeting sequence of a cytochrome oxidase subunit (Rizzuto et al.,

1992). In these cells therefore the expressed aequorin was segregated into mitochondria and, under appropriate conditions, could be used to measure $[Ca^{2+}]_m$. From the very beginning of these studies $[Ca^{2+}]_m$ was shown to be not stable but variable following the changes of $[Ca^{2+}]_i$ (Rizzuto et al., 1992). Moreover, the aequorin-expressing mitochondria, distributed all around within the cell, were found to serve an unexpected and precious function: that of $[Ca^{2+}]$ sensors in the microdomains of the cytosol immediately adjacent to the mitochondria themselves. The results obtained were exciting. By following in parallel the aequorin (mitochondria) and fura-2 (cytosol) signals it was in fact revealed that increases of average $[Ca^{2+}]_i$ triggered by Ca^{2+} influx from the extracellular medium induced effects on the $[Ca^{2+}]_m$ much smaller than those induced by apparently identical $[Ca^{2+}]_i$ increases generated however from the release of Ca^{2+} from intracellular stores. Even when the "cytosolic" $[Ca^{2+}]$ was clamped at 2 μM by bathing permeabilized cells in a Ca-EGTA buffer, exposure to IP_3 trigged the generation of robust $[Ca^{2+}]_m$ signals. Release of Ca^{2+} from IP_3 sensitive stores located adjacent to the mitochondria was in fact shown to induce a rapid accumulation of the cation into the matrix, across the internal membrane of the mitochondria, in spite of the low affinity of the uptake system into those organelles. These results clearly indicate that the cytoplasmic area around the mitochondria can reach $[Ca^{2+}]$ values well in the μM range, even when the average $[Ca^{2+}]_i$ signal does not exceed 0.5-0.6 μM. The existence of the cytosolic microdomains has thus been directly demonstrated (Rizzuto et al., 1993 and 1994). Their functional significance remains however only partially known. Do microdomains only serve to supply Ca^{2+} to mitochondria? Alternatively they might turn out to be highly important also for a variety of additional cell functions, which however have not been identified yet.

CONCLUSION

The four examples of Ca^{2+} microdomains in the cytosol of eukaryotic cells we have reported in some detail have modified profoundly our appreciation of the $[Ca^{2+}]_i$ dynamics, and changed also the approach of many experiments. Until recently it was common in the field to attempt straight correlations between average $[Ca^{2+}]_i$ and activities observed in a cell after stimulation. Correlations of this kind should now be considered with caution, taking into account not only the existence of microdomains of high $[Ca^{2+}]$ that have only moderate impact on average $[Ca^{2+}]_i$, but also their different localization within the cell: at the surface in case of stimulated influx; in the depth of the cytoplasm (where precisely?) when are the stores to be activated. Two other concepts need to be more accurately focussed in the microdomain problem. First, nothing excludes that intracellular Ca^{2+} stores move in the cytoplasm depending on the physiological activity of the cell, thus changing the site where microdomains are localized. So far the problem has been given only limited attention. The exclusion from dendritic spines of the calsequestrin-rich calciosomes of Purkinje neurons (Volpe et al., 1991) demonstrates however that specific localizations exist. Nothing in a cell is built to stay immobile for ever. The second concept concerns the composition of the cytosol. Expression of the various cytosolic high affinity Ca^{2+} binding proteins changes very much among cells. For example, Purkinje neurons express large amounts of at least two high affinity, E-F hand cytosolic proteins, calbindin and parvalbumin, whereas their neighbours, the granule neurons, express another such protein, calretinin. Moreover, some pyramidal cells of the hippocampus express calbindin, other don't (Villa et al., 1994). The reason of these heterogeneities is not clear. What is clear however is that the function attributed to the proteins, Ca^{2+} buffering in the cytosol, can change considerably in extent form cell to cell and even in the course of the life of a single cell. Although direct experimental evidence is still limited, these changes are expected to modify considerably the rate of Ca^{2+} diffusion in the cytosol, and thus the ability of a cell to establish and maintain Ca^{2+} microdomains. The type of cell heterogeneity discussed here is certainly quite different from those revealed by classical cytology. Still, it is certainly sufficient to contribute significantly to the control of numerous and fundamental cell functions.

REFERENCES

Allbritton NL, Meyer T and Stryer L (1992) Range of messenger action of calcium ion and inositol 1,4,5-trisphosphate. Science 258:1812-1815

Bezprozvanny I, Watras J and Ehrlich BE (1991) Bell-shaped calcium response curves of Ins(1,4,5)P_3- and calcium gated channels from endoplasmic reticulum of cerebellum. Nature 351:751

Brose N, Petrenko AG, Südhof TC and Jahn R (1992) Synaptotagmin: a calcium sensor on the synaptic vesicle surface. Science 256:1021-1025

Carafoli E (1987) Intracellular calcium homeostasis. Ann. Rev. Biochem. 56:395

Celio MR (1990) Calbindin D-28K and parvalbumin in the rat nervous system. Neuroscience 35:375-475

D'Andrea P, Zacchetti D, Meldolesi J, and Grohovaz F (1993) Mechanisms of $[Ca^{2+}]_i$ oscillations in rat chromaffin cells. J. Biol. Chem. 268:15213-15220

Grynkiewikz G, Poenie M, and Tsien RY (1985) A new generation of Ca^{2+} indicators with greatly improved fluorescence properties J. Biol. Chem. 260:3440-3450

Hunziker W (1994) Ca^{2+} binding proteins. Biochem. Biophys. Acta, Special Issue, in press.

Kasai H, Xin Li Y, and Miyashita T (1993) Subcellular distribution of Ca^{2+} release channels underlying Ca^{2+} waves and oscillations in exocrine pancreas. Cell 74:669-677

Llinas R, Sugimori M, and Silver RB (1992) Microdomains of high calcium concentration in a presynaptic terminal. Science 256:677

Lorenzon P, Zacchetti D, Codazzi F, Fumagalli G, Meldolesi J, and Grohovaz F (1994) Ca^{2+} waves in neurites: a bidirectional, receptor-orientated form of Ca^{2+} signalling. Submitted

Meldolesi J, and Ceccarelli B (1981) Exocytosis and membrane recycling. Trans. Royal Soc. London B 296:55-65

Pietrobon D, Di Virgilio F, and Pozzan,T (1990) Microdomains and functional aspects of calcium homeostasis in eukaryotic cells. Eur. J. Biochem. 193:599

Pozzan T, Rizzuto R, Volpe P, and Meldolesi J (1994) Molecular and cellular physiology of intracellular Ca^{2+} stores Physiol. Rev. In press

Rizzuto R, Simpson AWM, Brini M, and Pozzan, T (1992) Rapid changes of mitochondrial Ca^{2+} revealed by specifically targeted recombinant aequorin. Nature (Lond.) 358:325-327

Rizzuto R, Brini M, Murgia M, and Pozzan T (1993) Microdomains with high Ca^{2+} close to IP_3-sensitive channels that are sensed by neighboring mitochondria. Science 262:744-747

Rizzuto R, Bastianutto C, Brini M, Murgia M, and Pozzan T (1994) Mitochondrial Ca^{2+} homeostasis inintact cells. J. Biol. Chem. In press

Sitia R, and Meldolesi J (1992) The endoplasmic reticulum: a dynamic patchwork of specialized subregions Mol. Biol. Cell. 3:1067-1072

Thomas P, Wong JG, Lee AK, and Almers W (1993) A low affinity Ca^{2+} receptor controls the final steps inpeptide secretion from pituitary melanotrophs. Neuron 11:93-104

Torri Tarelli F, Valtorta F, Villa A, and Meldolesi J (1990) Functional morphology of the nerve terminal at the frog neuromuscolar junction: recent insights using immunocytochemistry. Prog. in Brain Res. 84:83

Tsien RW and Tsien RY (1990) Calcium channels, stores and oscillations. Annu Rev. Cell Biol. 6:715

Valtorta F, and Meldolesi J (1994) The presynaptic compartment: signals and targets. Sem. in Cell Biol. In press

Villa A, Podini P, Clegg DO, Pozzan T, and Meldolesi J (1991) Intracellular Ca^{2+} stores in chicken Purkinje neurons. J. Cell Biol. 133:779

Villa A, Podini P, Panzeri MC, Racchetti G, and Meldolesi J (1994) Cytosolic Ca^{2+} binding proteins during rat brain ageing: loss of calbindin and calretinin in teh hippocampus, with no change in the cerebellum. Eur. J. Neurosci. In press

Volpe P, Villa A, Damiani E, Sharp AH, Podini P, Snyder S, and Meldolesi J (1991) Heterogeneity of microsomal Ca^{2+} stores in chicken Purkinje neurons. EMBO J. 10:3183-3189

Versatile functions of Ca^{2+}-binding proteins in signal transduction and Ca^{2+} homeostasis

T.L. Pauls
Institute of Histology and General Embryology
University of Fribourg
Pérolles
1705 Fribourg
Switzerland

Ca^{2+} signaling and cytosolic Ca^{2+}-binding proteins

Transient changes in the intracellular Ca^{2+} concentration play an important role in triggering signal transduction pathways. Depending on the stimulus and mode of Ca^{2+} entry into the cytosol, distinct signaling pathways (e. g. via Ca^{2+}/calmodulin-dependent kinase and phosphatase, cAMP-dependent protein kinase A or protein kinase C) are chosen, leading to activation or inactivation of various enzymes and alterations in gene expression through distinct DNA regulatory elements (Bading *et al.*, 1993; Jain *et al.*, 1993; Peunova and Enikolopov, 1993). Many of the Ca^{2+}-dependent cellular functions are mediated by cytosolic Ca^{2+}-binding proteins which are known to influence Ca^{2+} levels and/or mediate Ca^{2+} signals within the cell (for a review, see Carafoli, 1987). Most of these proteins belong to a homologous class of small, acidic proteins, which constitute the superfamily of EF-hand type helix-loop-helix Ca^{2+}-binding proteins. Over 260 members of this family have been discovered so far (for reviews, see Moncrief *et al.*, 1990; Nakayama *et al.*, 1992; Nakayama and Kretsinger, 1993). They all share as a common structural motif a Ca^{2+}-binding loop which is flanked by two perpendicularly oriented α-helices (Fig. 1). Based on the crystal structure of parvalbumin (Fig. 2), which was the first Ca^{2+}-binding protein to be crystallized, this structure was named the "EF-hand" after the C-terminal E-helix-loop-F-helix Ca^{2+}-binding domain in this protein (Kretsinger and Nockolds, 1973). The Ca^{2+}-binding motif enables reversible binding of Ca^{2+} with dissociation constants in the submicromlar to micromolar range (0.01 - 10 μM), *i.e.* at Ca^{2+} concentrations as found in the cytoplasm under resting conditions and after stimulation of the cell.

NATO ASI Series, Vol. H 92
Signalling Mechanisms – from Transcription Factors to Oxidative Stress
Edited by L. Packer, K. Wirtz

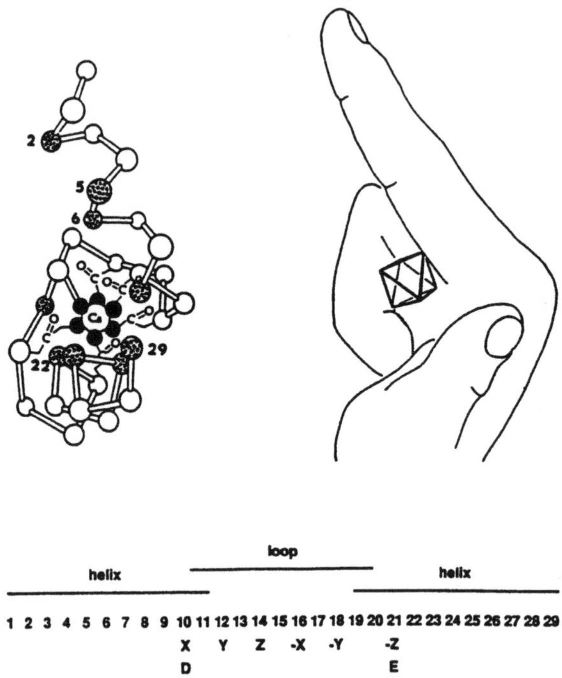

Fig. 1: The EF-hand Ca^{2+}-binding domain: In each EF-hand two α-helices, oriented at approximately 90°, flank a 12 amino acid containing loop, which provides 6 Ca^{2+}-coordinating oxygen ligands, defined as X, Y, Z, -X, -Y and -Z. The three-dimensional structure of the EF-hand domain can be visualized by the right hand with the stretched thumb and index finger representing the α-helices and the bent middle finger showing the Ca^{2+}-binding loop. As indicated, Asp (D) is usually found at position 10, and Glu (E) is often found at position 21 [taken from Moncrief *et al.*, 1990].

Two different roles of EF-hand Ca^{2+}-binding proteins

Two major functional roles have been assigned to EF-hand Ca^{2+}-binding proteins: the mediation of Ca^{2+} signals and the buffering of the cytosolic Ca^{2+} concentration.

A first group of proteins can interact with target proteins in a Ca^{2+}-dependent manner, thereby modulating or regulating enzyme activation. The proteins best characterized in this group are calmodulin, troponin C and calcineurin. The binding of Ca^{2+} to calmodulin induces marked conformational changes, which enable the protein to interact with and regulate a variety of enzymes (Tanaka and Hidaka, 1980; Veigl *et al.*, 1984; Cohen and Klee, 1988). For example, a Ca^{2+}-dependent activation and inhibition of transcription factors by calmodulin has been described (Bading *et al.*, 1993 and Corneliussen *et al.*, 1994, respectively). In contrast, troponin C, which shows a very similar overall structure to calmodulin (Herzberg and James, 1985), plays its sole role in the Ca^{2+}-dependent regulation of striated and heart muscle contraction. Calcineurin is a Ca^{2+}- and calmodulin-dependent protein phosphatase with two subunits, of which the EF-hand subunit is regulatory (Liu and Storm, 1989; Jain *et al.*, 1993).

For a second group of EF-hand Ca^{2+}-binding proteins no interaction with other proteins has been found. This type of Ca^{2+}-binding proteins is thought to constitute important Ca^{2+} buffers inside the cell, thereby influencing the cytosolic free Ca^{2+} levels. Parvalbumin (Fig. 2) and sarcoplasmic calcium-binding protein in fast skeletal muscle fibres of vertebrates and invertebrates, respectively, belong to this group (for reviews, see Wnuk *et al.*, 1982; Berchtold, 1989; Cox, 1990). They are thought to represent an abundant Ca^{2+}/Mg^{2+}-exchange system in these cells.

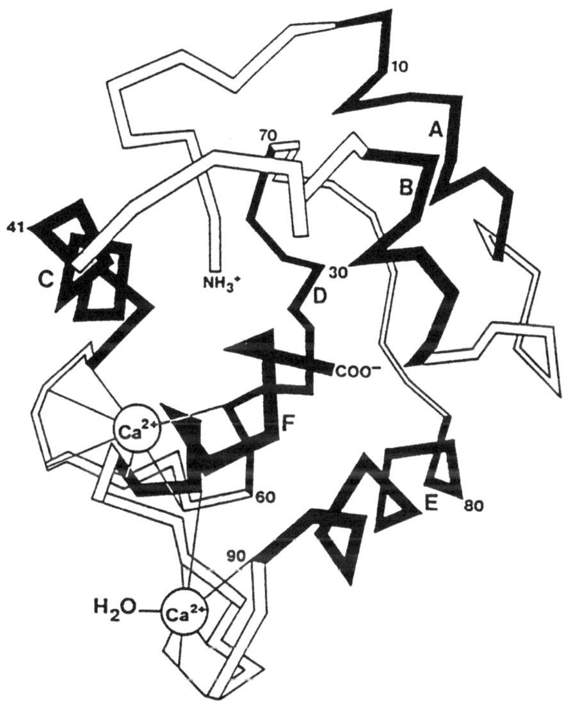

Fig. 2: Structure of carp PV: This protein contains three helix-loop-helix Ca^{2+}-binding domains, with helices AB, CD, EF indicated in black. Only the CD and the EF domains are functional, whereas the AB domain does not bind Ca^{2+}, due to a deletion of 2 amino acids. COO- refers to the C-terminus and NH_3^+ to the acetylated N-terminus. The numbers indicate amino acid positions in the protein [taken from Heizmann and Berchtold, 1987].

Two types of EF-hand Ca^{2+}-binding sites

Different modes of functions for EF-hand Ca^{2+}-binding proteins can be proposed according to the different types of Ca^{2+}-binding sites (Haiech *et al.*, 1979; Wnuk *et al.*, 1982). Two types of Ca^{2+}-binding sites can be distinguished, a Ca^{2+}/Mg^{2+}-mixed type and a Ca^{2+}-specific type, based on the differences in selectivity and affinity for Ca^{2+} and Mg^{2+} (Table 1).

The Ca^{2+}/Mg^{2+}-mixed sites bind Ca^{2+} with high affinity ($K_{Ca} = 10^7 - 10^9\ M^{-1}$) and Mg^{2+} with moderate affinity ($K_{Mg} = 10^3 - 10^5\ M^{-1}$) in a competitive fashion. Under physiological conditions at low resting Ca^{2+} levels they are saturated with Mg^{2+}. Elevated levels of intracellular Ca^{2+} cause a replacement of Mg^{2+} by Ca^{2+} with slow kinetics. These sites are thought to play an important role in intracellular Ca^{2+}-buffering.

Ca^{2+}-specific sites bind divalent cations with a reduced affinity for Ca^{2+} ($K_{Ca} = 10^5 - 10^7\ M^{-1}$) and a drastically reduced affinity for Mg^{2+} ($K_{Mg} = 10^1 - 10^2\ M^{-1}$) when compared to Ca^{2+}/Mg^{2+}-mixed sites (Milos *et al.*, 1986). Therefore, Mg^{2+} has only little effect on Ca^{2+} binding. In the resting cell Ca^{2+}-specific sites are thought to be in a metal-free state from which they can directly bind Ca^{2+} with fast kinetics when intracellular Ca^{2+} levels are increased. These sites are thought to confer Ca^{2+}-dependent regulatory functions of enzyme activation (Means *et al.*, 1988).

	Ca^{2+}/Mg^{2+}-mixed site	**Ca^{2+}-specific site**
K_{Ca}	$10^7 - 10^9\ M^{-1}$	$10^5 - 10^7\ M^{-1}$
K_{Mg}	$10^3 - 10^5\ M^{-1}$	$10^1 - 10^2\ M^{-1}$
Ca^{2+}/Mg^{2+} antagonism	strong	weak
binding status (at rest)	Mg^{2+}-loaded	metal free
functions	***Ca^{2+}-buffering*** - stress protection? - modulation of Ca^{2+} levels and transients?	***intracellular signaling*** - Ca^{2+}-dependent regulation of enzyme activation and gene expression

Table 1: Two types of EF-hand Ca^{2+}-binding domains: a Ca^{2+}/Mg^{2+}-mixed site and a Ca^{2+}-specific site. K_{Ca} and K_{Mg} are the affinity constants for Ca^{2+} and Mg^{2+} binding, respectively.

The group of Ca^{2+}-binding proteins with Ca^{2+}-specific regulatory sites has been intensively studied and a large variety of cellular functions have been described. On the other hand, Ca^{2+}-binding proteins which contain only Ca^{2+}/Mg^{2+}-mixed sites have mainly been thought to play a role in Ca^{2+}-buffering within the cell. For parvalbumin (containing two Ca^{2+}/Mg^{2+}-mixed sites) and calbindin D-28K (containing 1 high affinity and 2-3 low affinity Ca^{2+}-binding sites) a role as protective buffers against cytotoxic high Ca^{2+} concentrations was proposed for neuronal cells (Nitsch *et al.*, 1989; Christakos *et al.*, 1989).

A possible role of Ca^{2+}-buffering proteins in signal transduction

Recently, defined changes in the cytosolic Ca^{2+} levels were found to trigger expression of immediate early genes like c-fos and c-jun, which encode proteins of a family of transcription factors that mediate changes in patterns of cellular expression (Werlen *et al.*, 1993). These findings may indicate a possible function of Ca^{2+}-buffering proteins in transcellular signaling: Different Ca^{2+}-binding proteins with different kinetics (affinity constants, on and off rates) for Ca^{2+} binding may differentially affect the incoming Ca^{2+} signal, thereby modulating the cellular Ca^{2+} response. Data from a calbindin D-28K transfected pituitary cell-line showed that both T- and L-type Ca^{2+} currents were inactivated more rapidly in calbindin D-28K expressing than in non-transfected wild-type cells (Lledo *et al.*, 1992). Loading of rat dorsal root ganglia cells with parvalbumin or calbindin D-28K using microinjection did not significantly alter the basal cytosolic Ca^{2+} concentration when compared to untreated control cells (Chard *et al.*, 1993). Nevertheless, both proteins significantly affected depolarization-induced changes in the cytosolic Ca^{2+} concentration ($[Ca^{2+}]_i$) such as the rate of rise of $[Ca^{2+}]_i$, the maximal $[Ca^{2+}]_i$ peak and the rate of decay of $[Ca^{2+}]_i$ from its peak value. Calbindin D-28K caused an 8-fold decrease in the rate of rise in $[Ca^{2+}]_i$ and in addition altered the kinetics of decay of $[Ca^{2+}]_i$ to a pronounced slow component, with loss of the fast component, when compared to the biphasic decay in control cells. PV also slowed down the rate of rise in $[Ca^{2+}]_i$ but to a lesser degree, and increased a fast component of the $[Ca^{2+}]_i$ decay without affecting the slow component. Therefore, both proteins not only efficiently buffered the intracellular Ca^{2+} concentration but also modified the cytosolic Ca^{2+} transients in different ways which may be relevant for regulating or modulating Ca^{2+}-dependent aspects of cellular signaling. Further evidence for this hypothesis is given by experimental stimulation of the central nervous system in gerbils (Zuschratter *et al.*, 1994). Epileptic seizure-inducing stimuli caused a c-fos expression preferentially in neurons that lacked PV, whereas in PV-expressing neurons a time-delayed c-fos signal was found.

Wild-type and mutant Ca^{2+}-binding proteins with different metal-binding properties

In order to investigate the possible role of Ca^{2+}-buffering proteins in modulating Ca^{2+}-dependent signaling we have produced, by recombinant technology, a variety of wild-type and mutant Ca^{2+}-binding proteins with different metal-binding properties. In an initial step the recombinant wild-type EF-hand Ca^{2+}-binding protein parvalbumin (PV_{WT}) was generated and subsequently compared to recombinant oncomodulin (OM_{WT}), another EF-hand Ca^{2+}-binding protein with 50% amino acid sequence identity to PV (for a review, see Berchtold, 1989). PV_{WT} contains three Ca^{2+}-binding motifs, two of them (*i. e.* CD and EF domains) being responsible for high affinity Ca^{2+} binding, whereas the N-terminal (*i. e.* AB) domain is incapable of Ca^{2+} binding (Fig. 2). Both functional domains are of the Ca^{2+}/Mg^{2+}-mixed type, and are thought to play an important role in the relaxation of fast twitching muscle fibres (Briggs, 1975; Pechère *et al.*, 1977) and in the protection of neuronal cells against cytotoxic Ca^{2+} concentrations (Nitsch *et al.*, 1989). In contrast, OM_{WT} contains a single Ca^{2+}/Mg^{2+}-mixed site (*i. e.* EF domain) and a Ca^{2+}-specific site (*i. e.* CD domain), which is thought to confer calmodulin-like regulatory functions of enzyme activation (MacManus, 1981; Mutus *et al.*, 1985; Palmer *et al.*, 1990). Both, OM_{WT} and PV_{WT} were expressed and purified in high quantities (MacManus *et al.*, 1989; Pauls and Berchtold, 1993) and their metal-binding properties have been characterized (Cox *et al.*, 1990; Pauls *et al.*, 1993). Subsequently, the recombinant PV_{WT}, which lacks Trp and Tyr residues, was mutated by replacing a Phe at position 102 with a Trp in order to introduce an optical probe for structural studies into the protein. The cation-binding properties of this mutant protein (PV_{F102W}) were very similar to those of the recombinant unmodified PV_{WT} and of rat wild-type parvalbumin isolated from muscle (Fig. 3; Pauls *et al.*, 1993). Trp-fluorescence and UV-difference spectroscopy of PV_{F102W} indicated that the Trp residue is confined to a strong hydrophobic core which shows only small conformational changes in going from the metal-free to the metal-bound forms.

This PV_{F102W} was further modified in order to change essential features of metal-binding (Fig. 3, Table 2). In a first step, PV_{F102W} mutant proteins containing alterations essential for Ca^{2+}-binding in either the CD (*i. e.* PV_{-CD}), the EF (*i. e.* PV_{-EF}) or both (*i. e.* $PV_{-CD/-EF}$) Ca^{2+}-binding domains were produced (Pauls *et al.*, 1994). In each domain the highly conserved amino acids Asp (position 10) and Glu (position 21), which provide essential oxygen ligands for Ca^{2+} binding (Fig. 1), were replaced by the two non-polar amino acids Ala and Val, respectively. PV_{-CD} has a single Ca^{2+}/Mg^{2+}-mixed site with high affinity for Ca^{2+} (2.5-fold lower than PV_{F102W}) and moderate affinity for Mg^{2+} (2-fold higher than PV_{F102W}). Moreover, this mutant displayed a conformation and metal-induced conformational changes which are very similar to those in PV_{F102W}. In contrast, PV_{-EF} binds one Ca^{2+} or Mg^{2+} with 10-fold reduced affinity and weak Ca^{2+}/Mg^{2+} antagonism similar to a Ca^{2+}-specific site as

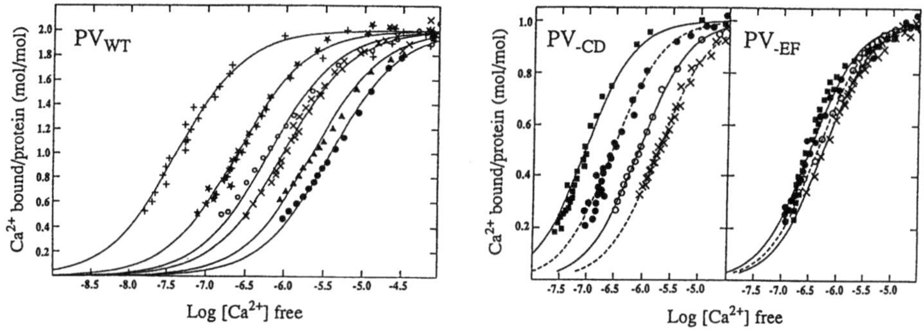

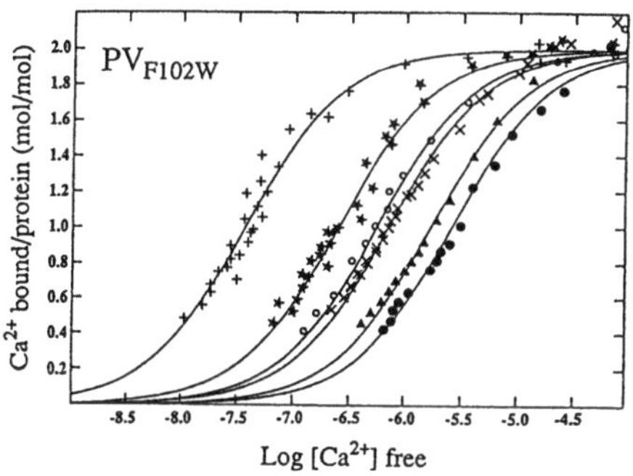

Fig. 3: Ca^{2+}-binding isotherms of PV$_{WT}$[a], PV$_{F102W}$[a], PV$_{-CD}$[b] and PV$_{-EF}$[b] in the absence and presence of Mg^{2+}: Ca^{2+} binding was measured by the flow dialysis method (Colowick and Womack, 1969) at 25°C. Protein concentrations were 10-30 µM. Mg^{2+} concentrations were 0 mM (✚), 2 mM (★), 5 mM (○), 10 mM (×), 30 mM (▲) or 50 mM (●) for **PV$_{WT}$** and **PV$_{F102W}$**; 0 mM (■), 1 mM (●), 5 mM (○) or 10 mM (×) for **PV$_{-CD}$** and **PV$_{-EF}$**. [a]taken from Pauls *et al.*, 1993; [b]taken from Pauls *et al.*, 1994.

found in OM$_{WT}$. Conformational studies suggest that PV$_{-EF}$ is much less structured in the metal-free form and recovers the native structure of PV$_{F102W}$ only upon binding Ca^{2+}. PV$_{-CD/-EF}$ binds neither Ca^{2+} nor Mg^{2+} and never establishes a native structure as found in PV$_{F102W}$.

Perspectives

In the future, we plan to overexpress PV and OM wild-type and mutant proteins with different metal-binding properties in cultured cells in order to study their influence on Ca^{2+} homeostasis and Ca^{2+}-mediated intracellular signaling. Previous data on PV expression in C127 cells (Rasmussen and Means, 1989) indicated an increase in cell cycle duration via prolongation of G_1 phase and mitosis for PV-transfected cells when compared to untransfected control cells. In order to investigate whether the observed changes in cell cycle duration are due to Ca^{2+}-dependent properties of PV, transfection of cells with our mutant PVs with modified metal-binding properties should be very helpful. The mutant PV$_{-CD/-EF}$, however, with defective metal-binding sites, constructed as a negative control protein, was

	Ca^{2+} bound per protein	Mg^{2+}bound per protein	K_{Ca} In 0 mM Mg^{2+}	K_{Mg}	Ca^{2+}/Mg^{2+} antagonism
PV_{WT}[a]	2	2	K' mean 2.4 10^7 M^{-1}	K' mean 2.9 10^4 M^{-1}	strong
PV_{F102W}[a]	2	2	K' mean 2.7 10^7 M^{-1}	K' mean 4.4 10^4 M^{-1}	strong
PV_{-CD}[b]	1	1	K' mean 1.1 10^7 M^{-1}	K' mean 8.0 10^4 M^{-1}	strong
PV_{-EF}[b]	1	1	K' mean 3.2 10^6 M^{-1}	K' mean 3.0 10^3 M^{-1}	weak
$PV_{-CD/-EF}$[b]	0	0	K' mean ---	K' mean ---	---
OM_{WT}[c]	2	1	K' 1 2.2 10^7 M^{-1} K' 2 1.7 10^6 M^{-1}	K' 1 4.0 10^3 M^{-1} K' 2 ---	strong weak

Table 2: Metal-binding properties of PV and OM wild-type and mutant proteins: Affinity constants K_{Ca} and K_{Mg} were obtained by Scatchard analysis of data derived from flow dialysis (Colowick and Womack, 1969; see Fig. 2) and equilibrium gel filtration (Hummel-Dryer, 1962), respectively. [a]taken from Pauls *et al.*, 1993; [b]taken from Pauls *et al.*, 1994; [c]taken from Cox *et al.*, 1990.

not able to attain the same conformation as found in PV_{F102W}. We, therefore, have started to generate new mutant PVs with single amino acid substitutions in each Ca^{2+}-binding site with the aim of creating mutant proteins which lack the ability to bind Ca^{2+} but still retain the physiological conformation of PV_{WT}. Such mutant proteins should yield indications whether the function of PV is restricted to metal binding or includes divalent cation independent functions. Furthermore, the generation of chimeric proteins with exchanges of complete Ca^{2+}-binding motifs between PV and OM is in progress. Expression of these wild-type and mutant Ca^{2+}-binding proteins in cultured cells should clarify whether Ca^{2+}-buffering proteins may play a role in modifying Ca^{2+} gradients, thereby influencing/regulating the incoming Ca^{2+} signal. Our questions for the future are: Do Ca^{2+}-binding proteins with different affinities and kinetics produce distinct temporal Ca^{2+} gradients? And if so do these changes result in a modulation of the different Ca^{2+}-dependent signal transduction pathways?

Conclusions

Two different types of EF-hand helix-loop-helix Ca^{2+}-binding proteins have been described: Proteins with Ca^{2+}-specific metal-binding sites possess direct Ca^{2+}-dependent regulatory capabilities of enzyme activation, whereas proteins with only Ca^{2+}/Mg^{2+}-mixed sites are thought to buffer the intracellular Ca^{2+} concentration. Recent data, however, indicate a possible role of the latter type of Ca^{2+}-buffering proteins also in modulating Ca^{2+}-dependent signal transduction, via protein specific modification of temporal changes in the cytosolic Ca^{2+} concentration. In order to test this hypothesis a variety of wild-type and mutant Ca^{2+}-binding proteins with modified metal-binding properties have been generated. The characterization of these proteins *in vivo* will show if Ca^{2+}-buffering proteins can modulate Ca^{2+}-specific signaling.

Aknowledgements:

I would like to thank Dr. M. W. Berchtold, Dr. M. Celio and Dr. B. Schwaller for their critical reading of my manuscript and Dr. J. A. Cox for providing me with his results on the metal-binding properties of wild-type and mutant PVs. This work was supported by Swiss National Foundation Grant 31-28847.90 (to M. W. Berchtold) and Grant 31-28637.90 (to J. A. Cox).

References

Bading, H., Binty, D. D., & Greenberg, M. E. (1993) *Regulation of gene expression in hippocampal neurons by distinct calcium signaling pathways*. Science **260**, 181-186.

Berchtold, M. W. (1989) *Structure and expression of genes encoding the three-domain Ca^{2+}-binding proteins parvalbumin and oncomodulin*. Biochem. Biophys. Acta **1009**, 201-215.

Briggs, N. (1975) *Identification of the soluble relaxation factor as a parvalbumin*. Fedn. Proc. **34**, 540.

Carafoli, E. (1987) *Intracellular calcium homeostasis*. Ann. Rev. Biochem. **56**, 395-433.

Chard, P. S., Bleakman, D., Christakos, S., Fullmer, C. S., & Miller, R. J. (1993) *Calcium buffering properties of calbindin D28k and parvalbumin in rat sensory neurones* . Journal of Physiology **472**, 341-357.

Christakos (1989) *Vitamin D-dependent calcium binding proteins: chemistry, distribution, functional considerations, and molecular biology*. Endocrine Rev. **10**, 3-26.

Cohen, P., & Klee, C. B. (1988) *Calmodulin*, Vol. 5: Molecular Aspects of Cellular Regulation, Elsevier, Amsterdam - New York - Oxford.

Colowick, S. P., & Womack, F. C. (1969) *Binding of diffusable molecules by macromolecules: Rapid measurement by rate of dialysis.* J. Biol. Chem. **244**, 774-777.

Corneliussen, B., Holm, M., Waltersson, Y., Onions, J., Hallberg, B., Thornell, A., & Grundström, T. (1994) *Calcium/calmodulin inhibition of basic helix-loop-helix transcription factor domains.* Nature **368**, 760-764.

Cox, J. A. (1990) *The role of intracellular calcium binding proteins.* **In:** Stimulus Response coupling (J.-R. Dedman and Smith, V.L. ed.) pp. 266-269, CRC Press, Boca Raton, Ann Arbor, Boston.

Cox, J. A., Milos, M., & MacManus, J. P. (1990) *Calcium- and magnesium-binding properties of oncomodulin.* J. Biol. Chem. **265**, 6633-6637.

Haiech, J., Derancourt, J., Pechère, J.-F., & Demaille, J. G. (1979) *Magnesium and calcium binding to parvalbumins: Evidence for differences between parvalbumins and an explantion of their relaxing function.* Biochemistry **18**, 2752-2758.

Heizmann, C. W., & Berchtold, M. W. (1987) *Expression of parvalbumin and other Ca^{2+}-binding proteins in normal and tumor cells: a topical review.* Cell Calcium **8**, 1-41.

Herzberg, O., & James, M. N. G. (1985) *Structure of the calcium regulatory muscle protein troponin-C at 28 Å resolution.* Nature **313**, 653-659.

Hummel, J. P., & Dryer, W. J. (1962) *Measurement of protein-binding phenomena by gel filtration.* Biochim. Biophys. Acta **63**, 530-532.

Jain, J., McCaffrey, P. G., Miner, Z., Kerppola, T. K., Lambert, J. N., Verdine, G. L., Curran, T., & Rao, A. (1993) *The T-cell transcription factor $NFAT_p$ is a substrate for calcineurin and interacts with Fos and Jun.* Nature **365**, 352-355.

Kretsinger, R. H., & Nockolds, C. E. (1973) *Carp muscle calcium-binding protein. II. Structure determination and general description.* J. Biol. Chem. **248**, 3313-3326.

Lledo, P.-M., Somasundaram, B., Morton, A. J., Emson, P. C., & Mason, W. T. (1992) *Stable transfection of calbindin-D28k into the GH3 cell line alters calcium currents and intracellular calcium homeostasis.* Neuron **9**, 943-954.

Liu, Y., & Storm, D. R. (1989) *Dephosphorylation of neuromodulin by calcineurin.* J. Biol. Chem. **264**, 12800-12804.

MacManus, J. P. (1981) *The stimulation of cyclic nucleotide phosphodiesterase by a Mr 11500 calcium binding protein from hepatoma.* FEBS Lett. **126**, 245-249.

MacManus, J. P., Hutnik, C. M. L., Sykes, B. D., Szabo, A. G., & Williams, T. C. (1989) *Characterization and site-specific mutagenesis of the calcium-binding protein oncomodulin produced by recombinant bacteria.* J. Biol. Chem. **264**, 3470-3477.

Means, A. R. (1988) *Molecular mechanisms of action of calmodulin.* Recent Prog. Horm. Res. **44**, 223-262.

Milos, M., Schaer, J.-J., Comte, M., & Cox, J. A. (1986) *Calcium-proton and calcium-magnesium antagonisms in calmodulin: microcalorimetric and potentiometric analyses.* Biochemistry **25**, 6279-6287.

Moncrief, N. D., Kretsinger, R. H., & Goodman, M. (1990) *Evolution of EF-hand calcium-modulated proteins. I. Relationships based on amino acid sequences.* J. Mol. Evol. **30**, 522-562.

Mutus, B., Karuppiah, N., Sharma, R. K., & MacManus, J. P. (1985) *The differential stimulation of brain and heart cyclic-AMP phosphodiesterase by oncomodulin.* Biochem. Biophys. Res. Commun. **131**, 500-506.

Nakayama, S., & Kretsinger, R. H. (1993) *Evolution of EF-hand calcium-modulated proteins. III. Exon sequences confirm most dendrograms based on protein sequences: calmodulin dendrograms show significant lack of parallelism.* J. Mol. Evol. **36**, 458-476.

Nakayama, S., Moncrief, N. D., & Kretsinger, R. H. (1992) *Evolution of EF-hand calcium-modulated proteins. II. Domains of several subfamilies have diverse evolutionary histories.* J. Mol. Evol. **34**, 416-448.

Nitsch, C., Scotti, A., Sommacal, A., & Kalt, G. (1989) *GABAergic hippocampal neurons resistant to ischemia-induced neuronal death contain the Ca^{2+}-binding protein parvalbumin.* Neurosci. Lett. **105**, 263-268.

Palmer, E. J., MacManus, J. P., & Mutus, B. (1990) *Inhibition of glutathione reductase by oncomodulin.* Arch. Biochem. Biophys. **74**, 149-154.

Pauls, T. L., & Berchtold, M. W. (1993) *Efficient complementary DNA amplification and expression using polymerase chain reaction technology.* Methods Enzymol. **217**, 102-122.

Pauls, T. L., Durussel, I., Cox, J. A., Clark, I. D., Szabo, A. G., Gagné, S. M., Sykes, B. D., & Berchtold, M. W. (1993) *Metal binding properties of recombinant rat parvalbumin wild-type and F102W mutant.* J. Biol. Chem. **268**, 20897-20903.

Pauls, T. L., Durussel, I., Berchtold, M. W., & Cox, J. A. (1994) *Inactivation of individual Ca^{2+}-binding sites in the paired EF-hand sites of parvalbumin reveals asymmetrical metal-binding properties.* Biochemistry **33**, 10393-10400.

Pechère, J.-F., Derancourt, J., & Haiech, J. (1977) *The participation of parvalbumins in the activation-relaxation cycle of vertebrate fast skeletal muscle.* FEBS Lett. **75**, 111-114.

Peunova, N., & Enikolopov, G. (1993) *Amplification of calcium-induced gene transcription by nitric oxide in neuronal cells.* Nature **364**, 450-453.

Rasmussen, C. D. & Means, A.R. (1989) *The presence of parvalbumin in a nonmuscle cell line attenuates progression through mitosis.* Mol. Endocrinol. **3**, 588-596.

Tanaka, T., & Hidaka, H. (1980) *Hydrophobic regions function in calmodulin-enzyme(s) interactions.* J. Biol. Chem. **255**, 11078-11080.

Veigl, M. L., Vanaman, T. C., & Sedwick, W. D. (1984) *Calcium and calmodulin in cell growth and transformation.* Biochim. Biophys. Acta **738**, 21-48.

Werlen, G., Berlin, D., Conne, B., Roche, E., Lew, D. P., & Prentki, M. (1993) *Intracellular Ca^{2+} and the regulation of early response gene expression in hl-myeloid leukemia cells.* J. Biol. Chem. **268**, 165906-16601.

Wnuk, W., Cox, J. A., & Stein, E. A. (1982) *Parvalbumins and other soluble high-affinity calcium-binding proteins from muscle.* **In:** Calcium and Cell Function (W. Y. Cheung ed.) 2, pp. 243-278, Academic Press, New York

Zuschratter, W., Gass, P., Herdegen, T. & Scheich, H. (1994) *Parvalbumin-immunoreactive neurons withstand c-fos expression following chemically-induced seizures.* Neurosci. Abstra. **20**, p. 397.

Purification of a neurite growth inhibitor (NI-250) from bovine CNS myelin

Adrian A.A. Spillmann, Christine E. Bandtlow, Flavio Keller* and Martin E. Schwab
Brain Research Institute
August Forel Strasse 1
8029 Zürich
Switzerland

Inhibitors of Neurite Growth

In the central nervous system (CNS) the ability of nerve fibres to regenerate upon injury is very limited. In contrast, peripheral nerve fibres have been observed to regrow over long distances. Transplantation experiments with CNS and PNS nerve explants have shown that the local microenvironment is crucial for regenerating nerve fibers: CNS neurons can extend their lesioned neurites into the peripheral nerve environment, but cease growth at the transition between PNS and CNS (Aguayo et al 1990; David and Aguayo 1981; Ramon y Cajal 1928; Tello 1911). To test if a possible lack of trophic factors in the adult CNS might be responsible for this observation, perinatal rat sensory, sympathetic, or retinal neurons were cocultured with explants of adult rat sciatic (PNS) or optic nerves (CNS) in the presence of nerve growth factor (NGF) or brain-derived neurotrophic factor (BDNF) (Schwab & Thoenen 1985). The pattern of fiber growth corresponded exactly to that found *in vivo*: many axons grew into the sciatic nerve explants, but none grew into the optic nerves. These results suggested that there is a neurite growth inhibitory activity present in adult CNS tissue that cannot be overcome by the stimulatory effects of neurotrophic factors (Schwab & Thoenen, 1985).

Furthermore, video time-lapse studies of dorsal root ganglion (DRG) neurites growing on a laminin substrate in the presence of NGF showed that contact with oligodendrocytes led to a rapid arrest and collapse of growth cones (Fawcett 1992; Bandtlow et al 1990). Contact of the growth cone filopodia with the processes of oligodendrocytes was sufficient to induce these phenomena (contact with astrocytes resulted in an unchanged or slightly reduced growth velocity) . Growth arrest was long lasting (hours) and strictly local, which means that neurites arising from the same cell and not in the contact with the oligodendrocytes continued to grow normally (figure 1).

---

*Present address: Libero Istituto Universitario Campus Bio-Medico, I-00186 Roma

NATO ASI Series, Vol. H 92
Signalling Mechanisms – from Transcription Factors to Oxidative Stress
Edited by L. Packer, K. Wirtz

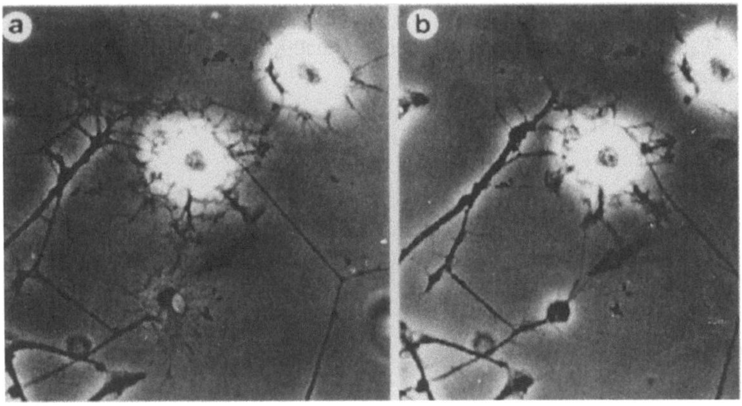

Figure 1: Phase-contrast photomicrographs showing DRG growth cones approching an oligodendrocyte. A single filopodial contact is sufficient to arrest growth cone motility and to collapse growth cone structure. Picture taken from Bandtlow et al 1990.

Testing membranes from oligodendrocyte enriched cultures and CNS myelin, a strong inhibitory effect was observed (Caroni & Schwab 1988a). These results indicated that neurite growth inhibitory factors are expressed on the cell surface of oligodendrocytes. Separation of spinal cord myelin proteins by SDS polyacrylamide gel electrophoresis (PAGE) and subsequent elution of the proteins and their reconstitution into liposomes allowed the screening for inhibitory activity. Two molecular weight regions at 35-kDa and 250-kDa were found in CNS myelin whereas in corresponding PNS regions these inhibitory active region for neurite outgrowth were absent (Caroni & Schwab 1988a). These two proteins were subsequently called neurite growth inhibitors NI-35 and NI-250. Immunological and recent biochemical data suggest that NI-250 and NI-35 are closely related. A monoclonal antibody, called IN-1, has been raised against gel purified NI-250 protein (Caroni & Schwab 1988b). It could be shown, both *in vitro* and *in vivo*, that this antibody neutralizes the neurite growth inhibition exerted by differentiated oligodendrocytes, CNS myelin and NI-35/NI-250 (Bandtlow et al 1990; Caroni & Schwab 1988b). It also allows regeneration of lesioned nerve fiber tracts in the rat spinal cord and brain (Schnell et al. 1994; Weibel et al 1994; Cadelli & Schwab 1991; Schnell & Schwab 1990).

Role of Intracellular Calcium in NI-35-Evoced Collapse of Neuronal Growth Cones

To further analyse the effect of NI-35 on neuronal growth cones, liposomes containing NI-35 were tested on cultured dorsal root ganglion (DRG) neurons. These DRG neurons arrested their growth upon application of NI-35 and collapsed within seconds to minutes (figure 2). The morphological changes of the growth cones observed upon contact with oligodendrocytes (figure 1) and with NI-35 (figure 2) were very similar and suggested the involvement of a second messenger leading to a reorganization of cytoskeletal proteins. For NI-35 it could be shown that preceding the morphological changes a large, transient increase in intracellular free calcium was detected (Bandtlow et al 1993). Neither a rise in intracellular free calcium nor collapse of growth cones was observed, when the NI-35 containing liposomes were preincubated with the neutralizing antibody IN-1. Interestingly, Dantrolene, which is an inhibitor of calcium release from caffeine-sensitive intracellular calcium stores, could protect growth cones from collapse in the presence of NI-35. Depletion of these calcium stores could be produced by caffeine pulses which evoked a modest rise in intracellular calcium concentration. When NI-35 liposomes were subsequently applied, they failed to induce the expected large increase in intracellular free calcium and no collapse occurred. These results suggest that the intracellular release of calcium is a crucial step in mediating the growth cone collapse of DRG neurons when confronted with NI-35 liposomes.

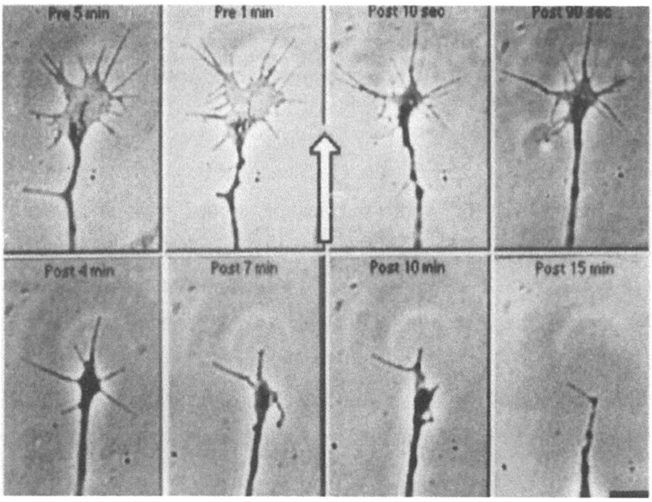

Figure 2: Phase-contrast photomicrographs of response of a DRG growth cone to the application of liposomes containing NI-35. The first two frames indicate the degree of spontaneous movement of a DRG neuron occurring over a 4-min interval before the addition of NI-35 liposomes (Pre). 4 min after (Post) application of liposomes (open arrow), unmistakable signs of collapse were seen. Collapse was complete within 1- to 15 min and withdrawal of the neurite was also observed. Photograph taken from Bandtlow et al 1993.

Purification of neurite growth inhibitor (NI-250) from bovine CNS myelin

The strategy for purification using crude bovine myelin involves the use of various chromatographic techniques in combination with two bioassays. The first bioassay tests for the effect of different column fractions on the spreading of 3T3 fibroblasts (spreading-assay; Caroni & Schwab 1988a). To further verify the characteristics of a specific neurite growth inhibitor the collapsing activity on growth cones of primary rat DRG neurons (figure 2) and the neutralizing effect of IN-1 was tested. When a fraction had collapsing activity for DRG neurons and could be neutralized by IN-1 it was regarded as an active fraction. By means of anion exchange (Q-Sepharose), reverse phase (C4-Reversed Phase) and size exclusion chromatography (TSK3000), a highly active protein fraction was obtained (table 1).

Table 1: Summary of the purification of a neurite growth inhibitory factor

Samples	Protein (mg)	Activity* (U)	spec. Act. (U/mg)	Purification (fold)	Recovery (%)
Extract	400	20000	50	1	100
Q-Pool	40	8000	200	4	40
C4-Pool	2	4000	2000	40	20
S-Pool	0.02	1333	66667	1200	6.7

*One unit (U) of activity is defined as the concentration of protein solution by which 50% of the 3T3 fibroblast cells do not spread.

The enrichment factor over the whole purification procedure is about 1200; the specific activity of the S-Pool fraction is about 15 ng/ml. Analysing at the protein pattern of the different active pools by SDS-PAGE (figure 3) a high molecular protein band at 250kd is enriched and is the only band seen in the S-Pool. As mentioned above this fraction could be totally neutralized with IN-1. Using this approach, the NI-35 protein could not be obtained. All these results suggest that we have obtained a candidate for the postulated NI-250 protein. In order to get enough material for amino acid sequencing we are currently optimizing the purification procedure.

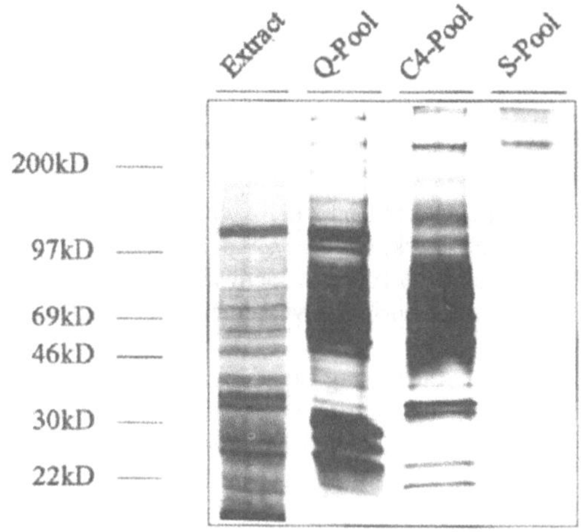

Figure 3: Identification of a NI-250kd candidate, which is responsible for the nonpermissive properties of CNS myelin, by SDS-PAGE under non-reducing conditions. Proteins from the CHAPS-solubilized extract of adult bovine spinal cord myelin (10 μg), Q-Pool (10 μg), C4-Pool (5 μg), and S-Pool (0.2 μg) were run in a 4–20% SDS-PAG and silver-stained. Molecular masses are indicated on the left. A 250kd band is visible in the S-Pool and is absent in the inactive fractions.

Literature references

Aguayo AJ, Carter DA, Zwimpfer TJ, Vidal-Sanz M, Bray GM (1990) Axonal regeneration and synapse formation in the injured CNS of adult mamals. In Brain Repair, ed.A Björklund, A Aguayo, D Ottoson, p. 251. New York: Stockton

Bandtlow CE, Schmidt MF, Hassinger TD, Schwab ME, Kater SB (1993) Role of itracellular clcium in NI-35-evoked collapse of neuronal growth cones. Science 259: 80-83

Bandtlow CE, Schwab ME (1991) Purification and biochemical characterization of rat and bovine CNS myelin associated neurite growth inhibitors NI-35 and NI-250. Soc. Neurosci. 17: 1495 (Abstr.)

Bandtlow CE, Zachleder T, Schwab ME (1990) Oligodendrocytes arrest neurite growth by contact inhibition. J. Neurosci. 10: 3937-48

Cadelli D, Schwab ME (1991) Regeneration of lesioned septohippocampal acetylcholinesterase-positive axons is improved by antibodies against the myelin-associated neurite growth inhibitors NI-35/250. Eur. J. Neurosci. 3: 825-32

Caroni P, Schwab ME (1988a) Two membrane protein fractions from rat central myelin with inhibitory properties for neurite growth and fibroblast spreading. J. Cell Biol. 106: 1281-88

Caroni P, Schwab ME (1988b) Antibody against myelin-associated inhibitor of neurite growth neutralizes nonpermissive substrate properties of CNS white matter. Neuron 1: 85-96

David S, Aguayo AJ (1981) Axonal elongation into peripheral nervous system "bridges" after central nervous system injury in adult rats. Science 214: 931-33

Ramon y Cajal S (1928. 1959) Degeneration and Regeneration of the Nervous System, ed. and transl. R. M. May 2 vols. New York: Hafner. 769 pp. (Reprint)

Schnell L, Schneider R, Kolbeck R, Barde Y-A, Schwab ME (1994) Neurotrophin-3 enhances sprouting of corticospinal tract during development and after adult spinal cord lesion. Nature 367: 170-173

Schnell L, Schwab ME (1990) Axonal regeneration in the rat spinal cord produced by an antibody against myelin-associated neurite growth inhibitors. Nature 343: 269-72

Schwab ME, Thoenen H (1985) Dissociated neurons regenerate into sciatic but not optic nerve explants in culture irrespective of neurotrophic factors. J. Neurosci. 5: 2415-23

Tello F (1911) La influencia del neurotropismo en la regeneracion de los centros nerviosos. Trab. Lab. Invest. Biol. 9: 123-59

Weibel D, Cadelli D, Schwab ME (1994) Regeneration of lesioned rat optic nerve fibers is improved after neutralization of myelin-associated neurite growth inhibitors. Brain Research 642: 259-266.

PROTEIN KINASE C AND LIPID SIGNALLING FOR CELLULAR REGULATION

Shun-ichi Nakamura†, Yoshinori Asaoka*, Kouji Ogita†, Françoise Hullin*, and Yasutomi Nishizuka†*
†Department of Biochemistry, Kobe University School of Medicine, Kobe 650, and *Biosignal Research Center, Kobe University, Kobe 657, Japan

It is well documented that protein kinase C (PKC) is activated by increased levels of diacylglycerol (DG) in the membrane that is produced as a result of hydrolysis of inositol phospholipids (PI) by signal-induced activation of phospholipase C (PLC) (Nishizuka, 1992). Recently, attention has been paid to the hydrolysis of other phospholipids, particularly phosphatidylcholine (PC). This reaction produces DG at a relatively later phase in cellular responses, and possible involvement of phospholipase D (PLD) in PKC activation has been postulated (see for reviews: Billah & Anthes, 1990; Exton, 1990; Cockcroft, 1992). Sustained activation of PKC is a prerequisite essential for long-term responses such as cell proliferation and differentiation. In addition, phospholipase A_2 (PLA_2) is activated by most of the signals which induce PI hydrolysis. Arachidonic acid regulates many physiological processes after its conversion to various eicosanoids, but products of PC hydrolysis catalyzed by PLA_2 potentiate PKC activation, thereby contributing to the signal transduction (Asaoka *et al.*, 1992a). However, the biochemical mechanism of activation of phospholipases D and A_2 remains largely unknown. PKC comprises a large family with multiple isoforms exhibiting individual characteristics (Nishizuka, 1992). Presently, a possibility may not be ruled out that some isoforms represent simply a redundancy, but plausible evidence suggests that the members of this enzyme family are activated in specific intracellular compartments in different ways, depending on various membrane lipid mediators, and play distinct roles for the control of major cellular functions. The present article will discuss some aspects of the generation of multiple lipid

NATO ASI Series, Vol. H 92
Signalling Mechanisms – from Transcription Factors to Oxidative Stress
Edited by L. Packer, K. Wirtz

mediators, which have potentials to prolonging PKC activation, that is needed for long-term cellular responses.

ELEVATION OF DIACYLGLYCEROL LEVELS

Upon stimulation of cell surface receptors, DG is immediately produced from inositol phospholipids, most rapidly from phosphatidylinositol-4,5-bisphosphate (PIP_2), as a result of PLC activation. This DG molecule disappears quickly. The level of DG, however, often increases again with a relatively slow onset, and it persists longer for minutes, occasionally more than hours (Fig. 1) (Liscovitch, 1992; Asaoka *et al.*, 1992a). It is generally thought that this second phase of DG comes from the hydrolysis of PC, because its fatty acid composition matches that of the PC molecule. This sustained elevation of DG is frequently observed in response to various long-acting signals such as growth factors, cytokines, and phorbol esters. Mitogenic signals sometimes cause only this second phase of DG elevation. Early experiments have shown

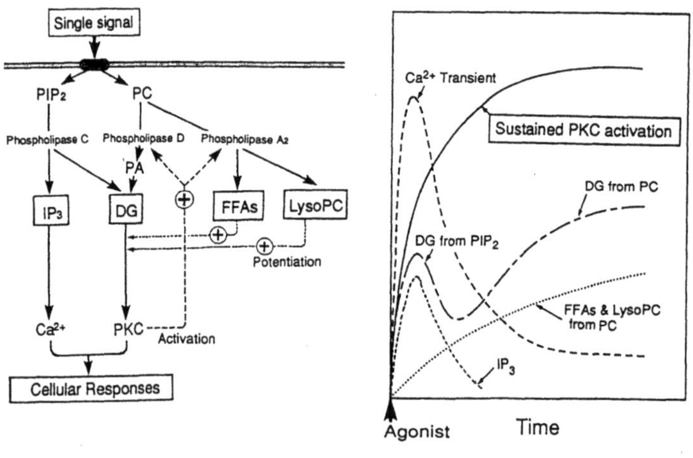

Fig. 1. Schematic representation of signal-induced phospholipid degradation. Left: potentiation of PKC activation by various lipid mediators. Right: time course of generation of lipid mediators. IP_3, inositol-1,4,5-trisphosphate; FFA, *cis*-unsaturated fatty acid. Other abbreviations, see text. Adapted from Nishizuka (1992)

that sustained elevation of DG levels for several hours is a prerequisite essential for various long-term cellular responses such as cell growth and differentiation (Asaoka *et al.*, 1991).

PLC which is reactive with PC to produce DG (PC-PLC) has been often proposed to occur in mammalian tissues. This proposal is based primarily on the observation that phosphorylcholine (P-choline) is produced quickly after cell stimulation. A second messenger role of P-choline for mitogenicity of growth factors has also been proposed (Cuadrado *et al.*, 1993). However, no unequivocal enzymological evidence is available presently indicating the existence of mammalian PC-PLC, which cleaves specifically PC to produce directly DG and P-choline. P-choline is produced rapidly by phosphorylation of choline which is an immediate product of PC hydrolysis by PLD, and choline kinase is very active in mammalian tissues. P-choline may also be produced from PC through lysophosphatidylcholine (lysoPC) and glycerophosphorylcholine by consecutive reactions initiated by PLA_2, followed by lysophospholipase and phosphodiesterase. In addition, sphingomyelinase may contribute to produce P-choline. Instead, several lines of evidence currently available favor a possibility that, at a later phase of cellular responses, PLD plays a role in the production of DG from PC (Billah & Anthes, 1990; Exton, 1990; Cockcroft, 1992). Phosphatidic acid (PA), a product of PC hydrolysis by PLD, is dephosphorylated to produce DG, which is capable of activating PKC.

MECHANISM OF ACTIVATION OF PHOSPHOLIPASE D

PLD is found in particulate fractions of many tissues and cell types. The particulate fractions *per se* are inactive, and cytosol and GTP-γ-S are necessary for the enzymatic activity (Fig. 2) (Anthes *et al.*, 1991; Olson *et al.*, 1991; Kusner *et al.*, 1993). This cytosolic factor appears to be a small GTP-binding protein, which is recently identified as ADP-ribosyla-

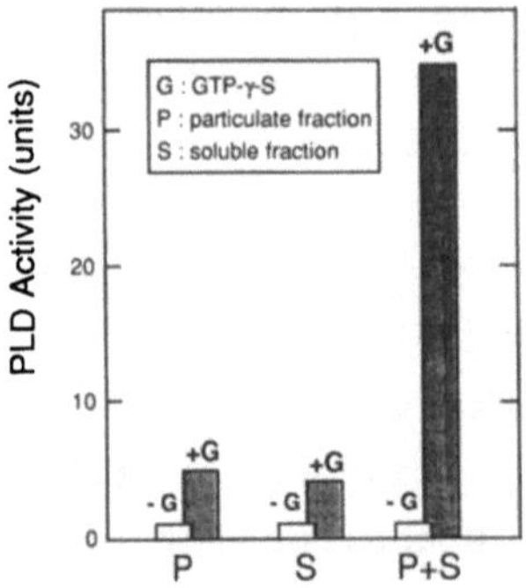

Addition	PLD activity (units)
Exp 1	
ATP	1
ATP + GTP-γ-S	35
ATP + GTP-γ-S + GDP-β-S	5
ATP + aluminium fluoride	2
Exp 2	
none	1
GTP-γ-S	13
ATP	5
GTP-γ-S + ATP	44
GTP-γ-S + ATP + genistein	18

Fig. 2. Activation of PLD. The PLD activity was assayed in a cell-free system with crude extracts of HL-60 cells. PLD is associated with particulate fractions. For full activation, soluble fraction, GTP-γ-S, and ATP were needed. Detailed experimental conditions will be described elsewhere.

tion factor (Brown *et al.*, 1993; Cockcroft *et al.*, 1994). The activation of PLD by GTP-γ-S in cell-free systems is further enhanced by ATP, and this enhancement is blocked by genistein, implying that tyrosine phosphorylation is involved in this PLD activation. With permeabilized cell systems, the enzyme is activated further by the addition of ATP and phorbol ester or a membrane-permeant DG, and this activation is sensitive to PKC inhibitors such as staurosporin and calphostin C. It seems likely, then, that PLD may be regulated by multiple mechanisms (see for reviews: Asaoka *et al.*, 1992a; Nishizuka, 1992). PKC once activated by DG, that is initially derived from PI hydrolysis, plays a role in the subsequent PC hydrolysis by PLD to produce additional DG to prolong PKC activation. The detailed mechanism of this receptor-mediated activation of PLD is a subject of great interest.

PHOSPHOLIPASE A_2 AND PKC ACTIVATION

PLA_2 is widely distributed in mammalian tissues, and several secretary and cytosolic PLA2 have been identified

(Dennis *et al.*, 1991). The enzymes are divided roughly into two groups. One cleave arachidonic acid preferentially, whereas the other cleave various fatty acids non-selectively that are esterified at the *sn*-2-position of phospholipids. Thus far, much attention has been paid to the arachidonic acid-selective type of enzymes because this fatty acid is the rate-limiting precursor to prostaglandins and leukotrienes, and most studies on the activation of PLA_2 have been made using cell preparations pre-labelled with radioactive arachidonic acid. The signals that provoke PI hydrolysis frequently release arachidonic acid, and hydrolysis of PC by PLA_2 generates two additional molecules, free *cis*-unsaturated fatty acids and lysophosphatidylcholine (lysoPC), which are both effective to enhance subsequent cellular responses (Asaoka *et al.*, 1992a).

Several *cis*-unsaturated fatty acids such as oleic, linoleic, and linolenic, arachidonic, and docosahexaenoic acids greatly enhance the DG-dependent activation of PKC in cell-free enzymatic systems. In the presence of both DG and fatty acid, the enzyme exhibits nearly full activity at the basal level of Ca^{2+} concentration (Shinomura *et al.*, 1991; Chen &

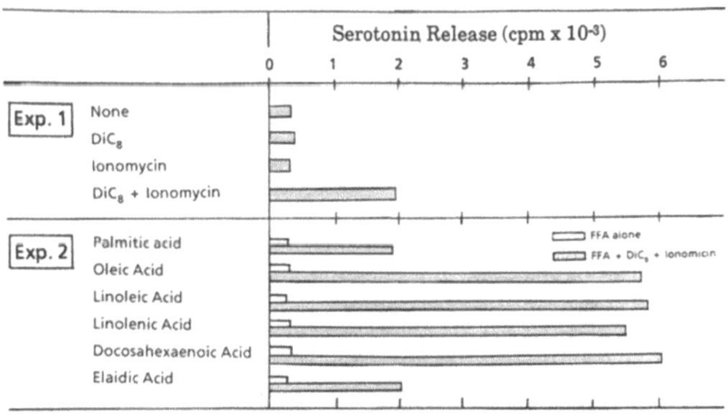

Fig. 3. Effect of various fatty acids on serotonin release from human platelets. [^{14}C]Serotonin-loaded platelets were stimulated for one min in the presence of various fatty acids with (shaded bars) or without (open bars) 1,2-dioctanoylglycerol plus ionomycin. Experiment 1 represents the serotonin release induced by 1,2-dioctanoylglycerol or ionomycin or both. Detailed experimental conditions were described elsewhere (Yoshida *et al.*, 1992). FFA, fatty acid; DiC_8, 1,2-dioctanoylglycerol.

Murakami, 1992). Neither saturated fatty acids such as palmitic and stearic acids nor *trans*-unsaturated fatty acids such as elaidic acid were capable of enhancing the PKC activation.

It is also shown that several *cis*-unsaturated fatty acids greatly enhance the DG-dependent activation of intact cells such as platelet release reaction (Fig. 3) (Yoshida *et al.*, 1992). For such cellular responses, a membrane-permeant DG and Ca^{2+}-ionophore are both essential. Consistent with the *in vitro* enzymatic reactions, the release reaction of platelets was enhanced by various *cis*-unsaturated fatty acids but not by *trans*-unsaturated and saturated fatty acids. The DG-induced PKC activation in intact platelets is indeed enhanced by the fatty acids as measured by the phosphorylation of a specific endogenous substrate, 47 kDa protein. Fatty acid *per se* is inactive if DG or phorbol ester is not added. Kinetic analysis with the Ca^{2+}-sensitive fluorescent dye fura-2 indicates that *cis*-unsaturated fatty acids markedly increase an apparent affinity of PKC activation to Ca^{2+}, thereby causing nearly full cellular responses at the basal level of Ca^{2+} concentration. An analogous synergistic action of *cis*-unsaturated fatty acid and DG or phorbol ester is observed for the differentiation of HL-60 cells to macrophages (Asaoka *et al.*, 1993). It is possible, then, that PKC once activated initially by PI hydrolysis may intensify its enzymatic activity even after the Ca^{2+} concentration returns to the basal level, when DG and *cis*-unsaturated fatty acids become available.

Fig. 4. Potentiation of phorbol ester-dependent HL-60 cell differentiation to macrophages by lysoPC, linoleic acid, or venom PLA_2. The differentiation was quantitated by measuring the expression of CD11b antigen. Detailed experimental conditions were described elsewhere (Asaoka *et al.*, 1993). TPA, tumour-promoting phorbol ester; LA, linoleic acid.

Lysophosphatidylcholine (lysoPC), the other product of PC hydrolysis by PLA_2, once produced, is rapidly metabolized further or acylated to produce PC again. LysoPC is known to exert detergent-like cytotoxic effects. However, in the presence of a membrane-permeant DG or a phorbol eater, lysoPC also greatly enhances cellular responses, particularly those in long-term such as cell growth and differentiation. For instance, lysoPC dramatically enhances the activation of human resting T-lymphocytes that is induced by a membrane-permeant DG and ionomycin, as determined by interleukin-2 receptor-α expression and thymidine incorporation (Asaoka *et al.*, 1992b). Similarly, lysoPC exerts its profound stimulatory effect on HL-60 cell differentiation to macrophage as measured by CDllb expression and appearance of phagocitic activity (Fig. 4) (Asaoka *et al.*, 1993). In either case, lysoPC is active only when both DG and ionomycin are present, indicating that this lysophospholipid interacts with the PKC pathway. Other lysophospholipids including lysophosphatidic acid (lysoPA) are practically inactive, except for lysophosphatidylethanolamine (lysoPE), which is slightly active as lysoPC. In a cell-free enzymatic reaction, lysoPC enhances the DG-dependent PKC activation at micromolar concentrations, but the mechanism underlying the biological action of lysoPC observed in intact cells remains unknown. A possibility may not be ruled out, however, that lysoPC facilitates intercalation of DG and phorbol ester into the membrane.

LysoPA has been proposed to be mitogenic by itself for various cell types. This action has been postulated to be mediated by activating tyrosine kinases and by Ras activation through stimulation of a specific cell surface G-protein-coupled receptor (Hordijk *et al.*, 1994). However, the putative receptor of this intriguing lipid action remains to be identified.

MECHANISM OF PHOSPHOLIPASE A_2 ACTIVATION

It was first suggested that, in cultured rat thyroid cells, PLA_2 may be activated by stimulation of α1-adrenergic receptor that is coupled to a pertussis toxin-sensitive G-protein (Axelrod *et al.*, 1988). Similarly, in rabbit platelets, histamine stimulates PLA_2 probably through the H1-receptor. A cytosolic PLA_2 with 85 kDa molecular size, which is found in many tissues, has been isolated, and its structure is elucidated. This enzyme is phosphorylated by mitogen-activated protein kinases (MAP kinases) at a specific threonyl residue, and subsequently bound to the membrane in a Ca^{2+}-dependent manner to reveal its catalytic activity (Lin *et al.*, 1993; Kramer *et al.*, 1993). This enzyme cleaves preferentially arachidonic acid. Since MAP kinases are located at the downstream of tyrosine kinases as well as most likely of PKC, this high molecular weight cytosolic PLA_2 is regulated indirectly by receptor stimulation.

A major question that remains to be answered, however, is the mechanism underlying receptor-mediated activation of Ca^{2+}-independent cytosolic PLA_2 enzymes that cleave various *cis*-unsaturated fatty acids non-selectively. Several Ca^{2+}-independent cytosolic PLA_2 with about 40 kDa molecular size have been described, but regulatory mechanism of this class of enzymes is unknown (Ackermann *et al.*, 1994).

The cells of HL-60 and U-937, when electro-permeabilized, release spontaneously various *cis*-unsaturated fatty acids, most abundantly oleic and arachidonic acids, especially in the acidic and alkaline medium (Table I) (Tsujishita *et al.*, 1994). This release reaction is accompanined by the formation of some lysoPC and lysoPE. The reaction in an alkaline medium is considerably accelerated by the addition of GTP-γ-S as well as aluminum fluoride, suggesting a potential role of a trimeric G-protein (Table II). The reaction is activated also by vanadate, and inhibited by genistein, implying that tyrosine phosphorylation is involved. This fatty acid release reaction is further affected by both PKC activators and inhibitors, but

		Peak I (acidic pH)	Peak II (alkaline pH)
		(nmoles/2x107 cells/hour)	
Exp 1	Palmitic acid	7.90	0.34
	Palmitoleic acid	2.55	2.50
	Stearic acid	2.05	3.13
	Oleic acid	13.20	14.80
	Linoleic acid	3.23	2.94
	Linolenic acid	2.95	2.62
	Arachidonic acid	6.62	8.86
Exp 2	Lyso PC	4.27	1.13
	Lyso PE	4.56	1.10

Table I. Release of various fatty acids and formation of lysophospholipids in permeabilized U937 cells. The cells were suspended in a buffer medium at pH 5.0 or pH 9.0, and electro-permeabilized. After incubation for one hour, fatty acids were extracted and determined with an HPLC system using ADAM fluorescent probe. Lysophospholipids were determined in a separate experiment with radioactive choline and ethanolamine. Detailed experimental conditions were described elsewhere (Tsujishita *et al.*, 1994). LysoPC, lysophosphatidylcholine; lysoPE, lysophosphatidylethanolamine.

Regulation	PLA_2 (Peak I) (acidic pH)	PLA_2 (Peak II) (alkaline pH)
Ca^{2+}	2 20 x 10^{-8} M	–
G-protein mechanism		
GTP-γ-S	–	+
Alminium fluoride	–	++
Phosphorylation mechanism		
Tyrosine-Kinase ?		
Va + ATP	–	++
Va + ATP + GTP-γ-S	–	+++
Va + ATP + genistein	–	–
Protein kinase C ?		
TPA + ATP	– *	+
DiC_8 + ATP	+	+

* DiC8 but not TPA enhances fatty acid release

Table II. Potential regulatory mechanism of putative PLA_2 enzymes in U937 cells. The detailed explanations were described elsewhere (Tsujishita *et al.*, 1994).

is insensitive to Ca^{2+} concentrations. The evidence available presently is all indirect, but it is suggestive that several Ca^{2+}-insensitive cytosolic, arachidonic acid-nonselective type of PLA_2 enzymes are linked to receptor stimulation (Tsujishita *et al.*, 1994). On the other hand, the fatty acid release reaction in an acidic medium is sensitive to low Ca^{2+} concentrations, but does not appear to be coupled directly to receptor stimulation. The reaction is stimulated by a membrane-permeant DG but not by phorbol ester. This is presumably due to membrane perturbation by DG.

HETEROGENEITY OF PROTEIN KINASE C ACTIVATION

Eleven isoforms of PKC have been presently identified in mammalian tissues, and these isoforms are divided into three groups; classical or conventional PKC isoforms (cPKC), new PKC isoforms (nPKC), and atypical PKC isoforms (aPKC) (Fig. 5) (see for reviews: Dekker & Parker, 1994; Newton 1993; Hug & Sarre, 1993; Bell & Burn, 1991, Nishizuka, 1988). The known members of the PKC family all required phosphatidylserine (PS). The membrane phospholipids *per se*, however, are inert for the activation of PKC. When cells are stimulated, the PKC isoforms exhibit their catalytic activities in subtly

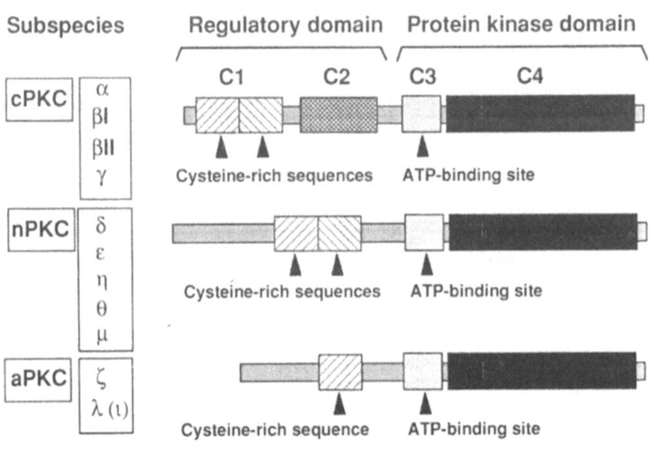

Fig. 5. Common structure of the PKC family. Detailed explanation is given elsewhere (Nishizuka, 1992).

different ways, in response to various lipid metabolites which are newly produced in membranes.

The amino terminal half of the regulatory domain of cPKC isoforms contains two conserved regions, C1 and C2, which play a critical role in the regulation of enzyme activation. The nPKC and aPKC isoforms lack the C2 region. The carboxyl terminal half of the enzyme contains additional two highly conserved regions, C3 and C4. The C3 region contains the ATP-binding consensus sequence, and the C4 region is responsible for protein substrate binding.

The C1 region comprises a preudosubstrate sequence followed by two tandem repeats of a cysteine-rich zinc finger-like motif. This structure is designated as a zinc butterfly because it coordinates two zinc atoms (Quest *et al.*, 1994). This zinc finger-like motif is responsible for DG and phorbol ester binding. The aPKC isoforms, like Raf-1, contain only one cysteine-rich sequence. The C2 region in the regulatory domain of the cPKC (α, βI, βII, and γ) isoforms is apparently needed for Ca^{2+}-sensitivity. This region, called CalB, is also noted in cytosolic PLA_2, GAP, PLCγ, and synaptotagmin. The region interacts with phospholipids in a Ca^{2+}-dependent fashion, and is implicated in translocation to membranes upon increase in the Ca^{2+} concentration. DG activates the cPKC isoforms in such a way that it increases greatly an apparent affinity of the enzyme for Ca^{2+} to the micromolar range. However, this range of Ca^{2+} concentrations is physiologically still high, and *cis*-unsaturated fatty acids, together with DG, further increase the affinity of the enzyme for Ca^{2+}, thereby causing full enzyme activation at the basal level of Ca^{2+} as described above.

The nPKC (δ, ε, η, θ, and μ) isoforms, which lack the C2 region, do not require Ca^{2+} for their activation. The μ-isoform is tentatively disclosed to be a member of the nPKC group, based on its primary structure (Johannes *et al.*, 1994). It contains two cysteine-rich zinc finger-like motifs, but has a long N-terminal sequence with a potential transmembrane domain. The ε-isoform is activated by *cis*-unsaturated fatty acids. On the other hand, the δ-isoform is activated slightly by *cis*-unsaturated fatty acids in the absence of DG, whereas the DG-dependent activation of this isoform is makedly inhibited by the fatty acids. The δ-, ε-, and η-isoforms are activated significantly by PIP_3, a product of PI-3 kinase, and to lesser extent by PIP_2, although selective activation of the ζ-isoform by PIP_3 has been reported. The η-isoform is activated by cholesterol sulfate, but the second messenger role of this unique lipid remains to be explored (Ikuta *et al.*, 1994).

The δ- and ε-isoforms exist in phosphorylated forms in native tissues, and appear as doublet or triplet bands upon gel electrophoresis. The δ-isoform is shown to be phosphorylated at tyrosyl residues, and this phosphorylation is enhanced by treatment of cells with phorbol esters (Denning *et al.*, 1993; Li *et al.*, 1994). The enzymatic activity of this isoform does not appear to be significantly affected by this tyrosine phosphorylation.

The aPKC (ζ and λ(ι)) isoforms also lack the C2 region, and contain only one cysteine-rich zinc finger-like motif. The ζ-isoform is dependent on PS for its catalytic activity, and is activated by *cis*-unsaturated fatty acids, and by equally PIP_2 and PIP_3. This isoform does not respond to DG nor to phorbol esters. The ι-isoform and the λ-isoform appear to be variants of a single entity, and show the highest amino acid sequence identity with the ζ-isoform (Selbie *et al.*, 1993; Akimoto *et al.*, 1994). This isoform does not bind phorbol esters. Presently, the physiological signal pathway to activate the aPKC isoforms is unknown.

Based on the observations outlined above, it is suggestive that the members of the PKC family respond differently to various combinations of lipids, including several phospholipids particularly PS, DG, fatty acids, lysoPC, and probably many other lipid metabolites in membranes, and hence the patterns of activation of the PKC isoforms may vary in extent, duration, and intracellular localization (Table III) (Nishizuka, 1992). In fact, the PKC isoforms so far examined show distinct tissue distribution and specific intracellular

	Subspecies	Amino acid residues	Lipid mediators	Tissue expression
cPKC	α	672	Ca^{2+}, DG, PS, FFA, LysoPC	Universal
	βI	671	"	Some tissues
	βII	673	"	Many tissues
	γ	697	"	Brain only
nPKC	δ	673	DG, PS	Universal
	ε	737	DG, PS, FFA, PIP3	Brain and others
	η	683	DG, PS, PIP3, Cholesterol sulfate	Skin, lung, heart
	θ	707	?	muscle, T cell etc.
	μ	912	?	NRK cells
aPKC	ζ	592	PS, FFA, PIP3?	Universal
	λ (ι)	587	?	Many tissues

Table III. Members of the PKC family from mammalian tissues. The activators for each subspecies are determined with calf thymus H1 histone and bovine myelin basic protein as model phosphate acceptors. The detailed enzymological properties of the θ-, μ-, and λ-subspecies have not yet been clarified. PS, phosphatidylserine; FFA, *cis*-unsaturated fatty acid. Other abbreviations, see text.

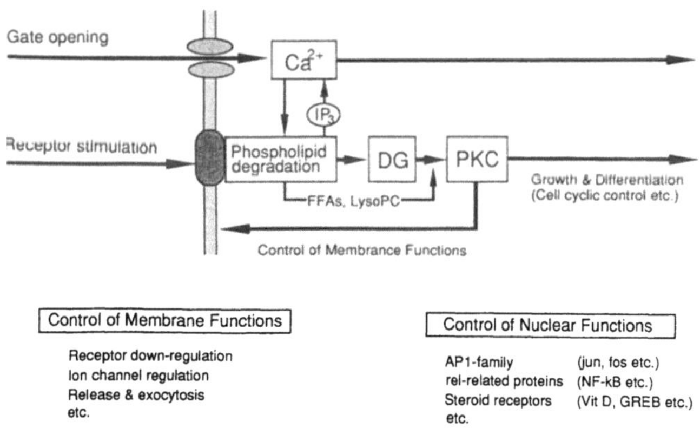

Fig. 6. Potential roles of PKC pathway in cellular regulation. IP_3, inositol-1,4,5-trisphosphate; FFA, *cis*-unsaturated fatty acid. Other abbreviations, see text. Adapted from Nishizuka (1992)

localization as documented repeatedly (see for a reviews: Nishizuka, 1988).

ROLE OF PKC AND PERSPECTIVES

Extensive studies have been made to identify specific roles of each isoform. Some isoforms may play crucial roles in regulating membrane functions, such as down-regulation of receptors, modulation of properties of ion channels and pumps, and metabolic regulation of membrane phospholipids that generate lipid mediators for cell signalling (Fig. 6) (see for a review; Nishizuka, 1986). Release and exocytosis are well known to be under control of the PKC signalling pathway. PKC itself as well as a protein kinase cascade leading to the activation of MAP kinases plays pivotal roles in cellular regulation (Blenis, 1993; Nishida & Gotoh, 1993; Lange-Carter *et al.*, 1993). For signalling to the nucleus, tyrosine kinases associated with growth factor receptors activate MAP kinases through a series of provocative protein-protein interactions

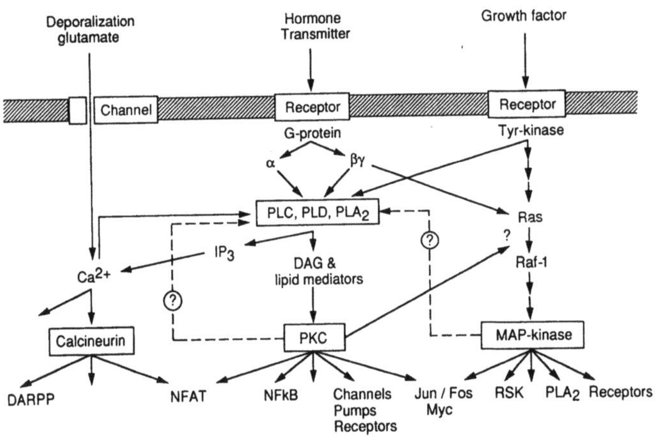

Fig. 7. Possible intracellular signalling network involving PKC. IP3, inositol-1,4,5-trisphosphate; DAG, diacylglycerol; RSK, cell cycle-regulated S6 protein kinase; DARPP, dopamine and cAMP regulated phosphoprotein. Other abbreviations, see text.

involving Ras and Raf-1 (Fig. 7). It is almost certain that stimulation of G-protein-coupled receptors also activates MAP kinases through PKC signalling pathway. The α-isoform has been proposed to phosphorylate and activate Raf-1 (Kolch *et al.*, 1993) directly, which in turn activates MAP kinases though a protein kinase cascade. However, the precise mechanism of the PKC signalling pathway or network to cause the MAP kinase activation has not been unequivocally established (MacDonald *et al.*, 1993). It has been proposed that, in T- and B-lymphocytes, PKC links with MAP kinases at the point of Ras activation (Downward *et al.*, 1990; Harwood *et al.*, 1993).

When PKC was first found to be activated by DG, it was thought that this DG may be derived solely from the hydrolysis of PI that is induced by hormones and neurotransmitters. The pertinent question that followed was obviously how PKC activation evokes cellular responses. Since then, extensive studies in numerous laboratories have given answers to a part of this question. Today, it is becoming clearer that several other membrane phospholipids may also be involved in transmitting information of a variety of extracellular signals across the membrane. However, it remains to be explored whether most of the lipid metabolites discussed herein indeed act as the signal mediators or messengers in physiological processes, because there is till a large distance between observations in *in vitro* studies and those in biological responses.

The principal architecture of major signalling pathways is, nevertheless, begins to emerge. Interactions among various pathways together with the multiplicity of the PKC family produce enormous variation of signalling network and cellular responses. Presumably, the function of each member of the PKC family may well coordinate with dynamic intracellular membrane lipid metabolism (Nishizuka, 1992). Cellular responses often persist longer, and sustained activation of PKC may be essential in this processes. However, unusually persistent activation of PKC such as that caused by phorbol esters, stable overexpression, or by alteration of lipid metabolism may all result in pathological responses, such as tumorigenesis. Spatiotemporal aspects and switch-off mechanism of the PKC activation are inevitable to be explored for understanding more the biochemical basis of cellular regulation.

ACKNOWLEDGEMENT:

Skillful secretarial assistance of Mrs. s. Nishiyama is cordially acknowledged. This work was supported in part by research grants from the Special Research Fund of the Ministry of Education, Science, and Culture, Japan; Merck sharp & Dohyme Research Laboratories; Yamanouchi Foundation for Research on Metabolic Disorders; Sankyo Foundation of Life Sciences; new Lead Research Laboratories of Sankyo Company; and Terumo Life Science Foundation.

REFERENCES:

Ackermann EJ, Kempner ES and Dennis EA (1994) Ca^{2+}-independent cytosolic phospholipase A_2 from macrophage-like $P388D_1$ cells. J. Biol. Chem. 269: 9227-9233

Akimoto K, Mizuno K, Osada S, Hirai S, Tanuma S, Suzuki K and Ohno S (1994) A new member of the third class in the protein kinase C family, PKCλ, expressed dominantly in an undifferentiated mouse embryonal cartinoma cell line and also in many tissues and cells. J. Biol. Chem. 269: in press

Anthes JC, Wang P, Siegel MI, Egan RW and Billah MM (1991) Granulocyte phospholipase D is activated by a guanine nucleotide dependent protein factor. Biochem. Biophys. Res. Commun. 175: 236-243

Asaoka Y, Oka M, Yoshida K and Nishizuka Y (1991) Metabolic rate of membrane-permeant diacylglycerol and its relation to human resting T-lymphocyte activation. Proc. Natl. Acad. Sci. USA 88: 8681-8685

Asaoka Y, Nakamura S, Yoshida K and Nishizuka Y. (1992a) Protein kinase C, calcium and phospholipid degradation. Trends Biochem. Sci. 17: 414-417

Asaoka Y, Oka M, Yoshida K, Sasaki Y and Nishizuka Y (1992b) Role of lysophosphatidylcholine in T-lymphocyte activation: Involvement of phospholipase A_2 in signal transduction through protein kinase C. Proc. Natl. Acad. Sci. USA 89: 6447-6451
Asaoka Y, Yoshida K, Sasaki Y and Nishizuka Y (1993) Potential role of phospholipase A_2 in HL-60 cell differentiation to macrophages induced by protein kinase C activation. Proc. Natl. Acad. Sci. USA 90: 4917-4921
Axelrod J, Burch RM and Jelsema CL (1988) Receptor-mediated activation of phospholipase A_2 via GTP-binding proteins: arachidonic acid and its metabolites as second messengers. Trends Neurosci. 11: 117-123
Bell RM and Burns DJ (1991) Lipid activation of protein kinase C. J. Biol. Chem. 266: 4661-4664
Billah MM and Anthes JC (1990) The regulation and cellular functions of phosphatidylcholine hydrolysis. Biochem. J. 269: 281-291
Blenis J (1993) Signal transduction via the MAP kinases: Proceed at your own RSK. Proc. Natl. Acad. Sci. USA 90: 5889-5892
Brown HA, Gutowski S, Moomaw CR, Slaughter C and Sternweis PC (1993) ADP-ribosylation factor, a small GTP-dependent regulatory protein, stimulates phospholipase D activity. Cell 75: 1137-1144
Chen SG and Murakami K (1992) Synergistic activation of type III protein kinase C by *cis*-fatty acid and diacylglycerol. Biochem. J. 282: 33-39
Cockcroft S (1992) G-protein-regulated phospholipases D, D and A_2-mediated signalling in neutrophils. Biochim. Biochys. Acta 1113: 135-160
Cockcroft S, Thomas GMH, Fensome A, Geny B, Cunningham E, Gout I, Hiles I, Totty NF, Truong O and Hsuan JJ (1994) Phospholipase D: a downstream effector of ARF in granulocytes. Science 263: 523-526
Cuadrado A, Carnero A, Dolfi F, Jiménez B and Lacal JC (1993) Phosphorylcholine: a novel second messenger essential for mitogenic activity of growth factors. Oncogene 8: 2959-2968
Dekker LV and Parker PJ (1994) Protein kinase C —a question of specificity. Trends Biochem. Sci. 19: 73-77
Denning MF, Dlugosz AA, Howett MK and Yuspa SH (1993) Expression of an oncogenic ras^{Ha} gene in murine keratinocytes induces tyrosine phosphorylation and reduced activity of protein kinase C δ. J. Biol. Chem. 268: 26079-26081
Dennis EA, Rhee SG, Billah MM and Hannun YA (1991) Role of phospholipases in generating lipid second messengers in signal transduction. FASEB J 5: 2068-2077
Downward J, Graves JD, Warne PH, Rayter S and Cantrell DA (1990) Stimulation of $p21^{ras}$ upon T-cell activation. Nature (London) 346: 719-723
Exton JH (1990) Signaling through phosphatidylcholine breakdown. J. Biol. Chem. 265: 1-4

Harwood AE and Cambier J (1993) B cell antigen receptor cross-linking triggers rapid protein kinase C independent activation of $p21^{ras1}$. J. Immunol. 151: 4513-4522
Hordijk PL, Verlaan I, van Corven EJ and Moolenaar WH (1994) Protein tyrosine phosphorylation induced by lysophosphatidic acid in rat-1 fibroblasts. Evidence that phosphorylation of MAP kinase is mediated by the Gi-$p21^{ras}$ pathway. J. Biol. Chem. 269: 645-651
Hug H and Sarre TF (1993) Protein kinase C isoenzymes: divergence in signal transduction? Biochem. J. 291: 329-343
Ikuta T, Chida K, Tajima O, Matsuura Y, Iwamori M, Ueda Y, Mizuno K, Ohno S and Kuroki T (1994) Cholesterol sulfate, a novel activator for the η isoform of protein kinase C. Cell Growth & Differentiation: in press
Johannes F-J, Prestle J, Eis S, Oberhagemann P and Pfizenmaier K (1994) PKCμ is a novel, atypical member of the protein kinase C family. J. Biol. Chem. 269: 6140-6148
Kolch W, Heidecker G, Kochs G, Hummel R, Vahidi H, Mischak H, Finkenzeller G, Marmé D, and Rapp UR (1993) Protein kinase Cα activates Raf-1 by direct phosphorylation. Nature (London) 364: 249-252
Kramer RM, Roberts EF, Manetta JV, Hyslop PA and Jakubowski J.A. (1993) Thrombin-induced phosphorylation and activation of Ca^{2+}-sensitive cytosolic phospholipase A_2 in human platelets. J. Biol. Chem. 268: 26796-26804
Kusner DJ, Schomisch SJ and Dubyak GR (1993) ATP-induced potentiation of G-protein-dependent phospholipase D activity in cell-free system from U937 promonocytic leukocytes. J. Biol. Chem. 267: 19973-19982
Lange-Carter CA, Pleiman CM, Gardner AM, Blumer KJ, and Johnson GL (1993) A divergence in the MAP kinase regulatory network defined by MEK kinase and Raf. Science 260: 315-319
Li W, Mischak H, Yu J-C, Wang L-M, Mushinski JF, Heidaran MA and Pierce JH (1994) Tyrosine phosphorylation of protein kinase C-δ in response to its activation. J. Biol. Chem. 269: 2349-2352
Liscovitch M (1992) Crosstalk among multiple signal-activated phospholipases. Trends Biochem. Sci. 17: 393-399
Lin L-L, Wartmann M, Lin AY, Knopt JL, Seth A and Davis RJ (1993) $cPLA_2$ is phosphorylated and activated by MAP kinase. Cell 72: 269-278
MacDonald SG, Crews CM, Wu L, Driller J, Clark R, Erikson RL and McCormick F (1993) Reconstitution of the Raf-1—MEK—ERK signal transduction pathway *in vitro*. Mol. Cell. Biol. 13: 6615-6620
Newton AC (1993) Interaction of proteins with lipid head groups: lessons from protein kinase C. Annu. Rev. Biophys. Biomol. Struct. 22: 1-25
Nishida E and Gotoh Y (1993) The MAP kinase cascade is essential for diverse signal transduction pathways. Trends Biochem. Sci. 18: 128-131

Nishizuka Y (1986) Studies and perspectives of protein kinase C. Science 233: 305-312

Nishizuka Y (1988) The molecular heterogeneity of protein kinase C and its implications for cellular regulation. Nature (London) 334: 661-665

Nishizuka Y (1992) Intracellular signaling by hydrolysis of phospholipids and activation of protein kinase C. Science 258: 607-614

Olson SC, Bowman EP and Lambeth JD (1991) Phospholipase D activation in a cell-free system from human neutrophils by phorbol 12-myristate 13-acetate and guanosine 5'-*O*-(3-thiotriphosphate) J. Biol. Chem. 266: 17236-17242

Quest AFG, Bardes ESG and Bell RM (1994) A phorbol ester binding domain of protein kinase Cγ. Deletion analysis of the cys2 domain defines a minimal 43-amino acid peptide. J. Biol. Chem. 269: 2961-2970

Selbie LA, Schmitz-Peiffer C, Sheng Y and Biden TJ (1993) Molecular cloning and characterization of PKCι, an atypical isoform of protein kinase C derived from insulin-secreting cells. J. Biol. Chem. 268: 24296-24302

Shinomura T, Asaoka Y, Oka M, Yoshida K and Nishizuka Y (1991) Synergistic action of diacylglycerol and unsaturated fatty acid for protein kinase C activation: Its possible implications. Proc. Natl. Acad. Sci. USA 88: 5149-5153

Tsujishita Y, Asaoka Y and Nishizuka Y (1994) Regulation of phospholipase A_2 in human leukemia cell lines: its implication for intracellular signaling. Proc. Natl. Acad. Sci. USA 91: 6274-6278

Yoshida K, Asaoka Y and Nishizuka Y (1992) Platelet activation by simultaneous actions of diacylglycerol and unsaturated fatty acids. Proc. Natl. Acad. Sci. USA 89: 6443-6446

PHOSPHOLIPASE D REGULATION IN RELATIONSHIP WITH SEVERAL PROTEIN KINASES ACTIVATION IN HUMAN NEUTROPHIL

Valérie Planat, Jean-Marc Lanau, Michel Record et Hugues Chap

INSERM Unité 326
Hôpital Purpan
31059 Toulouse, France

INTRODUCTION

The hydrolysis of phosphatidylcholine by phospholipase D (PLD) is now recognized as an important source of second messengers in many different signal transduction. The mechanisms by which agonists couple to PLD activity are not clearly understood, but there are a number of evidences suggesting that a variety of different pathways may occur. During the activation process of human polymorphonuclear neutrophil (PMN), PLD activity can be involved, and is regulated by many proteins -G proteins, tyrosine-kinase, serine/threonine kinase-, and also by lipids.
Stimulation of PLD activity by guanine nucleotide in permeabilized neutrophils, as well as in G protein-linked fMLP receptor, suggest a regulation by heterotrimeric G proteins. This is confirmed with the inhibition of PLD by pertussis toxin in rabbit peritoneal neutrophils (Kanaho et al., 1991). *In vitro* PLD studies have shown that this phospholipase activity required the presence of both membrane and cytosolic components for either activation. A cytosolic factor, needed for full activation of PLD has been proposed to be a GTP-binding protein, the ADP ribosylation factor, Arf (Brown et al., 1993 - Cockcroft et al.,1994). PLD activation is also associated with an increase in protein tyrosine phosphorylations, since the activity is decreased with a treatment by tyrosine kinase inhibitor. Many candidates are possible as substrate. Gomez-Cambronero et al. (1989) described phosphorylation of a p40kDa and p92kDa proteins, identified as $Gi_{2\alpha}$ and GM-CSF receptor respectively. Others proteins tyrosine phosphorylated in PMN have also been discovered, like MAP-kinases p42 and p44, under fMLP stimulation (Torres et al., 1993). However, phosphorylation of these proteins have not yet been directly related to PLD activation. PLD has been shown to be activated by PKC in particulate fractions, or its activators like phorbol esters (Nakamura et al., 1994), and we found similar result on PMN. By contrast, we also observed that fMLP

NATO ASI Series, Vol. H 92
Signalling Mechanisms – from Transcription Factors to Oxidative Stress
Edited by L. Packer, K. Wirtz

induced-PLD activity is PKC independent. It has recently been repported that phosphatidylinositol 4,5-*bis*phosphate (PIP_2) and phosphatidylinositol 3,4,5-*tris*phosphate (PIP_3) potently stimulate a brain membrane PLD activity *in vitro*, indicating that PIP_2 cycle could be important and determinant in regulating PLD activation (Liscovitch et al., 1994). However, there are little informations on the biochemical and molecular properties of signal-activated PLDs, and to date none of the mammalian PLD has been purified or cloned.

EXPRIMENTAL PROCEDURE

Neutrophil preparation and activation

Alkyl-phosphatidylcholine pool of human PMN (1µCi/2.10^7cells) was labelled *in vivo* by conversion of the precursor, 1-O[^{3}H]octadecyl-2-lyso-GPC, into phosphatidylcholine for 15min at 37°C in Hepes buffer. Labelled cells (2.10^7cells/ml) were stimulated by agonist (10^{-6}M fMLP or 10^{-7}M PMA). Reaction was arrested at indicated time, and lipids were extracted according to Bligh and Dyer (Bligh et al., 1959). Then lipids were separated on TLC using the solvant system chloroforme/methanol/acetic acid (90:10:10 - v/v). Products of PLD activity; phosphatidic acid, or phosphatidylethanol (in the presence of 0,5% ethanol during stimulation) were scrapped and counted for radioactivity. For checking the role of different kinases, neutrophils were treated with the corresponding inhibitor for 15min at 37°C, just before stimulation.

Immunoprecipitation

Total PMN lysate was obtained upon treatment of the cells with 1% Triton X100 and 0,1% SDS in a lysis buffer. Immunoprecipitation of protein phosphorylated on tyrosine residue was realised for 180min at 4°C in the presence of a specific agarose-conjugated antibody (anti P-Tyr (Ab1-A)). Tyrosine phosphorylated proteins were then eluted by competition with phenylphosphate 10mM, separated on 11% SDS-PAGE and visualised by silver stain method.

Western blotting

Proteins separated on 11% SDS-PAGE were transfered onto nitrocellulose membrane, and immunoblotted for 2 hours at 4°C with a MAP-kinase antibody (anti ERK1). MAP-kinase were detected by enhanced chimiluminescence system.

RESULTS AND DISCUSSION

Stimulation of human neutrophils by a variety of agonists leads to the activation of a PLD, hydrolising phosphatidylcholine into phosphatidic acid (PA) and choline. In the presence of 0,5% ethanol, phosphatidylethanol are produced instead of PA, by a specific PLD transphosphatidylation reaction. In the present work, we have studied the involvement of different agonists and protein kinases in the regulation of PLD activity.

Stimulation of PLD activity by two different agonists.

PLD can be activated by a pannel of agonists in various cell type. We used two of them, which act through different mechanisms. The first one is fMLP, formyl-Methionyl-Leucyl-Phenylalanine, a chemottractant peptide by which PMN activation is receptor-mediated. The second one is PMA, Phorbol-Myristate-Acetate, which is able to penetrate into the cell, through the plasma membrane. We observed that fMLP and PMA both stimulate PLD activity, but in different time-frame as shown in figure1.

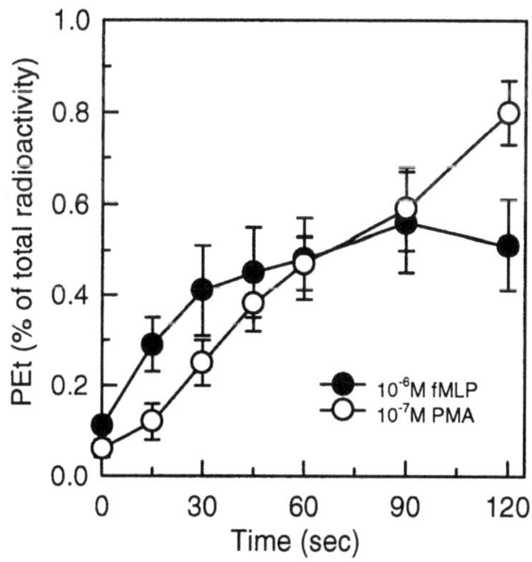

Figure 1. Time course activation of PLD activity. Labelled cells (1µCi/2.10^{7}cells) were stimulated with fMLP or PMA in the presence of 0,5% ethanol. Labelled products were separated on T.L.C. and counted for radioactivity. Phosphatidylethanol (PEt) produced are expressed in % of total radioactivity per sample +/- S.E.M. (n=3).

The maximum of PEt production reach 0,5% with 10^{-6}M fMLP, and 0,8% with 10^{-7}M PMA. In addition, the time course of PC hydrolysis is also different with PMA and fMLP. The maximum PLD activity is obtained at 1min with fMLP as agonist, whereas with PMA, PLD activity levelled off beyong 2min (Fig.1).

Involvement of protein kinase C

The point was to investigate whether or not protein kinase C is involved in the pathway activated by each of these agonists. PMA is known to act through direct stimulation of PKC. fMLP binds to a seven transmembrane-G protein linked receptor, before stimulating a PI-PLC activity, producing diglycerides which finally induce a PKC activation. So, each of the two agonists could stimulate PLD activity via PKC activation. To check for this hypothesis, we used a specific PKC inhibitor, a bisindolylmaleimide component, GF 109203X (Toullec et al., 1991). Following preincubation with different concentrations of GF 109203X, PLD activity was stimulated by 10^{-6}M fMLP for 1min or 10^{-7}M PMA for 5min. Results indicate that fMLP induced-PLD activity is independent of PKC activation, since PLD activity was unchanged in the presence of GF 109203X. By contrast, PMA induced-PLD activity required PKC activation (Fig.2). This is in agreement with the existence of several pathways leading to PLD activation.

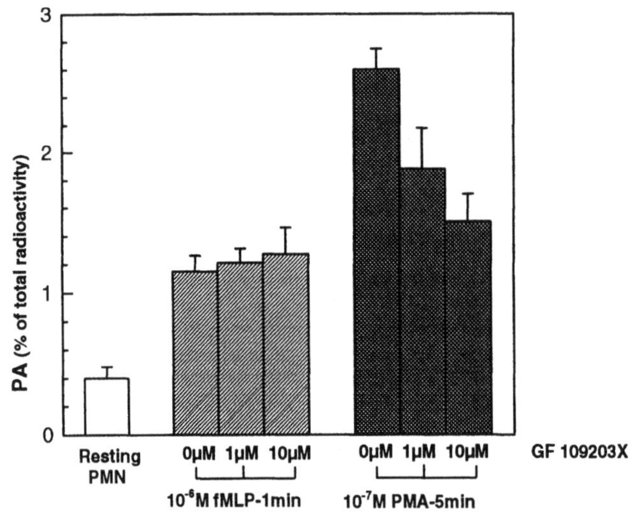

Figure 2. Dose-effect of PKC inhibitor (GF 109203X) on PLD activity stimulated with fMLP or PMA. Labelled cells (1µCi/2.10^7cells) were incubated for 15min at 37°C with or without different concentrations of GF 109203X and then stimulated for 1min with fMLP or 5min with PMA. Phosphatidic acid (PA) produced were expressed in % of total radioactivity per sample +/-S.E.M. (n=5).

Involvement of protein tyrosine kinases

Both fMLP and PMA increase the phosphotyrosine content of several neutrophil proteins (Thompson et al., 1991 - Uings et al., 1992). In order to determine the importance of these tyrosine phosphorylation for PLD activity, a tyrosine kinase inhibitor, genistein has been used (Akiyama et al., 1987). Cells were preincubated with 300μM genistein, then stimulated with fMLP or PMA. In each case PLD activity is inhibited but in different extend when stimulated with PMA or fMLP (around 25% and 80% inhibition respectively). Thereby, it appears that tyrosine kinases are involved in the PLD pathways triggered by fMLP and PMA. The further aim was to identify the type of protein tyrosine kinase involved and their targets. As previously described (Grinstein et al., 1992 - Torres et al., 1993), MAP-kinases are phosphorylated in fMLP-stimulated PMN. According to with this result we observed a time dependent increase of MAP-kinases p42 and p44 tyrosine phosphorylation under fMLP stimulation. Moreover, we found that PMA stimulation triggers only p42 MAP-kinase phosphorylation, in agreement with previous report on U937 cells (Adams et al., 1991). In both cases, phosphorylation were decreased by 300μM genistein or 200μM tyrphostin (a different tyrosine kinase inhibitor). Immunodetection methods clearly show that time course of PLD activation is similar to that of MAP-kinases phosphorylation. These first results let us suggest that MAP-kinases may be good candidates to be required in PLD activation. Moreover, recent evidences reveal that the Ras/Raf/MEK cascade is a primary response to fMLP leading to MAP-kinases phosphorylation and activation (Worthen et al., 1994). But up to now, other tyrosine kinases involved and most of their target proteins have not been identify, and relationship between these proteins and PLD remain to be established.

Requirement of calcium

The effect of calcium on PLD activation has been investigated in cells stimulated in the presence, or the absence of calcium (upon addition of 25μM Quin 2-AM and 1mM EGTA). We found that the presence of calcium was necessary for PLD activity triggered by fMLP. But by contrast, PMA-induced PLD activity is calcium independent. The same treatments were realised before checking for MAP-kinases phosphorylation. Results indicate that MAP-kinase phosphorylation induced by fMLP is still obtained in the absence of calcium, but is inhibited when the agonist is PMA (Fig.3). According to these results, PLD activation and MAP-kinases phosphorylation induced by these two agonists are differently regulated by calcium, providing the evidence that PLD and MAP-kinases activations are not related.

	1min fMLP stimulation		5min PMA stimulation	
	1mM $CaCl_2$	25µM Quin 1mM EGTA	1mM $CaCl_2$	25µM Quin 1mM EGTA
PLD activitation	+	-	+	+
MAP-kinase-phosphorylation	+ (p42, p44)	+	+ (p42)	-

Figure 3. Differential regulation by calcium of PLD activity compared to MAP-kinases phosphorylation.

In conclusion, our experiments suggest that protein kinases differently regulate PLD activation, depending upon the agonist. Using the receptor agonist fMLP, PLD activation is regulated by tyrosine kinases, and not by the serine/threonine kinases PKC. Instead, with the non-receptor agonist PMA, PKC is required for *in situ* PLD activation. In that respect, cell treatment with genistein has a weak effect on PMA-induced PLD activation. Results relative to calcium requierement show a dissociation between the PLD activation and MAP-kinases phosphorylation induced by fMLP and PMA.

REFERENCES

Adams P.D., Parker P.J. (1991) TPA-induced activation of MAP kinase. *F.E.B.S.* 290:77-82.

Akiyama T., Ishida J., Nakagawa S., Ogawara H., Watanabe S., Itoh N., Shibuya M., Fukami Y. (1987) Genistein, a Specific Inhibitor of Tyrosine-specific Protein Kinases. *J. Biol. Chem.* 262:5592-5595.

Bligh E.G., Dyer W.J. (1959) A rapid method of total lipid extraction and purification. *Can. J. Biochem. Physiol.* 37:911-917.

Brown H.A., Gutowski S., Moomaw C.R., Slaughter C., Sternweis P.C. (1993) ADP-Ribosylation Factor, a Small GTP-Dependent Regulatory Protein, Stimulates Phospholipase D Activity. *Cell* 75:1137-1144.

Cockcroft S., Thomas G.M.H., Fensome A., Geny B., Cunningham E., Gout I., Hiles I., Totty N.F., Truong O., Hsuan J.J. (1994) Phospholipase D: A Downstream Effector of ARF in Granulocytes. *Science* 263:523-526.

Gomez-Cambronero J., Huang C.K., Bonak V., Wang E., Casnellie J.E., Shiraishi T., Sha'afi R.I. (1989) Tyrosine phosphorylation in human neutrophil. *Biochem. Biophys. Res. Comm.* 162:1478-1485.

Grinstein S., Furuya W. (1992) Chemoattractant-induced Tyrosine Phosphorylation and Activation of Microtubule-associated Protein Kinase in Human Neutrophils. *J. Biol. Chem.* 267:18122-18125.

Kanaho Y., Kanoh H., Nozawa Y. (1991) Activation of phospholipase D in rabbit neutrophils by fMet-Leu-Phe is mediated by a pertussis toxin-sensitive GTP-binding protein that may be distinct from a phospholipase C-regulating protein. *F.E.B.S.* 279:249-252.

Liscovitch M., Chalifa V., Pertile P., Chen C.S., Cantley L.C. (1994) Novel Fonction of Phosphatidylinositol 4,5-Bisphosphate as a Cofactor for Brain Membrane Phospholipase D. *J. Biol. Chem.* 269:21403-21406.

Nakamura S., Nishizuka Y. (1994) Lipid mediators and Protein Kinase C Activation for the Intracellular Signalling Network. *J. Biochem.* 115:1029-1034.

Thompson H.L., Shiroo M., Saklatvala J. (1993) The chemotactic factor N-formylmethionyl-leucyl-phenylalanine activates microtubule-associated protein 2 (MAP) kinase and a MAP kinase kinase in polymorphonuclear leucocytes. *Biochem. J.* 290:483-488.

Thompson N.T., Bonser R.W., Garland L.G. (1991) Receptor-coupled phospholipase D and its inhibition. *T.rends in Pharmacol..Sci.* 12:404-408.

Torres M., Hall F.L., O'Niel K. (1993) Stimulation of Human Neutrophils with Formyl-Methionyl-Leucyl-Phenylalanine Induces Tyrosine Phosphorylation and Activation of Two Distinct Mitogen-Activated Protein-Kinases. *J. Immunol.* 150:1563-1577.

Toullec D., Pianetti P., Coste H., Bellevergue P., Grand-Perret T., Ajakane M., Baudet V., Boissin P., Boursier E., Loriolle F., Duhamel L., Charon D., Kirilovsky J. (1991) The Bisindolylmaleimide GF 109203X Is a Potent and Selective Inhibitor of Protein Kinase C. *J. Biol. Chem.* 266:15771-15781.

Uings I.J., Thompson N.T., Randall R.W., Spacey G.D., Bonser R.W., Hudson A.D., Garland L.G. (1992) Tyrosine phosphorylation is involved in receptor coupling to phospholipase D but not phospholipase C in the human neutrophil. *Biochem. J.* 281:597-600.

Worthen G.S., Avdi N., Mette Buhl A., Suzuki N., Johnson G.L. (1994) FMLP Activates Ras and Raf in Human Neutrophils. *J. Clin. Invest.* 94:815-823.

Phosphatidic Acid Phosphohydrolase: Its Role in Cell Signalling

Ian N. Fleming and Stephen J. Yeaman
Department of Biochemistry and Genetics
University of Newcastle-Upon-Tyne
Newcastle-Upon-Tyne
NE2 4HH, U.K.

Introduction

Phosphatidic acid phosphohydrolase [PAP; E.C.3.1.3.4] catalyses the dephosphorylation of phosphatidic acid (PA) into diacylglycerol. Regulation of PAP activity is thought to play a prominent role in controlling the rate of glycerolipid biosynthesis by limiting the supply of diacylglycerol, the precursor for triacylglycerol, phosphatidylcholine and phosphatidylethanolamine (Brindley, 1984; Martin *et al*, 1986). PAP can be regulated chronically by hormonal action, mediated at the level of enzyme synthesis and breakdown (Pittner *et al*, 1985), and in the short-term by translocation from the cytosol to the endoplasmic reticulum (Brindley, 1984) where it is thought to be metabolically active. Translocation onto the endoplasmic reticulum is induced by fatty-acids and their CoA esters (Brindley, 1984; Martin-Sanz *et al*, 1984). The enzyme can be displaced from the membranes *in vitro* by albumin and the amphiphilic amine chlorpromazine (Brindley, 1984), suggesting that this is a reversible mechanism. This regulatory mechanism has been suggested as a way in which the cell can coordinate the rate of lipid biosynthesis with the fatty-acid supply (Brindley, 1984).

More recently, PAP has been shown to play an important role in the phospholipase D signalling pathway (Exton, 1990; Billah and Anthes, 1990) by dephosphorylating the PA formed by cleavage of phosphatidylcholine. PA and its derivative lysophosphatidic acid (lysoPA) act as mitogens on fibroblast and non-fibroblast cell lines, through stimulation of polyphosphoinositide hydrolysis, activation of protein kinase C, inhibition of adenylyl cyclase and stimulation of DNA synthesis (Moolenaar *et al*, 1986; Murayama and Ui, 1987; Van Corven *et al*, 1989). Moreover, PA can also directly modulate the activity of several key functional proteins *in vitro*, including protein kinase C (Epand and Stafford, 1990), GTPase-activating protein (Tsai *et al*, 1989) phosphatidylinositol-4-phosphate kinase

NATO ASI Series, Vol. H 92
Signalling Mechanisms – from Transcription Factors to Oxidative Stress
Edited by L. Packer, K. Wirtz

(Moritz *et al*, 1992) and a novel phospholipid dependent protein kinase (Khan *et al*, 1994). The reaction product, diacylglycerol, has a major role as a second messenger, being an activator of protein kinase C (Nishizuka, 1984). Therefore, by controlling the relative levels of these two second messengers PAP could play a pivotal role in cell signalling.

The two functions of PAP appear to be performed by two different enzymes. In rat liver, there is one form of PAP in the plasma membrane thought to be involved in cell signalling and a second isoform, present on the microsomes and cytosol, which is believed to be involved in glycerolipid synthesis (Jamal *et al*, 1991; Day and Yeaman, 1992). These two enzymes differ in their size and charge (Day and Yeaman, 1992), have different Mg^{2+} requirements and can be readily distinguished in cell extracts by a variety of inhibitors including N-ethylmaleimide (NEM) (Jamal *et al*, 1991); PAP in the cytosol and microsomes is completely inhibited by NEM, whereas the plasma membrane enzyme is totally insensitive to this inhibitor.

Despite a number of attempts over the years, neither isoform of PAP has been purified to homogeneity from rat liver. Hence little information exists on the enzyme polypeptides, and the molecular basis of their regulation. Therefore, the aim was to purify and characterise the NEM-insensitive form of PAP from rat liver, to aid understanding of its role in the phospholipase D signalling pathway.

Methods

PAP activity was measured by following the release of [^{32}P] inorganic phosphate from [^{32}P]-labelled PA. [^{32}P]-labelled PA was prepared using the method of Walsh and Bell (1986), and delivered as a uniform mixed micelle with Triton X-100, in a Tris-maleate buffer, pH 7.0 (Day and Yeaman, 1992).

Fractions were incubated in the presence and absence of 4mM NEM for 15 minutes at 37°C before assaying. The NEM-insensitive activity was the activity measured following preincubation with NEM. The assay was carried out at 30°C, with one Unit of PAP being defined as the amount of enzyme which catalyses the release of 1nmol of inorganic phosphate per minute.

After each chromatography step, aliquots of PAP were subjected to SDS-polyacrylamide gel electrophoresis on 10% resolving gels under reducing and non-reducing conditions. The protein samples were precipitated in chloroform/methanol before electrophoresis to remove Triton X-100 (Wessel and Flugge, 1984). Proteins were detected by silver-staining (Heukeshoven and Dernick, 1985). Protein concentrations were determined by the method of Bradford (1976), using bovine serum albumin as the standard.

Results and Discussion

NEM-insensitive PAP was purified from rat liver essentially as by Kanoh *et al* (1992), but with several modifications. All operations were carried out at 4°C. Rat livers were thawed and homogenised in 2 volumes of buffer A (25mM Tris-HCl pH 7.4, 10%(w/v) glycerol, 50mM NaCl, 1mM DTT, 1mM EDTA, 0.2mM EGTA, 1mM benzamidine, 1mM PMSF, and 5μg/ml each of pepstatin A, leupeptin, antipain and soyabean trypsin inhibitor) for approximately 1 min. The homogenate was spun at 6000g for 20 min and the pellet discarded. The supernatant was centrifuged at 100,000g for 1 hour to collect the membranes, which were salt-washed twice by resuspending in 10mM Tris-HCl, pH 7.4, containing 1mM EDTA, 1mM DTT and 3M KCl, and recentrifuged at 100,000g. The membranes were resuspended in buffer A at 5 mg/ml protein.

The enzyme was solubilised by stirring the membranes gently for 1 hour in the presence of 1% (w/v) Octylglucoside. The solution was then centrifuged at 100,000g for 1 hour, and the solubilised PAP recovered in the supernatant. The supernatant was made 1%(w/v) with Triton X-100 before further purification.

The solubilised PAP preparation was slowly made to 40% saturation with solid ammonium sulphate, and stirred for 45 minutes, before centrifugation at 10,000g for 20 min. The floating pellet was discarded, and the supernatant brought up to 70% saturation with ammonium sulphate. After stirring for 30 min, the solution was centrifuged at 10,000g for 20min and the pellet resuspended in approximately 6 ml of buffer B (buffer A containing 1% (w/v) Triton X-100).

The resuspended ammonium sulphate pellet was loaded onto a Sephacryl S300 gel filtration column (2.5 x 65cm), equilibrated with buffer B. The column was run at 0.5 ml/min in the same buffer and 5ml fractions collected. NEM-insensitive PAP eluted at a Ve/Vo of approximately 1.33. The active fractions were pooled and loaded onto hydroxylapatite (2.5 x 8 cm), equilibrated with buffer B. The enzyme passed through the column, where it was collected and the active fractions pooled. The PAP was then loaded onto a heparin-Sepharose column (1.5 x 8 cm), equilibrated with buffer B. The column was washed with 45 ml of buffer B, before the enzyme was eluted with 120 ml of a linear NaCl gradient (0.05-1M) in buffer B at 0.5 ml/min; the PAP was eluted at 250mM NaCl.

Finally, the active enzyme was loaded onto Affi-Gel Blue F3GA (0.5 x 4 cm), equilibrated with buffer B. The column was washed with 20ml of buffer B, before the enzyme was eluted with 20ml of a linear NaCl gradient (0.05-3M) at 0.5 ml/min. The active PAP was eluted by approximately 1.75M NaCl. The fractions containing PAP activity were pooled (10ml total volume) and stored at -80°C until used.

A 5900-fold purification was produced, with a 2.4% yield of PAP activity (not shown). Silver-stained SDS-PAGE indicated that PAP is an 83kDa polypeptide, since it is the only protein band which elutes with the enzyme activity (Figure 1). Superose 6 gel-filtration of the purified PAP in the presence of Triton X-100 indicated a 265kDa native enzyme (not shown). Since PAP is in a Triton X-100 micelle of approximately 90kDa (Leonard and Chrambach, 1984), the native enzyme is 175kDa in size, and is likely to be a dimer.

Interestingly, S300 gel filtration also produced a second peak of PAP activity (in the void) which was also insensitive to inhibition by NEM. This activity eluted from the subsequent purification columns at a different position to the purified enzyme and is not an aggregated form of that enzyme, suggesting that there may be two isozymes of NEM-insensitive PAP in rat liver (not shown).

Whilst purified PAP required detergent for activity, it was not activated by fatty-acids or phospholipids. In contrast, amphiphilic amines, Mn^{2+} and Zn^{2+} all inhibited the purified enzyme, whereas Mg^{2+} and NEM had no effect (not shown), confirming that it corresponds to the enzyme described previously in rat liver membranes (Jamal *et al*, 1991).

A number of substrate analogues were tested for their ability to modulate the activity of purified NEM-insensitive PAP. Glycerol-3-phosphate (0-5mM; not shown), or triacylglycerol (0-2mM; Figure 2) had no effect on rat liver PAP activity. In contrast, lysoPA was an inhibitor of rat liver NEM-insensitive PAP (Figure 2), suggesting that the enzyme may dephosphorylate lysoPA. Indeed, using [^{32}P]-lysoPA we were able to confirm that purified PAP dephosphorylates this lipid in a linear manner with respect to both time and protein concentration (not shown). The purified enzyme was only 50% more active against PA than against the lysoPA substrate (not shown). This suggests that NEM-insensitive PAP may well have a role in destroying the signalling properties of lysoPA (Van Corven *et al*, 1989), as well as dephosphorylating PA. Moreover, the reaction products diacylglycerol and monoacylglycerol were able to significantly inhibit the rat liver enzyme (Figure 2), suggesting that the enzyme may be regulated, at least in part, by feed-back inhibition.

In view of its likely role in signal transduction, the possibility of the regulation of NEM-insensitive PAP by phosphorylation was investigated. Four different protein kinases were tested for their ability to phosphorylate and regulate the NEM-insensitive PAP (cAMP

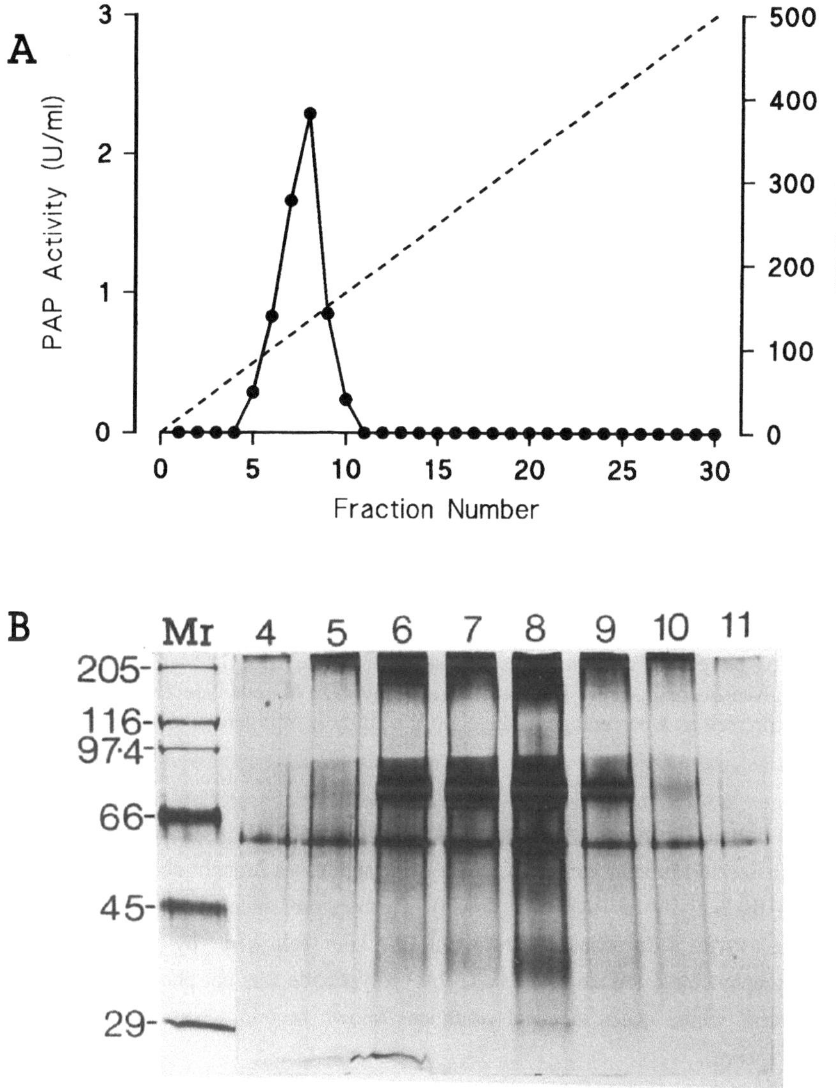

Figure 1: Analysis of purified PAP by Q-Sepharose chromatography and SDS-PAGE. 8 Units (1.5ml) of purified PAP were dialysed against 1litre of buffer C (buffer B containing no NaCl) for 3 hours at 4°C. The enzyme was loaded onto a 1ml Q-Sepharose Hitrap column, washed with 10ml of buffer C and eluted with a 30ml of a linear NaCl gradient (0-500mM) at 1ml/min. 1ml fractions were collected, and assayed for PAP activity (**A**). 100μl from each fraction was analysed by SDS-PAGE (**B**) as described in methods. The 57kDa doublet observed in each lane of the polyacrylamide gel is due to keratin contamination (Ochs, 1983).

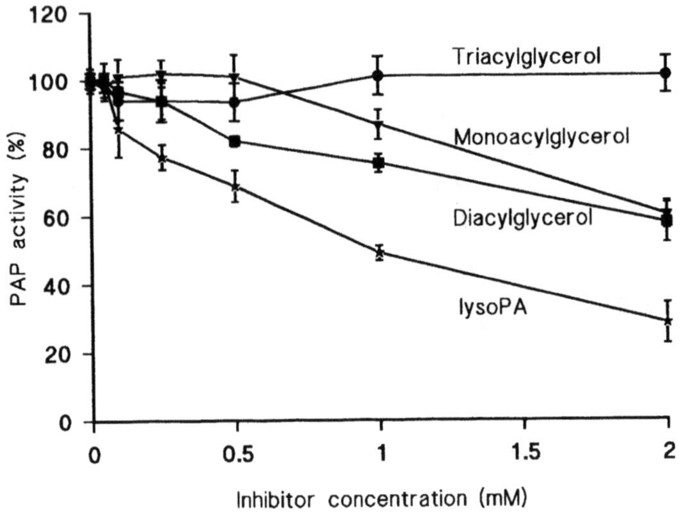

Figure 2: Effect of triacylglycerol, diacylglycerol, monoacylgycerol and lysophosphatidic acid on PAP.
Purified NEM-insensitive PAP was assayed against 0.5mM [^{32}P]-PA substrate in the presence of increasing concentrations of non-radioactive triacylglycerol, diacylglycerol, monoacylglycerol, or lysoPA. The results are the average of four independent experiments, and are expressed as a percentage of the control activity ± standard error.

dependent-protein kinase, MAP kinase [42kDa isozyme], Ca^{2+}-calmodulin-dependent protein kinase, and protein kinase C). None of these kinases brought about any observable change in the activity of PAP (not shown). Furthermore, incubation of purified NEM-insensitive PAP with kinase and [γ^{32}P]-ATP, followed by SDS-PAGE and autoradiography, indicated that the 83kDa PAP polypeptide was not phosphorylated by any kinase tested, whilst under identical conditions known protein substrates of each kinase were (not shown).

It has been suggested that NEM-insensitive PAP could play an important role in regulating the activity of the MAP kinase pathway by mediating the level of PA in the cell, so controlling the activity of the GTPase-activating protein and thereby the GTP:GDP ratio of ras (Tsai *et al*, 1989). These results suggest that MAP kinase does not self-regulate by altering the cellular PA levels, through modulation of NEM-insensitive PAP activity. Our results also suggest that protein kinase C does not regulate diacylglycerol levels by directly phosphorylating NEM-insensitive PAP. It is possible however, that protein kinase C could indirectly control diacylglycerol levels through sphingosine metabolism (Lavie *et al*, 1990).

Although the results presented here indicate that NEM-insensitive PAP activity is not regulated by any of the kinases tested, each of which has a relatively broad substrate specificity, it remains possible that its activity is controlled by a specific protein kinase.

NEM-insensitive PAP is present in the plasma membrane of rat liver. However it is not known whether the enzyme is located in the outer or inner leaflet of the membrane, or is present in both. It is important that the precise location of the enzyme is determined, since this may help to reveal whether the enzyme degrades extracellular PA and lysoPA, intracellular PA and lysoPA, or both. The availability of purified preparations of NEM-insensitive PAP will allow further biochemical studies into the regulation of the enzyme and the preparation of antibodies to help locate the precise location of the enzyme in the plasma membrane. Moreover, it will now be possible to carry out molecular studies on the enzyme, including cloning of the gene. Isolation of the PAP gene will hopefully confirm whether or not there are two isozymes of NEM-insensitive PAP in rat liver and will perhaps help to elucidate the relationship between NEM-insensitive PAP and the NEM-sensitive PAP believed to play a key role in regulating lipid biosynthesis.

We thank the Wellcome Trust for a Prize Studentship to fund I.N.F.

References

Billah, MM and Anthes, JC (1990) The regulation and cellular function of phosphatidylcholine hydrolysis. Biochem. J. 269, 281-291.

Bradford, MM (1976) A rapid and sensitive method for the quantitation of microgram amounts of protein utilising the principle of protein-dye binding. Anal Biochem. 72, 248-254.

Brindley, DN (1984) Intracellular translocation of phosphatidate phosphohydrolase and its possible role in the control of glycerolipid biosynthesis. Prog. Lipid Res. 23, 115-133.

Day, CP and Yeaman, SJ (1992) Physical evidence for the presence of two forms of phosphatidate phosphohydrolase in rat liver. Biochim. Biophys. Acta 1127, 87-92.

Epand, RM and Stafford, AR (1990) Counter-regulatory effects of phosphatidic acid on protein kinase C activity in the presence of calcium and diolein. Biochem. Biophys. Res. Commun. 171, 487-490.

Exton, JH (1990) Signalling through phosphatidylcholine breakdown. J. Biol. Chem. 265, 1-4.

Heukeshoven, J and Dernick, R (1985) Simplified method for silver-staining of proteins in polyacrylamide gels and the mechanism of silver-staining. Electrophoresis. 6, 103-112.

Jamal, Z, Martin, A, Gomez-Munoz, A and Brindley, DN (1991) Plasma membrane fractions from rat liver contain a phosphatidate phosphohydrolase distinct from that in the endoplasmic reticulum and cytosol. J. Biol. Chem. 266, 2988-2996.

Kanoh, H, Imai, S, Yamada, K and Sakane, F (1992) Purification and properties of phosphatidic acid phosphatase from porcine thymus. J. Biol. Chem. 267, 25309-25314.

Khan, WA, Blobe, GC, Richards, LA, and Hannun, YA (1994) J. Biol. Chem. 269, 9729-9735. Identification, partial purification, and characterisation of a novel phospholipid-dependent and fatty acid-activated protein kinase from human platelets.

Lavie, Y, Piterman, O and Liscovitch, M (1990) Inhibition of phosphatidate phosphohydrolase activity by sphingosine. FEBS Lett. 277, 7-10.

Leonard, M and Chrambach, A (1984) Solubilisation of functional membrane proteins. Methods in Enzymology, 104, 305-328.

Martin, A, Hopewell, R, Martin-Sanz, P, Morgan, JE and Brindley, DN (1986) Relationship between the displacement of phosphatidate phosphohydrolase from the membrane-associated compartment by chlorpromazine and the inhibition of synthesis of triacylglycerol and phosphatidylcholine in rat hepatocytes. Biochim. Biophys. Acta 876, 581-591.

Martin-Sanz, P, Hopewell, R and Brindley, DN (1984) Long-chain fatty-acids cause the translocation of phosphatidate phosphohydrolase from the cytosolic to the microsomal fraction of rat liver. FEBS Lett. 175, 284-288.

Moolenaar, WH, Kruijer, W, Tilley, BC, Verlaan, I, Bierman, AJ, and Delatt, SW (1986) Growth-factor like action of phosphatidic acid. Nature. 323, 171-173.

Moritz, A, DeGraan, PNE, Gispen, WH and Wirtz, KWA (1992) Phosphatidic acid is a specific activator of phosphatidylinositol-4-phosphate kinase. J. Biol. Chem. 267, 7207-7210.

Murayama, I and Ui, M (1987) Phosphatidic acid may stimulate membrane receptors mediating adenylate cyclase inhibition and phospholipid breakdown in 3T3 cells. J. Biol. Chem. 262, 5522-5539.

Nishizuka, Y (1984) The role of protein kinase C in cell surface signal transduction and tumour promotion. Nature 308, 693-698.

Ochs, D (1983) Anal. Biochem. 135, 470-474. Protein contaminants of Sodium Dodecyl Sulphate Polyacrylamide gels.

Pittner, RA, Fears, R and Brindley, DN (1985) Interactions of insulin, glucagon and dexamethasone in controlling the activity of glycerol phosphate acyltransferase, and the activity and subcellular distribution of phosphatidate phosphohydrolase in cutured rat hepatocytes. Biochem. J. 230, 525-534.

Tsai, MH, Yu, CI, Wei, FS and Stacey, DW (1989) The effect of GTPase activating protein upon Ras is inhibited by mitogenically responsive lipids. Science 243, 522-525.

Van Corven, EJ, Groenik, A, Jalink, K, Eichholtz, T and Moolenaar, WH (1989) Lysophosphatidic acid-induced cell proliferation: identification and dissection of signalling pathways mediated by G-proteins. Cell. 59, 45-54.

Walsh, JP, and Bell, RM (1986) Sn-1,2-Diacylglycerol kinase of *Escherichia coli*. J. Biol. Chem. 261, 6239-6247.

Wessel, D and Flugge, UI (1984) A method for the quantitative recovery of protein in dilute solution in the presence of detergents and lipids. Anal. Biochem. 138, 141-143.

A Novel Form of Type-IV$_A$ cAMP Phosphodiesterase found in rat brain.

Linda Pooley & Miles D. Houslay.
Molecular Pharmacology Group,
Department of Biochemistry,
University of Glasgow,
Glasgow,
G12 8QQ,
SCOTLAND,
UK.

Introduction

Cyclic AMP (cAMP) plays a pivotal role in controlling cell functioning mediating the action of a number of hormones (Sutherland, 1972; Dumont *et al*, 1989). In such systems binding of a hormone to its receptor stimulates adenylate cyclase to increase synthesis of cAMP. Cyclic AMP then elicits the appropriate cellular response via activation of cAMP-dependent protein kinase (Scott, 1991). Alternatively it may be extruded from the cell (reviewed by Barber and Butcher, 1983 and Brunton and Heasley, 1988) or inactivated. The sole means of its inactivation is through the action of cAMP phosphodiesterases (PDEs) which hydrolyse cAMP to 5'AMP (Beavo, 1990; Houslay and Kilgour, 1990; Conti and Swinnen, 1990). Therefore, cAMP phosphodiesterases play a key role in cAMP metabolism and hence the associated signalling systems.

Cyclic AMP phosphodiesterases form a large family of enzymes encoded by multiple genes with the added complexity of multiple splicing (Beavo, 1990; Conti and Swinnen, 1990; Davis, 1990). Such cAMP specific PDEs exhibit distinct biochemical characteristics and can be divided into at least five functional classes on the basis of their substrate specificity, sensitivity to

NATO ASI Series, Vol. H 92
Signalling Mechanisms – from Transcription Factors
to Oxidative Stress
Edited by L. Packer, K. Wirtz

specific inhibitors and response to known effectors. This classification is summarised in Table 1.

Isoenzyme Family	Regulatory Characteristics	Selective Inhibitors
I.	Stimulated by Ca^{2+}/ calmodulin Isoforms located within and between tissue types with differing K_m values for cAMP and cGMP.	vinpocetine
II.	Stimulated by low µM [cGMP]. High K_m for both cAMP and cGMP.	MEP 1
III.	cAMP activity is inhibited by low µM [cGMP]. Low K_m for both cAMP and cGMP.	milrinone cilostamide piroximone
IV.	Specific for cAMP No known regulator	rolipram Ro20-1724
VII.	Specific for cAMP Insensitive to cGMP	None reported Insensitive to rolipram and IBMX

Table 1: The isoenzyme families of the cyclic AMP specific phosphodiesterases.

This diversity would appear to place such species in a unique position where they are poised to tailor cAMP metabolism to the requirements of specific cell types and also to mediate cross-talk between signalling systems.

The Type-IV PDE family specifically hydrolyse cAMP, are insensitive to low concentrations of cGMP and Ca^{2+}/calmodulin and are specifically and selectively inhibited by rolipram (Beavo, 1990; Conti and Swinnen, 1990). It is now generally accepted that there are four Type-IV PDE isoform families, called IV_A, IV_B, IV_C and IV_D, (Beavo, 1990), being the products of four different genes (Davis, 1990; Conti and Swinnen, 1990; Conti *et al*, 1991).

RD1 is a Type-IV_A PDE which was cloned by homology from a rat brain cDNA library using the drosophila *dunc* PDE cDNA as a probe (Davis *et al*, 1989). It was the first mammalian Type-IV PDE cDNA to be isolated. We have characterised this species in COS-cells showing that the unique N-terminal domain of RD1 is responsible for membrane association and enhanced thermostability of this enzyme, but does not affect its substrate affinity or sensitivity to rolipram and IBMX (Shakur *et al*, 1993). In order to see whether the protein predicted from the cDNA is expressed *in vivo* , we generated an antiserum against a C-terminal dodecapeptide of RD1 and used this to probe various regions of the rat brain. Native RD1 was detected in cerebellum membranes as a 70kDa species that co-migrates on SDS PAGE with the species observed by expression of the cDNA in COS-cells (Shakur *et al*, 1994). RD1 was also detected in other regions of the brain as a membrane associated species (Lobban *et al*, 1994).

However, the antiserum originally believed to be specific for RD1, identified an additional immunoreactive species of higher molecular weight which occurs in both soluble and membrane fractions (Pooley *et al*, 1994). This novel species is particularly abundant in the olfactory lobe of the rat brain and we have termed it 'olf-RD1'. In this study we show that olf-RD1 exhibits Type-IV cAMP specific phosphodiesterase activity and suggest that it is likely to reflect an IV_A splice variant which is highly related to RD1.

Materials and Methods

Preparation of soluble and membrane fractions from the olfactory lobe and cerebellum region of rat brain

All operations were performed at 4 ^{o}C.

The olfactory lobes and cerebellum regions of the brain of Sprague Dawley rats were removed by dissection and immediately transferred to glass homogenisers containing 0.5ml/brain TES buffer [0.25M sucrose, 10mM Tris, 1mM EDTA pH7.4] supplemented with protease inhibitor cocktail [a 1000-fold stock was prepared by dissolving 2mg aprotonin, 2mg antipain, 2mg pepstatin A, 2mg leupeptin, 313mg benzamidine and 80mg of PMSF in 2ml of dimethyl sulfoxide (DMSO)]. The preparations were homogenised using ten strokes of a mechanical homogeniser and cell debris was removed by low speed centrifugation. The supernatants were transferred to Beckmann Ti50 tubes and spun at 48,000rpm for 90 minutes to separate the membrane and soluble fractions.

Determination of protein concentration

The protein concentration of the two fractions was determined by the method of Bradford (1976) using bovine serum albumin as a standard. Membranes were solubilised on ice with 0.2% Triton X-100 before assay and Triton X-100 included in the standards.

Western Blot Analysis

Western blotting was performed according to the method of Sambrook, Fritsch and Maniatis (1989). Membrane fraction and cytosolic fraction proteins (100μg) were separated by 10% SDS PAGE and then transferred to nitrocellulose. The nitrocellulose was then probed using antibody raised against RD1 and the blots visualised using ECL detection reagents (Amersham International, Amersham, Bucks, U.K.).

Determination of Phosphodiesterase Activity

Phosphodiesterase activity was determined as previously described by Marchmont and Houslay (1980). All assays were conducted at 30 ^{o}C.

Immunoprecipitation

Sample was diluted two-fold in immunoprecipitation buffer, [1% Triton X-100, 10mM EDTA, 100mM $NaH_2PO_4.2H_2O$, 50mM HEPES pH7.2], containing protease inhibitor cocktail. Antibody was then added(5µl antibody/100µl volume) and the precipitation mixture left overnight at 4 ^{o}C. If using membranes, the sample was stored on ice for 30 minutes before the addition of the antibody to allow the Triton to solubilise the membranes. Next, 10% Pansorbin (Calbiochem) was added (25µl/100µl sample) followed by mixing at 4 ^{o}C for at least 2 hours. The precipitate was pelleted in a bench top microfuge and washed two times with immunoprecipitation buffer before resuspension in a buffer suitable for subsequent assay.

Determination of K_m, IC_{50}, and relative V_{max} values

The K_m and V_{max} values were determined using seven concentrations of cyclic nucleotide. IC_{50} values for the general PDE inhibitor iso-butyl-1-methyl-xanthine (IBMX) and the Type-IV specific inhibitor rolipram were determined as described by Lavan *et al* (1990) at the K_m concentration of the substrate.

Results

1. Antiserum against a C-terminal dodecapeptide of RD1 identifies a novel immunoreactive species.

All preparations were subjected to Western blot analysis using antibody raised against a C-terminal dodecapeptide of RD1(Shakur *et al* , 1994). The results are shown in Figure 1 (Pooley *et al* , 1994).

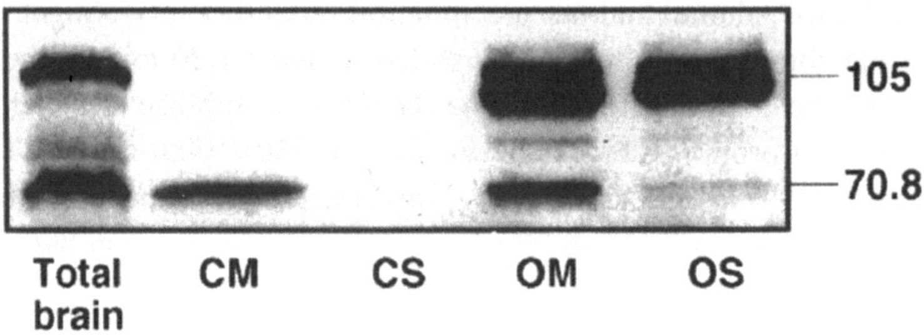

Figure 1: Membranes and soluble fractions were prepared from the cerebellum and olfactory lobe regions of the rat brain as detailed in the methods. A total brain (lacking olfactory lobe) homogenate was also prepared. All preparations, total brain, cerebellum membranes (CM), cerebellum soluble fraction (CS), olfactory lobe membranes (OM) and olfactory lobe soluble fraction (OS) were subjected to Western blotting and probed using an C-terminal anti-RD1 antiserum.

RD1 was found to be present in olfactory lobe and cerebellum membranes but absent from both soluble fractions. Indeed, in cerebellum membranes, RD1 was the only immunoreactive species. However, in both the membrane and soluble fraction of the olfactory lobe, the RD1 antibody, previously believed to be specific for RD1, also detected a protein which migrated as a doublet with apparent molecular weights of 108kDa and 101kDa. This species could also be detected in a total brain preparation lacking olfactory lobe, but interestingly only the higher molecular weight species of this doublet was observed.

2. Determination of the kinetic properties of olf-RD1

The cytosolic species in the soluble fraction of the olfactory lobe was specifically immunoprecipitated allowing us to determine its kinetic properties in the absence of RD1. This activity exhibited a low K_m for cAMP (Table 2 and Figure 2a), was dose-dependently inhibited by low concentrations of rolipram (Table 2 and Figure 2b) and could also be inhibited by IBMX (Table 2 and Figure 2c). By separately immunoprecipitating RD1 from cerebellum membranes and then assessing the relative amounts of RD1 and olf-RD1 using immunoblotting, we were able to calculate the relative V_{max} values for the two enzymes (Table 2).

	K_m cAMP (μM)	K_i rolipram (μM)	EC_{50} IBMX (μM)	V_{max} pmoles cAMP hydrolysed/min/ unit protein
OLF	2.3 +/- 1.9	0.19 +/-.03	5.7 +/-4.3	103
RD1	4.1 +/- 2.3	0.4 +/- 0.1	26 +/- 8	1541

Table 2: Kinetic properties of Olf-RD1 and RD1 (Pooley *et al,* 1994).

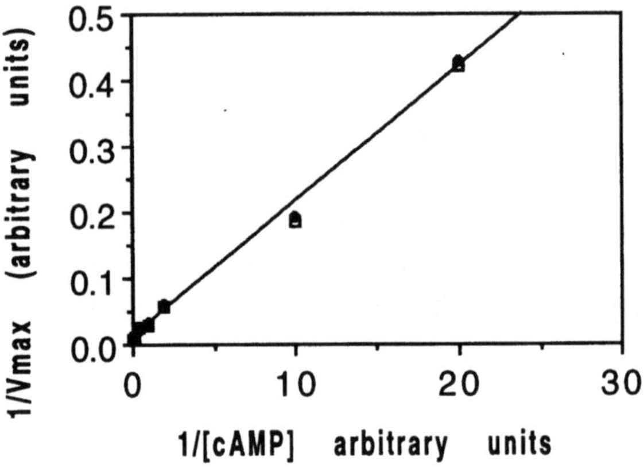

Figure 2a: Determination of K_m for olf-RD1 (Pooley *et al,* 1994).

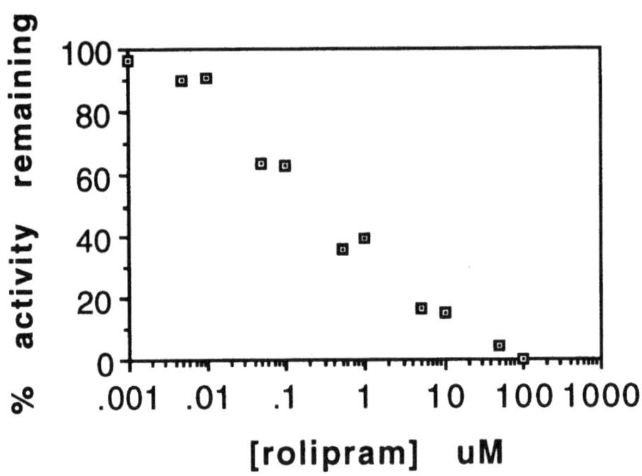

Figure 2b: Determination of the rolipram sensitivity of olf-RD1 (Pooley *et al,* 1994).

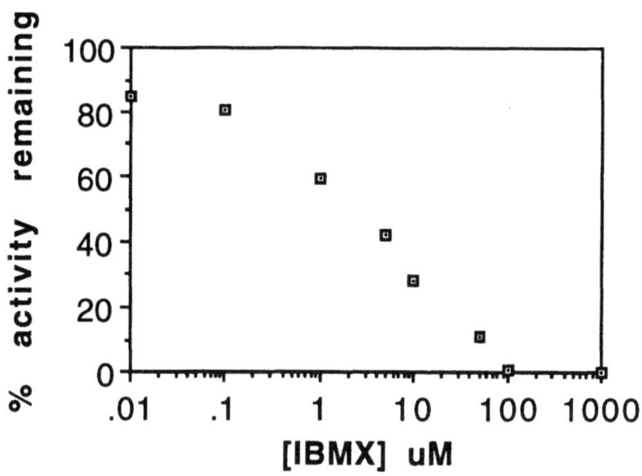

Figure 2c: Determination of IBMX sensitivity of olf-RD1 (Pooley *et al,* 1994).

Discussion

Using the antiserum raised against a C-terminal dodecapeptide of the rat brain rolipram-inhibited cAMP phosphodiesterase RD1, we have identified a novel immunoreactive species that is particularly abundant in the olfactory lobe of the rat brain. Unlike RD1 which is always membrane associated (Lobban *et al,* 1994), this species which we have termed olf-RD1 was also evident in the soluble fraction.

In order to determine whether this species was indeed a phosphodiesterase, we specifically immunoprecipitated the soluble species as this could be characterised in the absence of RD1. We were able to immunoprecipitate phosphodiesterase activity which could be >90% inhibited by 10μM rolipram assayed in the presence of 1μM cAMP. This cAMP hydrolysing activity was insensitive to the addition of Ca^{2+}/calmodulin and cGMP and no cGMP hydrolysing activity was immunoprecipitated. Activity analyses of this activity showed that olf-RD1 has a low K_m value for cAMP and a low K_i for rolipram similar to RD1. However, olf-RD1 appears to be more sensitive to

IBMX and exhibits a markedly different V_{max} (approximately 6% of that exhibited by RD1). We also undertook an examination of the thermostability of olf-RD1. This activity decayed with a half life similar to that observed for N-terminally-truncated soluble RD1 in COS-cells (Shakur *et al*, 1993) and as a single exponential suggesting that a single species has been immunoprecipitated.

Considering the different subcellular localisation of olf-RD1 and RD1 together with the kinetic data presented here, the two PDE species recognised by our antiserum are clearly very different and we would like to suggest that olf-RD1 is in fact a splice variant of RD1. This form would of course be expected to share a similar if not identical C-terminal 12 amino acids with RD1 for it to be recognised by our antiserum. Indeed using RD1 cDNA as a probe we have isolated a clone which could encode a species larger than RD1. This species shows complete sequence homology with RD1 from its extreme C-terminus up until 24 amino acids from its N-terminus. This does suggest an alternative protein product of the IV_A gene which would be expected to be recognised by our antiserum. It is possible that a number of alternative products of the IV_A gene may exist. The doublet nature of olf-RD1 seen here appears to be unique to olfactory tissue and might reflect either a further splice variant or perhaps is due to anomalous migration of a covalently modified form.

Acknowledgement

This work was supported by a grant from the Medical Research Council(UK).

References

Barber, R. and Butcher, R.W. (1983) Adv. Cyclic Nucleotide Research **15** 119 - 135.

Beavo, J.A. (1990) in 'Cyclic Nucleotide Phosphodiesterases: Structure, Regulation and Drug Action' Molecular Pharmacology of Cell Regulation series Vol. 2. Beavo, J.A. and Houslay, M.D. eds. pp 3 - 15, John Wiley and Sons Ltd., pubs, Chichester and New York.

Bradford, M.M. (1976) Anal. Biochem. **72** 248 - 254

Brunton, L.L. and Heasley, L.E. (1988) Methods Enzymology **159** 83 - 93

Conti, M and Swinnen, J.V. (1990) in 'Cyclic Nucleotide Phosphodiesterases: Structure, Regulation and Drug Action' Molecular Pharmacology of Cell Regulation series Vol. 2. Beavo, J.A. and Houslay, M.D. eds. pp 243 - 266, John Wiley and Sons Ltd., pubs, Chichester and New York.

Conti, M., Jin, C., Monaco, L., Repaske, D.R. and Swinnen, J.V. (1991) Endocrine Reviews **12** 218 - 234.

Davis, R.L., Takayusu, H., Eberwine, M and Myres J. (1989) PNAS **86** 3604 - 3608.

Davis, R.L. (1990) in 'Cyclic Nucleotide Phosphodiesterases: Structure Regulation and Drug Action' Molecular Pharmacology of Cell Regulation series Vol. 2. Beavo, J.A. and Houslay, M.D. eds. pp 227 - 241, John Wiley and Sons Ltd., pubs, Chichester and New York.

Dumont, J.E., Jauniaux, J-C. and Roger, P. (1989) TIBS **14** 67 - 71

Houslay, M.D. and Kilgour, E. (1990) in 'Cyclic Nucleotide Phosphodiesterases: Structure, Regulation and Drug Action' Molecular Pharmacology of Cell Regulation series Vol. 2. Beavo, J.A. and Houslay, M.D. eds. pp 185 - 226, John Wiley and Sons Ltd., pubs, Chichester and New York.

Lavan, B.E., Lakey, T. and Houslay, M. D. (1990) Biochem. Pharmacol. **38** 4123 - 4136

Lobban, M., Shakur, Y., Beattie, J. and Houslay, M.D. (1994) Submitted.

Marchmont, R.J. and Houslay, M.D. (1980) Biochem J. **187** 381 - 392

Pooley, L., Lobban, M., Shakur, Y., Horton,Y., Wilson,M., Beattie, J., Reed,R., and Houslay M.D. (1994). In preparation.

Sambrook, J., Fritsch, E.F. and Maniatis, T. (1989) Molecular cloning. A laboratory manual pp 18.47 - 18.45

Scott, J.D. (1991) Pharmacol. Ther. **14** 1 - 24.

Shakur, Y., Pryde, J.G. and Houslay, M.D. (1993) Biochem J. **292** 677 - 686.

Shakur, Y., Wilson, M., Pooley, L, Lobban, M., Campbell, A.M., Beattie, J., Daley, C. and Houslay, M.D. (1994). In press.

Sutherland, E.W. (1972) Science **17** 401 - 408

INTERACTION OF THE RETINAL CYCLIC GMP-PHOSPHODIESTERASE INHIBITOR WITH TRANSDUCIN (G_t) AND WITH OTHER G-PROTEINS (G_i, G_o AND G_s)

Annie Otto-Bruc, Bruno Antonny, Marc Chabre, Pierre Chardin, Eva Faurobert and T.Minh Vuong.
CNRS - Institut de Pharmacologie Moléculaire et Cellulaire.
660 route des Lucioles, Sophia Antipolis,
F-06560 Valbonne, France.

INTRODUCTION

Heterotrimeric G-proteins act as signal transducers at the cytoplasmic face of the cell's plasma membrane by interacting first with an activated receptor and then with an effector such as an enzyme or a channel (Gilman, 1987; Ross, 1989). The first interaction leads to the activation of the G protein, i.e., the switching of its α-subunit from an inactive, GDP-bound state to an active, GTP-bound state. The G-protein α-subunit (Gα), now bearing GTP, binds to and turn on an effector. G-protein effector coupling is commonly studied by monitoring changes in the levels of a second messenger, which reflects effector activity. However, such measurements are generally too indirect to allow accurate determinations of the kinetic parameters that describe the G-protein effector interaction. Here, we summarize recent studies where intrinsic tryptophan fluorescence was used as a biophysical means to monitor the interaction between transducin (G_t), the G-protein of the visual system and its effector target, the retinal cGMP-phosphodiesterase (PDE).

PDE is a membrane-bound enzyme and is composed of a large catalytic dimer, PDEαβ, whose activity is controlled by two small inhibitory subunits, PDEγ. G_tαGTP binds to PDEγ and relieves its inhibition on PDEαβ. (Fung et al., 1981; Wensel & Stryer, 1986; for review Chabre & Deterre, 1989). PDEγ is thus the central element: It regulates the phosphodiesterase activity by interacting with both PDEαβ and G_tαGTP. This 87 amino acids polypeptide can be expressed in *Escherichia coli* (Brown & Stryer, 1989). To directly monitor the binding of PDEγ to PDEαβ, Wensel and Stryer (1990) were first in applying a physical technique. The fluorescence anisotropy of fluorescein-labelled PDEγ is increased upon binding to PDEαβ. This signal was used to determine the on and off rates of the PDEγ-PDEαβ interaction and to study

NATO ASI Series, Vol. H 92
Signalling Mechanisms – from Transcription Factors to Oxidative Stress
Edited by L. Packer, K. Wirtz

the effect of certain mutations of PDEγ on this interaction (Brown, 1992). We recently showed that the interaction of PDEγ with its other partner, $G_t\alpha$GTP, can be monitored via changes in the tryptophan fluorescence of the two proteins (Otto-Bruc et al., 1993). The fluorescence signal depends on the conformation of $G_t\alpha$ and permits accurate determination of the affinity of PDEγ for $G_t\alpha$. PDEγ interacts tightly with $G_t\alpha$GTPγS ($K_d < 0.1$ nM) but retains a substantial affinity for the inactive $G_t\alpha$GDP ($K_d \approx 3$ nM). By site directed mutagenesis, we identified the two tryptophan residues that are responsible for this fluorescence signal (Otto-bruc et al., 1993; Faurobert et al., 1993). One tryptophan (W70) belongs to PDEγ. The other tryptophan (W207) is in one of the three regions of $G_t\alpha$ whose conformations are dramatically affected upon activation of the α-subunit.

Finally, we showed that other Gα-subunits, and particularly those sharing a large sequence homology with $G_t\alpha$, can interact quite well with PDEγ (Otto-Bruc et al., 1994). More interestingly, this interaction depends on the conformations of the Gα-subunits, as with $G_t\alpha$. The implication of this observation on the sites on $G_t\alpha$ that are involved in the binding to PDEγ will be discussed.

1.INTRINSIC FLUORESCENCE CHANGES ARE CORRELATED TO THE FORMATION OF THE $G_t\alpha$GTPγS-PDEγ COMPLEX.

During the activation of PDE by the α-subunit of transducin, the $G_t\alpha$GTP-PDEγ complex remains membrane-bound through weak interactions with PDEαβ as well as with phospholipids (Clerc & Bennett, 1992; Catty et al., 1992). However, this complex can be easily solubilized and purified (Deterre et al., 1986). Hence, the *in vitro* extractable, soluble $G_t\alpha$-PDEγ complex is a valid minimalist model that can shed some light on the molecular details of the interaction between transducin and PDE.

Fluorescence emission spectra of PDEγ and of $G_t\alpha$GTPγS are shown in figure 1A (dotted traces). If PDEγ did not interact with $G_t\alpha$GTPγS, the spectrum of a mixture of the two proteins should equal the arithmetic sum of the individual spectra of PDEγ and $G_t\alpha$GTPγS measured separately. This is not the case: the spectrum of the mixture (thick trace) exhibits a blue shift with respect to the sum of the invidual spectra (thin trace). Consequently, the differential spectrum is biphasic: for $\lambda_{em} < 330$ nm, interaction between PDEγ and $G_t\alpha$GTPγS gives a fluorescence increase, while a decrease is observed for $\lambda_{em} > 330$ nm. A fluorescence change is also observed when PDEγ is added to the inactive GDP-bound form of $G_t\alpha$ (figure 1B). Association of PDEγ with $G_t\alpha$GDP induces a near doubling of the fluorescence yield without much spectral shift.

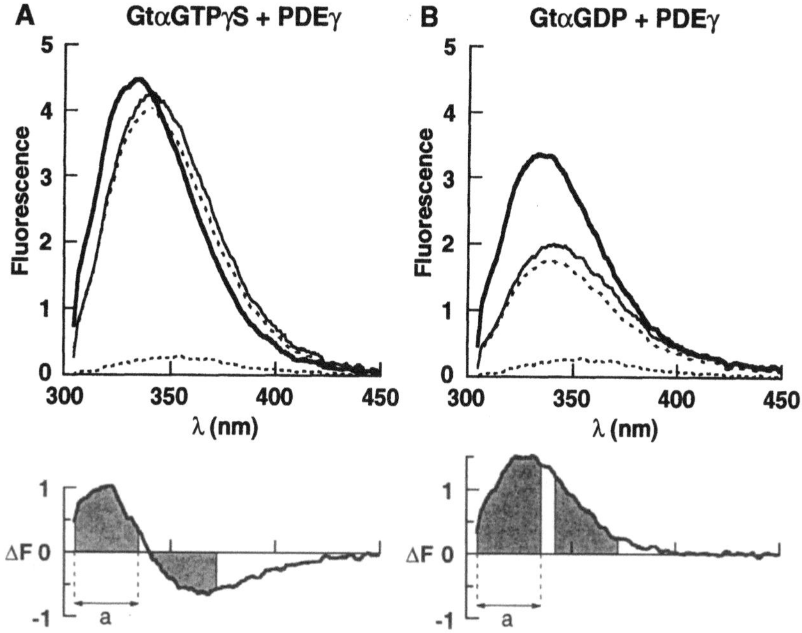

Figure 1: Fluorescence emission spectra the $G_t\alpha$–PDEγ complexes. **A**. $G_t\alpha$GTPγS + PDEγ. **B**. $G_t\alpha$GDP + PDEγ. The $G_t\alpha$GTPγS (or $G_t\alpha$GDP) and PDEγ concentrations are 100 nM. The buffer (TKM) contains 20 mM Tris, pH 7,5, 120 mM KCl, 2 mM $MgCl_2$. Excitation wavelength: 292 ± 2.5 nm. Emission bandwidth: 5 nm. The lower curve in each panel corresponds to the difference between the spectrum of the mixture and the arithmetic sum of the individual spectra. The stippled area (a) delimit the emission bandwidth chosen for the time scan measurements (320 ± 15 nm).

This fluorescence association signal can be conveniently used to determine the affinity of the interaction between PDEγ and $G_t\alpha$GTPγS or $G_t\alpha$GDP. Fluorescence intensity changes are recorded as successive PDEγ aliquots (25 nM) are added to a solution containing 100 nM $G_t\alpha$GTPγS (figure 2A). The fluorescence increase that occurs on the first addition of PDEγ, is much larger than when the same concentration of PDEγ is injected into buffer containing no $G_t\alpha$. When the total concentration of PDEγ exceeds that of $G_t\alpha$ during the final injections, the fluorescence jumps become more comparable to those from free PDEγ alone. A binding curve is obtained from these data as the fluorescence increases can be simply correlated to the concentration of the $G_t\alpha$-PDEγ complex formed (see legend of figure 2). $G_t\alpha$GTPγS interacts tightly with PDEγ (K_d < 0.1 nM). The affinity of PDEγ for $G_t\alpha$GDP is substantial (3 nM) but is at least 30 times weaker than for $G_t\alpha$GTPγS (figure 2B).

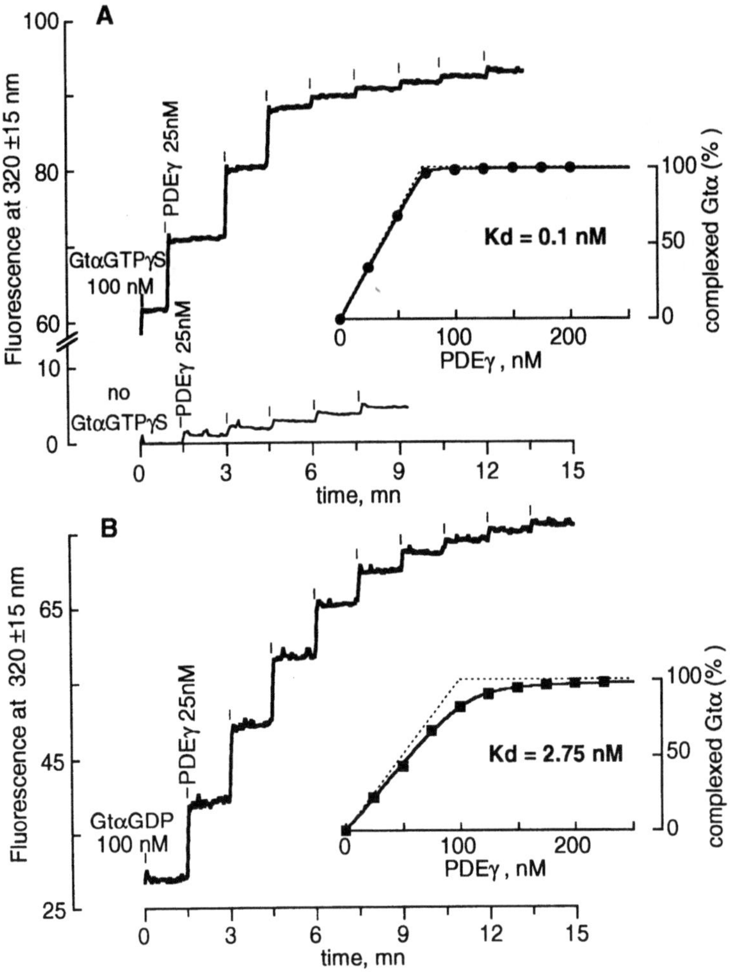

Figure 2: The change in fluorescence emission from a solution containing 100 nM $G_t\alpha GTP\gamma S$ (A) or 100 nM of $G_t\alpha GDP$ (B) is continuously monitored while equal amounts of PDEγ (25 nM) are added every 90 s to the fluorescence cuvette (thick trace). For control, an identical experiment is performed without $G_t\alpha$ (thin trace). At each concentration of PDEγ, the difference between the fluorescence change in the presence of $G_t\alpha$ and that of the control is directly proportional to the concentration of the $G_t\alpha$-PDEγ complex. The inset shows the molar fraction of $G_t\alpha$ complexed to PDEγ as a function of [PDEγ]. The fitting equation is:

$$[\mathrm{PDE}\gamma\text{-}\mathrm{G}\alpha] = \frac{1}{2}\{(C_0 + n\Delta x + K_d) - \sqrt{(C_0+n\Delta x+K_d)^2 - 4.C_0.n\Delta x}\}$$

It describes a simple bimolecular association scheme between $G_t\alpha$ and PDEγ. C_0 is the total concentration of Gα and n.Δx is the total concentration of PDEγ after the n-th injection. The K_d value from the best fit is given. For the $G_t\alpha$-PDEγ complex the K_d value obtained can only be considered as an upper limit. The dotted line correspond to the case with $K_d = 0$ and indicate the stoichiometries of the complexes.

Site directed mutagenesis was used to localize the tryptophan residues that are responsible for the fluorescence signals described above (Otto-Bruc et al., 1993; Faurobert et al., 1994). Mutation of either W70 in PDEγ or W207 in $G_t\alpha$ to a phenylalanine has two effects: the fluorescence change that accompanies the binding of the two proteins is modified and the affinity is reduced by a factor of at least 100. On the other hand, the mutation W70F on PDEγ does not prevent it from inhibiting PDEαβ suggesting that this residue is only implicated in the binding of PDEγ to $G_t\alpha$. W207 on $G_t\alpha$ belongs to the switch II domain, a region that is among the most conserved within the Gα's sequences and whose conformation depends on the nature of nucleotide bound to the protein. The recent reports on the structure of transducin α-subunit showed that W207 is not exposed when $G_t\alpha$ is in the active, GTPγS-bound form (Noel et al., 1993; Lambright et al., 1994). The effect of its replacement by a phenylalanine on the ability of $G_t\alpha$ to bind to PDEγ is thus more complex than originally thought (Faurobert et al., 1994).

2. INTERACTION OF PDEγ WITH VARIOUS Gα SUBTYPES

In order to investigate the ability of PDEγ to bind to different Gα subtypes, the same spectroscopic approach as in section 1 is used. When PDEγ is mixed with $G_i\alpha$, $G_o\alpha$ or $G_s\alpha$, either in the GTPγS form or in the GDP form, the resulting fluorescence change closely resembles that observed with $G_t\alpha$ (figure 3) and reflect the interaction between PDEγ and the Gαs. This fluorescence change depends on the conformation of the Gα subunit. For the GαGTPγS form, the differential spectra show up as biphasic traces. $G_{i1}\alpha$ and $G_{i3}\alpha$ are nearly identical to $G_t\alpha$ in their interaction with PDEγ. $G_o\alpha$ and $G_s\alpha$ also interact to some extent with PDEγ. From the amino-acid sequences, these two α-subunits are more distantly related to G_t than the G_i's. Yet as shown in figure 3, a fluorescence blue shift is also observed for mixtures containing PDEγ and $G_s\alpha$GTPγS or $G_o\alpha$GTPγS. For the interaction between PDEγ and GαGDP, differential spectra were detectable for only three subtypes, G_{i1}, G_t and G_{i3}.

Affinity for PDEγ of each Gα subtype were determined from experiments similar to those shown for Gtα (figure 2). For the GTPγS forms, the emission window was set at 360 ±15 nm. In this way, the fluorescence jumps start out negative as PDEγ interacts with GαGTPγS, cross zero upon saturation and become positive when the signal only comes from excess, free PDEγ (Otto-Bruc et al., 1994). The results are summarized in figure 4A. The interaction is strong (K_d ≈ 10 nM) with the GTPγS form of $G_{i1}\alpha$ and $G_{i3}\alpha$ and weaker (K_d ≈ 1 μM) with $G_o\alpha$ and $G_s\alpha$. As expected, binding tightness is related to the similarity in the amino acid sequence of the Gα subtype tested and that of $G_t\alpha$.

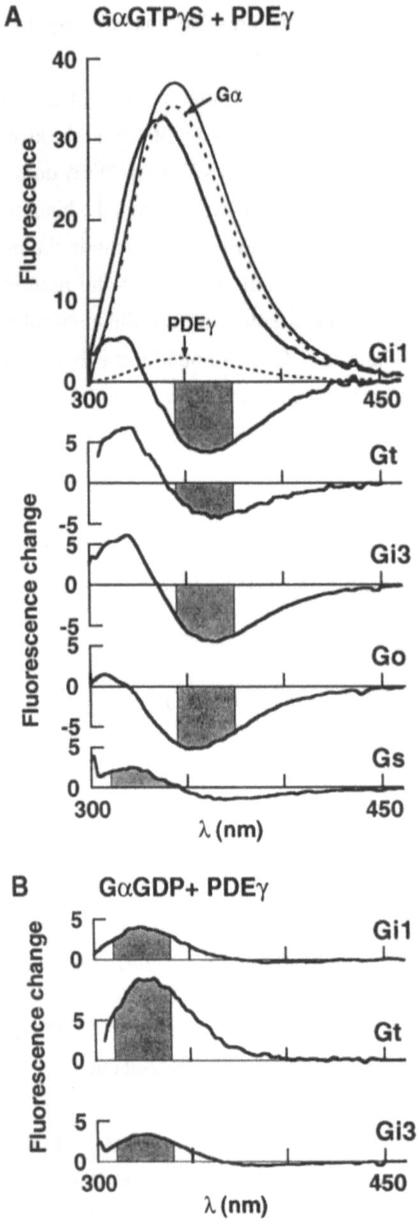

Figure 3: Fluorescence changes induced by the binding of PDEγ to various Gα-subtypes. **A.** Individual fluorescence emission spectra of PDEγ (500 nM) or G_{i1}GTPγS (500 nM) are measured separately (dotted traces). The sum of the two spectra (thin trace) is markedly different from the spectrum of a mixture containing the two proteins (thick trace). The differential spectrum is obtained by subtracting the sum of individual spectra from the mixture spectrum. This treatment is applied to all five GαGTPγS subunits, resulting in the five differential spectra shown here. Protein concentrations were: [G_{i1}]= 500 nM; [G_t]= 100 nM; [G_{i3}]= 750 nM; [G_o]= 500 nM; and [G_s]= 500 nM. For each Gα-subtype, [PDEγ] is equal to [Gα]. The differential spectra were normalized according the amount of protein used. **B.** For the interaction between PDEγ and GαGDP, differential spectra were detectable for only three subtypes, G_{i1}, G_t and G_{i3}. The stippled areas show the two emission windows used in quantifying the binding of PDEγ to the five Gα subunits.

When the GDP forms of $G_{i1}\alpha$, $G_{i3}\alpha$ or $G_t\alpha$ are used in lieu of the GTPγS forms, the relative decreases in affinity for PDEγ are comparable : In all cases the ratios of affinities between GTPγS and GDP states are ≥ 30. Interaction between $G_o\alpha$GDP or $G_s\alpha$GDP with PDEγ were not detectable.

A

Equilibrium constants of the Gα-PDEγ complexes

	Gt	Gi1	Gi3	Go	Gs
GTPγS form	≤ 0.1 nM	11 nM	6 nM	2 μM	1,7 μM
GDP form	3 nM	0,36 μM	0,22 μM	?	?

B

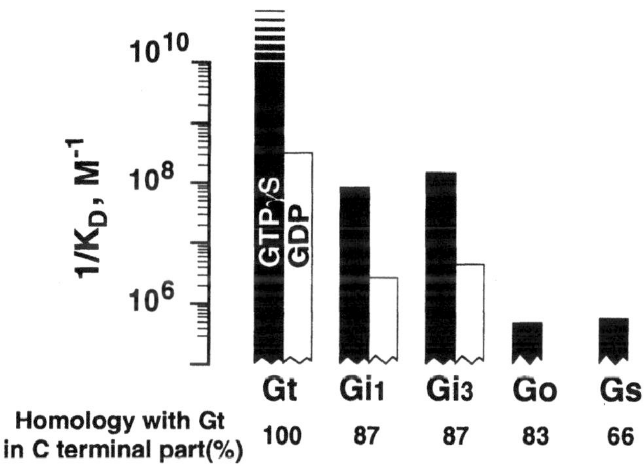

Figure 4: **A.** Affinities of PDEγ for the various Gα subtypes were determined from fluorescence assays as illustrated in figure 2. Each value represents the mean of duplicate determinations. The maximal variation in the K_d value was 20%. **B.** Affinities ($1/K_d$) of PDEγ for the five Gα-subtypes are shown in log scale. Given the logarithmic relationship between free energy and affinity ($\Delta G_0 = -2.3.RT \log_{10}[1/K_d]$), this figure shows that the decrease in free energy of the PDEγ-Gα complex (or the increase in PDEγ binding energy) as GDP is replaced by GTP in the nucleotide site is the same for $G_t\alpha$, $G_{i1}\alpha$ and $G_{i3}\alpha$. The affinity of $G_t\alpha$GTPγS for PDEγ is only a lower limit, as illustrated by the vanishing top end of the bar. Homologies in the C-terminal of the primary structure (starting from D196 of $G_t\alpha$) between each Gα-subtype and $G_t\alpha$ are shown.

CONCLUSION

In 1987 Higashijima and coworkers showed that a large intrinsic fluorescence change occurs upon activation of G-protein α subunits. This signal permitted numerous time-resolved studies of the off/on cycle of these proteins. Here we demonstrate that tryptophan fluorescence can be also used to monitor the event that follows the activation of $G_t\alpha$: its interaction with the γ-subunit of the retinal phosphodiesterase. A large change in the intrinsic tryptophan fluorescence of $G_t\alpha$ and PDEγ is observed when the two purified subunits are mixed together. Thanks to this signal, the affinities of PDEγ for the GDP- and the GTPγS-bound states of $G_t\alpha$ and of various Gα subtypes could be measured. The affinity of PDEγ for $G_t\alpha$GTPγS is extremely high ($K_d < 10^{-10}$ M) and remains substantial for $G_t\alpha$GDP ($K_d = 3 \times 10^{-9}$ M). In another study using PDEγ labelled with a fluorescence probe, a much lower affinity for $G_t\alpha$GTPγS was found ($K_d = 3.6 \times 10^{-8}$ M) and no interaction with $G_t\alpha$GDP was detected (Artemyev et al., 1992). However, the probe probably affects the interaction of PDEγ with $G_t\alpha$ since it was attached to Cys68, a residue close to Trp70, which is involved in the interaction with $G_t\alpha$.

One drawback of using intrinsic tryptophan fluorescence is that the signal-to-noise ratio is drastically reduced by the addition of other proteins and of membranes or vesicles. Thus, the technique cannot be used to study the formation of the whole complex between $G_t\alpha$GTPγS and PDE$\alpha\beta\gamma_2$ at the surface of a plasma membrane. Heck and Hoffman (1993) recently described a real time, light scattering assay for this interaction.

By comparing the amino acid sequence of $G_t\alpha$ with those of the other Gα's, one expects that $G_i\alpha$ should be better than $G_s\alpha$ and $G_o\alpha$ in the interaction with PDEγ, provided such interaction can be measured. Beyond confirming this obvious expectation, the studies summarized here reveal a more significant feature: Whenever binding of PDEγ to both the GDP and GTPγS conformations is measurable, the affinity for the latter is always $\geq$ 30 times higher than for the former. This suggests that the side chains that are involved in both conformational change and interaction with PDEγ are well conserved among the Gα-subtypes. Our studies with the W207F mutant of $G_t\alpha$ seems to indicate that the switch II domain might well be this nucleotide-sensing effector interaction domain: It harbors the conserved residue W207 and its conformation changes as GTP replaces GDP. But by the very fact of its high level of conservation (figure 5), this domain cannot confer the specificity that each Gα must have with respect to its cognate effector. Another domain of $G_t\alpha$ (293-304) is involved in PDEγ binding (Rarick et al., 1992) but is not affected by the GDP-to-GTP conformational change (Lambright et al., 1994). We suggest that it is this second domain which makes $G_t\alpha$ interact specifically with its native effector, the γ-subunit of PDE. Presumably, this domain would play the same role in other Gα subunits, namely to make each one of them specific for one effector.

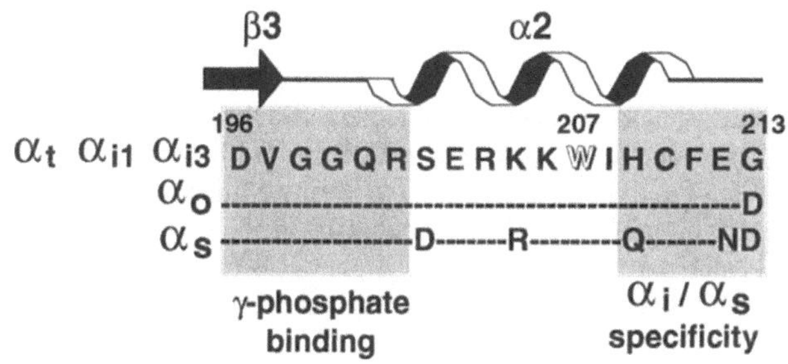

Figure 5: Sequence alignment of the switch II domain of G-protein α-subunits. The sequence is conserved among G_t, G_{i1} and G_{i3}. $G_s\alpha$ and $G_o\alpha$ differ from this conserved sequence at the residues shown.

REFERENCES

Artemyev, N O, Rarick, H M, Mills, J S, Skiba, N P and Hamm, H E (1992) Sites of interaction between rod G-protein α-subunit and cGMP-phosphodiesterase γ-subunit - Implications for the phosphodiesterase activation mechanism. J. Biol. Chem. 267: 25067-25072.

Brown, R L and Stryer, L (1989) Expression in bacteria of functional inhibitory subunit of retinal rod cGMP phosphodiesterase. Biochemistry 86: 4922-4926.

Brown, R L (1992) Functional regions of the inhibitory subunit of retinal rod cGMP phosphodiesterase identified by site-specific mutagenesis and fluorescence spectroscopy. Biochemistry 31: 5918-5925.

Catty, P, Pfister, C, Bruckert, F and Deterre, P (1992) The cGMP phosphodiesterase-transducin complex of retinal rods - Membrane binding and subunits interactions. J. Biol. Chem. 267: 19489-19493.

Chabre, M and Deterre, P (1989) Molecular mechanisms of visual transduction. Eur. J. Biochem. 179: 255-266.

Clerc, A and Bennett, N (1992) Activated cGMP phosphodiesterase of retinal rods - A complex with transducin α subunit. J. Biol. Chem. 267: 6620-6627.

Deterre, P, Bigay, J, Robert, M, Pfister, C, Kühn, H and Chabre, M (1986) Activation of retinal rod cyclic GMP phosphodiesterase by transducin: Characterization of the complex formed by phosphodiesterase inhibitor and transducin α-subunit. Proteins Struct. Funct. Genetics 1: 188-193.

Faurobert, E, Otto-Bruc, A, Chardin, P and Chabre, M (1993) Tryptophan W207 in transducin Tα is the fluorescence sensor of the G protein activation switch and is involved in the effector binding. EMBO Journal 12: 4191-4198.

Fung, BKK, Hurley, JB and Stryer, L (1981) Flow of information in the light-tiggered nucleotide cascade of vision. Proc. Natl. Acad. Sci. USA 78: 152-156.

Gilman, A G (1987) G proteins: Transducers of receptor-generated signals. Ann. Rev. Biochem. 56: 615-649.

Heck, M and Hofmann, K P (1993) G-protein effector coupling - A real-time light-scattering assay for transducin-phosphodiesterase interaction. Biochemistry 32: 8220-8227.

Higashijima, T, Ferguson, K M, Sternweis, P C, Ross, E M, Smigel, M D and Gilman, A G (1987) The effect of activating ligands on the intrinsic fluorescence of guanine nucleotide-binding regulatory proteins. J. Biol. Chem. 262: 752-756.

Lambright, D G, Noel, J P, Hamm, H E and Sigler, P B (1994) Structural determinants for activation of the α-subunit of a heterotrimeric G protein. Nature 369: 621-628.

Noel, J P, Hamm, H E and Sigler, P B (1993) The 2.2 angstrom crystal structure of transducin-α complexed with GTPγS. Nature 366: 654-663.

Otto-Bruc, A, Antonny, B, Vuong, T M, Chardin, P and Chabre, M (1993) Interaction between the retinal cyclic GMP phosphodiesterase inhibitor and transducin - Kinetics and affinity studies. Biochemistry 32: 8636-8645.

Otto-Bruc, A, Vuong, T M and Antonny, B (1994) GTP-dependent binding of G(i), G(o) and G(s) to the γ-subunit of the effector of G(t). FEBS Lett. 343: 183-187.

Rarick, H M, Artemyev, N O and Hamm, H E (1992) A site on rod G-protein α-subunit that mediates effector activation. Science 256: 1031-1033.

Ross, E M (1989) Signal sorting and amplification through G protein-coupled receptors. Neuron 3: 141-152.

Wensel, T G and Stryer, L (1986) "Reciprocal control of retinal rod cyclic GMP phosphodiesterase by its γ subunit and transducin". Proteins Struct. Funct. Genet. 1: 90-99.

Wensel, T G and Stryer, L (1990) Activation mechanism of retinal rod cyclic GMP phosphodiesterase probed by fluorescein-labeled inhibitory subunit. Biochemistry 29: 2155-2161.

Identification of stable opioid receptor Go-protein complexes using GTP-binding protein selective antisera

Zafiroula Georgoussi, Graeme Milligan* and Christine Zioudrou

Institute of Biology,
National Centre of Scientific Research "Demokritos",
153 10 Ag. Paraskevi, POB 60228,
Athens,Greece.

INTRODUCTION

Agonist interaction of all three opioid receptor subtypes μ,δ and can result in the inhibition of adenylate cyclase and/or the regulation of ion channels by activation of one or more pertussis toxin sensitive guanine nucleotide binding proteins (G-proteins) acting as signal transducers (Childers, 1991). Although there has been extensive analysis of opioid receptor function and pharmacology our understanding of the molecular basis of these properties is limited. Recently all subtypes of opioid receptors have been cloned (Evans et al.,1992, Reisine and Bell, 1993 and references therein). Comparison of amino acid sequences of the three cloned opioid receptors reveals high sequence similarity of the intracellular loops suggesting that they might interact with similar G-proteins.

The nature of the pertussis toxin sensitive G-proteins that interact with opioid receptors in brain is less well established. Reconstitution experiments using G-proteins and the μ-opioid receptor from rat brain have indicated the potential of this receptor to interact with Gi or Go (Ueda et al., 1988), whereas in NG108-15 hybrid cells the δ-opioid receptor appears to activate Gi2 (McKenzie and Milligan, 1990), Go (Offermanns et al., 1991) and possibly Gi3 (Roerig et al., 1992). Using selective G-protein antisera in native brain membranes we have previously demonstrated that both the carboxyterminal and aminoterminal regions of Goα play a key role in opioid receptor Go-protein interaction (Georgoussi et al.,1993).

*Molecular Pharmacology Group, Departments of Biochemistry and Pharmacology, University of Glasgow, Glasgow G12, 8QQ, Scotland United Kingdom.

NATO ASI Series, Vol. H 92
Signalling Mechanisms – from Transcription Factors to Oxidative Stress
Edited by L. Packer, K. Wirtz

Attempts to solubilize opioid receptors which retain binding characteristics similar to those in intact membranes has been reported in numerous laboratories. In most of the cases occupancy of the receptor with a ligand before solubilization or high concentrations of sodium chloride were required for the receptor in order to displace high affinity guanine nucleotide sensitive binding (Simonds et al., 1980, Ofri et al., 1992, Cote et al., 1993, Wong et al., 1989).

In this study we report the nature of interaction of μ and δ opioid receptors with the guanine nucleotide binding protein Go using a new immunoprecipitation protocol with selective Go protein antisera. It is shown that the binding of the carboxyterminal and the aminoterminal directed antisera that uncouple opioid receptor-Go protein complexes in native membranes (Georgoussi et al., 1993) can not immunoprecipitate soluble opioid receptors, while an antiserum directed against an internal region of Goα can. Moreover antiserum βN3 which is directed against the aminoterminal ten amino acids of Gβ also immunoprecipitated solubilized opioid receptors. This allowed for the first time the observation of stable soluble μ and δ opioid receptor Go-protein complexes derived from rat brain membranes.

METHODS

Solubilization of membranes

Rat brain cortical membranes were prepared as described by Georgoussi et al., 1993, in the presence of 0.1 mM EGTA and 50 mM Tris-HCl pH 7.5. Plasma membranes were diluted with buffer containing 50 mM Tris-HCl pH 7.5, 50μg/ml trypsin inhibitor and 0.1mM PMSF (Buffer A) and spun for 15 min at 165,000 x g. The pellet was solubilized according to Simonds et al., 1980, in Buffer A containing 10 mM CHAPS for 1hr at 4°C with gentle stirring. The solubilized preparation was centifuged at 100,000 x g for 1hr at 4°C. The clear supernatant was removed diluted 10 fold in 50 mM Tris-HCl pH 7.5, 10 mM $MgCl_2$, 0.1 mM PMSF and 50 μg/ml trypsin inhibitor (Buffer B) and concentrated to half volume using a CentriCell . The solubilized material was then used for binding or immunoprecipitation experiments. The concentration of protein in each sample was approximately 250-350μg/ml and 1ml samples were used for binding experiments which were performed as described by Georgoussi et al., 1993.

Immunoprecipitation of opioid receptor-G-protein complexes

To immunoprecipitate opioid receptor-G-protein complexes, a

sample of solubilized receptor (1 ml) was incubated with the appropriate G-protein antiserum (20µl) under constant rotation at 4°C. After 6hr 100 µl of 50% (w/v) protein A-Agarose beads were added to each sample and incubated overnight. Finally another aliquot of antiserum was added to bring the final antiserum dilution to 1:30. These samples were further incubated for 3-4 additional hours and then centrifuged in an Eppendorf microcentrifuge for 2 min. The supernatant was removed, the immune complex was washed by addition of 1 ml buffer B and recentrifuged. The immunoprecipitate was resuspended in 1 ml buffer B and the opioid receptors were detected with binding assays, or the immune complex was resuspended in 30 µl of Laemmli sample buffer and resolved by SDS-PADE [10% (w/v) acrylamide].

Immunological experiments

The generation and specificity of the antisera used in this study are described extensively by McKenzie et al., 1990, Georgoussi et al,.1993, Mullaney and Milligan, 1990, Goldsmith et al., 1987.

RESULTS

Active, high affinity opioid receptors were solubilized from brain cortical membranes with 10 mM CHAPS in the absence of NaCl according to Simonds et al., 1980. The CHAPS extracts displayed high affinity, guanine nucleotide sensitive agonist binding as assessed by displacement experiments of [^{3}H]diprenorphine by DADLE, performed in the presence and absence of the poorly hydrolyzed analog of GTP, GppNHp. As indicated in Fig. 1, 500nM DADLE displaced 45% of the specific binding of [^{3}H]diprenorphine (1.8 nM) in the absence of GppNHp, whereas in its presence the same concentation of DADLE was able to displace only 8% of the specific binding. Similar results were obtained with 100nM DADLE (Fig.1). These data indicate that even in the absence of agonist it was possible to solubilize a complex of opioid receptors and G-protein(s) which remained stable in the presence of CHAPS.

Abbreviations : DAMGO, [D-Ala^2,N-Me-Phe^4,Gly^5-ol]enkephalin; DSLET,[D-Ser^2,Leu^5,Thr^6]enkephalin; DADLE,[D-Ala^2,D-leu^5] enkephalin; GppNHp,guanosine 5'-(βγ-imido)triphosphate; CHAPS, 3,[(3-cholamidopropyl)dimethylammonio]1-propanesulfonate;PMSF, methylsulfonylfluoride.

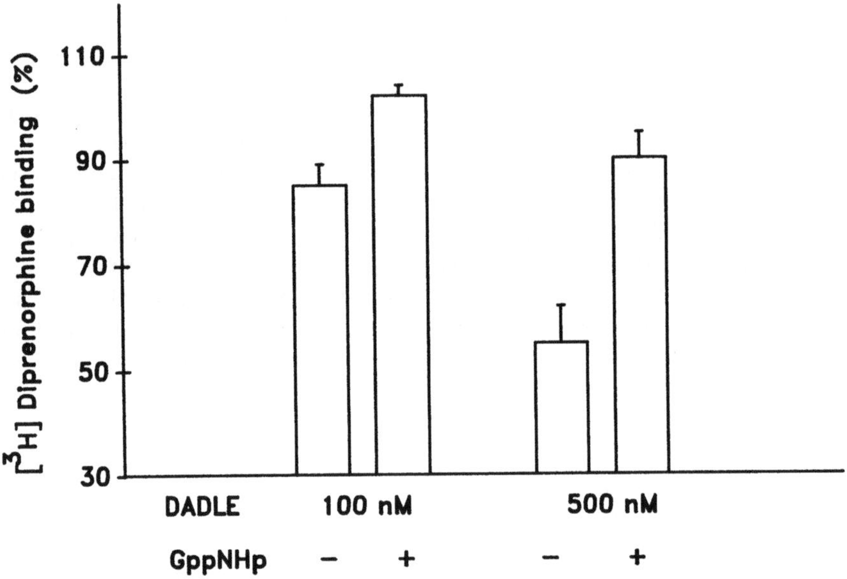

Fig. 1.The effect of GppNHp on displacement of [^{3}H]diprenorphine by DADLE from solubilized opioid receptors.

Rat brain cortical membranes were solubilized as described in Methods. The CHAPS extracts were incubated with 100μM GppNHp for 45 min at 30° C prior to binding which was performed as described by Georgoussi et al., 1993. Non specific binding was assessed in the presence of 10μM diprenorphine. One hundred percent specific [^{3}H]diprenorhpine binding (1.8 nM) was 55±5 fmoles/mg of protein for solubilized membranes (control) and 49±4 fmoles/mg of protein for the membranes that have been incubated in the presence of 100μM GppNHp. Results are the mean ± SD of three independent experiments carried out in triplicates.

Antipeptide antisera IM1, OC2 and ON1 generated against peptides common to polypeptides corresponding to products from both the Go1 and Go2 splice variants of the Goα gene have been previously characterized (Mullaney and Milligan,1990). Both antisera IM1 and OC2 were able to immunoprecipitate a 39kDa polypeptide in the solubilized material which could be identified as Goα by immunoblotting these immunoprecipitates with antiserum ON1 (Fig 2). These results indicated that each of these antisera could bind Goα in the CHAPS extracts of rat cortical membranes and therefore if the opioid receptors were coupled to Go it should be possible to co-immunoprecipitate the receptors using the Goα antisera.

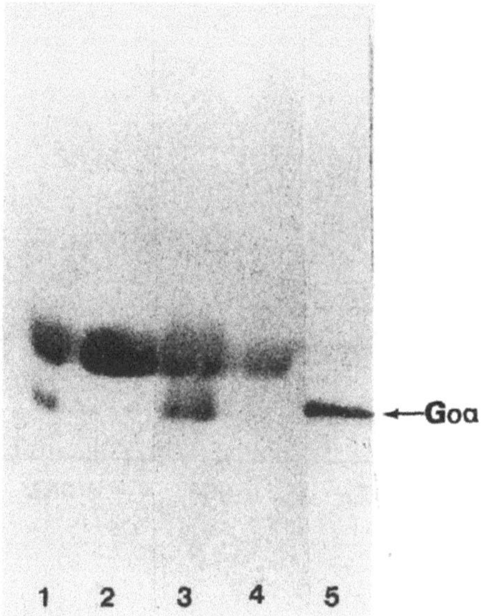

Fig. 2. Anti-Goα antisera immunoprecipitate Goα from solubilized rat cortical membranes.

Rat brain cortical membranes were solubilized and immunoprecipitated at final antiserum dilution of 1:30 with either antisera OC2 (lane 1), IM1 (lane 3) or preimmune serum (lanes 2,4) as described in Materials and Methods. Both the solubilized membranes (lane 5) and the immunoprecipitates were subjected to SDS-PAGE (10% w/v acrylamide) and immunoblotted using antiserum ON1 as the primary antiserum.

Antiserum OC2, which is directed against the carboxyterminal region of Goα failed to immunoprecipitate specific [^{3}H]diprenorphine binding sites from the CHAPS extract when compared to a preimmune serum control (Fig.3a). A similar lack of opioid receptor immunoprecipitation was observed with antiserum ON1 which is directed against the aminoterminal hexadecapeptide of isoforms of Goα (Fig.3). By contrast, antiserum IM1 generated against aminoacids 22-35 of Goα was able to immunoprecipitate effectively Goα-opioid receptor complexes from solubilized cortical rat brain membranes, measured by specific [^{3}H]diprenorphine binding in the immunoprecipitate. Immunoprecipitation of the opioid receptors with Goα was specific, as pretreatment of the membranes with GppNHp prior to solubilization and immunoprecipitation blocked the ability of antiserum IM1 to co-immunoprecipitate Goα-opioid receptor complexes (Fig 3b).

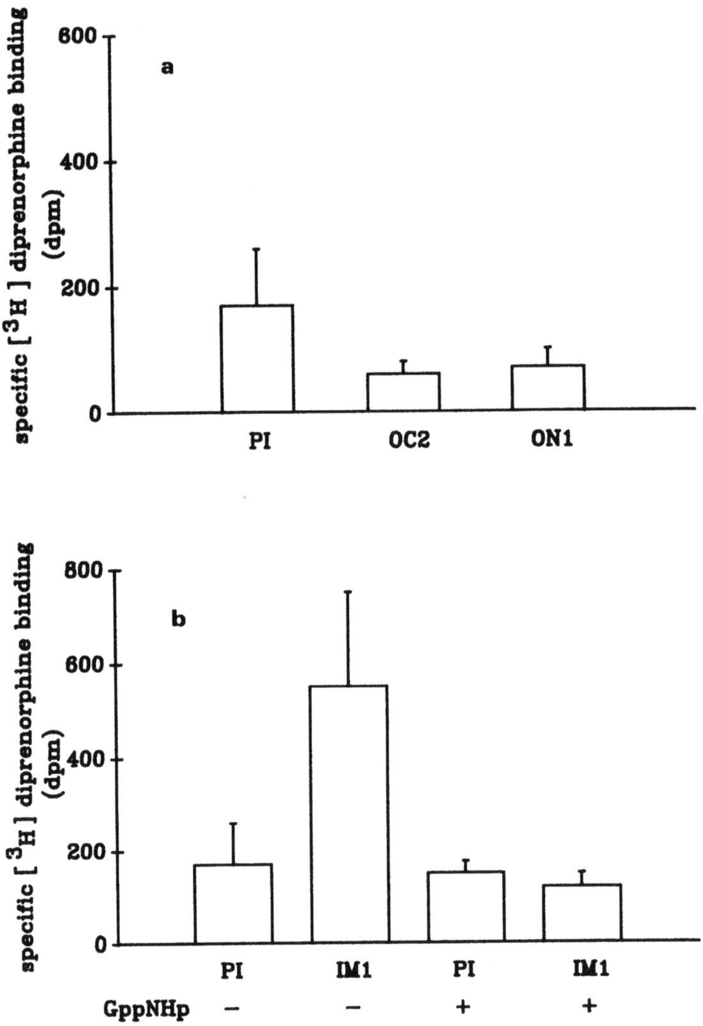

Fig. 3. Immunoprecipitation of soluble opioid receptors from rat cortical membranes with anti-Go protein antisera.

Soluble opioid receptors from rat cortical membranes were immunoprecipitated with either a) OC2, ON1 and preimmune serum (PI), or b) IM1 and preimmune serum (PI). The presence of opioid receptors in the immunoprecipitate was detected by the specific binding of [^{3}H]diprenorphine (2.3nM). In experiments whereas GppNHp was present the membranes were preincubated with 100µM GppNHp for 45 min at 30°C prior to solubilization. Values are presented as the amount of specific binding (dpm) and are averages of mean ± SD of four different experiments performed in triplicate for the OC2 and ON1 antisera and seven experiments for the IM1 antiserum.

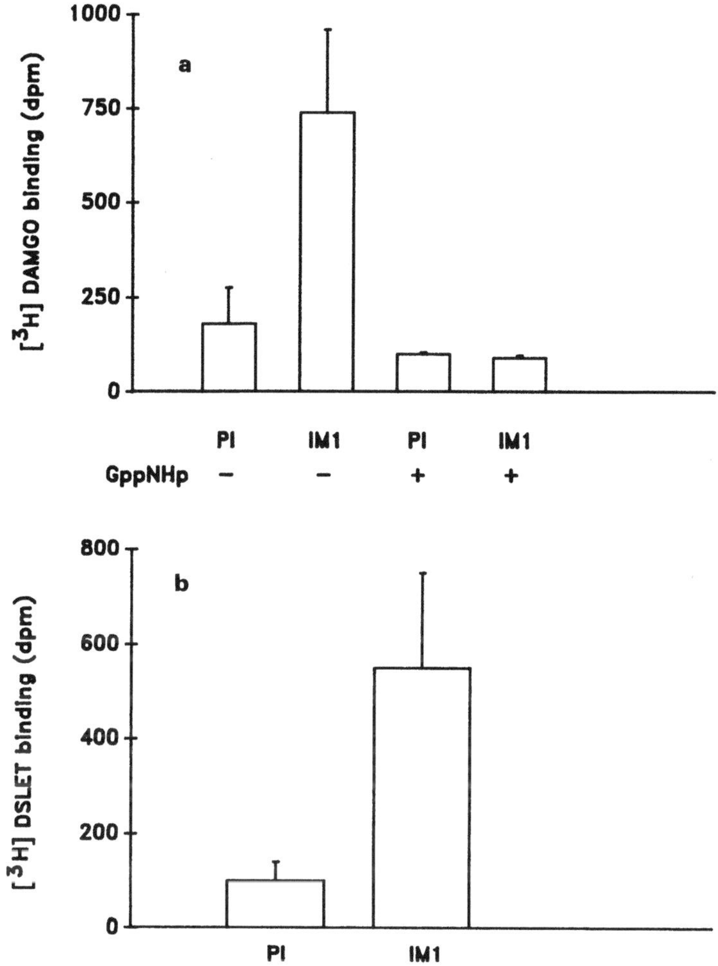

Fig. 4. Immunoprecipitation of μ-and δ-opioid receptors from soluble rat cortical membranes with the IM1 antiserum. Concentrated solubilized membranes were immunoprecipitated as described in Methods with IM1 antiserum at final dilution of 1:30. Specific [^{3}H]DAMGO (3 nM) binding and [^{3}H]DSLET (3.5 nM) binding in the immunoprecipitate was carried out as described by Georgoussi et al., 1993. In experiments where GppNHp was included cortical membranes were incubated with 100 μM GppNHp prior to solubilization. Values are presented as the amount of specific binding (dpm) and are the averages of mean ± SD from four independent experiments.

In order to define which opioid receptors interact with Goα we tested for the presence of both μ and δ opioid receptors in

IM1 immunoprecipitates by using the highly selective opioid ligands [^{3}H]DAMGO (μ-agonist) and [^{3}H]DSLET (δ-agonist). As shown in Figure 4, a significant amount of [^{3}H]DAMGO binding was present in the IM1 immunoprecipitate as compared with that of preimmune control, indicating that IM1 antiserum can immunoprecipitate μ-opioid-Goα protein complexes. This binding was selectively abolished by pretreatment of the membranes with the nonhydrolyzable GTP analog GppNHp, suggesting that [^{3}H]DAMGO specifically labels Go-protein coupled μ-opioid receptors in the immunoprecipitate (Fig.4a). Similar experiments using [^{3}H]DSLET have demonstrated that δ-opioid receptors-Goα complexes could be determined in the IM1 immunoprecipitates of CHAPS solubilized rat cortical membranes (Fig.4b).

Table 1.

	[^{3}H]diprenorphine binding
Preimmune serum	137±10
βN3 antiserum	459±180

Table 1. Gβ directed antisera selectively immunoprecipitate solubilized opioid receptors from rat cortical membranes.
CHAPS extracts were immunoprecipitated with either antiserum βN3 or preimmune serum at dilution 1:30 as described in Methods.The presence of opioid receptors in the immunoprecipitate was detected by [^{3}H]diprenorphine binding (2.5 nM). Data are representative of an individual experiment which was repeated three times (mean±SD).

To determine whether the β-subunits are associated with the opioid receptors, antiserum βN3 was used to immunoprecipitate opioid receptors-G-protein complexes from solubilized brain cortical membranes. Indeed, antiserum βN3 , which is directed against amino acids 1-10 of Gβ effectively immunoprecipitated solubilized opioid receptors as indicated by the binding in specific [^{3}H]diprenorphine binding (2.4nM) sites in contrast to preimmune control (Table 1). These results suggest that identification of functional interactions between Goα and Gβ with opioid receptors can be observed in solubilized rat cortical membranes following immunoprecipitation with selective antipeptide antisera.

DISCUSSION

Biochemical data to support a physical association of opioid receptors and G-proteins and the nature of these interactions is limited. The work presented herein describes an immunoprecipitation based approach using a series of selective antipeptide antisera directed against amino acids 22-35 of forms of Goα (IM1), and βN3 antiserum generated against amino acids 1-10 of Gβ, which can be used to identify functional opioid receptor-G-protein interactions following solubilization of rat cortical membranes. Solubilization was carried out in the presence of 10mM CHAPS as described by Simonds et al., 1980, without altering the binding characteristics of the receptor. Pretreatment of the solubilized preparation with the poorly hydrolyzed analogue of GTP, GppNHp, reduced the affinity of DADLE to displace specific [^{3}H] diprenorphine, indicative of interaction between the receptor and G-protein(s) (Fig.1). These observations suggested to us that occupancy of the receptor with a ligand before solubilization is not required to stabilize the receptor and the formation of receptor-G-protein complex. Similar putative receptor-G-protein complexes have been reported following solubilization of the vasopressin (Georgoussi et al., 1990) the muscarinic acetylcholine (Matesic et al., 1989) and the D_2-dopamine receptors (Senogles et al.,1990).

The important finding in our studies is the observation that both μ and δ opioid receptors form functional complexes with one or more variants of Goα following solubilization of cortical membranes with CHAPS. Antiserum IM1 which was raised against a region of Goα not believed to play a key role in receptor-G-protein interactions does not uncouple Goα from opioid receptors in cortical membranes (Georgoussi et al., 1993) but as demonstrated herein IM1 was able to co-immunoprecipitate opioid receptors along with Go (Fig.3). The inability of both, the antiserum against the carboxyterminal region (OC2) and the aminoterminal region (ON1) of Goα, to immunoprecipitate opioid receptor-Goα complexes may be a reflection that both of these antisera identify regions of Goα which could represent important contact regions between receptors and G-proteins. It has been reported that the carboxyterminal region of Gα plays a crucial role for receptor G-protein interaction and for the activation of the α subunit following agonist binding to receptors (Masters et al., 1988) whereas the amino-terminal domain of G-protein is responsible for interaction of the α subunit with βγ subunits (Schmidt and Neer, 1991). The lack of ability of antisera OC2 and ON1 to immunoprecipitate soluble opioid receptor-Go complexes may be attributed, either to uncoupling of the complexes by preventing the formation of the heterotrimer, or to recognition of these antisera with epitopes of Goα which are in direct contact to opioid receptors. In support to this explanation, we have previously demonstrated that both antisera ON1 and OC2 uncouple opioid receptors from Go based on the ability of each antiserum to reduce the affinity of the opioid peptide DADLE to compete for the specific [^{3}H]diprenorphine binding in rat cortical membranes (Georgoussi et al., 1993) In contrast to antisera OC2 and ON1, antiserum IM1 was unable to produce a reduction in affinity of DADLE for the opioid receptors, indicating that binding of antiserum IM1 to Go does not interfere with receptor-G protein coupling (Georgoussi et al., 1993). Interestingly, opioid receptors appear to associate with Gβ, as indicated by the ability of antiserum βN3 directed against 1-10 amino acids of the β-subunit of G-proteins, to immunoprecipitate solubilized rat cortical brain opioid receptors.

Recent work has utilised antisera directed against the C-terminal region of Giα and Goα to immunoprecipitate somatostatin (Law et al., 1991), α2-adrenergic (Okuma and Reisine, 1992) and muscarinic (Matesic et al., 1989) receptor-G-protein complexes. Such studies might have been predicted to be ineffectual as antisera of this type have been widely used to interfere with receptor signaling pathways (Simonds et al., 1989, McFadzean et al., 1989). While it has been assumed that such antisera act to uncouple G-proteins from their receptors, the effect of the antisera may only be to uncouple receptors and G-proteins functionally rather than physically.

Various studies revealed profound differences between μ and δ opioid receptors in coupling to pertussis toxin sensitive G-proteins (McKenzie and Milligan, 1990, Offermanns et al., 1991, Seward et al., 1991, Laugwitz et al., 1993). However since activated μ and δ opioid receptors have identical effects on several known effector systems, it is tempting to speculate that both receptors might also interact with identical G-protein(s). In this study we demonstrate that both μ and δ opioid receptors form stable complexes with one or more variants of Go employing a coimmunoprecipitation approach using selective G-protein subtype antisera. Such information could allow the elucidation of the molecular mechanisms underlying the intracellular events induced by opioid receptors in brain.

REFERENCES

Childers, S.R. (1991) Opioid receptor-coupled second messenger systems. Life Sci.48: 1991-2003

Cote TE, Gosse ME, and Weems HB (1993) Solubilization of high affinity, guanine nucleotide-sensitive μ-opioid receptors from 7315c cell membranes. J.Neurochem. 61:973-978

Evans, C.J., Keith, D.E., Morrison, H., Magendzo, K. and Edwards RH (1992) Cloning of a delta opioid receptor by functional expression. Science 258:1952-1955

Georgoussi Z, Taylor SJ, Bocckino SB and Exton JH (1990) Purification of the hepatic vasopressin receptor using a novel affinity column. Biochim. Biophys. Acta 1055:69-74

Georgoussi Z, Carr C and Milligan G (1993) Direct measurements of in situ interactions of rat brain opioid receptors with the guanine nucleotide-binding protein Go. Mol.Pharmacol. 44:62-69

Goldsmith PK, Gierschik P, Milligan G, Unson C G, Vinitsky R, Malech HL, and Spiegel AM (1987) Antibodies directed against synthetic peptides disdinguish between GTP-binding proteins in neutrophil and brain. J. Biol. Chem. 262:14683-14688

Laugwitz K-L, Offermanns S, Spicher K and Schultz G (1993) μ and δ opioid receptors differentially couple to G protein subtypes in membranes of human neuroblastoma SH-SY5Y cells. Neuron 10:233-242

Law SF, Manning DR and Reisine T (1991) Identification of the subunits of GTP-binding proteins coupled to somatostatin receptors. J. Biol. Chem. 266: 17885-17897

Masters SB, Sullivan KA, Miller RT, Beiderman B, Lopez, NG, Ramachandran J and Bourne HR (1988) Carboxyl terminal domain of Gsα specifies coupling of receptors to stimulation of adenylate cyclase. Science 241: 448-451

Matesic DF, Manning DR, Wolfe BB and Luthin GR (1989) Pharmacological and biochemical characterization of complexes of muscarinic acetylcholine receptor and guanine nucleotide-binding protein. J. Biol. Chem. 264: 21638-21645

McFadzean I, Mullaney I, Brown DA and Milligan G (1989) Antibodies to the GTP binding protein, Go, antagonize noradrenaline-induced calcium current inhibition in NG108-15 hybrid cells. Neuron 3: 177-182

McKenzie, FR and Milligan G (1990) δ-opioid-receptor-mediated inhibiton of adenylate cyclase is transduced specifically by the guanine-nucleotide-binding protein Gi2. Biochem. J. 267: 391-398

Mullaney I and Milligan G (1990) Identification of two distinct isoforms of the guanine nucleotide binding protein Go in neuroblastoma X glioma hybrid cells: Independent regulation during cyclic AMP-induced differentiation. J. Neurochem.55: 1890-1898

Offermanns S, Schultz G and Rosenthal W (1991) Evidence for

opioid receptor mediated activation of the G proteins, Go, Gi2 and in membranes of neuroblastoma X glioma (NG108-15) hybrid cells. J.Biol.Chem. 266: 3365-3368

Ofri D, Ritter AM, Liu Y, Gioannini TL, Hiller JM, and Simon EJ (1992) Characterization of CHAPS-solubilized opioid receptors:Reconstitution and uncoupling of guanine nucleotide-sensitive agonist binding. J. Neurochem. 58: 628-635

Okuma Y and Reisine T (1992) Immunoprecipitation of α2a-adrenergic receptor-GTP-binding protein complexes using GTP-binding protein selective antisera. J. Biol. Chem. 267: 14826-14831

Reisine T and Bell GI (1993) Molecular biology of opioid receptors. Trends Neurosci. 16: 506-510

Roerig SC, Loh HH and Law PY (1992) Identification of three separate guanine nucleotide-binding proteins that interact with the δ-opioid receptor in NG108-15 neuroblastoma X glioma hybrid cells Mol. Pharmacol. 41: 822-831

Senogles SE, Spiegel AM, Padrell E, Iyengar R and Caron, MG (1990) Specificity of receptor-G preotein interactions: Discrimination of Gi subtypes by the D2 dopamine receptor in a reconstituted system. J. Biol. Chem. 265: 4507-4514

Schmidt CJ and Neer EJ (1991) In vitro synthesis of G-proteins βγ dimers. J. Biol. Chem. 266: 4538-4544

Seward E, Hammond C and Henderson G (1991) μ-Opioid-receptor-mediated inhibition of the N-type calcium-channel current. Proc. R. Soc. Lond. B 244: 129-135

Simonds WF, Koski G, Streaty RA, Hjelmeland LM and Klee WA (1980) Solubilization of active opiate receptors. Proc. Natl. Acad. Sci. USA 77: 4623-4627

Simonds WF, Goldsmith PK, Codina J, Unson CG and Spiegel AM (1989) Gi2 mediates α2- adrenergic inhibition of adenylate cyclase in platelet membranes: In situ identification with Gα C-terminal antibodies. Proc. Natl. Acad. Sci. USA. 86: 7809-7813

Ueda H, Harada H, Nozaki M, Katada T, Ui M, Satoh M and Takagi H (1988) Reconstitution of rat brain μ opioid receptors with purified guanine nucleotide-binding regulatory proteins, Gi and Go. Proc. Natl. Acad. Sci. USA 85: 7013-7017

Wong YH, Demoliou-Mason CD and Barnard EA (1989) Opioid receptors in magnesium-digitonin-solubilized rat brain membranes are tightly coupled to a pertussis toxin-sensitive guanine-nucleotide binding protein. J. Neurochem. 52: 999-1009

THE USE OF STABLY TRANSFECTED CELL LINES IN THE ANALYSIS OF FUNCTIONAL INTERACTIONS BETWEEN Goα1 & Giα2 AND THE α2C10 ADRENOCEPTOR

Morag A. Grassie and Graeme Milligan
Molecular Pharmacology Group
Division of Biochemistry and Molecular Biology
Institute of Biomedical and Life Sciences
University of Glasgow
Glasgow G12 8QQ
Scotland, U.K.

INTRODUCTION

Heterotrimeric guanine nucleotide-binding proteins (G-proteins) play a central role in the transmission of information inside the cell from extracellular signals.The signal crosses the plasma membrane via serpentine receptors resulting in the activation of heterotrimeric G-proteins which can then in turn interact with a number of effector systems. Both heterotrimeric G-proteins and G-protein-linked receptors exist as large multi-member families. Therefore the specificity of G-protein-receptor interaction is clearly very important as depending on which G-protein the receptor couples to, a different effector system can activated.

The family of α2 adrenoceptors are typical examples of receptors which mediate inhibition of adenylyl cyclase via agonist induced stimulation of one or more members of the pertussis toxin-sensitive G-proteins (Limbird, 1989).This occurs in both platelets (Simonds *et al.* 1989) and NG108-15 neuroblastoma x glioma hybrids (McClue & Milligan 1990) by Gi2. However, a number of other reports of effector regulation by the α2 adrenoceptors exist. This includes the rapid inhibition of Ca^{2+} currents (McFadzean *et al.* 1989), stimulation of adenylyl cyclase (Eason et *al.* 1992), and stimulation of phospholipase A2 (Cotecchia *et al.* 1990), C (Jones *et al.* 1991) and D (McNaulty *et al.* 1992).

As the fast inhibition of calcium influx by α2 adrenoceptors is most probably a consequence of activated Goα inhibiting N-type Ca^{2+} channels in both NG108-15 (Claufield *et al.* 1992) and rat sympathetic neurons (Mathie *et al.* 1992), this pathway may represent an important mechanism by which presynaptic α adrenoceptors inhibit neurotransmitter release. This inhibition of calcium influx may represent a more important role for the α adrenoceptor in the central nervous system than that of agonist mediated inhibition of adenylyl cyclase which occurs via Giα2.

NATO ASI Series, Vol. H 92
Signalling Mechanisms – from Transcription Factors to Oxidative Stress
Edited by L. Packer, K. Wirtz

We have used stably transfected rat-1 cells to investigate the relative ability of two G-proteins Goα1 and Giα2 to interact with the α2C10 adrenoceptor (pharmacologically defined as an α2A adrenoceptor). By stably transfecting the cDNA of Goα1 and the α2A receptor into rat-1 cells (which endogenously express Giα2) we were able to quantitatively and qualitatively analyse the ability of these G-proteins to functionally interact with the α2A adrenoceptor.

MATERIALS AND METHODS

The generation of clone 1C cells (Rat 1 cells expressing the α2C10 adrenoceptor) has previously been described in detail (Milligan *et al.* 1991). Briefly, this involved transfection of Rat 1 fibroblasts, using calcium phosphate precipitation, with genomic DNA encoding the human α2C10 adrenoceptor (Kobilka *et al.* 1987) subcloned into the mammalian expression vector pDOL. Further clones including Clone 3 which is analysed in this study, were generated from clone 1C cells by co-transfection in a 10:1 ratio of a cDNA encoding rat Go1α in the mammalian expression vector pcEXV-3 (Parenti *et al.* 1993) and the plasmid pBABE hygro, which is able to direct expression of the hygromycin B resistance marker, in the presence of DOTAP (Boehringer Mannheim), used according to the manufacturers instructions. Following transfection the cells were maintained in hygromycin B(200 µg/ml) for 7 days following which individual colonies were selected and subsequently expanded. Cells were grown in DMEM supplemented with 5% newborn calf serum, penicillin (100units/ml) and streptomycin (100µg/ml) in 5% CO_2 at 37°C. Membranes from the cells were prepared by homogenization with a teflon on glass homogenizer and differential centrifugation as described for a variety of cells (Milligan, 1987).

For immunoblotting studies following SDS-PAGE, proteins were transferred to nitrocellulose (Costar) and blocked for 2 hours in 5% gelatin in phosphate buffered saline, pH 7.5 (PBS). Primary antisera (the generation of antisera SG1 (Green *et al,* 1990), ON1 and OC1 (Mullaney & Milligan 1990) has been described previously), were added in 1% gelatin in PBS containing 0.2 % Nonidet P40 (NP 40) and incubated for at least 2 hours. The primary antiserum was then removed and the blot washed extensively with PBS containing 0.2% NP40. Secondary antiserum (donkey anti-rabbit IgG coupled to horseradish peroxidase) (Scottish Antibody Production Unit, Wishaw, Scotland) was added (1:200 dilution in 1% gelatin in PBS containing 0.2% NP40) and incubated with the nitrocellulose for 2 hours. The antiserum was then removed and following extensive washing of the blot with PBS containing 0.2% NP40 and finally with PBS alone, the blot was developed using o-dianisidine hydrochloride (Sigma) as the substrate for horseradish peroxidase as previously described (McKenzie & Milligan 1990). Relative quantitation of immunoblots was performed using a Biorad Model GS-670 imaging densitometer.

Cholera toxin-catalysed [^{32}P] ADP-ribosylation of cell membranes of clones 1C and 3 was performed in the absence of added guanine nucleotide basically as in (Milligan *et al.* 1991). However, in the immunoprecipitation assays sodium phosphate, pH 7.0 replaced potassium phosphate, pH 7.0. Further additions to the assays were as detailed in the text. Dried gels were autoradiographed for up to fourteen days using Kodak X-O-Mat X-Ray film or were exposed overnight on a Canberra Packard Instant Imager model 2024. The immunoprecipitation conditions were as follows; membranes were suspended in 200μl 1% (w/v) SDS and boiled for 4 minutes. Subsequently 800μl of ice cold buffer 1 (1.25% (w/v) Triton X-100, 190mM NaCl, 6mM EDTA, 50mM Tris-HCl, pH 7.5) containing protease inhibitors (0.5mM PMSF and 0.15μM aprotinin) was added. After 1 h. between 5 and 10 μl of the relevant antiserum was added and incubated with rotation for 12h at 4°C. Subsequently, 50 μl of Pansorbin (Calbiochem) was added and the incubation continued for a further 3h. Immuno-complexes were collected by centrifugation (14,000g, 3 min, 4°C). These pellets were washed 3 times with 1ml of a buffer consisting of a 4:1 mixture of buffer 1 and 1% (w/v) SDS and once subsequently with 50mM Tris-HCl, pH 6.8. The final pellets were resuspended in Laemmli sample buffer and placed in a boiling water bath for 5 minutes. Samples were then centrifuged (14,000g, 3 min) and the supernatant then resolved by SDS-PAGE (10% (w/v) acrylamide) at 60V. Gels were stained with Coomassie blue, destained, dried and then autoradiographed using Kodak X-O-Mat X ray film and intensifying screens or were exposed to a phosphor storage plate and developed using a FUJIX Bio-imaging analyzer.

Binding experiments were performed at 30°C for 30 minutes in 10mM Tris HCl, 50 mM sucrose, 20 mM $MgCl_2$, pH 7.5 (Buffer B). In saturation experiments using [^{3}H] yohimbine the concentration of ligand was varied between 0.5 and 20nM. Non-specific binding was defined in all cases by parallel assays containing 100μM idozoxan. Non-specific binding increased with [^{3}H] ligand concentration in a linear manner. Binding experiments were terminated by rapid filtration through Whatman GF/C filters followed by three washes of the filter with ice cold buffer B (5ml).

Curve Fitting and Data Analysis was performed using the Kaleidograph curve fitting package driven by an Apple Macintosh computer.

GENERATION AND ANALYSIS OF STABLE CELL LINES

Cells of Rat 1 fibroblasts which had been previously transfected to express the human α2C10 adrenoceptor (clone 1C) (Milligan *et al.* 1991) were further transfected with the plasmids pBABE hygro, which expresses the product of the hygromycin B resistance gene, and pcEXV-3 into which a cDNA encoding the rat Go1α splice variant had been ligated (Parenti *et al.* 1993) (Figure 1). Clones which displayed resistance to hygromycin B were selected, expanded and tested for maintained expression of the α2C10 adrenoceptor and novel

RAT-1 fibroblast transfected with α2C10 adrenoceptor genomic DNA (expression driven by β-actin promoter in pDol mammalian vector). Clones were selected in medium containing 700ug/ml geneticin.

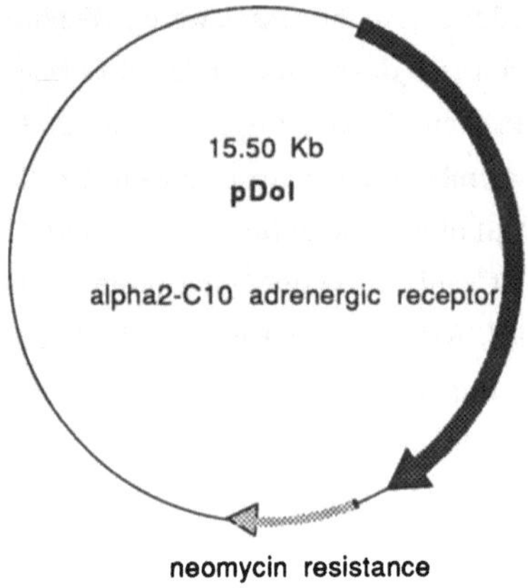

Generation of 1C cell line expressing the α2-C10 adrenoceptor

1C cell line co-transfected with pEXV-3 containing Go1α + a second plasmid pBABE, conferring hygromycin resistance.

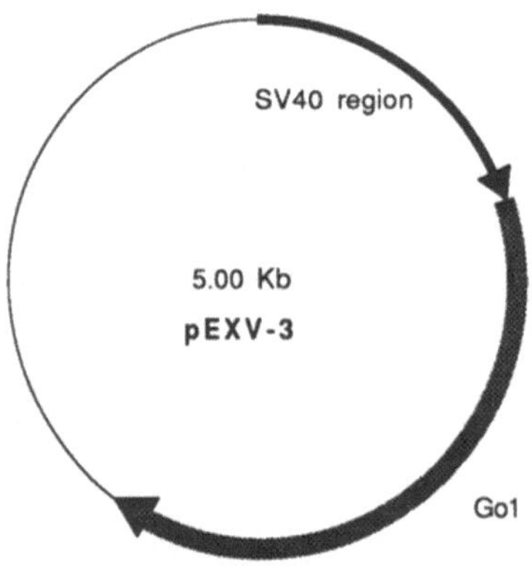

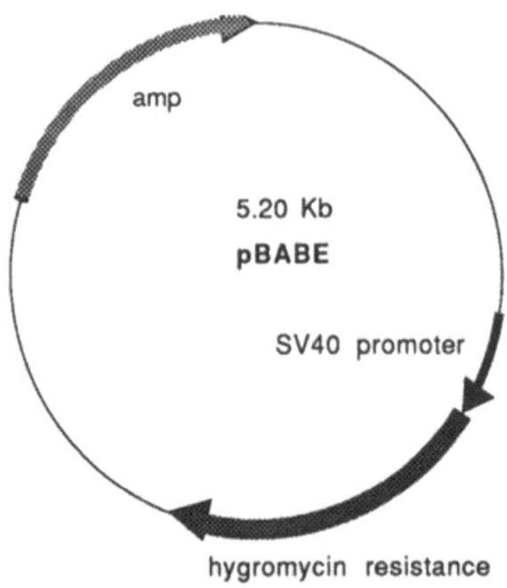

Generation of Clone 3 cell line expressing α2-C10 adrenoceptor and Go1α

Figure 1 : Construction of clones 1C and clone 3 from rat 1 fibroblasts

Rat 1 fibroblasts were transfected with the plasmid pDol (containing human α2C10 adrenoceptor) and geneticin resistant clones selected (clone 1C). Clone 1C was further transfected with pBabe and pEXV-3 (containing Goα1). Hygromycin resistant clones were selected and analysed (see Table 1).

expression of Go1α. A number of clones co-expressing these proteins were analysed and the range of Go1α and receptor expression measured (Table 1). Clone 3 was selected for detailed study. This clone expressed 2.23 ± 0.23 pmol/mg membrane protein of the α2C10 adrenoceptor which bound the α2 adrenoceptor antagonist [^{3}H] yohimbine with high affinity (Kd 2.8 ± 0.5 nM). Figure 2 shows a representative Scatchard analysis of clone 3 cell membranes (20ug). Estimation of the levels of Goα1 in clone 3 has been previously assessed by quantitative immunoblotting of clone 3 membranes vs differing amounts of purified recombinant Go1α (Grassie & Milligan, 1994). Construction of standard curves from such immunoblots allowed the levels of Go1α in clone 3 membranes to be determined. Clone 3 was found to express 94.3 ± 13.2 pmol Go1α/mg membrane protein (mean ± S.E.M. n =3).

AGONIST STIMULATED CHOLERA TOXIN ADP [^{32}P] RIBOSYLATION

To assess the ability of Go1α and Giα2 to functionally interact with the α2A adrenoceptor agonist driven cholera toxin [^{32}P] ADP-ribosylation was performed in the absence of guanine nucleotides (Milligan *et al.* 1991). Under normal conditions cholera toxin

Clone number	Go1α Immunoreactivity	α2C10 adrenoceptor (fmol/mg membrane protein)
1	+++	224
3	+++	2302
6	-	504
10	++	15
11	++	631
12	-	977
13	-	1097
14	+	426
15	++	311
16	-	604

Table 1: Characterization of clones derived from clone 1C following co-transfection with plasmid pEXV-3 containing rat Goα1 and plasmid pBABE hygro.

α2C10 adrenoceptors were measured by the specific binding of a single concentration (10nM) of [^{3}H] yohimbine with subsequent correction for receptor occupancy based on a measured Kd of 3.3nM for [^{3}H] yohimbine in membranes of clone 3 cells. Relative amounts of Goα1 were assessed by immunoblotting with the Goα antiserum OC1. Clone 3 expresses some 100pmol Goα1/mg membrane protein (Grassie & Milligan, 1994).

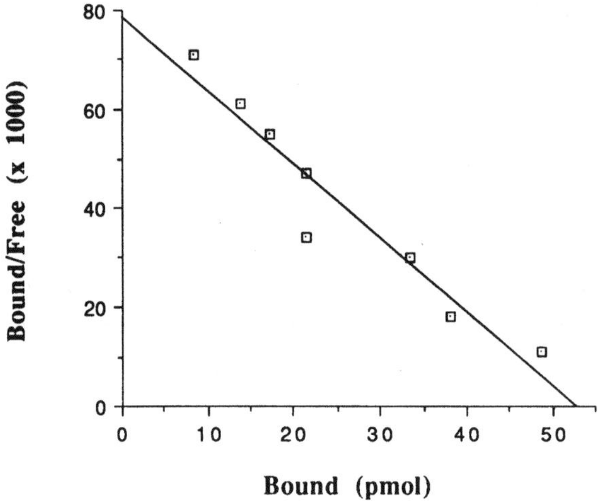

Figure 2 : Quantitation of levels of the α2C10 adrenoceptor in clone 3 membranes.
Membranes of clone 3 cells (20ug) were analysed for the presence of α2 adrenoceptors in saturation assays using varying concentrations of [3H] yohimbine as described in Methods. Data from a representative experiment is presented as a Scatchard plot for visual examination but all quantitative data were calculated using a non-linear least squares curve fit program as described in Methods. In the example displayed Bmax = 2.5pmol/mg membrane protein.

ADP-ribosylates Gs, but not Gi or Go-like G-proteins. It has however been shown that cholera toxin catalysed ADP-ribosylation of "inappropriate" pertussis toxin sensitive G-proteins can occur if the guanine nucleotide-binding pocket of these G-proteins is empty. It has been further observed that the addition of agonist for a receptor known to interact with a pertussis toxin sensitive G-protein causes a marked enhancement of the incorporation of [^{32}P]ADP-ribose into that G-protein. This strategy has been useful in defining which pertussis toxin sensitive G-proteins interact with which receptors.

By generating a stable cell line expressing the α2a adrenoceptor, Goα1 and endogenous Giα2, we have been able to compare in detail the relative ability of both of these G-proteins to interact with the α2A adrenoceptor using this technique of agonist stimulated cholera toxin ADP ribosylation. Cholera toxin-catalysed [^{32}P]ADP-ribosylation was performed on membranes of each of clone 3 cells, clone 1C cells and parental untransfected Rat 1 cells in the absence of guanine nucleotides and in the presence and absence of the selective α2 adrenoceptor agonist UK14304 (10μM). Resolution of membrane proteins from each cell type by SDS-PAGE (10%

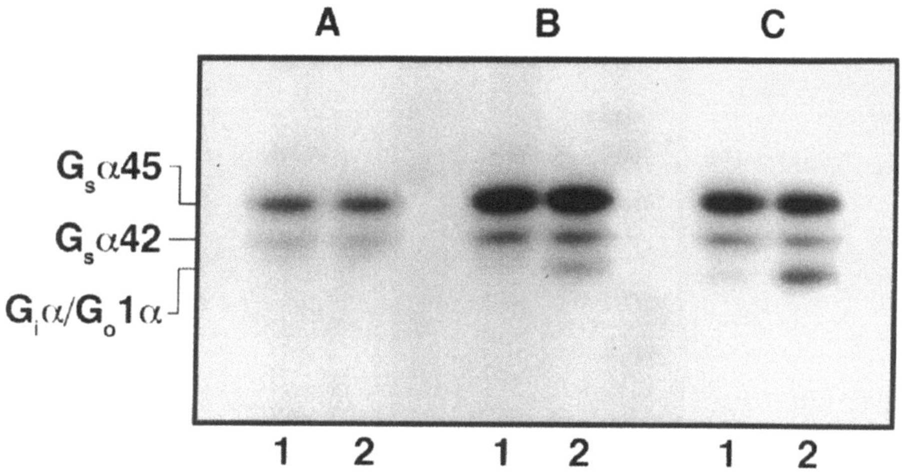

Figure 3. Cholera toxin catalysed [32P]ADP-ribosylation of G-proteins. The effect of agonist. Membranes (60μg) of (A) Rat 1 fibroblasts, (B) clone 1C or (C) clone 3 were incubated with [^{32}P]NAD and thiol-activated cholera toxin in the absence of guanine nucleotides and in the absence (1) or presence (2) of UK14304 (10μM) for 2 hours as described in Methods. Samples were precipitated, resolved by SDS-PAGE (10% (w/v) acrylamide) and subsequently exposed to a phosphor storage plate and developed (Grassie & Milligan, 1994).

(w/v) acrylamide) and subsequent autoradiography following such treatment led to the incorporation of radioactivity into the 45 and 42 kDa splice variants of Gsα in both the absence and presence of UK14304 in all three membrane preparations (Figure 3). In both clone 1C and clone 3 membranes, but not in membranes of Rat 1 cells, there was also incorporation of radioactivity into an apparently single band of some 40kDa primarily when the experiments were performed in the presence of the receptor agonist (Figure 3) as noted previously for clone 1C (Milligan *et al* 1991). Incorporation of radioactivity into this 40kDa polypeptide was greater in membranes of clone 3 than in clone 1C. Such data defines the interaction of the agonist-occupied adrenoceptor with "Gi-type" G-protein(s) (Milligan *et al.* 1991), however, as the subtypes of Giα and Go1α essentially comigrate in such gels this does not allow discrimination between interactions of the receptor with Go1α or other endogenous Gi-like G-proteins .

To explore in detail the molecular identity of the pertussis toxin-sensitive G-protein(s) activated by the α2C10 adrenoceptor in these clones we performed cholera toxin-catalysed

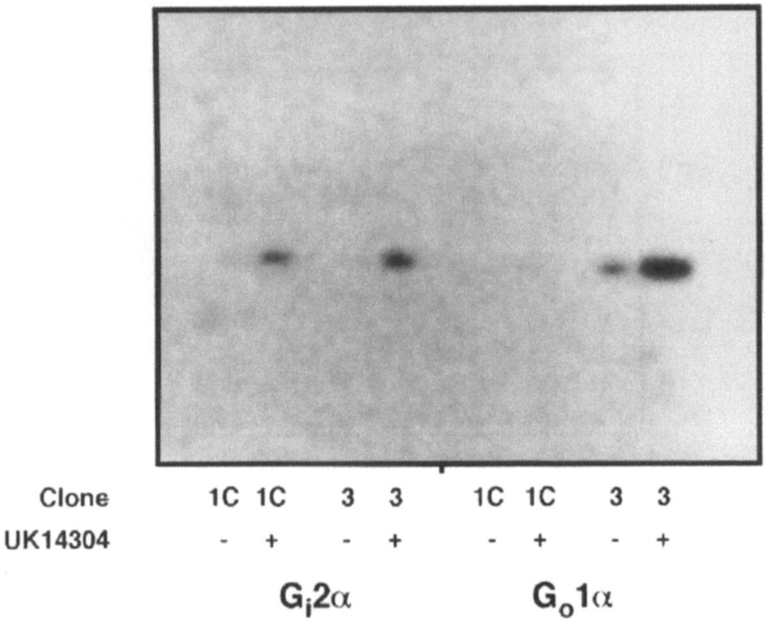

Figure 4 : Immunoprecipitation of agonist-induced cholera toxin catalysed [^{32}P]ADP-ribosylation of Gi2α and Go1α in membranes from clones of Rat 1 fibroblasts.
Cholera toxin catalysed [32P] ADP ribosylation was performed in the absence of guanine nucleotides and in the presence or absence of UK14304 (10μM) on membranes (60 μg) of clone 1C and clone 3 cells as described in Methods. Subsequently samples were immunoprecipitated using either the anti-Gi2α antiserum SG or the anti Goα antiserum ON. Immunoprecipitates were resolved by SDS-PAGE (10% (w/v) acrylamide), dried and radioactivity detected over a 17h period by electron capture using a Canberra Packard Instant Imager model 2024 as described in Methods (Grassie & Milligan, 1994).

[^{32}P]ADP-ribosylation of membranes of both clone 1C and clone 3 in the absence and presence of UK14304 and subsequently immunoprecipitated these samples with antisera which identify Gi2α (antiserum SG) or Go1α (antiserum ON). SDS-PAGE and autoradiography of these immunoprecipitates demonstrated UK14304-stimulated incorporation of radioactivity into Gi2α in both clone 1C and clone 3 and also incorporation of radioactivity into Go1α in clone 3 but not in clone 1C (Figure 4). A small degree of cholera toxin-catalysed incorporation of radioactivity into Go1α in membranes of clone 3 cells was also recorded in the absence of UK14304 (Figure 4). Immunoprecipitation of such experiments with a Gsα specific antiserum (CS) confirmed that the 45 and 42 kDa polypeptides radiolabelled by cholera toxin treatment in both the absence and presence of UK14304 were indeed isoforms of Gsα (data not shown). Although it may seem from the data of Figure 3 that the α2C10 adrenoceptor activated more Go1 than Gi2 in membranes of clone 3 as there is greater amount of radiolabel in the Go1α immunoprecipitates than in the Gi2α immunoprecipitates such conclusions must be drawn with caution as this will be dependent to a large extent on the immunoprecipitation efficiencies of respective antibodies.

To investigate the relative interaction between Goα1 or Giα2 with the α2A adrenoceptor, a dose response to agonist stimulated cholera toxin ribosylation was performed as before with the exception that all samples were immunoprecipitated with either ON antiserum (directed against the N'terminus of Goα) or SG1 (directed against the C'terminus of Giα2) before SDS/PAGE (Grassie & Milligan, 1994). Analysis of these dose effect curves indicated that agonist driven cholera toxin catalysed [^{32}P]ADP-ribosylation of Goα1 and Giα2 were highly similar, generating an EC_{50} value for Goα1 of 34±7nM with the EC50 value for Giα2 being 22±6nM (mean ± S.E.M. n= 4) (Grassie & Milligan, 1994).

AGONIST DRIVEN SELECTIVE G-PROTEIN DOWNREGULATION

In many circumstances sustained exposure of cells expressing a G-protein-coupled receptor to agonist results in a selective down-regulation of the α subunit of the G-protein(s) activated by the receptor (Milligan, 1993). To further confirm the ability of both Giα2 and Goα1 to functionally interact with the α2A adrenoceptor we exposed clone 3 to 10^{-5}M UK14304 for 17 hours. Membranes were then prepared from treated and control cells and immunoblotted for both Goα1 and Giα2. A marked downregulation of both Go1α and Gi2α was observed whereas membrane-associated levels of Gqα/G11α and Gsα were unaltered by such treatment. By using a range of concentration of UK14304 a dose-response for agonist-mediated down-regulation of both Go1α and Gi2α were carried out. (Figure 5). Densitometric scanning of the immunoblots indicated the half maximal concentration of UK14304 for Goα1 and Giα2 downregulation to be 176±67nM and 58±40 nM (mean±S.E.M., n= 4) respectively (Grassie & Milligan, 1994).

DISCUSSION

The use of stably transfected cell lines has certain advantages over a number of alternative systems. Unlike transient transfection, the generation of a stable cell line allows a homologous population of cells to be studied, where the number of α subunits of heterotrimeric G-proteins and binding sites for the receptor in question can be accurately determined. This differs quite dramatically from transient systems, where after transfection a heterogeneous population of cells exists, with a wide range of the levels of proteins encoded by the transfected cDNAs. A previous study has reported on the relative affinity of interaction of the α2C10 adrenoceptor with various G-proteins following transient co-expression in HEK293 cells (Chabre *et al.* 1994). This study, whilst demonstrating that the α2 adrenoceptor interacted more effectively with Gi over Gs or Gq, was unable to assess as reported herein either the levels of

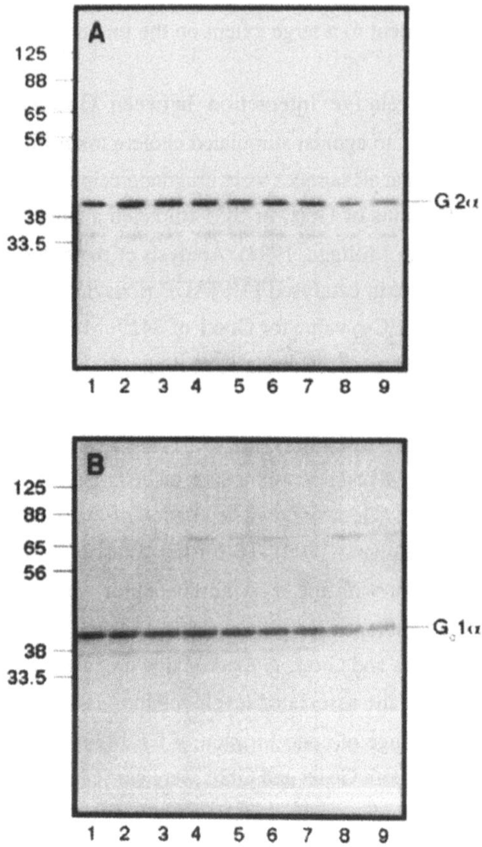

Figure 5 : Dose-response for UK14304-induced down-regulation of Go1α and Gi2α in clone 3
Clone 3 cells in tissue culture were treated with varying concentration of UK14304 for 16 hr, the cells harvested and membranes prepared. Samples of these membranes were resolved by SDS-PAGE (10% (w/v) acrylamide) transferred to nitrocellulose and immunoblotted for the presence of either Gi2α (panel A) or Go1α (panel B) (1, control, 2, 100pM, 3 = 1nM, 4 = 10nM, 5 = 30nM, 6 = 100nM, 7 = 300nM, 8 =1μM, 9 =10μM). Markers represent apparent molecular mass (x10-3). Data is presented from a representative experiment.

expression of the various polypeptides or the relative ability of the receptor to activate two G-proteins when they were co-expressed. In reconstitution assays using phospholipid vesicles, quantitation of the signalling components of interest is possible and also protein-protein interactions can be investigated. However, the lack of intracellular components in this system can limit and potentially influence the results obtained. We have therefore used the stable transfection of Rat 1 fibroblasts to analyse the relative ability of Giα2 and Goα1 to functionally interact with the human α2C10 adrenoceptor.

α2A adrenoceptors are classically defined in terms of second messenger regulation as inhibitors of adenylyl cyclase. For the α2A (molecularly defined as the α2C10) adrenoceptor of human platelets (Simonds *et al*, 1989) this effector function has been demonstrated to result from a selective activation of the G-protein Gi2. However, it is clear that in a number of other systems, expression of an α2 adrenoceptor can result in the activation of multiple G-proteins and effector cascades. For example, in both Rat 1 fibroblasts and CHO cells transfected to express the α2C10 adrenoceptor, agonist activation of both Gi2 and Gi3 has been recorded (Milligan *et al.* 1991, Gerhardt & Neubig 1991) while following transfection of this receptor into LLC-PK1-O cells a more complex pattern has been reported with the receptor appearing to be able to interact with each of Gi1, Gi2, Gi3 and Go (Okuma & Reisine 1992).

The question of potential interactions of α2 adrenoceptors with Go is of particular interest as α2 adrenoceptors in the central nervous system are known to be able to inhibit flux through N-type Ca2+ channels and receptor regulation of these channels has been established to be mediated via the individual splice variant subtypes of Go (Kleuss *et al.* 1991). Furthermore, regulation of such ionic conductance may be a more significant function in the central nervous system than inhibition of adenylyl cyclase. Although the potential interaction of α2A adrenoceptors with multiple G-proteins has been noted above, there is little information on the relative selectivity of interactions of the receptor with various G-proteins in a membrane or whole cell. We have therefore examined the interactions of the α2C10 adrenoceptor with Go1α and compared this to the interaction of the receptor with Gi2α following construction of a Rat 1 fibroblast cell line expressing each of these polypeptides. This was achieved by initially transfecting with the adrenoceptor, identifying a clone which expressed appropriate levels of the receptor and then further transfecting this clone with the cDNA for Goα1. In order to select successfully transfected cells for a second time, an additional antibiotic resistance was required. To achieve this, the Goα1 cDNA was co-transfected with a plasmid containing the hygromycin B resistance gene.

As the α2 adrenoceptors have been demonstrated to have the potential to interact weakly with Gs to stimulate adenylyl cyclase (Eason *et al.* 1992) we wished to examine if there was a substantial difference in the efficiency of the α2C10 adrenoceptor in clone 3 cells to activate Gi2α and Go1α. We (Milligan & McKenzie, 1988, Milligan *et al.* 1991) and others (Gierschik *et al.* 1989, Roerig *et al.* 1992, Dell'Acqua *et al.* 1993, Remaury *et al.* 1993) have previously made use of the ability of agonist to promote cholera toxin-catalysed [^{32}P] ADP-ribosylation to define pertussis toxin-sensitive G-proteins which interact with receptors. Using this technique we have been able to compare relative interactions of Giα2 and Goα1 with the adrenoceptor. No large difference in dose-effect curves for the covalent modification of the two G-proteins by agonist stimulated cholera toxin ADP ribosylation was recorded indicating that the receptor is able to activate both G-proteins at similar degrees of receptor occupancy and that activation of Go1α did not occur only at high receptor occupancy.

To further demonstrate the interaction of the receptor with both Gi2 and Go1 we took advantage of the growing literature which indicates that maintained exposure of a receptor to agonist can result in downregulation of the G-protein(s) which are activated by that receptor without alteration in levels of G-proteins which are not activated by the receptor (see Milligan 1993, for review). We noted that maintained exposure of clone 3 cells to UK14304 resulted in a substantial and essentially equal degree of down-regulation of membrane associated levels of Gi2α and Go1α but that there was no parallel down-regulation of the phosphoinositidase C-linked G-proteins Gqα/G11α. Whereas it has been noted following transfection that α2 adrenoceptors can in some instances cause the stimulation of a phosphoinositidase C via a pertussis toxin-insensitive mechanism (Jones *et al.* 1991) and thus presumably via activation of Gq and/or G11, we have previously noted in clone 1C that UK14304 is unable to stimulate the production of inositol phosphates (McNaulty *et al.* 1992). It is thus not surprising that there was no down-regulation of Gq/G11 upon agonist occupation of the α2C10 adrenoceptor.

It is likely that the absolute stoichiometry of activation of two different G-proteins by a single receptor will be dependent upon the relative levels of expression of the G-proteins. In clone 3 cells Go1α (approx 100 pmol/mg membrane protein) is expressed at somewhat higher levels than Gi2α (approx 50 pmol/mg membrane protein) (McClue *et al.* 1992), however, these levels are not dissimilar to the measured levels of 'Go' and 'Gi' in mammalian adult brain (Milligan *et al.* 1987) although it is obviously not yet possible to estimate levels of these G-proteins in individual neurons. This suggests that in the absence of physical constraints α2 adrenoceptors are likely to be able to activate both Giα2 and Goα1 in individual neurones and that the integration of signal information following activation of an α2 adrenoceptor will depend on both the absolute levels of expression of each G-protein and what mole fraction of the G-protein must be activated to significantly alter the activity of its effector.

Acknowledgement

These studies were supported by a project grant from the Medical Research Council (U.K.) to G.M.

REFERENCES

Caulfield, M.P., J. Robbins & D.A. Brown (1992). Neurotransmitters inhibit the omega-conotoxin-sensitive component of Ca current in neuroblastoma x glioma hybrid (NG108-15) cells, not the nifedepine-sensitive component. Pflueger's Arch. 420: 486-492.

Chabre, O., B.R. Conklin, S. Brandon, H.R. Bourne & L.E. Limbird (1994). Coupling of the α2A-adrenergic receptor to multiple G-proteins. A simple approach for estimating receptor-G-protein coupling efficiency in a transient expression system. J. Biol. Chem. 269: 5730-5734.

Cotecchia, S., B.K. Kobilka, K.W. Daniel, R.D. Nolan, E.Y. Lapetina, M.G.Caron, R.J. Lefkowitz & J.W. Regan (1990). Multiple second messenger pathways of α adrenergic receptor subtypes expressed in eukaryotic cells. J. Biol. Chem. 265: 63-69.

Dell'Acqua, M.L., R.C. Carroll & E.G. Peralta (1993). Transfected m2 muscarinic acetylcholine receptors couple to Gαi2 and Gαi3 in Chinese hamster ovary cells. Activation and desensitization of the phospholipase C signaling pathway. J. Biol. Chem. 268: 5676-5685.

Eason, M.G., H. Kurose, B.D. Holt, J.R. Raymond & S.B. Liggett (1992) Simultaneous coupling of α2-adrenergic receptors to two G-proteins with opposing effects. Subtype-selective coupling of α2C10, α2C4 and α2C2 adrenergic receptors to Gi and Gs. J. Biol. Chem. 267: 15795-15801.

Gerhardt, M.A. & R.R.Neubig (1991). Multiple Gi protein subtypes regulate a single effector mechanism. Mol. Pharmacol. 40: 707-711.

Gierschik. P., D. Sidiropoulos & K.-H. Jakobs (1989). Two distinct Gi-proteins mediate formyl peptide receptor signal transduction in human leukemia (HL-60) cells. J. Biol. Chem. 264: 21470-21473.

Grassie, M.A. & G.Milligan. (1994). Analysis of the relative interactions between the α2C10 adrenoceptor and the guanine nucleotide binding proteins Go1α and Gi2α following co-expression of these polypeptides in Rat-1 fibroblasts. Biochem. J. (in press)

Green, A., J.L.Johnson & G. Milligan (1990). Down-regulation of Gi sub-types by prolonged incubation with an adenosine receptor agonist. J. Biol. Chem. 265: 5206-5210.

Jones, S.B., S.P. Halenda & D.B.Bylund (1991). α2-adrenergic receptor stimulation of phospholipase A2 and of adenylate cyclase in transfected Chinese hamster ovary cells is mediated by different mechanisms. Mol. Pharmacol. 39:239-245.

Kleuss C., J. Hescheler, C. Ewel, W. Rosenthal, G. Schultz & B. Wittig (1991). Assignment of G-protein subtypes to specific receptors inducing inhibition of calcium currents. Nature (Lond.) 353:43-48.

Kobilka, B.K., H. Matsui, T.S. Kobilka, T.L. Yang-Feng, U. Francke, M.G. Caron, R.J. Lefkowitz & J.W. Regan (1987). Cloning, sequencing, and expression of the gene coding for the human platelet α2-adrenergic receptor subtype. Science 238: 650-656.

Limbird, L. (ed) (1988) The alpha 2 adrenergic receptors, Humana Press, New York .

MacNulty, E.E., S.J.McClue, I.C.Carr, T.Jess, M.J.O. Wakelam & G. Milligan (1992). α2C10 adrenergic receptors expressed in Rat 1 fibroblasts can regulate both adenylyl cyclase and phospholipase D-mediated hydrolysis of phosphatidylcholine by interacting with pertussis toxin-sensitive guanine nucleotide binding proteins. J. Biol. Chem. 267: 2149-2156.

McClue, S. & G. Milligan (1990). The α2B adrenergic receptor of undifferentiated neuroblastoma x glioma hybrid NG108-15 cells interacts directly with the guanine nucleotide binding protein Gi2. FEBS Lett. 269: 430-434.

McClue S. J., E. Selzer, M. Freissmuth & G. Milligan (1992). Gi3 does not contribute to the inhibition of adenylate cyclase when an α2- adrenergic receptor causes activation of both Gi2 and Gi3. Biochem. J. 284: 565-568.

McFadzean, I., I. Mullaney, D.A. Brown & G Milligan (1989). Antibodies to the GTP binding protein Go antagonize noradrenaline-induced calcium current inhibition in NG108-15 hybrid cells. Neuron 3: 177-182.

McKenzie, F.R & G. Milligan (1990). δ-opioid receptor mediated inhibition of adenylate cyclase is transduced specifically by the guanine nucleotide binding protein Gi2. Biochem. J. 267: 391-398.

Mathie, A., L.Bernheim & B. Hille (1992). Inhibition of N-and L-type calcium channels by muscarinic receptor activation in rat sympathetic neurons. Neuron 8: 907-914.

Milligan, G., R.A. Streaty, P. Gierschik, A.M. Spiegel & W.A. Klee (1987). Development of opiate receptors and GTP-binding regulatory proteins in neonatal rat brain. J. Biol. Chem. 262: 8626-8630.

Milligan, G (1987). Foetal-calf serum stimulates a pertussis-toxin-sensitive high affinity GTPase activity in rat glioma C6 BU1 cells. Biochem. J. 245: 501-505.

Milligan, G. & F.R. McKenzie (1988). Opioid peptides promote cholera-toxin-catalysed ADP-ribosylation of the inhibitory guanine-nucleotide-binding protein (Gi) in membranes of neuroblastoma x glioma hybrid cells. Biochem. J. 252: 369-373.

Milligan, G., C. Carr, G.W.Gould, I. Mullaney & B.E. Lavan (1991). Agonist-dependent, cholera toxin-catalysed ADP-ribosylation of pertussis toxin-sensitive G-proteins following transfection of the human α2-C10 adrenergic receptor into Rat 1 fibroblasts. Evidence for the direct interaction of a single receptor with two pertussis toxin-sensitive G-proteins, Gi2 and Gi3. J. Biol. Chem. 266: 6447-6455.

Milligan, G (1993). Agonist regulation of cellular G-protein levels and distribution: mechanisms and functional implications. Trends Pharmacol. Sci. 14: 413-418.

Mullaney, I. & G. Milligan (1990). Identification of two distinct isoforms of the guanine nucleotide binding protein Go in neuroblastoma x glioma hybrid cells: Independent regulation during cyclic AMP-induced differentiation. J. Neurochem. 55: 1890-1898.

Parenti, M., M.A.Vigano, C.M.H.Newman, G. Milligan & A.I. Magee (1993). A novel N-terminal motif for palmitoylation of G-protein α subunits. Biochem. J. 291: 349-353.

Okuma, Y. & T (1992). Reisine.Immunoprecipitation of α2A-adrenergic receptor-GTP-binding protein complexes using GTP-binding protein selective antisera. Changes in receptor/GTP-binding protein interaction following agonist binding. J. Biol. Chem. 267:14826-14831.

Remaury, A., D. Larrouy, D. Daviaud, B. Rouot & H. Paris (1993). Coupling of the α2-adrenergic receptor to the inhibitory G-protein Gi and adenylate cyclase in HT29 cells. Biochem. J. 292: 283-288.

Roerig S. C., H.H. Loh & P.Y. Law (1992). Identification of three separate guanine nucleotide binding proteins that interact with the δ-opioid receptor in NG108-15 cells. Mol Pharmacol. 41: 822-831.

Simonds, W.F, P. K. Goldsmith , J. Codina , C.G.Unson & A.M. Spiegel (1989). Gi2 mediates α2 -adrenergic inhibition of adenylate cyclase in platelets membranes : in situ identification with Gα C-terminal antibodies. Proc. Natl. Acad. Sci (USA) 86: 7809-7813.

MODULATION OF THE AMPA/KAINATE RECEPTORS BY PROTEIN KINASE C

Carlos B. Duarte[1], Ana L. Carvalho[2] and Arsélio P. Carvalho[1]
Center for Neurosciences of Coimbra
Department of Zoology[1] and Department of Biochemistry[2]
University of Coimbra
3049 Coimbra Codex
Portugal

Glutamate is the major excitatory neurotransmitter in the brain, and glutamate receptors mediate rapid excitatory synaptic transmission in the CNS. Glutamate activates ionotropic receptors and receptors coupled to guanine nucleotide-binding proteins (Schoepp and Conn, 1993; Schoepfer et al., 1994). Based on ligand binding assays, and on molecular biological and electrophysiological properties, the mammalian ionotropic glutamate receptors were subdivided into three classes: the N-methyl-D-aspartate (NMDA), S-α-amino-3-hydroxy-5-methyl-4-isoxazole-propionic acid (AMPA) and kainate receptors. The activity of these receptors is modulated during neuronal plasticity, most likely by phosphorylation (Raymond et al., 1993a). In this chapter we will review the basic structure of the non-NMDA ionotropic glutamate receptors, and the modulation of the channel activity by protein kinase C-dependent mechanisms. Finally, the role of these receptors and PKC in long term depression in the cerebellum is also discussed.

Subunit composition of the AMPA/kainate glutamate receptors

Cloning studies revealed at least three non-NMDA glutamate receptor classes in mammals: GluRA-GluRD, GluR5-GluR7, and KA1 and KA2 (Seeburg, 1993; Schoepfer et al., 1994). GluRA-GluRD subunits display high affinity for AMPA (in the nM range) and low affinity for kainate (in the μM range), and exhibit fast desensitizing current responses to AMPA and steady-state current-responses to kainate (Hollmann et al., 1989; Boulter et al., 1990; Keinänen et al., 1990; Nakanishi et al. 1990). Similarly, the chicken kainate-binding protein, which is expressed exclusively by the cerebellar Bergmann glia (Gregor et al., 1989;

NATO ASI Series, Vol. H 92
Signalling Mechanisms – from Transcription Factors
to Oxidative Stress
Edited by L. Packer, K. Wirtz

Somogyi et al., 1990), has an affinity for kainate in the μM range (Henley and Barnard, 1991). In mammals the GluRB subunit dominates the ion flow properties of the receptors. Channels formed from GluRA, GluRC and GluRD subunits show a high Ca^{2+} permeability, whereas heteromeric receptors containing the GluRB subunit, as appears to be the case in most of the native AMPA receptors, have low divalent ion permeabilities (Hollmann et al., 1991; Verdoorn et al., 1991).

The GluR5-GluR7 subunits give rise to receptors with high affinity for kainate (in the nM range), and exhibiting fast desensitizing current responses to the agonist (Egebjerg et al., 1991; Bettler et al., 1992; Sommer et al., 1992: Lomeli et al., 1992). KA1 and KA2 bind kainate with very high affinity (low nM range), but these subunits form functional receptors only when co-expressed with GluR-5 or -6 (Werner et al., 1991; Herb et al., 1992; Sakimura et al., 1992). The KA2/GluR6 channels can be activated by AMPA, which generates a non-desensitizing current component (Herb et al., 1992; Sakimura et al., 1992). In cells expressing the GluR6 subunits the influx of Ca^{2+} through the receptor-associated channels may be modulated by the extent to which GluR6 is structurally altered by RNA editing (Seeburg, 1993). The above mentioned electrophysiological studies, showing that the high affinity kainate receptors can be gated by AMPA and that receptors formed by GluRA-GluRD can be activated by kainate, lead to the AMPA/kainate classification.

Mechanisms of $[Ca^{2+}]_i$ regulation by AMPA/kainate receptors

The activation of the ionotropic glutamate receptors depolarizes the cells and, in some cases, increases Ca^{2+} conductances through the receptor associated channels (Seeburg et al., 1993; Schoepfer et al., 1994). Therefore, the $[Ca^{2+}]_i$ responses to glutamate receptor agonists are likely to be a combination of Ca^{2+} entry through the receptor-associated channels and through voltage-sensitive Ca^{2+} channels (VSCCs) activated by membrane depolarization. The contribution of the former pathway can be evaluated by stimulating the cells with a specific agonist in the presence of the non-permeable cation N-methyl-D-glucamine (NMG; Carvalho et al., 1994; Duarte et al., 1994).

The activation of the AMPA/kainate receptors, generally regarded as low-Ca^{2+} permeability receptors, is thought to increase the $[Ca^{2+}]_i$ mainly via depolarization-induced opening of VSCCs (Mayer and Westbrook, 1987; Mayer and Miller, 1990). However, some Ca^{2+} permeable AMPA/kainate receptors have been described (e.g. Brorson et al., 1992; Burnashev et al., 1992; Carvalho et al., 1994). Indeed, the stimulation of cultured neuronal chick retina cells with kainate, in Na^+ medium, increases transiently the $[Ca^{2+}]_i$. After the initial peak the $[Ca^{2+}]_i$ decreases to a plateau at about 74% of the peak concentration (Fig. 1A). In NMG medium, kainate evokes a more sustained increase of the $[Ca^{2+}]_i$, about 60% of the control (Fig. 1A), indicating that neuronal retina cells are endowed with Ca^{2+} permeable AMPA/kainate receptors.

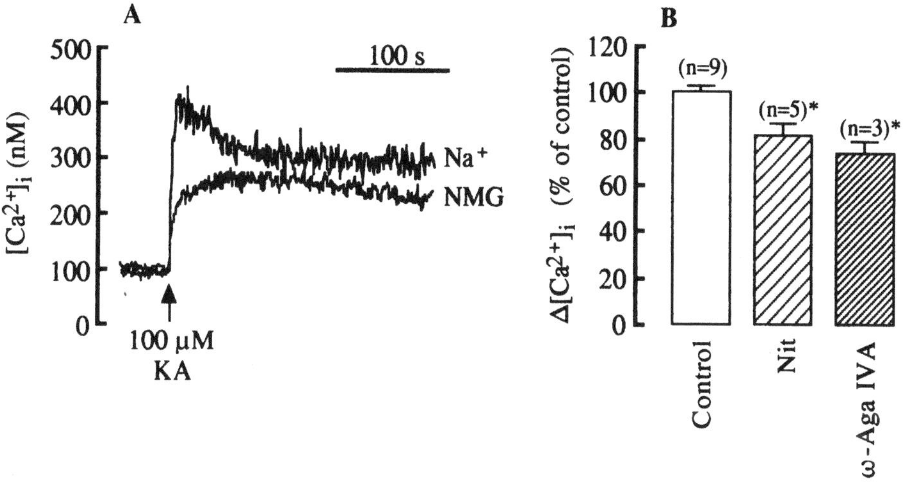

Fig. 1. $[Ca^{2+}]_i$ responses to kainate (KA) in cultured chick retina cells, in Na^+ (Na^+) and N-methyl-D-glucamine (NMG) media (A). In (B) the control initial responses to kainate (100 µM) in Na^+ medium are compared with those observed in the presence of nitrendipine (Nit; 1.5 µM) or ω-Aga IVA (200 nM), antagonists of the L- and P-type VSCCs, respectively. Cultures of chick retina cells, enriched in amacrine cells, were prepared as described (Duarte et al., 1992), and the $[Ca^{2+}]_i$ was measured using the fluorescent indicator Indo-1 (Duarte et al., 1992).

The contribution of the L- and P-type VSCCs to the $[Ca^{2+}]_i$ transients evoked by kainate in Na^+ medium can be determined using the antagonists nitrendipine (McCleskey et al., 1986) and ω-Agatoxin IVA (ω-Aga IVA; Mintz et al., 1992), respectively. After pre-incubation with the antagonists of the L- or P-type VSCCs, the initial $[Ca^{2+}]_i$ responses due to kainate stimulation is reduced by about 20% and 25% (Fig. 1B), respectively, and the effects of ω-Aga IVA are due mainly to an inhibition of the inactivating component (Duarte et al., 1994). Recently, ω-Aga IVA was shown also to interact with the Q-type VSCCs at the concentration used (Sather et al., 1993; Wheeler et al., 1994) and, therefore, a possible contribution of these channels to the responses to kainate can not be ruled out.

Modulation of the AMPA/kainate receptor responses by protein kinase C

Phosphorylation of voltage-gated and ligand-gated ion channels is important in the regulation of their activity, and in modulating neuronal excitability (Huganir and Greengard, 1990; Soderling et al., 1993; Hescheler and Schultz, 1993). Molecular cloning studies have shown that the high affinity- AMPA and -kainate receptors contain consensus sequences for both protein kinase C (PKC) and calcium/calmodulin-dependent kinase II, located in the intracellular loop (loop 3) between the third and forth transmembrane domains (Boulter et al., 1990; Keinänen et al., 1990; Raymond et al., 1993b). The GluR6, but not the GluRA-D subunits, also contains a protein kinase A (PKA) consensus sequence, and was shown to be directly phosphorylated by PKA (Raymond et al., 1993a,b; Wang et al., 1993).

The activation of protein kinase C was shown to increase the phosphorylation of GluRA receptors (McGlade-McCulloh et al., 1993) and to inhibit both the desensitization of GluRA+C receptors and the peak current amplitude (Dildy-Mayfield and Harris, 1994). Similar results were reported for the effect of phorbol esters on the activity of GluRC, GluRB+C and GluR6 receptors (Dildy-Mayfield and Harris, 1994). The introduction of mutations at the PKC consensus sites will allow determining whether the receptor subunits are directly phosphorylated by PKC or indirectly by phosphorylation of regulatory proteins.

Studies in neurons and using oocytes expressing rat or chick brain RNA lead to conflicting results regarding the role of PKC in the modulation of AMPA/kainate receptors. Indeed, the activation of PKC with phorbol esters produced either no effect, inhibition, or potentiation of the responses to kainate or AMPA in oocytes injected with brain RNA (Sigel and Baur, 1988; Moran and Dascal, 1989; Kelso et al., 1992). Moreover, PKC was without effect on the AMPA/kainate responses in trigeminal neurons (Chen and Huang, 1992), and inhibited the responses to AMPA in cultured cerebellar neurons (Linden and Connor, 1991; Ito and Karachot, 1992) and the currents evoked by low kainate concentrations in hippocampal neurons (Wang et al., 1994). In the latter preparation, however, the responses to near maximal concentrations of kainate or AMPA were potentiated by the intracellular injection of the catalytically active fragment of PKC (Wang et al., 1994).

In agreement with the observations in cultured hippocampal neurons, the activity of the Ca^{2+}-permeable AMPA/kainate receptors, in cultured chick retina cells, is potentiated by activation of PKC with phorbol esters (Carvalho et al., 1994). In that preparation, the $[Ca^{2+}]_i$ rise evoked by kainate in NMG medium, due to Ca^{2+} entry through the receptor associated channel (see above), is increased to about 130% of the control by activation of PKC with phorbol 12-myristate 13-acetate (PMA; Fig. 2). However, in Na^+ medium, PMA decreases the effect of kainate to about 80% of the control, most likely due to an inhibition of Ca^{2+} entry through VSCCs (Carvalho et al., 1994). Indeed, in cultured chick retina cells PMA decreases, by about 25%, the $[Ca^{2+}]_i$ rise evoked by K^+-depolarization (Fig. 3), which is mainly due to Ca^{2+} influx through the nitrendipine-sensitive L-type VSCCs (Fig. 3C; see also Duarte et al., 1992). Thus, although the activity of the Ca^{2+} permeable AMPA/kainate receptors is increased by PKC activation, the overall $[Ca^{2+}]_i$ response in Na^+ medium is inhibited due to the effect of the phorbol esters on the VSCCs. In agreement with our observations in cultured chick retina cells, PKC activation has also been shown to reduce L-type calcium channel activity in different cell types (e.g. Messing et al., 1986; Rane and Dunlap, 1988).

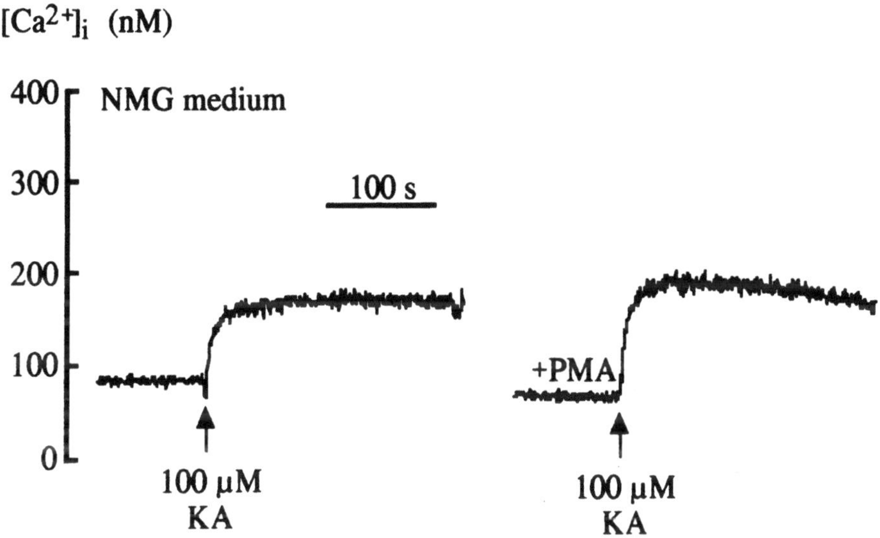

Fig. 2. Effect of PKC activation on the $[Ca^{2+}]_i$ rise due to Ca^{2+} entry through the AMPA/kainate receptor associated channel. The cultured chick retina cells were stimulated with kainate (KA) in NMG medium, in the absence (left) or after pre-incubation with 200 nM PMA, for 4 min (right). The $[Ca^{2+}]_i$ was determined with the fluorescent dye Indo-1 (Duarte et al., 1992).

The Ca^{2+}-insensitive protein kinase C subspecies δ and ξ are not expressed in the chick amacrine cells (Osborne et al., 1994), the most abundant cell type in the preparation (Duarte et al., 1992). Therefore, the effects of PKC on the chick retinal $[Ca^{2+}]_i$ responses to kainate or K^+-depolarization are due to the activation of one or more of the classical PKC isoforms, known to be present in the amacrine cells (Osborne et al.,1992). The modulatory effects might not arise, totally or partially from the direct phosphorylation of the receptor or channel proteins. For example, PKC may increase the activity of other kinases, such as PKA and/or calcium/calmodulin-dependent kinase II, which in turn may have modulated the AMPA/kainate receptor channel and the VSCCs (Browning and Dudek, 1992). Alternatively, PKC activation may affect the aggregation and assembly of the AMPA/kainate receptor subunits (Swope et al., 1992) and/or affect the channel activity by phosphorylating proteins that regulate the integrity of the cytoskeleton (Prat et al., 1993).

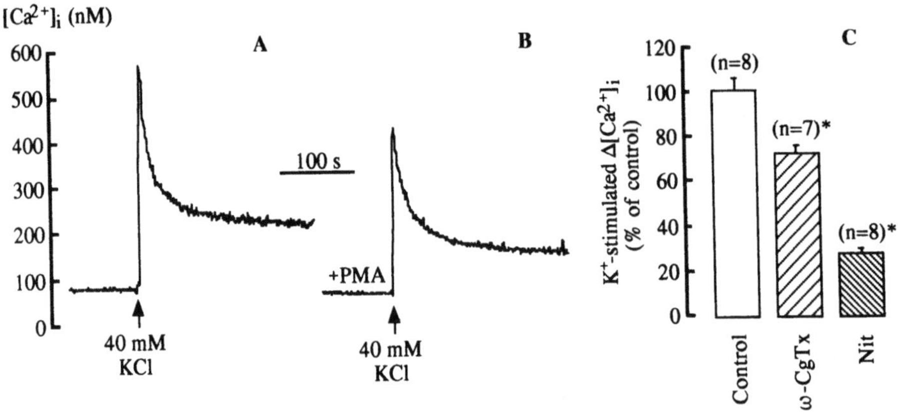

Fig. 3. Effect of the phorbol ester PMA on the $[Ca^{2+}]_i$ response to K^+-depolarization, in cultured chick retina cells. The cells were stimulated with 40 mM (A,B) or with 50 mM (C) KCl, in the presence (B) or in the absence (A,C) of 200 nM PMA. The pre-incubation time with the phorbol ester or nitrendipine was 4 min, and the $[Ca^{2+}]_i$ was determined with the fluorescent indicator Indo-1 (Duarte et al., 1992). When ω-Conotoxin GVIA (ω-CgTx; 0.5 μM), an antagonist of the N-type VSCCs, was tested (C) the cells were pre-incubated with the toxin for 1h.

Cultured chick neuronal retina cells were shown to possess metabotropic glutamate receptors, coupled to the hydrolysis of inositol phosphates (Duarte et al., 1994). The activation of these receptors, called ACPD receptors (Schoepp and Conn, 1993), does not lead to the mobilization of Ca^{2+} from intracellular stores, suggesting that the important signal may be diacylglycerol, the other second messenger produced by the phosphoinositides turnover (Carvalho et al., 1994; Duarte et al., 1994). In the brain, it has been suggested that one or the other of the two branches of this signalling pathway may be biased in a given location (Worley et al., 1987). If the ionotropic and the metabotropic glutamate receptors are expressed by the same cells, PKC activation by ACPD receptors may modulate the responses of the ionotropic glutamate receptors. Accordingly, in chick Purkinje neurons, the metabotropic glutamate receptor agonist (1S,3R)-1-aminocyclopentane-1,3-dicarboxylic acid ([1S,3R]-ACPD) supresses the AMPA/kainate receptor responses, and the effect could be mimicked by activation of PKC and/or protein kinase G (Mori-Okamoto et al., 1993). In the rat dorsal horn neurons (1S,3R)-ACPD potentiates the responses due to activation of the NMDA and non-NMDA receptors, but the interactive effect was relatively short-lived, suggesting that phosphorylation mechanisms may not be involved (Bleakman et al., 1992).

Considering that the AMPA/kainate receptors are probably oligomeric assemblies of subunits (Wenthold et al., 1992), different combinations of receptor subunits may generate subspecies of receptors with different selectivities for permeating ions and with distinct properties regarding the modulation by protein kinases. Although it is still not known whether the isoforms assemble randomly to form functional receptors, the differences regarding to the effect of PKC activation, by phorbol esters and by agonists of the metabotropic glutamate receptors, may result from the functional polimorphism.

Modulation of the AMPA/kainate receptors by PKC in long term depression

Long-term depression (LTD) in the intact cerebellum is induced after co-activation of parallel fiber (PF) and climbing fiber (CF) inputs to a Purkinje neuron, and leads to a persistent reduction in the amplitude of excitatory postsynaptic potentials (EPSP) at the PF-PC synapse (Ito, 1989). These synapses are enriched in AMPA/kainate receptors, and their desensitization is thought to cause LTD (Ito, 1989). The depression of glutamate induced currents is completely blocked by antibodies against the mGluR1 metabotropic glutamate receptors, which are coupled to the hydrolysis of phosphoinositides (Schoepp and Conn, 1993; Shigemoto et al., 1994). The observation that the induction of LTD depends on the co-activation of metabotropic and AMPA/kainate glutamate receptors and on Ca^{2+} influx, lead to the proposal that PKC may play a major role in the process (Ito, 1989; Mori-Okamoto et al., 1993). Indeed, recent studies have shown that phorbol esters, together with AMPA, evoke LTD in cultured mouse PC neurons, and the development of LTD is inhibited by PKC inhibitors (Linden and Connor, 1991). Moreover, phorbol esters decrease the responses due to activation of AMPA/kainate receptors in cerebellar Purkinje cells (Crepel and Krupa, 1988; Ito and Karachot, 1992). In addition to PKC, protein kinase G, activated by the second messenger cGMP, may also play a role in the generation of LTD (Ito and Karachot, 1992; Linden and Connor, 1993). However, although a direct phosphorylating action of PKG on AMPA/kainate receptors can not be excluded, it seems unlikely (Ito and Karachot, 1992; Taverna and Hampson, 1994).

ACKNOWLEDGEMENTS

The authors greatefully acknowledge the fruitfull discussions with Professors Carlos Faro and Euclides Pires. Work from the author's laboratory was supported by research grants form the Portuguese Research Council (JNICT) and by the Human and Capital Mobility Program.

REFERENCES

Bettler B, Egejerg J, Harma G, Pecht G, Hermans-Borgmeyer I, Moll C, Stevens CF, Heinemann S (1992) Cloning of a putative glutamate receptor: a low affinity kainate-binding subunit. Neuron 8: 257-265.

Bleakman D, Rusin KI, Chard PS, Glaum SR, Miller RJ (1992) Metabotropic glutamate receptors potentiate ionotropic glutamate responses in the rat dorsal horn. Mol Pharmacol 42: 192-196.

Boulter J, Hollmann M, O'Shea-Greenfield A, Hartley M, Deneris E, Maron C, Heinemann S (1990) Molecular cloning and functional expression of glutamate receptor subunit genes. Science 249: 1033-1037.

Brorson JR, Bleakman D, Chard PS, Miller RJ (1992) calcium directly permeates kainate/α-amino-3-hydroxy-5-methyl-4-isoxazolepropionic acid receptors in cultured cerebellar Purkinje neurons. Mol Pharmacol 41: 603-608.

Browning MD, Dudek E (1992) Activators of protein kinase C increase the phosphorylation of the synapsins at sites of phosphorylation by cAMP-dependent and Ca^{2+}/calmodulin-dependent protein kinase in the rat hippocampal slice. Synapse 10: 62-70.

Burnashev N, Khodorova A, Jonas P, Helm PJ, Wisden W, Monyer H, Seeburg PH, Sakmann B (1992) Calcium permeable AMPA-kainate receptors in fusiform cerebellar glial cells. Science 256: 1566-1570.

Carvalho AP, Ferreira IL, Carvalho AL, Duarte CB (1994) Glutamate receptor modulation of [^{3}H]GABA release and intracellular calcium in cultured chick retina cells. Ann N Y Acad Sci. (in press)

Chen L, Huang L-Y (1992) Protein kinase C reduces Mg^{2+} block of NMDA-receptor channels as a mechanism of modulation. Nature 356: 521-523.

Crepel F, Krupa M (1988) Activation of protein kinase C induces a long-term depression of glutamate sensitivity of cerebellar Purkinje cells. Brain Res 458: 397-410.

Dildy-Mayfield JE, Harris RA (1994) Activation of protein kinase C inhibits kainate-induced currents in oocytes expressing glutamate receptor subunits. J Neurochem 62: 1639-1642.

Duarte CB, Ferreira IL, Santos PF, Oliveira CR, Carvalho AP (1992) Ca^{2+}-dependent release of [^{3}H]GABA in cultured chick retina cells. Brain Res 591: 27-32.

Duarte CB, Ferreira IL, Santos PF, Oliveira CR, Carvalho AP (1993) Glutamate increases the $[Ca^{2+}]_i$ but stimulates $[Ca^{2+}]_i$ independent release of [^{3}H]GABA in cultured chick retina cells. Brain Res 611: 130-138.

Duarte CB, Santos PF, Carvalho AP (1994) Mechanisms of $[Ca^{2+}]_i$ by glutamate receptor agonists in cultured chick retina cells. (submitted)

Egebjerg J, Bettler B, Hermans-Borgmeyer I, Heinemann S (1991) Cloning of a cDNA for a glutamate receptor activated by kainate but not by AMPA. Nature 351: 745-748.

Gregor P, Mano I, Maoz I, Mckeown M, Teichberg VI (1989) Molecular structure of the chick cerebellar kainate-binding subunit of a putative glutamate receptor. Nature 342: 689-692.

Henley JM, Barnard E (1991) Comparison of solubilised kainate and α-amino-3-hyroxy-5-methylisoxazolepropionate binding sites in chick cerebellum. J. Neurochem. 56: 702-705.

Herb A, Burnashev N, Werner P, Sakmann B, Wisden W, Seeburg PH (1992) The KA-2 subunit of excitatory amino acid receptors shows widespread expression in brain and forms ion channels wih distantly related subunits. Neuron 8: 775-785.

Hescheler J, Schultz G (1993) G-proteins involved in the calcium channel signalling system. Curr Opin Neurobiol 3: 360-367.

Hollmann M, O'Shea-Greenfield A, Rogers SW, Heinemann S (1989) Cloning by functional expression of a member of the glutamate receptor family. Nature 342: 643-648.

Hollmann M, Hartley M, Heinemann S (1991) Ca^{2+} permeability of KA-AMPA-gated glutamate receptor channels depends on subunit composition. Science 252: 1028-1031.

Huganir RL, Greengard P (1990) Regulation of neurotransmitter receptor desensitization by protein phosphorylation. Neuron 5: 555-567.

Ito M (1989) Long-term potentiation. Ann Rev Neurosci 12: 85-102.

Ito M, Karachot L (1992) Protein kinases and phosphatase inhibitors mediating long-term desensitization of glutamate receptors in cerebellar Purkinje cells. Neurosci Res 14: 27-38.

Keinänen K, Wisden W, Sommer B, Werner P, Herb A, Verdoorn TA, Sakmann B, Seeburg PH (1990) A family of AMPA-selective glutamate receptors. Science 249: 556-560.

Kelso SR, Nelson TE, Leonard JP (1992) Protein kinase C-mediated enhancement of NMDA currents in *Xenopus* oocytes. J Physiol (Lond) 449: 705-718.

Linden DJ, Connor JA (1991) Participation of postsynaptic PKC in cerebellar long-term depression in culture. Science 254: 1656-1659.

Linden DL, Connor JA (1993) Cellular mechanisms of long-term depression in the cerebellum. Curr Opin Neurobiol 3: 410-406.

Lomeli H, Wisden W, Kohler M, Keinanen K, Sommer B, Seeburg PH (1992) High-afinity kainate and domoate receptors in rat brain. FEBS Lett 307: 139-143.

Mayer ML, Miller RJ (1990) Excitatory amino acid receptors, second messengers and regulation of intracellular Ca^{2+} in mamalian neurons. Trends Pharmacol Sci 11: 254-260.

Mayer ML, Westbrook L (1987) Permeation and block of N-methyl-D-aspartic acid receptor channels by divalent cations in mouse cultured central neurons. J. Physiol (Lond) 394: 501-527.

McCleskey EW, Fox AP, Feldman D, Tsien RW (1986) Different types of calcium channels. J Exp Biol 124; 177-190.

McGlade-McCulloh E, Yamamoto H, Tan S-E, Brickey DA, Soderling TR (1993) Phosphorylation and regulation of glutamate receptors by calcium/calmodulin-dependent protein kinase II. Nature 362: 640-642.

Messing RO, Carpenter CL, Greenberg DA (1988) Inhibition of calcium flux and calcium channel antagonist binding in the PC-12 neural cell line by phosbol esters and protein kinase C. Biochem Biophys Res Commun 136: 1049-1056.

Mintz I, Venema V, Swiderek K, Lee T, Bean B, Adams M (1992) P-type calcium channels blocked by the spider toin ω-Aga IVA. Nature 355: 827-829.

Moran O, Dascal N (1989) Protein kinase C modulates neurotransmitter responses in *Xenopus* oocytes injected with rat brain RNA. Mol Brain Res 5: 193-202.

Mori-Okamoto K, Tatsuno J (1993) Intracellular mechanisms underlying the supression of AMPA responses by *trans*-ACPD in cultured chick Purkinje neurons. Mol Cell Neurosciences 4: 375-386.

Nakanishi N, Shneider NA, Axel R (1990) A family of glutamate receptor genes: evidence for the formation of heteromeric receptors with distinct channel properties. Neuron 5: 569-581.

Prat AG, Bertorello AM, Ausiello DA, Cantiello HF (1993) Activation of epithelial Na^+ channels by protein kinase A requires actin filaments. Am J Cell Physiol 265: C224-233.

Rane S, Dunlap K (1988) Kinase C activator 1,2-oleylacetyl-glycerol attenuates voltage-dependent calcium current in sensory neurons. Proc Natl Acad Sci USA 83: 184-188.
Raymond LA, Blackstone CD, Huganir RL (1993a) Phosphorylation of amino acid neurotransmitter receptors in synaptic plasticity. Trends Neurosci 16: 147-153.
Raymond LA, Blackstone CD, Huganir RL (1993b) Phosphorylation and modulation of recombinant GluR6 glutamate receptors by cAMP-dependent protein kinase. Nature 361: 637-641.
Sakimura K, Morita T, Kushiya E, Mishina M (1992) Primary structure and expression of the gamma 2 subunit of the glutamate receptor channel selective for kainate. Neuron 8: 267-274.
Sather WA, Tanabe T, Zhang J-F, Mori Y, Adams M, Tsien RW (1993) Distinctive biophysical and pharmacological poperties of class A (BI) calcium channel $\alpha 1$ subunits. Neuron 11: 291-303.
Schoepfer R, Monyer H, Sommer B, Wisden W, Sprengel R, Kuner T, Lomeli H, Herb A, Köhler M, Burnashev N, Günther W, Ruppersberg P, Seeburg P (1994) Molecular biology of glutamate receptors. Prog Neurobiol 42: 353-357.
Schoepp DD, Conn PJ (1993) Metabotropic glutamate receptors in brain function and pathology. Trends Pharmacol Sci 14: 13-20.
Seeburg PH (1993) The molecular biology of mammalian glutamate receptor chanels. Trends Neurosci 16: 359-365.
Shigemoto R, Abe T, Nomura S, Nakanishi S, Hirano T (1994) Antibodies inactivating mGluR1 metabotropic glutamate receptor block long-term depression in cultured Purkinje cells. Neuron 12: 1245-1255.
Sigel E, Baur R (1988) Activation of protein kinase C differentially modulates neuronal Na^+, Ca^{2+}, and γ-aminobutyrate type A channels. Proc Natl Acad Sci USA 85: 6192-6196.
Soderling TR, Tan SE, McGlade-McCulloh E, Yamamoto H, Fukunaga K (1993) Excitatory interactions between glutamate receptors and protein kinases. J Neurobiol 25: 304-311.
Somogyi P, Eshhar N, Teichberg VI, Roberts JD (1990) Subcellular localization of a putative kainate receptor in Bergman glial cells using a monoclonal antibody in the chick and fish cerebellar cortex. Neurosci. 35: 9-30.
Sommer B, Burnashev N, Verdoorn TA, Keinänen K, Sakmann B, Seeburg PH (1992) A glutamate receptor channel with high affinity for domoate and kainate. EMBO J 11:1651-1656.
Swope SL, Moss SJ, Blackstone CD, Huganir RL (1992) Phosphorylation of ligand-gated ion channels: a possible mode of synaptic plasticity. FASEB J 6: 2514-2523.
Taverna FA, Hampson DR (1994) Properties of a recombinant kainate receptor expressed in baculovirus-infected insect cells. Eur J Pharmacol 266: 181-186.
Wang L-Y, Taverna FA, Huang X-P, MacDonald JF, Hampson DR (1993) Phosphorylation and modulation of a kainate receptor (GluR6) by cAMP-dependent protein kinase. Science 259: 1173-1175.
Weeler D, Randall A, Tsien RW (1994) roles of N-type and Q-type Ca^{2+} channels in supporting hippocampal synaptic transmission. Science 264: 107-111.
Wenthold RJ, Yokotani N, Doi K, Wada K (1992) Immunochemical characterization of the non-NMDA glutamate receptor using subunit-specific antibodies. J Biol Chem 267: 501-507.
Werner P, Voigt M, Keinänen K, Wisden W, Seeburg PH (1991) Cloning of a putative high-affinity kainate receptor expressed predominantly in hippocampal CA3 cells. Nature 351: 742-744.
Worley PF, Baraban JM, Colvin JS, Snyder S (1987) Inositol trisphosphate receptor localization in brain: variable stoichiometry with protein kinase C. Nature 325: 159-161.
Verdoorn TA, Burnashev N, Monyer H, Seeburg PH, Sakmann B (1991) Structural determinats of ion flow through recombinant glutamate receptor channels. Science 252: 1715-1718.

Interaction of MARCKS, a major protein kinase C substrate, with the membrane[1].

Guy Vergères[2], Stéphane Manenti[3] and Thomas Weber[4]
Department of Biophysical Chemistry
Biocenter of the University of Basel
Switzerland

The cell has developed complex molecular machineries to transport the information contained in external stimuli to specific subcellular locations. An excellent example is the modulation of the interaction of MARCKS, a major substrate of protein kinase C, by activators of protein kinase C. Phosphorylation and dephosphorylation induce the cycling of MARCKS between membrane and cytosol. The dynamics of this interaction may regulate the function of downstream targets of MARCKS, such as calmodulin and actin filaments. The molecular aspects of the interaction of MARCKS with membranes are reviewed here.

1) Distribution of MARCKS

In 1982 Greengard and coworkers identified an 87kDa protein phosphorylated in a calcium/phospholipid-dependent manner upon activation of protein kinase C (PKC)[5] following depolarization-induced calcium influx in synaptosomes of rat cerebral cortex (Wu et al., 1982). The development of a purification procedure for the 87kDa protein (Patel and Kligman, 1987) has allowed the subsequent cloning and sequencing of the bovine brain gene (Stumpo et al., 1989). This protein turned out to be the first of a family of PKC substrates now commonly refered as to MARCKS (for "Myristoylated Alanine-Rich C Kinase Substrate")

Footnotes: [1]This work was supported in part by the Swiss National Foundation for Promotion of Science Grant 31.32188.91. [2]Address for correspondence: Guy Vergères, Department of Biophysical Chemistry, Biocenter of the University of Basel, Klingelbergstrasse 70, 4056 Basel, Switzerland. Tel +41 612672179 Fax +41 612672189 Email vergeres@urz.unibas.ch [3]Hopital PURPAN, Unité INSERM 326, Toulouse, France. [4]Dept. of Biochemistry Swiss Federal Institute of Technology (ETH), Zürich, Switzerland. Present address: Program of Cellular Biochemistry and Biophysics, Rockefeller Research Laboratories, Memorial Sloan-Kettering Cancer Center, New York, USA. [5]Abbreviations: protein kinase C (PKC), phosphorylation site domain (PS domain), bacterial lipopolysaccharides (LPS).

NATO ASI Series, Vol. H 92
Signalling Mechanisms – from Transcription Factors to Oxidative Stress
Edited by L. Packer, K. Wirtz

(Stumpo et al., 1989). MARCKS has a widespread distribution among mammalian species (Albert et al., 1986; Blackshear et al., 1986) and has also been detected in chicken (Graff et al., 1989a) as well as in the electric organ of *Torpedo californica* (Albert et al., 1986). Analysis of vertebrate tissues reveals that MARCKS is concentrated in the brain, but is also present at various concentrations in all other tissues examined (Albert et al., 1986; Blackshear et al., 1986). In the central nervous system, MARCKS has been detected in both neurons and glial cells (Ouimet et al., 1990; Byers et al., 1993). Among non-neuronal cells, the protein is also found in fibroblasts (Rozengurt et al., 1983), neutrophils (Thelen et al., 1990) and macrophages (Aderem et al., 1988a; Li and Aderem, 1992).

2) MARCKS structure

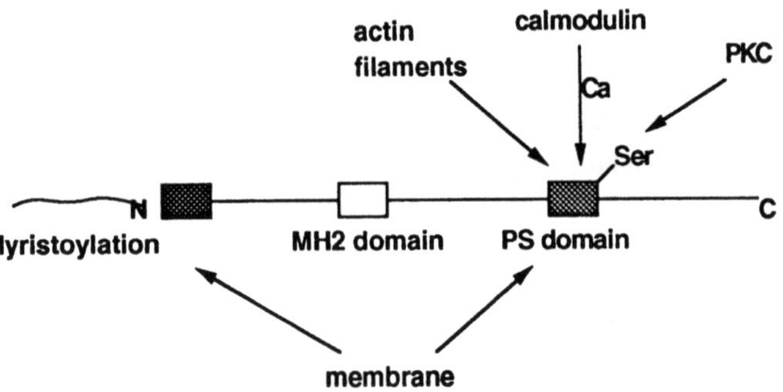

Figure 1. Structure of MARCKS. The sites of interaction of MARCKS with PKC, calmodulin, actin filaments and the membrane are shown.

The MARCKS family comprises two groups of proteins: the members of the first group are ubiquitous proteins, 288-322 residues in length, which have apparent molecular weights of 68-87kDa on SDS-polyacrylamide gels. A second group is composed of 42-45kDa proteins which contain 199-200 residues and are specifically expressed in macrophages. This group of proteins is refered as to MacMARCKS, MRP or F52 in the literature (Umekage and Kato, 1991; Li and Aderem, 1992; Blackshear et al., 1992; Chapline et al., 1993). MARCKS has three conserved domains (Figure 1). The N-terminus contains a consensus sequence, (Met)-Gly-X-X-X-Ser, for acylation by N-myristoyltransferase (Towler et al., 1988). A highly basic region (20-25 residues), the phosphorylation site (PS) domain, is present in the middle of the sequence and can be modeled as an amphipathic α-helix (Aderem, 1992a). These two domains are separated by a highly conserved MH2 domain, whose function is as yet unknown but which shows homology to the cytoplasmic domain of the mannose-6-phosphate receptor, a region thought to be involved in internalization. At the mRNA level, this region contains the site of intron splicing that accounts for different forms of mRNA (Li and Aderem, 1992; Blackshear, 1993). MARCKS is an acidic (calculated pI < 5.0), heat-stable protein which is soluble at low pH and unusually rich in alanine, glutamic acid, proline and glycine. Furthermore MARCKS does not contain significant hydrophobic regions and has a low content of secondary structure (Blackshear, 1993; Manenti et al., 1992). EM-studies show that MARCKS has a rod-like structure and that, interestingly, the length of this rod is dependent on the phosphorylation state of the protein, pointing to major conformational changes upon phosphorylation/dephosphorylation (Hartwig et al., 1992).

3) MARCKS phosphorylation

Treatment of intact cells with a variety of PKC activators such as phorbol esters, phospholipase C, diacylglycerol and growth factors induce the phosphorylation of MARCKS (Wu et al., 1982; Rozengurt et al., 1983; Rodriguez-Pena and Rozengurt, 1985; Blackshear et al., 1985; Blackshear et al., 1986; James and Olson, 1989). Bacterial lipopolysaccharides (LPS), tumor necrosis factor α, and chemotactic peptides specifically induce the phosphorylation of MARCKS in macrophages (Aderem et al., 1988a; Thelen et al., 1991) and in neutrophils (Thelen et al., 1990). PKC phosphorylates purified recombinant MARCKS and MacMARCKS with positive cooperativity and with substrate concentrations corresponding to the half-maximal reaction velocity, $S_{0.5}$, of 100 and 238 nM, respectively (Verghese et al., 1994). Whereas three serine residues are phosphorylated in the PS domain of MARCKS (Rosen et al., 1989), MacMARCKS can be phosphorylated at two serine residues only, the third serine being replaced by an alanine (Li and Aderem, 1992; Verghese et al., 1994).

In neutrophils, activation of PKC with a chemotactic peptide results in a rapid but transient phosphorylation of MARCKS. After 40 s stimulation, the equilibrium between kinase and phosphatase shifts to favour the dephosphorylation of MARCKS (Thelen et al., 1991). Dephosphorylation of MARCKS has also been reported in a cell-free system and in fibroblasts (Clarke et al., 1993).

Kinases other than PKC also phosphorylate MARCKS. In neutrophils, a peptide that corresponds to the PS domain of human MARCKS is a substrate for 69- and 63-kDa non-PKC kinases (Ding and Badwey, 1993). MARCKS is also phosphorylated on serine residues outside of the PS domain (Taniguchi et al., 1994). The sites of phosphorylation are consensus sequences for proline-directed protein kinases such as mitogen-activated protein (MAP) kinase or Cdk5 kinase.

4) MARCKS myristoylation

Aderem and coworkers have first demonstrated that MARCKS can be labeled with ^{3}H-myristate in activated macrophages (Aderem et al., 1986; Aderem et al., 1988a; Aderem et al., 1988b). The myristoylated protein is resistant to treatment with hydroxylamine, suggesting that the fatty acid is bound to MARCKS by an amide linkage. Cloning of MARCKS has confirmed the presence of a N-terminal consensus sequence for myristoylation (Stumpo et al., 1989). When expressed in COS cells (Graff et al., 1989b), or translated in a cell-free system (George and Blackshear, 1992), a mutant of MARCKS in which the N-terminal glycine is replaced by an alanine fails to be labeled with ^{3}H-myristate, demonstrating that MARCKS is acylated at the N-terminus (Graff et al., 1989b). Although MARCKS was proposed to be myristoylated posttranslationally upon activation of macrophages with LPS (Aderem et al., 1988a), pulse chase labeling of myocytes reveals no evidence for posttranslational stimulus-dependent myristoylation of a pre-existing pool of the protein (James and Olson, 1989). In addition, cycloheximide completely abolishes incorporation of ^{3}H-myristate by MARCKS showing that the protein is acylated cotranslationally (James and Olson, 1989). However, a pool of non-myristoylated MARCKS is present in brain cytosolic extracts (McIlhinney and McGlone, 1990; Manenti et al., 1993) and although myristoylation is thought to be irreversible (Towler et al., 1988), a demyristoylation enzymatic activity has been detected in cytosolic fractions of brain synaptosomes indicating that deacylation of MARCKS could potentially serve to regulate the function of the protein (Manenti et al., 1994).

5) MARCKS binding to calmodulin

In 1989 Blackshear and coworkers have shown that MARCKS binds to calmodulin in a calcium-dependent manner (Graff et al., 1989c). Both MARCKS and MacMARCKS bind to calmodulin with high affinity, with dissociation constants lower than 10 nM. The PS domain was identified as the calmodulin binding site. Phosphorylation of MARCKS by PKC prevents calmodulin from binding as well as disrupts the preformed MARCKS-calmodulin complex (Graff et al., 1989c; Verghese et al., 1994). In contrast to the myristoylated protein, non-myristoylated MARCKS can be eluted from a calmodulin affinity column by high ionic strength in the presence of calcium indicating that the myristic group directly or indirectly promotes the binding of MARCKS to calmodulin (Manenti et al., 1993). *In vivo* studies with the fresh water protozoan *Paramecium tetraurelia* indicate that the interaction of MARCKS with calmodulin is physiologically relevant (Hinrichsen and Blackshear, 1993). The calmodulin-dependent swimming behavior of this organism is altered by injecting a synthetic peptide corresponding to the PS domain of MARCKS. This behavioral response can be prevented by activation of endogenous PKC with a phorbol ester.

6) MARCKS binding to actin filaments

The association of MARCKS with microtubules in neurons (Ouimet et al., 1990) as well as the co-localization of MacMARCKS with talin and vinculin in macrophages (Rosen et al., 1990) suggest that MARCKS interacts with the cytoskeleton. Aderem and coworkers have shown that MARCKS is a filamentous actin crosslinking protein (Kd = 0.28 μM) (Hartwig et al., 1992). Phosphorylation by PKC inhibits the crosslinking activity of MARCKS but not its ability to bind actin filaments. In the presence of calcium, calmodulin also inhibits the crosslinking activity of MARCKS to actin filaments. Two models were proposed to explain these findings: 1) MARCKS has one actin binding site and can dimerize but phosphorylation or calmodulin monomerize MARCKS and consequently disrupt the crosslinking of actin filaments; 2) MARCKS has two binding sites for actin and only one is modulated by PKC and calmodulin, namely the PS domain (Hartwig et al., 1992). A distinction between these two models is not possible at this point and this question clearly awaits further investigation.

7) Dynamics of MARCKS subcellular localization

Immunocytochemistry reveals that MARCKS is primarily associated with microtubules, but is also present in the cytosol and at the plasma membrane of neurons (Ouimet et al., 1990). In brain extracts, 80-90% of total MARCKS can be purified from the particulate fraction (Thelen et al., 1991) whereas phosphorylated MARCKS is found in the particulate as well as in the cytosolic fractions (Wu et al., 1982; Walaas et al., 1983a; Walaas et al., 1983b; Patel and Kligman, 1987). In intact synaptosomes, activation of PKC results in increased phosphorylation of MARCKS and its translocation from the membrane to the cytosol. More than 80% phosphorylated MARCKS is cytosolic after phorbol ester treatment whereas only *ca* 20% of total MARCKS is translocated to the cytosol (Wang et al., 1989). Translocation is reversible since removal of stimulus results in reassociation of MARCKS with the membrane.

In macrophages, MARCKS has a punctate distribution at the cell-substratum interface of the membranes of pseudopodia and filopodia. The major portion of myristoylated MARCKS is in the membrane fraction, whereas the major portion of phosphorylated MARCKS is in the cytosolic fraction. Activation of PKC with phorbol esters results in the rapid disappearance of punctate staining and is concomitant with phosphorylation. 75% of myristoylated MARCKS is translocated into the cytosol after activation with PKC (Aderem et al., 1988; Rosen et al., 1990).

In neutrophils, MARCKS is mostly confined to the membrane. Treatment of neutrophils with phorbol esters results in the phosphorylation of MARCKS and its displacement from the plasma membrane. Transient phosphorylation of MARCKS with a chemotactic peptide is accompanied with the displacement of myristic acid-labeled protein from the membrane to the cytosol followed by dephosphorylation concomitant with reassociation of the myristoylated protein with membrane fractions (Thelen et al., 1991).

In glioma cells, MARCKS is mostly located in the membrane fraction. PKC-mediated phosphorylation induces translocation of MARCKS to the cytosol. However phosphorylation of MARCKS is increased to the same extent in cytosolic and particulate fractions (Byers et al., 1993).

In myocytes, the subcellular distribution of MARCKS is unchanged following stimulation with phorbol esters. The phosphorylation of both the membrane-bound and soluble forms is increased to the same extent upon stimulation of myocytes (James and Olson, 1989).

In neuroblastoma cells, phosphorylated MARCKS is confined to the membrane fraction only and no change in the distribution of the myristoylated protein is seen following activation of PKC with phorbol esters (Byers et al., 1993).

8) Interaction of MARCKS with membranes

The analysis of the subcellular localization of MARCKS prior and subsequent to PKC-mediated phosphorylation raises two major questions concerning the interaction of MARCKS with membranes: 1) how does MARCKS bind to membranes? and 2) how is this binding regulated? These aspects will be discussed in detail in the next sections.

8.1) Binding of MARCKS to membranes

Myristoylation. MARCKS can be released from brain synaptosomal membranes with nonionic detergents but not with NaCl (Albert et al., 1986) indicating a tight hydrophobic interaction of the protein with membranes. Since the protein is myristoylated at the N-terminus, the fatty acid is a likely candidate for promoting the binding of MARCKS to membranes. A mutant of MARCKS, in which the N-terminal glycine residue is changed to an alanine and therefore cannot be acylated by N-myristoyltransferase, is mostly cytosolic when expressed in COS cells (Graff et al., 1989b). In a cell-free system, the same mutant binds to fibroblast membranes to a much lower extent than the myristoylated wild type protein (George and

Blackshear, 1992). These experiments demonstrate that the myristic group mediates MARCKS binding to membranes. The cytosolic localization of a pool of non-myristoylated MARCKS further support these findings (McIlhinney and McGlone, 1990; Manenti et al., 1993). Although the myristic group is believed to anchor proteins *via* insertion of the carbon chain into the lipid bilayer no direct proof of such an interaction has been presented so far. We have therefore crosslinked MARCKS with (^{125}I)-TID-PC/16, a radioactive phosphatidylcholine trifluoromethyl-diazirine probe which has been developed to photolabel the apolar core of membrane proteins (Figure 2) (Weber et al., 1994). Photolysis of MARCKS bound to acidic lipid vesicles containing trace amounts of (^{125}I)-TID-PC/16 results in the labeling of the protein (lane 1). The labeling on MARCKS is almost completely lost upon subsequent deacylation (lane 2), demonstrating that the myristic group is inserted into the lipid bilayer.

Interaction of the basic PS domain with membranes. The Gibbs free binding energy of myristoylated glycine to membranes, 8kcal/mol, provides barely enough energy to attach proteins (Peitzsch and McLaughlin, 1993). The presence of a pool of myristoylated MARCKS

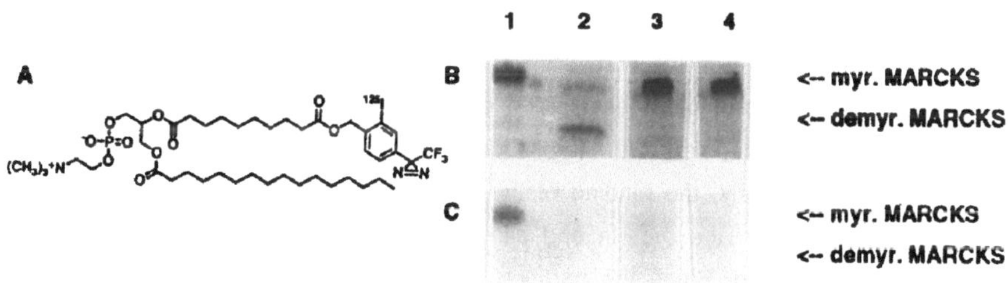

Figure 2. Photolabeling of MARCKS. Panel A: structure of (^{125}I)-TID-PC/16. Panels B,C: photolabeling of MARCKS. MARCKS (0.3 μM), purified from calf brain, was incubated with 100 nm vesicles (60 μM lipids) containing 5 μCi (^{125}I)-TID-PC/16. Following photolysis, lipids were removed by chromatography on DEAE in the presence of 0.1% Triton-X100 and the samples were analyzed on a 10% SDS-polyacrylamide gel (lanes 1, 3). In lane 2 MARCKS was demyristoylated enzymatically (Manenti et al., 1994). In a control experiment MARCKS was added to vesicles subsequent to photolysis (lane 4). Lipid composition of vesicles: 100% phosphatidylcholine (lane 3); 80% phosphatidylcholine / 20% phosphatidylserine (lanes 1, 2, 4). Panel B: Coomassie blue staining. Panel C: autoradiography.

in the cytosol of cells (James and Olson, 1989; Manenti et al., 1992) further indicates that the myristic group is not sufficient to bind MARCKS to membranes. On the other hand non-myristoylated MARCKS can, to some extent, bind to membranes indicating that the myristic group is not the only determinant of the binding of MARCKS to membranes (Nakaoda et al., 1992; George and Blackshear, 1992). *In vitro* studies with purified MARCKS show that the protein has a higher affinity for acidic than for neutral lipid vesicles suggesting that the basic PS domain might interact with the negatively charged membrane surface (Nakaoda et al., 1992; Kim et al., 1994a). Studies with peptides have confirmed this hypothesis: a synthetic peptide, corresponding to the PS domain of MARCKS, requires acidic lipids to bind to vesicles (Taniguchi and Manenti, 1993; Kim et al., 1994b). The binding of the PS domain peptide increases sigmoidally with the proportion of acidic lipids in the vesicles (Kim et al., 1994b).

Synergism of binding between the myristic group and the PS domain. A careful quantification of the individual contributions of the myristic group and of the PS domain to the binding of MARCKS to membranes reveals that neither hydrophobicity nor electrostatics alone can provide a strong anchoring of MARCKS to the membrane (Kim et al., 1994b). Based on these findings, McLaughlin and coworkers propose a two balls (myristic group and PS domain) and string (spacer residues) model in which there is a synergism in the two binding reactions. Binding of one domain to the membrane confines the second domain in the vicinity of the membrane. The probability that the second domain meets the membrane is therefore increased (Kim et al., 1994a). Our photolabeling experiments show that MARCKS is labeled with (^{125}I)-TID-PC/16 when incubated with acidic lipid vesicles but not with neutral vesicles (Figure 2, lanes 1 and 3, respectively). Binding of the PS domain to acidic lipids is therefore necessary to promote the insertion of the myristic group into the bilayer. This observation is in good agreement with the pseudocooperative two balls and string model (Kim et al., 1994a). Alternatively, the binding of the PS domain to acidic lipids could induce a conformational change in the protein and allow the subsequent insertion of the myristic group into the membrane. Since the partition coefficient for the binding of myristoylated MARCKS to phosphatidylcholine vesicles (2.6×10^3 M^{-1}) (Kim et al., 1994a), is four times smaller than the partition coefficients of myristoylated model peptides (1×10^4 M^{-1}) (George and Blackshear, 1992), MARCKS may be in a conformation which inhibits the insertion of the myristic group into the membrane. This inhibition could be released when the PS domain binds to acidic lipids. The hypothesis of a conformational change is supported by the observation that myristoylation of MARCKS modulates the affinity of the protein for calmodulin (Manenti et al., 1993). The PS domain and the myristoylated N-terminus might therefore interact with each other to regulate the interaction of MARCKS with the membrane as well as with calmodulin and other proteins. In this respect and in analogy to the calcium-myristoyl switch described for recoverin (Zozulya and Stryer, 1992), Gordon and coworkers have introduced the term myristoyl-phosphoryl switch to describe this aspect of the function of MARCKS (Johnson et al., 1994).

8.2) Regulation of the binding of MARCKS to membranes

Cellular studies indicate that the subcellular localization of MARCKS is a dynamic event, the protein being translocated reversibly between membrane and cytosol (section 7). In light of the mechanism by which MARCKS binds to membranes (section 8.1), the subcellular localization of MARCKS could potentially be regulated at two sites on the protein: removal of the myristic group or inhibition of the electrostatic interaction of the PS domain with membranes should both be sufficient to trigger the translocation of the protein.

Myristoylation/demyristoylation. Although N-myristoyl groups apparently do not turn over on proteins in a rapid acylation/deacylation/reacylation cycle (Towder et al., 1988), the detection of an enzymatic deacylation of MARCKS in the cytosol of cell extracts indicates that removal of the myristic group could potentially regulate the function of MARCKS (Manenti et al., 1994). Cohn and colleagues (Aderem et al., 1988a) proposed a model in which the phosphorylation of MARCKS induces the demyristoylation of the protein and its subsequent release to the cytosol. This model implies that the membrane-bound pool of MARCKS is the target for demyristoylation. The ability of cytosolic extracts to demyristoylate MARCKS has however been demonstrated for the free protein only (Manenti et al., 1994) and still awaits confirmation for the membrane-bound form. In addition, pulse chase experiments with ^{3}H-myristate in myocytes have demonstrated that there are no alterations in the myristoylation of MARCKS in either the particulate- or the cytosolic fractions following activation of PKC with phorbol esters (James and Olson, 1989). In view of these contradictory results, the importance of a deacylation reaction is unclear and further investigations are needed to estimate the contribution of this reaction to the dynamics of interaction of MARCKS with membranes.

Phosphorylation/dephosphorylation. In macrophages and neutrophils as well as in intact synaptosomes, activation of PKC is concomitant with the translocation of MARCKS to the cytosol (section 7) suggesting that the binding of MARCKS to membranes is regulated by phosphorylation. *In vitro* studies with synaptosomal membranes (Wang et al., 1989) or with the purified protein bound to lipid vesicles (Nakaoda et al., 1992; Taniguchi and Manenti, 1993; Kim et al., 1994a) demonstrate that phosphorylation by PKC drastically reduces the affinity of MARCKS for membranes. PKC-mediated phosphorylation inhibits the binding of synthetic peptides, corresponding to the PS domain of MARCKS, to acidic lipid vesicles (Taniguchi and Manenti, 1993; Kim et al., 1994b). Upon phosphorylation, this peptide rapidly dissociates ($t_{1/2}$ < 30 s) from the phospholipid vesicles (Kim et al., 1994b). In agreement with the pseudocooperative model proposed for the binding of MARCKS to membranes (Kim et al., 1994a), the above findings show that the phosphorylated PS domain cannot bind to membranes, leaving MARCKS with only one available binding site, namely the myristic group. Since the myristic group is insufficient for a stable anchoring of the protein (section 8.1), phosphorylation results in the translocation of the protein to the cytosol. Furthermore, dephosphorylation of the PS domain induces the reassociation of MARCKS with the membrane, thereby allowing a cycling of the protein between membrane and cytosol (Thelen et al., 1991). Phosphatases are therefore likely to play an important role in the regulation of the interaction of MARCKS with membranes.

In myocytes, glioma and neuroblastoma cells, MARCKS phosphorylation does not correlate with its translocation to the cytosol (section 7). In a cell-free expression system, the amount of MARCKS released from fibroblast membranes into the incubation medium following PKC activation does not result from phosphorylation but rather is the consequence of the addition of divalent ions such as magnesium and calcium to the medium (George and Blackshear, 1992). In addition, a non-phosphorylatable tetraglycine mutant, in which all serine residues in the PS domain are replaced by glycines, is detached from the membrane to the same extent as the wild type protein under the conditions used for phosphorylation by PKC (George and Blackshear, 1992). As these observations contradict the findings described in the previous paragraph, they suggest that the regulation of the interaction of MARCKS with membranes is complex, depends on the cell type investigated and is probably modulated by additional factors. Some of these aspects are discussed below.

Receptor protein. The punctate distribution of MacMARCKS in focal contacts at the substrate-adherent plasma membrane of macrophages suggests that MARCKS associates with the plasma membrane through a receptor (Thelen et al., 1991). However, in a cell-free system, treatment of fibroblast membranes with heat or trypsin does not alter the binding of MARCKS, indicating that the membrane association of MARCKS in this *in vitro* system does not appear to involve a specific cytoplasmic-face protein receptor (George and Blackshear, 1992). The balls and string model predicts that any protein, located at a specific site at the membrane and which could bind to MARCKS, even with a weak interaction, would be sufficient to target MARCKS to a specific location. The actin cytoskeleton is a candidate for this kind of interaction (Kim et al., 1994a). Although this model does not need to postulate the presence of a protein with a strong affinity for MARCKS to explain a specific location of the protein, the existence of a high affinity receptor for MARCKS cannot be excluded.

Lipid composition. The effects of various lipids on the binding of MARCKS to vesicles suggest that the charge rather than the structure of the lipid is the critical parameter determining the affinity of MARCKS for membranes (Kim et al., 1994a). Thus the concentration of acidic lipids in the membranes of different cell types and organelles could determine the affinity of MARCKS for membranes and therefore its subcellular distribution.

MARCKS sequence heterogeneity. Variations in the sequences of MARCKS from different species may result in different affinities of the protein for membranes. Within the same species, MARCKS and MacMARCKS might also bind differently to membranes. In this respect, the binding of MARCKS should be particularly sensitive to changes in the structure of the PS domain.

Non-PKC kinases. As for PKC, phosphorylation of the PS domain with non-PKC kinases could potentially alter the binding of MARCKS to the membrane (Ding and Badwey, 1993). Also, although the sites of phosphorylation of MARCKS by proline-directed kinases have been found outside the PS domain (Taniguchi et al., 1994), the effect of these phosphorylations on the interaction of MARCKS with membranes has not been investigated.

Calmodulin. When present in large excess, calmodulin inhibits the binding of MARCKS by competizing with the membrane for binding to the PS domain (Kim et al., 1994a). Calmodulin is however inefficient at removing MARCKS from vesicles at lower calmodulin/MARCKS ratios, which better reflect the physiological situation (concentrations of MARCKS and calmodulin in the brain: 12 μM and 60 μM respectively (Blackshear, 1993)).

9) Function of MARCKS

Changes in the expression, stucture and localization of MARCKS are concomitant with cellular events such as secretion, membrane trafficking, motility, mitosis and transformation (Aderem, 1992). Over the last decade, intensive work by several groups has led to some insights on how MARCKS is involved in these cellular events. Blackshear (1993) has proposed that MARCKS regulates calmodulin activity by reversibly sequestering calmodulin at the membrane. MARCKS might also regulate the plasticity of the actin cytoskeleton at the membrane as well as in the cytosol in a calcium/calmodulin-dependent manner (Aderem, 1992a; Aderem, 1992b). In order to assess the validity of these models, a quantitative analysis as well as an understanding of these interactions at the molecular level are necessary. Furthermore, it is unlikely that all factors acting downstream of MARCKS in the PKC-mediated signal transduction cascade have been identified. The development of reconstituted *in vitro* models systems should therefore complement the cellular approach and shed some light on the hitherto unknown function of MARCKS in signal transduction.

Acknowledgements

We thank Drs. J. Ramsden and G. Schwarz for critically reading the manuscript.

References

Aderem AA, Keum MM, Pure E and Cohn ZA (1986) Proc. Natl. Acad. Sci. USA 83: 5817-5821
Aderem AA, Albert KA, Keum MM, Wang JKT, Greengard P and Cohn ZA (1988a) Nature 332: 362-364
Aderem AA, Marratta DE and Cohn ZA (1988b) Proc. Natl. Acad. Sci. USA 85: 6310-6313
Ade em AA (1992a) Cell 71: 713-716
Aderem AA (1992b) Trends Biol. Sci. 17: 438-443
Albert KA, Walaas SI, Wang JKT and Greengard P (1986) Proc. Natl. Acad. Sci. USA 83: 2822-2826
Blackshear PJ, Witters LE, Girard PR, Kuo JF and Quamo SN (1985) J. Biol. Chem. 260: 13304-13315
Blackshear PJ, Wen L, Glynn BP and Witters LE (1986) J. Biol. Chem. 261: 1459-1469
Blackshear PJ, Verghese GM, Johnson JD, Haupt DM and Stumpo DJ (1992) J. Biol. Chem. 267: 13540-13546
Blackshear PJ (1993) J. Biol. Chem. 268: 1501-1504
Byers DM, Palmer FBSC, Spence MW and Cook HW (1993) J. Neurochem. 60: 1414-1421
Chapline C, Ramsey K, Klauck T and Jaken S (1993) J. Biol. Chem. 268: 6858-6861
Clarke PR, Siddhanti SR, Cohen P and Blackshear PJ (1993) FEBS 336: 37-42
Ding J and Badwey JA (1993) J. Biol. Chem. 268: 17326-17333
George DJ and Blackshear PJ (1992) J. Biol. Chem. 267: 24879-24885
Graff JM, Stumpo DJ and Blackshear PJ (1989a) Mol. Endocrinol. 3: 1903-1906
Graff JM, Gordon JI and Blackshear PJ (1989b) Science 246: 503-506
Graff JM, Young TN, Johnson JD and Blackshear PJ (1989c) J. Biol. Chem. 264: 21818-21823
Hartwig JH, Thelen M, Rosen A, Janmey PA, Nairn AC and Aderem AA (1992) Nature 356: 618-622
Hinrichsen RD and Blackshear PJ (1993) Proc. Natl. Acad. Sci. USA 90: 1585-1589
James G and Olson EN (1989) J. Biol. Chem. 264: 20928-20933
Johnson DR, Bhatnagar RS, Knoll LJ and Gordon JI (1994) Annu. Rev. Biochem. 63: 869-914
Kim J, Shishido T, Jiang X, Aderem AA and McLaughlin S (1994a) J. Biol. Chem. In press
Kim J, Blackshear PJ, Johnson JD and McLaughlin S (1994b) Biophys. J. 67: 227-237
Li J and Aderem AA (1992) Cell 70: 791-801
Manenti S, Sorokine O, Van Dorsselaer A and Tanigushi H (1992) J. Biol. Chem. 26/: 22310-22315
Manenti S, Sorokine O, Van Dorsselaer A and Tanigushi H (1993) J. Biol. Chem. 268: 6878-6881
Manenti S, Sorokine O, Van Dorsselaer A and Taniguchi H (1994) J. Biol. Chem. 269: 8309-8313
McIlhinney RAJ and McGlone K (1990) Biochem. J. 271: 681-685
Nakaoda T, Kojima N, Hamamoto T, Kurosawa N, Lee YC, Kawasaki H, Suzuki K and Tsuji S (1993) J. Biochem. 114: 449-452
Ouimet CC, Wang JKT, Walaas SI, Albert KA and Greengard P (1990) J. Neurosci. 10: 1683-1698
Patel J and Kligman D (1987) J. Biol. Chem. 262: 16686-16691
Peitzsch RM and McLaughlin S (1993) Biochemistry 32: 10436-10443
Rodriguez-Pena A and Rozengurt E (1985) EMBO J. 4: 71-76
Rosen A, Nairn AC, Greengard P, Cohn ZA and Aderem AA (1989) J. Biol. Chem. 264: 9118-9121
Rosen A, Keenan KF, Thelen M, Nairn AC and Aderem AA (1990) J. Exp. Med. 172: 1211-1215
Rozengurt E, Rodriguez-Pena M and Smith KA (1983) Proc. Natl. Acad. Sci. USA 80: 7244-7248

Stumpo DJ, Graff JM, Albert KA, Greengard P and Blackshear PJ (1989) Proc. Natl. Acad. Sci. USA 86: 4012-4016
Taniguchi H and Manenti S (1993) J. Biol. Chem. 268: 9960-9963
Taniguchi H, Manenti S, Suzuki M and Titani K (1994) J. Biol. Chem. 269: 18299-18302
Thelen M, Rosen A, Nairn AC and Aderem AA (1990) Proc. Natl. Acad. Sci. USA 87: 5603-5607
Thelen M, Rosen A, Nairn AC and Aderem AA (1991) Nature 351: 320-322
Towler DA, Gordon JI, Adams SP and Glaser L (1988) Ann. Rev. Biochem. 57: 69-99
Umekage T and Kato K (1991) FEBS 286: 147-151
Verghese GM, Johnson JD, Vasulka C, Haupt DM, Stumpo DJ and Blackshear PJ (1994) J. Biol. Chem. 269: 9361-9367
Walaas SI, Nairn AC and Greengard P (1983a) J. Neurosci. 3: 291-301
Walaas SI, Nairn AC and Greengard P (1983b) J. Neurosci. 3: 302-311
Wang JKT, Walaas SI, Sihra TS, Aderem AA and Greengard P (1989) Proc. Natl. Acad. Sci. USA 86: 2253-2256
Weber T, Paesold G, Galli C, Mischler R, Semenza G and Brunner J (1994) J. Biol. Chem. 269: 18353-18358
Wu WCS, Walaas SI, Nairn AC and Greengard P (1982) Proc. Natl. Acad. Sci. USA 79: 5249-5253
Zozulya S and Stryer L (1992) Proc. Natl. Acad. Sci. USA 89: 11569-11573

Phosphorylase Kinase: A Protein Kinase For All Seasons

Theodore G. Sotiroudis, George A. Maridakis, Anna-Maria G. Psarra and Athanassios E. Evangelopoulos

Institute of Biological Research and Biotechnology,
The National Hellenic Research Foundation,
48 Vassileos Constantinou Avenue,
Athens 116 35, Greece

If one protein kinase deserved the title of the kinase with the most long lasting and continuously renuable interest, then phosphorylase kinase (PhK), a key regulatory enzyme of glycogen metabolism, would definetely be the first candidate. In experiments done nearly 40 years ago Krebs and Fischer (1956) showed for the first time that the activity of an enzyme-glycogen phosphorylase-can be turned on and off by reversible phosphorylation, a process known today as one of the most prevalent mechanisms of cellular regulation. PhK, the enzyme responsible for phosphorylation and activation of glycogen phosphorylase was thus the first protein kinase discovered. It is today estimated that about one third of the proteins expressed in a typical mammalian cell contain covalently bound phosphate and perhaps 2-3% of the genes in the entire eucaryotic genome may code for protein kinases (Hubbard and Cohen, 1993).

By far, the predominant number of studies concerning PhK have focused on the rabbit skeletal muscle PhK isozymes. In this respect, it is generally accepted that PhK is a dedicated Ca^{2+}-dependent protein kinase, that is a protein kinase with an exclusive function, in this case the catalysis of glycogen phosphorylase phosphorylation and activation. Moreover, it is considered that within the cell PhK is either "soluble" in the cytosol or associated with glycogen particles, which contain all the enzymes of glycogen metabolism (reviewed by Pickett-Gies and Walsh, 1986). Nevertheless, increasing evidence suggests that several properties of PhK and the subcellular distribution of PhK activities in a number of cells are not compatible with an exclusive function and localization of this kinase and that this protein kinase may well have functions in addition to the phosphorylation of glycogen phosphorylase. From now on we will be terming the cytosolic "soluble", glycogen-associated PhK, conventional PhK and the phosphorylase *b* to *a* converting activity, of PhK conventional PhK activity. All other PhK forms, associated with subcellular fractions other than soluble cytosol or cytosolic glycogen particles will be termed unconventional forms and the PhK catalyzed reactions other than the phosphorylation of phosphorylase b, as unconventional reactions of PhK.

This short review briefly summarizes the unconventional characteristics of PhK together with recent results from our laboratory concerning the presence of PhK activity in rat liver smooth endoplasmic reticulum as well as in rat brain membranes and mitochondria. Moreover, some interesting amino acid sequence homologies between PhK subunits and

NATO ASI Series, Vol. H 92
Signalling Mechanisms – from Transcription Factors to Oxidative Stress
Edited by L. Packer, K. Wirtz

putative target locus proteins of specific cellular structures are presented. Putative roles of unconventional PhK activities are also discussed.

I. Conventional PhK today

With a mass of $1.3x10^6$ Da and four copies each of four different subunits $(\alpha\beta\gamma\delta)_4$, PhK is by far the biggest and most complex protein kinase known. It is regulated in a complex way by phosphorylation and Ca^{2+} and it is situated at the crossroad of glycogenolysis linking glycogen breakdown to both nervous and endocrine stimulation, while at the same time it receives information from metabolic pathways. The final level of phosphorylase *a* is thus accurately determined by a fine-tuning of PhK regulatory and catalytic properties, which is the result of the integration by PhK of all available information. The δ-subunit is identical to calmodulin and confers Ca^{2+} sensitivity to the enzyme. The γ-subunit has catalytic activity and is similar to other protein kinases. The two large subunits, α and β, are homologous proteins, they carry all phosphorylation sites and at least one of their function is regulatory.The primary structures of γ- and δ- subunits have been determined and cDNAs encoding α-, β- and γ- subunits have been cloned (for reviews see Pickett-Gies and Walsh, 1986, Heilmeyer, 1991). Furthermore, the active γ- subunit has been obtained by expresion of its cDNA in various systems and the roles of different residues and domains were analyzed by specific DNA manipulations and construction of truncated forms (Cawley et al., 1989; Chen et al., 1989; Lee et al., 1992; Cox and Johnson, 1992; Huang et al., 1993; Lanciotti and Bender, 1994; Huang et al., 1994).

Characterization of the gene for rat PhK γ- subunt revealed that this gene extends over 16 kb of DNA and contains eight introns within the coding region plus one 3.3-kb intron upstream in the 5-untranslated region (Cawley et al., 1993).The genetic heterogeneity of PhK reflects the structural complexity of this enzyme and all three specific subunits, α, β and γ have several isoforms. Tissue-specific isoforms encoded by distinct genes were identified for α- and γ- subunits: muscle α_M and γ_M, liver α_L and testis γ_T. In addition several subtypes of α and β subunits have been recently characterized by Kilimann's group and it has been shown that are generated by alternative mRNA splicing. The α_M and α_L genes, are X-chromosomal, while the β, γ_M and δ genes are located in autosomes (Schneider, et al., 1993; Davidson, et al., 1992 and references therein).

Although the functional properties and regulation of PhK have been studied extensively (reviewed by Pickett-Gies and Walsh, 1986; Heilmeyer, 1991) much less is known however, about the three dimensional structure of this huge multisubunit kinase. However, the large size of PhK is an advantage for its microscopic characterization, and indeed a variety of microscopy techniques have been used for studying the spatial arrangement of its subunits. A bilobal form of the enzyme has been reported described as "batterfly-like". It appears that the two large, elongated lobes are either connected by one or two bridges or separated by a central region of low mass (reviewed by Wilkinson et al., 1994). In such a model structure approximated by two parentheses-like lobes the dimensions of each lobe were 25 nm x 10.5 nm as determined by small-angle X-ray and neutron scattering (Henderson et al., 1992). Moreover an epitope of the α-subunit of PhK was localized to the tips of the bilobal kinase molecule by two types of immunoelectron microscopy (Wilkinson et al., 1994). In electron micrographs PhK shows also a second major molecular form termed the "chalice" form, which probably represents an alternative viewing orientation (Schramm and Jennissen, 1985). Recently, Louise Johnson's laboratory in Oxford has made progress with structural studies of the kinase domain of γ-subunit. Residues 1-298 of the kinase domain have been expressed in E.coli, purified and crystallized. The structure has been solved to 3A resolution and it already indicated the essential features of Mg-ADP recognition, domain orientation and provides an explanation as to why the kinase is active without the need for post-translational modification (L.N.Johnson, personal communication).

Some of the highlights in rabbit skeletal muscle PhK are listed below:

1956 - Discovery of PhK (Krebs and Fischer)
1964 - Purification of PhK, (Krebs et al.)
1968 - Activation by cAMP-dependent protein kinase (Walsh et al.)
1973 - Subunit structure of PhK (Hayakawa et al.; Cohen)
1974 - First electron micrographs of PhK (Cohen)
1978 - Identificatior of calmodulin as the fourth subunit of PhK (Cohen et al)
1983 - PhK-dephosphorylation as a criterion for the classification of protein phosphatases (Ingebritsen and Cohen)
1984 - Amino acid sequence of the catalytic γ-subunit (Reimann et al.)
1986 - Preparation of active γ-subunit (Kee and Graves)
1988 - cDNA cloning and primary structures of α and β subunits (Zander et al.; Kilimann et al)

II. Unconventional characteristics of PhK

1. How many catalytic actions?

In addition to phosphorylase b, PhK can, in vitro, phosphorylate a number of different proteins. These include: PhK itself, glycogen synthase, troponin I, troponin T, κ-casein, myelin basic protein, histone H1, myosin light chain, the sarcolemmal Na^+, K^+ ATPase phospholamban, soluble and microsomal fractions of rat liver and a 40 kDa brain protein from synaptic plasma membranes (reviewed by Chan and Graves, 1984; Pickett-Gies and Walsh, 1986) as well as hormone-sensitive lipase (Olsson et al., 1986) bovine cardiac C-protein (Schlender et al., 1988), β and δ subunits of nicotinic acetylcholine receptor of mouse myocytes (Smith et al., 1989) a 93 kDa protein of human erythrocyte membranes, possibly representing the anion transport band 3 polypeptide (Sotiroudis et al., 1990), an actin-like protein of bovine stomach smooth muscle (Sotiroudis et al., 1991), and the neuronal tissue-specific proteins B-50 (GAP-43, neuromodulin) and neurogranin (Paudel et al., 1993).

Although PhK was until now considered as a Ser/Thr protein kinase it has been recently demonstrated by Yuan et al. (1993) that the holoenzyme of PhK purified from rabbit skeletal muscle is a dual specificity protein kinase showing significant tyrosine kinase activity with angiotensin II in the presence of Mn^{2+} but not with Mg^{2+}. Using a recombinant truncated form of the catalytic subunit and site directed mutagenesis the same authors have proved that the tyrosyl kinase activity is a property of the γ-subunit and one site is most probably involved in both serine and tyrosine kinase activities.

With the exception of phosphorylase conversion, no physiological significance has yet been established for these reactions and in certain cases the observed phosphorylation reactions may be due to contaminanting foreign kinases especially when low rates of phosphorylation are observed. Nevertheless, it has been shown that myelin basic protein, κ-casein, troponin complex, isolated troponin T and troponin I can be phosphorylated by recombinant γ-subunit (Huang et al., 1993). Moreover, several other attributes of PhK suggest that it may have functions in addition to the regulation of phosphorylase:

a. PhK has a very complex subunit structure which it appears more complex than it would be necessary for the phosphorylation of a single substrate (phosphorylase). In this case the multisubunit structure of the muscle "soluble"-isoforms and the glycogen-associated liver isozyme are considered, which seems to be similar for both enzymes (Pickett-Gies and Walsh, 1986). Whether this type of subunit structure is common for other isozymes, is not known.

b. The amount of PhK present in skeletal muscle is far greater than is needed to phosphorylate and activate phosphorylase *b* (discussed by Pickett-Gies and Walsh, 1986).

c. We have shown that the α subunit of phosphorylase kinase can be labelled specifically by fluorescein isothiocyanate and the covalent modification is accompanied by partial inhibition of the enzyme activity (Sotiroudis and Nikolaropoulos, 1984). Kinetic studies revealed that both ATP and the ATP/Mg^{2+} complex (but not ADP and AMP) compete with fluorescein isothiocyanate binding suggesting that the holoenzyme may contain two substrate binding sites: one located on the γ- and the other on the α-subunit (Zaman et al., 1989). Only one fluorescent peptide of the α-subunit was isolated upon labelling of the holoenzyme and its sequence was determined and localized in the primary structure of the α-subunit (lysine residue 588 is modified) (Zaman et al., 1989). However, sequence comparison of this region of α-subunit or of the whole α- or β- sequence with that of other known members of the protein kinase family shows no apparent homology. Although subdomains are recognizable which are characteristic for catalytic centres of the protein kinase family, e.g., a glycine rich cluster, Gly-X-Gly-X-X-Gly, or Asp-Phe-Gly, Leu-Asp-Ala and other subdomains of ATP/Mg^{2+}-binding sites (Zaman, et al., 1989; Heilmeyer, 1991), these subdomains are located N-terminal to the glycine rich cluster but not C-terminal as in the protein kinase family. The above results suggest that α-subunit belongs to a different family of proteins than the known protein kinase family. Of course one cannot exclude the possibility that the α-subunit is indeed a protein kinase but without the typical characteristics of other protein kinases. In this respect, isocitrate dehydrogenase kinase/phosphatase is a good example of such as an atypical protein kinase molecule (La Porte et al., 1989). To explain the function of such a putative substrate binding domain on the α-subunit, Heilmeyer (1991) proposed that ATP/Mg^{2+} binding to this subunit triggers an ADP/ATP exhange on the catalytic γ- subunit by analogy with a signal transducing G-protein.

d. Whereas phosphorylase is totally associated with glycogen particles the bulk of PhK is cytosolic and only 20-30% is glycogen bound (Meyer et al., 1970). Moreover, a number of studies have shown specific localization or association of PhK activities with membrane compartments or components of the contractile apparatus (see sections 2 and 3 of this article). This is in accordance with the described above characteristic of PhK to prefer as non-phosphorylase substrates, principally membrane proteins and proteins of the contractile apparatus.

e. In brain, despite a low PhK activity, the activity ratio of PhK to phosphorylase is higher than in skeletal muscle (Drummond and Bellward, 1970; Gross and Mayer, 1974).

f. PhK has been found to have an intrinsic ATPase activity with characteristics similar to those of phosphorylase *b* to *a* converting activity. Experimental evidence suggests that the γ-subunit is a candidate for carrying out the ATPase reaction (Paudel and Carlson, 1991). Nevertheless, the involvement of the α-subunit in such a catalytic process is an intriguing possibility.

2. PhK activities associated with subcellular compartments other than soluble cytosol and glycogen particles

Electron micrographs of rabbit striated muscle incubated with polyclonal antibodies to PhK holoenzyme and labelled by immunoperoxidase indicate that the kinase is mainly localized in the intermyofibrillar space and inside the myofibrils in association with small glycogen particles. The second major labelled site is the sarcoplasmic reticulum (SR) (Thieleczek et al., 1987). In the latter case the association of PhK with this membrane compartment must be considered as functional and not a simple artefactual adsorption of the cytosolic enzyme since PhK activity is also present in SR membranes of I-strain mice that are genetically deficient in the cytosolic form of PhK (Varsanyi et al., 1978). These localizations of PhK activities generate a number of major questions: (i) which is the relationship between the conventional PhK, which is a "soluble" cytosolic enzyme and the PhKs associated with the contractile apparatus and SR membranes? (ii) Does glycogen mediate the association of "soluble" PhK with the specific cellular compartments or the above compartmentations is the result of a direct interaction of PhK molecules with myofibrillar and membrane components? (iii) Does conventional PhK interact in vitro with components of the contractile apparatus and membranes or membrane constituens? (v) Does PhK have structural features which could justify the above interactions? and (iv) Is PhK localized in similar or other cellular compartments of non-muscle cells?

In the following sections of this article we will try to briefly present experimental evidene and structural consideration which could shed light on the above questions.

a. PhK and the contractile apparatus

The localization of PhK (alone or as a component of glycogen particles) in the myofibrils (Thieleczek et al. 1987) suggests that within the muscle cell PhK and the contractile apparatus are highly suitable for one to be regulating the other. A variety of experimental results and a number of sequence characteristics of PhK subunits (see section c) are in favour of a direct interaction of PhK with myofibrilar components:

(i) In addition to its intrinsic calmodulin, skeletal muscle PhK can be specifically activated by extrinsic calmodulin but also with the homologous protein skeletal muscle troponin C. In this respect Cohen (1980) has argued that troponin C rather than calmodulin is the physiological activator of holoPhK, since it is present in the cell at a concentration greater than that required for activation and the Ca^{2+} concentration required for PhK activation appears to be in a more physiological range than that required of extrinsic calmodulin.

(ii) Cohen(1980), has also shown that artificial thin filaments composed of actin, tropomyosin and the troponin subunits are equally as effective as isolated troponin C in the activation of holo PhK. This finding greatly supports the above Cohen's proposal since most troponin C in the cell is bound as a component of the myofibrils.

(iii) The polymerized form of actin, F-actin, activates holoPhK through promotion of an increased affinity for phosphorylase and an increased Vmax (Livanova et al., 1983).

(iv) HoloPhK binds to F-actin filaments and the binding is influenced by tropomyosin and troponin as well as by Ca^{2+} and ATP (Zemskova et al., 1991).

(v) In addition to holoPhK, troponin C activates the isolated γ-subunit and binds with approximately the same affinity as calmodulin. Moreover actin inhibits the activity of γ-calmodulin and γ-troponin C complexes. Conversely the γ-subunit is able to inhibit actomyosin ATPase (Paudel and Carlson, 1990).

(vi) Fischer et al. (1978) indicated that the γ-subunit of dogfish, PhK preparations shows strong similarities with dogfish actin. They suggested that actin can not be present as a mere contaminant in the kinase preparations but that the γ-subunit could copolymerize with actin allowing thus PhK to interact with actin networks.

(vii) Although the above phenomenon is not observed in rabbit muscle PhK, we recently presented experimental evidence suggesting that PhK purified from bovine stomach smooth muscle is tightly bound to an aggregate of actin-like molecules (Zevgolis et al. 1991). Our finding is in accordance with the postulation by Paul (1989) that glycogenolysis in smooth musle is linked to contractile filaments.

(viii) In the course of the purification of either the rabbit or the dogfish PhK, activity always emerges in two peaks on Sepharose 4B gel filtration columns: a high Mr excluded fraction which is turbid and the included one with the standard 1.3 MDa molecular mass. In both cases the turbid fraction contains a higher proportion of an actin-like material and a 95-100 kDa protein which is most probably a-actinin (Malencik and Fischer, 1982).

b. PhK associated with membranes and cellular organelles

It is well documented that PhK activity is associated with SR and T-tubules (Thieleczek et al., 1987; Dombradi et al., 1984). Nevertheless, the relationship between the membrane-bound PhK(s) and the cytosolic enzyme is not well understood, although PhK subunits have been localized at the SR of rabbit skeletal muscle by monoclonal and polyclonal antibodies (Thieleczek et al., 1987). Studies in our laboratory were focused on the understanding of the hydrophobic properties of cytoplasmic PhK. We have addressed the question if hydrophobic interactions of PhK with membrane constituents may lead under certain intracellular conditions to the transformation of "soluble" PhK to a membrane bound form as in the case of protein kinase C. We have shown that: (i) acidic phospholipids, the neutral phospholipid lysophosphatidylcholine and gangliosides stimulated the pH 6.8 activity of PhK. Most important the ganglioside $G_{D1\alpha}$ dramatically increases the activity of the kinase at low Ca^{2+}. (ii) The sphingolipids psychosine and sphingosine effectively inhibit PhK activity at neutral pH. (iii) PhK binds to the innerface of the erythrocyte membrane in a Ca^{2+}- and Mg^{2+}- dependent manner (reviewed by Sotiroudis et al., 1990).

Heilmeyer et al. (1992) recently demonstrated the presence of farnesylcysteine residues in the carboxyl termini of the α- and β- subunits of PhK. This post-translational modification may be the critical cellular process leading PhK to membrane localization. In addition farnesyl groups may also be responsible for association of phosphatidylinositol 4-kinase with PhK (Georgoussi and Heilmeyer, 1986; Heilmeyer et al., 1992). Since this lipid kinase colocalized with PhK in muscles, it has been proposed that PhK might serve as a kind of carrier for this phosphatidylinositol kinase (Heilmeyer et al., 1992).

In a recent report Paudel et al. (1993) communicated that a relatively high specific activity of PhK is present in rat brain synaptosomes and that 32% of the kinase in synaptosomes is associated with membranes. It thus appears that the association of PhK with membranes is not restricted to muscle SR and T-tubule membranes but it might be a general phenomenon, where PhK can be bound to a variety of non muscle cellular membranes or membrane bearing cellular organelles. To test this hypothesis we examined the subcellular localization of PhK activities in two non-muscle tissues possessing significant amounts of PhK activity i.e. liver (Maridakis and Sotiroudis, submitted) and brain (Psarra and Sotiroudis, submitted).

Subcellular localization of liver PhK

We measured PhK activity in seven rat liver subcellular fractions nuclei (NUC) mitochondria (MIT), postmitochondrial supernatant (PMS), plasma membranes (PM), smooth endoplasmic reticulum (SER) and rough endoplasmic reticulum (RER) (Fig. 1A). Although no enzyme activity was detected in purified nuclei, plasma membranes and mitochondria (not shown), fractionation of PMS (prepared from the livers of fed or 48h fasted rats) by discontinuous sucrose gradient centrifugation showed that the recovered activity (100%) was almost totally located in the soluble fraction (53%) and the SER fraction (39%). In the final SER pellet that was obtained after 5-fold dilution of the SER band and recentrifugation, the remaing PhK activity was the same for both fasted and fed animals. Since SER glycogen is depleted after 48h starvation we can suggest that the presence of PhK in SER is not due to the asssociation of the kinase with glycogen particles but to a direct PhK-membrane interaction. This hypothesis was strengthened by the fact that treatment of glycogen depleted SER with α-amylase did not alter significantly associated PhK activity. SER associated PhK differed from the soluble liver enzyme in several aspects including a higher resistance to inhibition by antibodies against rabbit muscle holoenzyme or to inhibition by proteolytic attack.

PhK in brain subcellular fractions

Glycogen is the largest energy store in brain which may serve a protective function during ischemia and hypoglycemia or it can be subject to an active stimulus-responsive metabolism under normal conditions (Swanson, 1992). Although brain PhK activities bave been detected both in soluble and synaptic membrane fractions (Drummond and Bellward, 1970; Paudel et al., 1993) no detailed subcellular distribution studies have been performed. Because of the significance of glycogen metabolism in brain and the possible compartmentation of its metabolism (Knull and Khandelwall, 1982) we have undertaken to evaluate in detail the distribution of PhK in rat brain subcellular fractions (Psarra and Sotiroudis, submitted).

The highest amount of PhK activity is recovered with cytoplasmic fraction (79%) but significant amounts of the kinase activity are also found associated with nuclei (11%), microsomes (6%) and mitochondria (3%) (Fig. 1B), while the myelin fraction does not express PhK activity (not shown). Further submitochondrial and subsynaptic fractionation of PhK activities indicate that the enzyme is highly enriched in the intermembrane space of mitochondria and the synaptoplasmic fraction. Significant latent PhK activities can be released by Triton X-100 treatment of all fraction examined except soluble cytosol (Fig. 1B), including mitochondrial membranes, synaptic vesicles, synaptic membranes and synaptic mitochondria. Latent activities indicate PhK trapped in synaptosomes or associated with membranes and cell organelles. The highest latent activities, expressed as the ratio of PhK activity of Triton X-100 treated vs. untreated fractions are found in synaptic (15.6) and mitochondrial membranes (10), suggesting a strong binding of PhK to these membranes. Rat brain mitochondria-associated PhK differ from the soluble neuronal enzyme in several properties including Ca^{2+} and pH-dependency, inactivation by proteinase K and inhibition by antibodies against rabbit PhK.

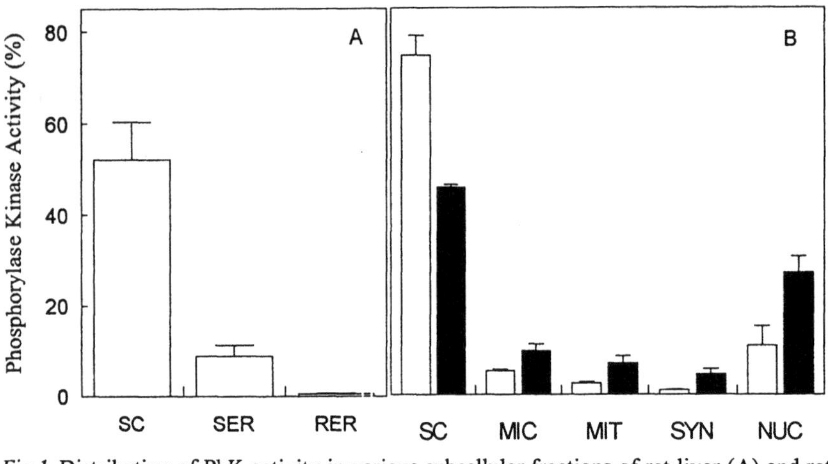

Fig 1. Distribution of PhK activity in various subcellular fractions of rat liver (A) and rat brain (B). PhK activity was determined before (open bars) or after preincubation of the fractions with 0.5% (v/v) Triton X-100(shaded bars).SC, soluble cytosol; SER, smooth endoplasmic reticulum; RER, rough endoplasmic reticulum; MIC, microsomal fraction; MIT, mitochondrial fraction; SYN, synaptosomal fraction; NUC, nuclear fraction. Each value represents the mean ± SEM, n=3. The PhK units present in rat liver post mitochondrial supernatant or the total PhK units of all rat brain subcellulars fractions were taken as 100% PhK activity.

3. Sequence similarities

With the aim of elucidating the biological function of PhK and understanding the mode of interaction of this kinase with contractile apparatus, cellular membranes and organelles we searched for homology between its subunits and amino acid sequence regions of proteins of the contractile machinery, membrane associated proteins and nuclear proteins.

We found a striking sequence homology between residues 748-786 of the α-subunit (domain T) and 124-164 of rabbit skeletal muscle α-tropomyosin (Sotiroudis and Geladopoulos, 1992). It is especially intriguing that residues 126-164 of α-tropomyosin are coded by exon V suggesting that the α-subunit of PhK may have evolved from the fusion of an α-subunit protogene with a progenitor of exon V of α-tropomyosin. Moreover, subsequence 125-136 and the region 150-180 overlapping the homologous segment represent putative actin and troponin binding domains respectively. We proposed that domain T of α-subunit participates in the interaction of holokinase with thin filaments (Sotiroudis and Geladopoulos, 1992).

Searching for homologies with other proteins of the contractile apparatus we unexpectedly found that a subsequence of the domain T (residues 750-770) was similar to a 21-amino acids region of cardiac α-myosin heavy chain (Kavinsky et al., 1984) (Fig. 2A). The alignment reveals 10 identities (48%) and 3 conservative replacements (14%) and the homology is significant (alignment score, A=3.1). In the same region, skeletal muscle myosin heavy chain (Saez and Leinwand, 1986) shows a slightly lower homology with 9 identities (43%) and 3 conservative replacements (not shown). The homologous domain in cardiac myosin lies immediately amino terminal to the hinge region of S2 which is important for force production (McNally et al, 1989).

In parallel, Paudel and Carlson (1990) have proposed that γ-subunit might mimic troponin I in interacting with actin and troponin C, as it possesses at its carboxyl quarter an amino acid sequence segment with significant homoly with the region of troponin I coded by exon VII, which is responsible for the interaction of troponin I with actin and troponin C. In this respect it is noteworthy that γ-subunit contains two regions showing sequence similarity with actin (Fig. 2A). The first one, spanning residues 105-120 is homologous (7 identities - 41%) with an actin segment (residues 180-194) which has been proposed to be an actin-actin interaction site and it shows similarities with actin binding proteins (Tellam et al., 1989). These similarities involve primarily the central actin pentapeptide motif LTDYL, a varient of which is present in the first homologous region of the γ-subunit. The second region of the catalytic subunit spanning residues 327-352 and overlaping the calmodulin binding domains of the γ-subunit (Dasgupta et al., 1989) is homologous (11 identities -41%, A=3.5) to an actin segment (residues 161-185) overlaping the N-terminal region of the actin-actin interaction site (Leu 180-Thr 194).

In relation to the association of PhK with membranes and membrane components we have identified a number of interesting sequence similarities concerning α-subunit:

A subsequence of the domain T (residues 750-773) is similar to a 24-residues region of the juxtamembrane cytoplasmic domain of human EGF receptor (residues 663-686) (Sotiroudis and Geladopoulos, 1992) and it could be involved in PhK-membrane interaction. A 30-amino acid segment of the α-subunit (residues 306-335) has 50% identity (A=6.5) with a skeletal muscle calsequestrin region (residues 124-152) (Fig. 2A). Calsequestrin is a high-capacity Ca^{2+}-binding protein localized in luminal spaces of the terminal cisternae of the SR of muscle cells (Fliegel et al., 1987). This membrane protein has been reported to be tightly associated with PhK activity (Varsanyi and Heilmeyer, 1979). The homologous segment of calsequestrin is located within a region (residues 86-191) responsible for the interaction of

A.

```
α    E V H L P R D Q S G E V D F Q A L V L Q L (750-770)
MHC  E F D I S Q L N S K I E D E Q A L V L Q L (648-668)

α    E D P N R L Y Y E P A E L K L F E N I E C E W P L F W T Y F (306-335)
CAL  E D P V E L I E G E R E L Q A F E N I E D E I K L I - G Y F (124-152)

α    Q V Q E Y R E A L E A V L I K G K N G V P L L P E L Y S V P P D K (347-379)
PKC  Q A H I D R E V L I V V V R D A K N L V P M D P N G L S D P Y V K (165-197)

γ    S V R I Y Y Q Y R R V K P V T R E I V I R D (314-335)
H4   S G L I Y E E T R G V L K V F L E N V I R D (47-68)

γ    L M K K G - E L F D Y L T E K V T (105-120)
AC   L D L A G R D L T D Y L M K I L T (180-196)

γ    V T R E I V I R D P Y A L - R P L R R L I D A Y A F R (327-352)
AC   V T H N V P I Y E G Y A L P H A I M R L - D - L A G R (161-185)
```

B.

β	RKlkgkygfkrflrdgyrtsledpKRRyy	(321-349)
β	KRflrdgyrtsledpKRRyy	(330-349)
β	KRegpnfitqegtvsdhieRlyRR	(748-771)
γ	KKgelfdyltekvtlseKetRK	(107-128)
γ	RRvkpvtreivirdpyalRplRR	(322-344)

Fig 2 A.Alignment of amino acid sequence segments of the α- or γ- subunits of rabbit muscle Phk with regions of the sequences of cardiac α-myosin heavy chain (MHC), skeletal muscle calsequestrin (CAL), protein kinase C (β1/β2 isozymes) (PKC), calf thymus histone H4 (H4) and actin (AC). Boxes and solid bars show identities and conservative replacements respectively. Statistical analysis of Gaps in the alignments are indicated as dashes. Boxes and solid bars show identities and conservative replacements respectively. Statistical analysis of homologies was performed according to Sotiroudis amd Geladopoulos (1992). The number in parentheses show residue numbers of starting and ending sites for the compared segments.
B. Amino acid sequences of β- and γ-subunits of Phk containing the bipartite nuclear targeting motif. Basic amino acids of the basic clusters are shown in upper case, all other amino acids are shown in lower case.

the Ca^{2+}-binding protein with the junctional face membrane of SR (Collins et al., 1990). Thus, it is possible that the above region of α-subunit which is homologous to calsequestrin may be involved in the interaction of PhK with SR membranes. Furthermore, a 33-residues region of the α-subunit (residues 347-379) shows a striking sequence homology with a segment of protein kinase C (residues 165-197) β_1/β_2 isozymes (Ohno et al., 1987; Parker et al., 1989) which are members of a family of cytosolic enzymes that translocate to the particulate fraction on activation. The alignment reveals 13 identities (39%) and 4 conservative replacements (12%), and the homology is significant (A=4.4) (Fig.2A). The homologous domain of protein kinase C overlaps the variable region V2 and part of the conserved region C2 located in the regulatory domain of this kinase (Parker et al., 1989). It has been shown that protein kinase C binds to specific protein, termed receptors for activated C-kinase or RACKS, in the praticulate fraction and that the C2 region is involved in this binding (Mochly-Rosen et al., 1992). We suggest that this segment of the α-subunit may mediate the translocation of PhK to analogous intracellular structures.

We have already shown that a significant amount of brain PhK activity is associated with the nuclear fraction (Fig. 1B). In addition, a number of studies have reported the association of glycogen with the nuclear envelope and the presence of this polysaccharide inside the nucleus, while very recently a novel role for glycogen has been presented, which involves the formation of a particular higher order chromatin intermediate (Hartl et al., 1994 and references therein). Therefore we searched the primary structure of PhK subunits for the existence of sequence homologies with nuclear proteins or nuclear targeting motifs. Concerning the later case, Dingwall and Laskey (1991) have identified a bipartite motif in a large number of proteins transported into the nucleus and proposed that this motif may serve as a consensus for nuclear targeting. This bipartite motif consists of a downstream cluster of two basic amino acids, a spacer region of any amino acids (usually 10-22) and an upstream cluster in which three out of five amino acids are basic. We found that a 22-amino acids segment of the γ-subunit (residues 314-335), overlaping part of its calmodulin binding domains (Dasgupta et al., 1989), has a 50% identity (A=3.9) with a region of histone H4 (residues 47-68) (Ogawa et al., 1969) (Fig 2A). Moreover we have identified two segments of the primary structure of γ-subunit and three of the β-subunit that mach the bipartite motif (Fig. 2B). Two of them, one of the γ- and the other of the β-subunit, spanning residues 322-344 and 748-771 respectively, overlap part of the calmodulin binding domains of γ- (Dasgupta et al., 1989) and β-subunit (Heilmeyer, 1991). Future research should reveal whether brain PhK, binds to any nuclear component and whether such binding involves the bipartite motifs or the region of γ-subunit homologous to histone H4 identified here.

III. Conclusions and future research targets

A vast body of information available today supports the hypothesis that PhK is not a dedicated kinase specified only for the regulation of glycogen metabolism in the cytosol, associated with glycogen particles, but it may well play some role(s) in addition to the catalysis of phosphorylase *b* to *a* conversion. Moreover, even in the case where its main role is to activate phosphorylase in order to supply the glycolytic pathway with a regulated amount of phosphorylated glucose units, this role could be perfectly performed in a variety of intracellular compartments not necessarily limited to the soluble cytosol or the cytosolic glycogen complex. If so, then one should expect that localization of PhK molecules as well as their substrate specificities will largely depend on the particular cell type which could express specific PhK species. In any case, the particular interactions between PhK molecules and components of the specific intracellular locus where PhK is bound, would drastically affect its substrate specificity. At the same moment, PhK could allosterically modify the conformation of locus proteins with which interacts. According to the model proposed by Hubbard and Cohen (1993), phosphatases and kinases with generalized specificity are converted to more selective enzymes by virtue of their binding to a targeting subunit (T_{sub}) which localizes the catalytic subunit (C_{sub}) of a phosphatase or a kinase to a particular target locus (T_{loc}) of a cellular structure such as glycogen particles, contractile apparatus, cytoskeleton, organelle membranes, chromosomes etc. The interaction between a T_{sub} and a C_{sub} or between T_{sub} and T_{loc} may alter (allosterically) the catalytic activity or specificity of the C_{sub} for nearby substrates. PhK can perfectly fit to this model if we assume that one or both of its regulatory subunits (α and β) play the role of a T_{sub} (although other targeting subunits cannot be excluded) for the catalytic γ-subunit. Such a function is analogous to that of glycogen-targetting subunit of protein phosphatase 1 (Hubbard and Cohen, 1993). We propose that α- rather than β-subunit acts as a T_{sub} of PhK to maintain the holoenzyme at specific target loci, for the following reasons:

(i) The α-subunit is oriented more peripherally and it is more exposed than the β-subunit, which occupies a more central and protected position in the holokinase (Daube et al., 1991; Wilkinson et al., 1994).

(ii) Glycogen can directly interact with PhK and the α-subunit plays an important role in this process (Chan and Graves, 1984).

(iii) Only the α-subunit possesses an ATP/Mg^{2+} binding domain which can be labelled specifically by fluorescein isothiocyanate. One might assume that binding of ATP/Mg^{2+} to the α-subunit is followed by hydrolysis of the nucleotide which triggers a signal transducing function by analogy with G protein-coupled receptor kinases (GRKs) (Inglese et al., 1993). In this respect, it is interesting that the GRK rhodopsin kinase has another important feature of α- (and β-) subunit, that of the isoprenylation of its protein, while it has been shown that farnesylation of rhodopsin kinase is required for light-dependent membrane translocation (Inglese et al., 1992).

(iv) Sequence comparisons of the subunits of PhK with putative T_{loc} protein components show homologies preferentially with segments of the α-subunit.

(v) A "multiphosphorylation loop" is uniquely present in the α-subunit but not in the homologous β- subunit (Meyer et al., 1990) containing all the phosphoserine residues of α-subunit. This domain may be responsible for the interaction of PhK with a variety of T_{loc} components through electrostatic interactions.

However we cannot exclude the possibility that in certain cases a direct interaction of the C_{sub} of PhK with T_{loc} may occur. This type of association may be involved in PhK localization in actin filaments. In this case, PhK either utilizes actin or an actin-binding protein as a regulatory subunit and/or it serves as an actin-binding protein to regulate the arrangement and turnover of actin.

The association of PhK with cellular membranes seems to follow a two-step procedure. First, at mobilizing Ca^{2+} conditions, PhK translocates to membrane compartments where it interacts with membrane components from which it can be easily released (Kyriakidis et al., 1988). The segments of α-subunit homologous to EGF receptor (Sotiroudis and Geladopoulos, 1992) or protein kinase C (Fig.2), as well as the farnesylcysteine residues of both α- and β- subunits (Heilmeyer et al., 1992) are probably involved in this interaction. We have already studied this type of reversible PhK-membrane association with erythrocyte membranes (Kyriakidis et al., 1988), rat liver SER membranes (Maridakis and Sotiroudis, submitted) and a fraction of rat brain microsomal membranes (Psarra and Sotiroudis, submitted). In several cases the reversibly membrane associated PhK is transformed to a stable membrane-anchored kinase solubilized only by detergents. This type of PhK was observed in SR (Thieleczek et al., 1987) and rat brain membrane fractions (Psarra and Sotiroudis, submitted). The molecular basis of this transformation may include the carboxy-terminal processing events which follow protein prenylation and have already been observed in ras proteins (Powers, 1991). Further studies are needed to determine the isoforms of PhK involved in the interaction with liver and brain membranes, and the exact sequence of events leading in membrane-anchoring of PhK in these cells.

In membranes, PhK should perform a different role than in soluble cytosol. This hypothesis is rather certain in the case of SR-associated PhK, since phosphorylase *b* is not present in these membranes and the PhK activity of the SR membranes of I-strain mice which are deficient in the cytoplasmic enzyme does not convert phosphorylase *b* upon muscle contraction (discussed by Thieleczek et al., 1987).

Particularly intriguing is the presence of PhK in brain mitrochondria, both in the intermembrane space and mitochondrial membranes (Psarra and Sotiroudis, submitted), in contrast to its absence from skeletal muscle, cardiac and liver mitochondria. In parallel, a significant latent glycogen phosphorylase activity always accompanies the mitochondrial PhK which increases several times after Triton X-100 treatment (unpublished results). Although intramitochondrial glycogen has been described by many investigators both in nerve cells (Ishikawa and Pei, 1965; Shabadash and Zelikina, 1970) as well as in a variety of other cells either in normal or in pathological states (Jones and Ferrans, 1973 and references therein), a detailed analysis of the enzyme machinery involved in mitochondrial glycogen processing is absent. Moreover, nothing is known about the metabolic availability and fate of this compartmentalized pool of glycogen or the place of its synthesis. By calculating the ratios of mitochondrial vs cytoplasmic specific activities for glycogen phosphorylase and synthase in brain, from the data of the subcellular fractionation study of Knull and Khandelwal (1982), we find that the ratio for phosphorylase is 4.5-fold higher than that of synthase. Although, the presence of glycogen phosphorylase and synthase in brain mitochondria was not appreciated in that publication, our data help us now to assume that glycogen degradation -and not glycogen synthesis- is the main metabolic route followed by this polysaccharide in neuronal mitochondria. Studies are in progress in our laboratory to investigate the role of glycogenolytic system in brain mitochondria and the possible involvement of PhK isozyme(s) in other neuronal events which may vary drastically according to the neuronal cell type and the subcellular and regional localization of this kinase.

References

Cawley KC, Akita CG, Walsh DA (1989) Expression of a cDNA for the catalytic subunit of skeletal muslce phosphorylase kinase in transfected 3T3 cells. Biochem J 263:223-229

Cawley KC, Akita CG, Angelos KL and Walsh DA (1993) Characterization of the gene for rat phosphorylase kinase catalytic subunit. J Biol Chem 268:1194-1200

Chan K-FJ, Graves DJ (1984) Molecular properties of phosphorylase kinase. In: Cheung WY (ed), Calcium and Cell Function, Vol 5, Academic Press, New York, pp 1-31

Chen L-R, Yuan C-J, Somasekhar G, Wejksora P, Peterson JE, Myers AM, Graves L, Cohen PTW, daCruz e Silva EF, Graves DJ (1989) Expression and characterization of the γ-subunit of phosphorylase kinase. Biochem Biophys Res Commun 161:746-753

Cohen P (1973) The subunit structure of rabbit skeletal muscle phosphorylase kinase and the molecular basis of its activation reactions. Eur J Biochem 34:1-14

Cohen P (1974) The role of phosphorylase kinase in the nervous and hormonal control of glycogenolysis in muslce. Biochem Soc Symp 39:51-73

Cohen P (1980) The role of calcium ions, calmodulin and troponin in the regulation of phosphorylase kinase from rabbit skeletal muscle. Eur J Biochem 111:563-574

Cohen P, Burchell A, Foulkes G, Cohen PTW (1978) Identification of the Ca^{2+}-dependent modulator protein as the fourth subunit of rabbit skeletal muscle phosphorylase kinase. FEBS Lett 92:287-293

Collins JH, Tarcsafalvi A, Ikemoto N (1990) Identification of a region of calsequestin that binds to the junctional face membrane of sarcoplasmic reticulum. Biochem Biophys Res Commun 167: 189:193

Cox S, Johnson LN (1992) Expression of the phosphorylase kinase gamma subunit catalytic domain in Escherichia coli. Protein Eng 5:811-819

Dasgupta M, Honeycutt T, Blumenthal DK (1989) The γ-subunit of phosphorylase kinase contains two noncontiguous domains that act in concert to bind calmodulin. J Biol Chem 264: 17156-17163

Davidson JJ, Ozcelik T, Hamacher C, Willems PJ, Francke U, Kilimann MW (1992) cDNA cloning of a liver isoform of the phosphorylase kinase α-subunit and mapping of the gene to Xp22.2-p22.1, the region of human X-linked liver glycogenosis. Proc Natl Acad Sci USA 89:2096-2100

Daube H, Billich A, Mann K, Schramm HJ (1991) Cleavage of phosphorylase kinase and calcium-free calmodulin by HIV-1 protease. Biochem Biophys Res Commun 178: 892-898

Dingwall C, Laskey RA (1991) Nuclear targeting sequences-a consensus? Trends Biochem Sci 16:478-481

Dombradi VK, Silberman SR, Lee EYC, Caswell AH, Brandt NR (1984) The association of phosphorylase kinase with rabbit muscle T-tubules. Arch Biochem Biophys 230:615-630

Drummond GI, Bellward G (1970) Studies on phosphorylase kinase from neural tissues. J Neurochem 17: 475-482

Fischer EH, Alaba JO, Brautigan DL, Kerrick WGL, Malencik DA, Moeschler HJ, Picton C, Pocinwong S (1978) Evolutionary aspects of the structure and regulation of phosphorylase kinase. In: Li CH (ed) Versatility of proteins, Academic Press, New York, pp 133-145

Fliegel L, Ohnishi M, Carpenter MR, Khanna VK, Reithmeier RAF, MacLennan DH (1987) Amino acid sequence of rabbit fast-twich skeletal muscle calsequestrin deduced from cDNA and peptide sequencing. Proc Natl Acad USA 84:1167-1171

Georgoussi Z, Heilmeyer LMG, Jr (1986) Evidence that phosphorylase kinase exhibits phosphatidylinositol kinase activity. Biochemistry 25:2867-3874

Gross SR, Mayer SE (1974) Characterization of the phosphorylase *b* to *a* converting activity in skeletal muscle extracts of mice with the phosphorylase *b* kinase deficiency mutation. J Biol Chem 249:6710-6718

Hartl P, Olson E, Dang T, Forbes DJ (1994) Nuclear assembly with λDNA in fractionated *Xenopus* egg extracts: An unexpected role for glycogen in formation of a higher order chromatin intermediate. J Cell Biol 124:235-248

Hayakawa T, Perkins JP, Krebs EG (1973) Studies on the subunit structure of rabbit skeletal muscle phosphorylase kinase. Biochemistry 12:574-580

Heilmeyer LMG, Jr (1991) Molecular basis of signal intergration in phosphorylase kinase. Biochim Biophys Acta 1094: 168-174

Heilmeyer LMG, Jr, Serwe M, Metzger J, Hoffmann-Posorske E, Meyer HE (1992) Farnesylcysteine, a constituent of the α- and β- subunits of rabbit skeletal muscle phosphorylase kinase: Localization by conversion to S-ethylcysteine and by tandem mass spectrometry. Proc Natl Acad Sci USA, 89:9554-9558

Henderson SJ, Newsholme P, Heidorn DB, Mitchell R, Seeger PA, Walsh DA, Trewhella J (1992) Solution structure of phosphorylase kinase studied using small-angle X-ray and neutron scattering. Biochemistry 31:437-442

Huang CYF, Yuan CJ, Livanova NB, Graves DJ (1993) Expression, purification, characterization and deletions mutations of phosphorylase kinase γ-subunit: identification of an inhibitory domain in the γ-subunit. Mol Cell Biochem 127/128:7-18

Huang CYF, Yuan CJ, Luo SQ, Graves DJ (1994) Mutational analyses of the metal ion and substrate binding site of phosphorylase kinase γ- subunit. Biochemistry 33:5877-5883

Hubbard MJ, Cohen P (1993) On target a new nechanism for the regulation of protein phosphorylation. Trends Biochem Sci 18:172-177

Ingebritsen TS and Cohen P (1983) The protein phosphatases involved in cellular regulation. 1. Classification and substrate specificities. Eur J Biochem 132:255-261

Inglese J, Glickman JF, Lorenz W, Caron MG, Lefkowitz RJ (1992) Isoprenylation of a protein kinase. Requirement of farnesylation/α-carboxyl methylation for full enzymatic activity of rhodopsin kinase. J Biol Chem 267: 1422-1425

Inglese J, Freedman NJ, Kock WJ, Lefkowitz RJ (1993) Structure and mechanism of the G protein-coupled receptor kinases. J Biol Chem 268: 23735-23738

Ishikawa T, Pei YF (1965) Intramitochondrial glycogen particles in rat retinal receptor cells. J Cell Biol 25: 402-407

Jones M, Ferrans VJ (1973) Intramitochondrial glycogen in hypertrophied infundibular muscle of patients with congenital heart diseases. Am J Pathol 70: 69-88

Kavinsky CJ, Umeda PK, Levin JE, Sinha AM, Nigro JM, Jakovcic S, Rabinowitz M (1984) Analysis of cloned mRNA sequences encoding subfragment 2 and part of subfragment 1 of α- and β- myosin heavy chains of rabbit heart. J Biol Chem 259:2775-2781

Kee SM and Graves DJ (1986) Isolation and properties of the active γ-subunit of phosphorylase kinase. J Biol Chem 261:4732-4737

Kilimann MW, Zander NF, Kuhn CC, Crabb JW, Meyer HE, Heilmeyer LMG, Jr (1988) The α- and β- subunits of phosphorylase kinase are homologous: cDNA cloning and primary structure of the β- subunit. Proc Natl Acad Sci USA 85:9381-9385

Knull HR, Khandelwall RL (1982) Glycogen metabolizing enzymes in brain. Neurochem Res 7:1307-1317

Krebs EG, Fischer EH (1956) The phosphorylase *b* to *a* converting enzyme of rabbit skeletal muscle. Biochim Biophys Acta 20:150-157

Krebs EG, Love DS, Bratvold GE, Trayser KA, Meyer WL, Fischer EH (1964) Purification and properties of rabbit skeletal muscle phosphorylase kinase. Biochemistry 3:1022-1033

Kyriakidis SM, Sotiroudis TG, Evangelopoulos AE (1988) Ca^{2+}- and Mg^{2+}- dependent association of phosphorylase kinase with human erythrocyte membranes. Biochim Biophys Acta 972: 347-352

Lanciotti RA, Bender PK (1994) Baculovirus-directed expression of the γ-subunit of phosphorylase kinase:purification and calmodulin dependence. Biochem J 299:183-189

LaPorte DC, Stueland CS, Ikeda TP (1989) Icocitrate dehydrogenase kinase/phosphatase. Biochimie 71:1051-1057

Lee J-H, Maeda S, Angelos KL, Kamita SG, Ramachandran C, Walsh DA (1992) Analysis by mutagenesis of the ATP binding site of the γ-subunit of skeletal muscle phosphorylase kinase expressed using a baculovirus system. Biochemistry 31, 10616-10625

Livanova NB, Silonova GV, Solovyeva NV, Andreeva IE, Ostrovskaya MV, Poglazov BF (1983) Regulation of muscle phosphorylase kinase by actin and calmodulin. Biochem Int 7:95-105

Malencik DA, Fischer EH (1982) Structure, function and regulation of phosphorylase kinase. In: Cheung WY (ed) Calcium and cell function, Vol 3, Academic Press, New York, pp 161-188

McNally EM, Kraft R, Bravo-Zehnder M, Taylor DA, Leinwand LA (1989) Full-length rat alpha and beta cardiac myosin heavy chain sequences. J Mol Biol 210:665-671

Meyer F, Heilmeyer LMG, Jr, Haschke RH, Fischer EH (1970) Control of phosphorylase activity in a muscle glycogen particle. I. Isolation and characterization of the protein-glycogen complex. J Biol Chem 245:6642-6648

Meyer HE, Meyer GF, Dirks H, Heilmeyer LMG, Jr (1990) Localization of phosphoserine residues in the α-subunit of rabbit skeletal muscle phosphorylase kinase. Eur J Biochem 188: 367-376

Mochly-Rosen D, Miller KG, Scheller RH, Khaner H, Lopez J, Smith BL (1992) p65 fragments homologous to the C2 region of protein kinase C, bind to the intracellular receptors for protein kinase C. Biochemistry 31:8120-8124

Ogawa Y, Quagliarotti G, Jordan J, Taylor CW, Starbuck WC, Busch H (1969) Structural analysis of the glycine-rich, arginine-rich histone. J Biol Chem 244: 4387-4392

Ohno S, Kawasaki H, Imajoh S, Suzuki K, Inagaki M, Yokokura H, Sakoh T, Hidaka H (1987) Tissue-specific expression of three distinct types of rabbit protein kinase C. Nature 325:161-166

Olsson H, Stralfors P, Belfrage P (1986) Phosphorylation of the basal site of hormone-sensitive lipase by glycogen synthase kinase-4. FEBS Lett 209: 175-180

Parker PJ, Kour G, Marais RM, Mitchell F, Pears C, Schaap D, Stabel S,Webster C (1989) Protein kinase C - a family affair. Mol Cell Endocrinol 65:1-11

Paudel HK, Carlson GM (1990) Functional and structural similarities between the inhibitory region of troponin I coded by exon VII and the calmodulin-binding regulatory region of the catalytic subunit of phosphorylase kinase. Proc Natl Acad Sci USA 87:7285-7289

Paudel HK, Carlson GM (1991) The ATPase activity of phosphorylase kinase is regulated in parallel with its protein kinase activity. J Biol Chem 266:16524-16529

Paudel HK, Zwiers H, Wang JH (1993) Phosphorylase kinase phosphorylates the calmodulin-binding regulatory regions of neuronal tissue-specific proteins B-50 (GAP-43) and neurogranin. J Biol Chem 268:6207-6213

Paul RJ (1989) Smooth muscle energetics. Ann Rev Physiol 51:331-349

Pickett-Gies CA, Walsh DA (1986) Phosphorylase kinase. In: Boyer P, Krebs EG (eds), The Enzymes, Vol.17, Academic Press, Orlando, pp 395-459

Powers S (1991) Protein prenylation: a modification that sticks. Curr Biol 1: 114-116

Reimann EM, Titani K, Ericsson LH, Wade RD, Fischer EH, Walsh KA (1984) Homology of the γ-subunit of phosphorylase *b* kinase with cAMP-dependent protein kinase. Biochemistry 23:4185-4192

Saez L, Leinwand LA (1986) Characterization of diverse forms of myosin heay chain expressed in adult human skeletal muscle. Nucl Acid Res 14:2951-2969

Schlender KK, Thysseril TJ, Hegazy MG (1988) Calcium-dependent phosphorylation of bovine cardiac C-protein by phosphorylase kinase. Biochem Biophys Res Commun 155:45-51

Schneider A, Davidson JJ, Wullrich A, Kilimann MW (1993) Phosphorylase kinase deficiency in I-strain mice is associated with a frameshift mutation in the α-subunit muscle isoform. Nature genetics 5:381-385

Schramm HJ, Jennissen HP (1985) Two-dimensional electron microscopic analysis of the chalice form of phosphorylase kinase. J Mol Biol 181: 503-516

Shabadash AL, Zelikina TI (1970) Histochemical detection of glycogen in mammalian nerve-cell mitochondria. Dokl Acad Nauk SSSR 192: 196-198

Smith MM, Merlie JP, Lawrence JC, Jr (1989) Ca^{2+}-dependent and cAMP-dependent control of nicotinic acetylcholine receptor phosphorylation in muscle cells. J Biol Chem 264:12813-12819

Sotiroudis TG, Geladopoulos TP (1992) A domain of the a-subunit of rabbit phosphorylase kinase shows homologies with regions of rabbit a-tropomyosin, human EGF receptor and the a-chain of bovine S-100 protein. Biosci Rep 12:313-317

Sotiroudis TG, Nikolaropoulos S (1984) Selective labelling of phosphorylase kinase with fluorescein isothiocyanate FEBS Lett 176:421-425

Sotiroudis TG, Zevgolis VG, Baltas LG, Kyriakidis SM (1990) Control of glycogen metabolism and phosphorylase kinase. A model system for studying signal transduction mechanism mediated by protein phosphorylation and Ca^{2+}. In: Ranjeva R and Boudet AM (eds) Signal Perception and Transduction in Higher Plants. NATO ASI Series, Vol H47,Springer-Verlag, Berlin, pp 201-212

Sotiroudis TG, Zevgolis VG, Evangelopoulos AE (1991) Control of cellular activity by protein phosphorylation-dephosphorylation: phosphorylase kinase from bovine stomach smooth muscle. In: Ross EMM and Wirtz KWA (eds) Biological Signal Transduction. NATO ASI Series, Vol H52, Springer-Verlag, Berlin, pp 309-320

Swanson RA (1992) Physiologic coupling of glial glycogen metabolism to neuronal activity in brain. Can J Physiol Pharmacol 70:S138-S144

Thieleczek R, Behle G, Messer A, Varsanyi M, Heilmeyer LMG, Jr, Drenckhahn D (1987) Localization of phosphorylase kinase subunits at the sarcoplasmic reticulum of rabbit skeletal muscle by monoclonal and polyclonal antibodies. Eur J Cell Biol 44: 333-340

Tellam RL, Morton DJ, Charke FM (1989) A common theme in the amino acid sequences of actin and many actin-binding proteins? Trends Biochem Sci 14: 130-133

Varsanyi M, Groschel-Stewart K, Heilmeyer LMG, Jr (1978) Characterization of a Ca^{2+}-dependent protein kinase in skeletal muscle membranes of I-strain and wild-type mice. Eur J Biochem 87:331-340

Varsanyi M, Heilmeyer LMG, Jr (1979) The protein kinase properties of calsequestrin. FEBS Lett 103:85-288

Walsh DA, Perkins JP, Krebs EG (1968) An adenosine 3,5-monophosphate-dependent protein kinase from rabbit skeletal muscle. J Biol Chem 243:3763-3774

Wilkinson DA, Marion TN, Tillman DM, Norcum MT, Hainfeld JF, Seyer JM, Carlson GM (1994) An epitope proximal to the carboxyl terminus of the α-subunit is located near the lobe tips of the phosphorylase kinase hexadecamer. J Mol Biol 235:974-982

Yuan CJ, Huang CYF, Graves DJ (1993) Phosphorylase kinase, a metal ion-dependent dual specificity kinase. J Biol Chem 268:17683-17686

Zaman N, Varsanyi M, Heilmeyer LMG, Jr, Sotiroudis TG, Johnson CM, Crabb JW (1989) Reaction of fluorescein isothiocyanate with an ATP-binding site on the phosphorylase kinase α- subunit. Eur J Biochem 182:577-584

Zander NF, Meyer HE, Hoffmann-Posorske E, Crabb JW, Heilmeyer LMG, Jr, Kilimann MW (1988) cDNA cloning and complete primary structure of skeletal muscle phosphorylase kinase (α- subunit). Proc Natl Acad Sci USA 85:2929-2933

Zemskova MA, Shur SA, Skolysheva LK, Vul'fson PL (1991) Interaction of phosphorylase kinase with proteins of the thin filaments of rabbit skeletal muscles. Biochimiya 56:100-108

Zevgolis VG, Sotiroudis TG, Evangelopoulos AE (1991) Phosphorylase kinase from bovine stomach smooth muscle: a Ca^{2+}-dependent protein kinase associated with an actin-like molecule. Biochim Biophys Acta 1091:222-230

SIGNAL, MESSENGER AND TRIGGER MOLECULES FROM FREE RADICAL REACTIONS AND THEIR CONTROL BY ANTIOXIDANTS

John M C Gutteridge
Oxygen Chemistry Laboratory
Unit of Critical Care
Department of Anaesthesia and Intensive Care
Royal Brompton Hospital NHS Trust
Sydney Street, London, SW3 6NP

1 Introduction to Free Radicals and Antioxidants

A. Free Radicals and Reactive Oxygen Species

The element oxygen (O) exists in air as a molecule (0_2) known as dioxygen or molecular oxygen. It was first isolated and characterised between 1772 to 1774 by the individual skills of the great European scientists Priestley, Lavoisier and Scheele. Dioxygen, hereafter referred to as oxygen, appeared in significant amounts on the surface of the Earth some 2.5×10^9 years ago, and geological evidence suggests that this was due to the photosynthetic activity of micro-organisms (blue-green algae).

The percentage of oxygen in dry air is now around 21% making it, after nitrogen (78%) the second most abundant element in the atmosphere. However, this amount of oxygen in the air is negligible when compared with that present as part of the water molecule in oceans, lakes and rivers, and that present as part of mineral reservoirs in the Earth's crust. When the Earth's atmosphere changed from a highly reducing state to the oxygen-rich state that we know today, anaerobic life forms ceased to exist or retreated to places where oxygen was excluded. The slow change from anaerobic to aerobic life necessitated the evolution of specialised antioxidants to protect against the toxic properties of oxygen. When we oxidise molecules with oxygen, the oxygen molecule itself becomes reduced and forms intermediates; two of which are free radicals (Equations 1-4)

1 $0_2 + e + H^+$ ---------> $H0_2^{\bullet}$ hydroperoxyl radical

$H0_2^{\bullet}$ --------> $H^+ + 0_2^{\bullet-}$ superoxide radical

NATO ASI Series, Vol. H 92
Signalling Mechanisms – from Transcription Factors to Oxidative Stress
Edited by L. Packer, K. Wirtz

2	$2O_2^{\cdot} + 2H^+ + e$	--------->	H_2O_2 hydrogen peroxide
3	$H_2O_2 + e$	--------->	$OH^- + {}^{\cdot}OH$ hydroxyl radical
4	$OH^{\cdot} + e + H^+$	--------->	H_2O

When oxygen is reduced by the stepwise addition of electrons (equation 1-4) two free radicals ($HO_2^{\cdot}$, $^{\cdot}OH$) are formed, together with hydrogen peroxide (H_2O_2). At a physiological pH value of 7.4 the hydroperoxyl radical ($HO_2^{\cdot}$) with a pKa of 4.8 (the pH value at which equal concentrations of both acid ($HO_2^{\cdot}$) and base ($O_2^{\cdot-}$) are present) dissociates to give the superoxide anion radical ($O_2^{\cdot-}$).

$$(5) \quad HO_2^{\cdot} \quad ----------> \quad H^+ + O_2^{\cdot-}$$

A free radical may be defined as, any chemical species capable of independent existence that contains one or more unpaired electrons. This definition (Halliwell and Gutteridge, 1989) is a broad biological one which does not specify exactly where the unpaired electron is. It is preferred because it allows us to classify most of the transition metal ions as free radicals, and so better understand the close inter-relationship between oxygen and reactive metal ions.

Superoxide ($O_2^{\cdot-}$) is a radical anion formed when one electron enters one of the II *2p orbitals of oxygen. The chemistry of superoxide differs greatly depending on its solution environment. In aqueous solution $O_2^{\cdot-}$ is a weak oxidising agent able to oxidise molecules such as ascorbic acid, and thiols. However, $O_2^{\cdot-}$ is a much stronger reducing agent, and is able to reduce several iron complexes such as cytochrome c and ferric-EDTA.

Any system producing superoxide will, as a result of the dismutation reaction, also produce hydrogen peroxide (H_2O_2)(Fridovich, 1974). Hydrogen peroxide is a weak oxidant and weak reducing agent that is relatively stable in the absence of transition metal ions. The molecule has an uncharged covalent structure. It readily mixes with water, and is treated as a water molecule by the body, rapidly diffusing across cell membranes. The hydroxyl radical ($^{\cdot}OH$) is a major product arising from the high energy ionization of water (radiolysis). The $^{\cdot}OH$ radical is an extremely aggressive oxidant that can attack most biological molecules at an almost diffusion controlled rate. Other reactive forms of oxygen such as singlet oxygen ($'\Delta gO_2$) ozone (O_3) nitric oxide ($N\dot{O}$), nitrogen dioxide ($N\dot{O}_2$) and hypochlorous acid (HOCl) are important components of oxidative stress leading to molecular damage.

B. Biological Antioxidants

The term 'antioxidant' is frequently used in the biomedical literature, but rarely defined, often implying that it refers to chemicals with chain-breaking properties such as vitamin E (α-tocopherol) and vitamin C (ascorbic acid). The author's view (Halliwell and Gutteridge, 1989) is much wider than this, and defines an antioxidant as :-

> "Any substance that when present at low concentrations, compared to those of the oxidizable substrate, significantly delays, or inhibits, oxidation of that substrate"

Antioxidants can act at many different stages in an oxidative sequence, such as :-

> Removing oxygen or decreasing local 0_2 concentrations, Removing catalytic metal ions, Removing key reactive oxygen species (ROS) such as $0_2^{\cdot}$ and $H_2 0_2$, Scavenging initiating radicals such as $^{\cdot}0H$, $RO^{\cdot}$, $RO_2^{\cdot}$, Breaking the chain of an initiated sequence, Quenching or scavenging singlet oxygen

Many antioxidants have more than one mechanism of action. Propyl gallate, for example, a partially water-soluble phenolic antioxidant used by the food industry, is a chain-breaking antioxidant, a powerful scavenger of $^{\cdot}OH$ radicals and an iron-binding agent. Cells have formidable defenses against oxidative damage, many of which may not at first sight be seen as antioxidants. Antioxidant protection can operate at several different levels within cells, for example by :-

> Preventing radical formation, Intercepting formed radicals, Repairing oxidative damage, Increasing elimination of damaged molecules, Non-repair-recognition of excessively damaged molecules in order to prevent mutations occurring.

The different antioxidant strategies used within cells, membranes, and extracellular fluids prompted the proposal in 1986 that these profound differences were important for humoral signalling (Halliwell and Gutteridge, 1986).

1. Antioxidants and Intracellular Signalling

Oxygen metabolism occurs within cells, and it is here we expect to find antioxidants evolved to deal speedily and specifically with reduced intermediates of oxygen. Enzymes such as the superoxide dismutases rapidly promote the dismutation of superoxide into hydrogen peroxide and oxygen at a rate considerably faster than it occurs uncatalysed (Fridovich, 1974). Hydrogen peroxide a product of the dismutation reaction can be destroyed by two different enzymes namely catalase and glutathione peroxidase (a selenium containing enzyme). During normal aerobic metabolism these enzymes function in concert to eliminate toxic reduction intermediates of oxygen inside the cell thereby allowing a small pool of low molecular mass iron to safely exist for DNA synthesis and the manufacture of iron-containing proteins.

1.1 Iron as a Signal Molecule

Cells normally accumulate iron via the binding of transferrin to high affinity surface receptors. However, there is a transferrin-independent pathway of cellular iron uptake that involves a membrane-based transport system (Kaplan, Jordan and Sturrock, 1991). When non-transferrin bound iron appears in plasma it triggers the induction of membrane transporters which remove low molecular mass iron from the extracellular environment. As already mentioned, the intracellular environment can better cope with low molecular mass iron than can the extracellular compartments. Delivery of iron to the cell is, however, normally achieved by the transferrin receptor (TfR), and iron is stored in the cell within the ferritin molecule (can store up to 4,500 iron atoms). The synthesis rate of TfR and ferritin is regulated at the post-transcriptional level by cellular iron and coordinated by the iron-dependent binding of a cytosolic protein called 'the iron responsive element binding protein (IRE-BP)' which binds to specific sequences on their mRNAs (Klausner, Rouault and Harford, 1993). It appears that low molecular mass iron is capable of acting as a signal to regulate ferritin and TfR synthesis in this way. Recent work has shown that IRE-BP is identical in sequence to the cytosolic enzyme aconitase (Kennedy et al, 1992). The protein functions as an active aconitase when it has an Fe-S cluster present or as an RNA-binding protein when iron is absent (Haile et al, 1992a). Switching between these two forms depends on cellular iron status (Haile et a,1992b) such that when iron is replete it is an active aconitase, whereas when deprived of iron it has only RNA-binding activity. Intracellular low molecular mass iron may also be regulated by the oxidative stress/heat shock protein haem oxygenase which increases intracellular levels of ferritin (Vile and Tyrrell, 1993).

2. Antioxidants and Membrane Signalling

Within the hydrophobic lipid interior of membranes, different types of lipophilic radicals are formed from those seen in the intracellular aqueous milieu, and lipophilic radicals require different types of antioxidants for their removal. Vitamin E (α-tocopherol), a fat-soluble vitamin, is a poor antioxidant outside a membrane bilayer but is extremely effective when incorporated into the membrane. Membrane stability and protection very much depends on the way in which the membrane is assembled from its lipid components. Structural organisation requires that the 'correct' ratios of phospholipids to cholesterol are present, and that the 'correct' types of phospholipids and their fatty acids are attached (reviewed in Gutteridge and Halliwell, 1988).

When a cell is damaged, or dies, it is highly likely that its lipid membrane will undergo peroxidation (Halliwell and Gutteridge, 1984), normal antioxidant defences being unable to cope with such extreme oxidative stress. Peroxidation of membrane polyunsaturated fatty acids produces a plethora of reactive primary peroxides and secondary carbonyls. It has been suggested that lipid oxidation products such as these, resulting from cell death, might act as triggers for new cell growth (Gutteridge and Stocks, 1976). So far, few of these numerous lipid oxidation products have been identified or biological characterised. Through the pioneering work of Hermann Esterbauer and colleagues (1988), however, we have considerable information on the biological reactivity of one typical oxidation product namely, '4-hydroxy-2-nonenal, which is formed from n=6 fatty acids. This molecule is a potent trigger for chemotaxis (Schaur et al, 1994), it can inactivate thiol containing molecules (Esterbauer, Schau and Zollner, 1991) as well as activate certain enzymes (Natarajan, Scribner and Taher, 1993).

3. Antioxidants and Extracellular Signalling

Body extracellular fluids contain little, or no, catalase activity, and extremely low levels of superoxide dismutase. Glutathione peroxidases, in both selenium-containing and non-selenium-containing forms, are present in plasma but there is little glutathione in plasma (< 1 μM) to satisfy an enzyme with a Km for GSH in the millimolar range. "Extracellular" superoxide dismutases (EC-SOD) have recently been identified (Marklund, Holme and Hellner,, 1982), and shown to contain copper and attached carbohydrate groups (glycosylated). By allowing the limited survival of $O_2^{\bullet-}$, H_2O_2, lipid peroxides (LOOH) and hypochlorous acid (HOCl) in extracellular fluids the body can utilise these molecules, and others such as nitric oxide ($NO^{\bullet}$), as useful messenger, signal or trigger molecules (Halliwell and Gutteridge, 1986: Saran and Bors, 1989: Saran and Bors, 1994). A key feature of such a proposal is that $O_2^{\cdot}$, H_2O_2, LOOH and HOCl do not meet with reactive iron or copper, and that extracellular antioxidant protection has evolved to keep iron and copper in poorly or non-reactive forms (Gutteridge, 1982: Halliwell and Gutteridge, 1990).

The iron transport protein transferrin is normally one third loaded with iron and keeps the concentration of 'free' iron in plasma at effectively nil. Iron bound to transferrin will not participate in radical reactions, and the available iron-binding capacity gives it a powerful antioxidant property towards iron-stimulated radical reactions (Gutteridge et al, 1981). Similar considerations apply to lactoferrin which like transferrin can bind two moles of iron per mole of protein, but hold onto its iron down to pH values as low as 4.0. Haemoglobin, myoglobin and haem compounds can accelerate lipid peroxidation by at least two different mechanisms. The haem ring can react with peroxides to form active iron-oxo species such as perferryl (iron oxidation state V) and ferryl (IV), and a molar excess of peroxide can cause fragmentation of the cyclic tetrapyrrol rings releasing chelatable iron (Gutteridge, 1986). Plasma also contains proteins such as haptoglobins and haemopexin specifically to bind and conserve haemoglobin and haem iron respectively. Binding to these proteins greatly diminishes the ability of haem proteins to accelerate lipid peroxidation (Gutteridge and Smith 1988: Gutteridge, 1987).

The major copper-containing protein of human plasma is caeruloplasmin, unique for its intense blue coloration. Apart from its known acute-phase reactant properties its biological functions remain an enigma. However, the author and his colleagues have pointed out that the protein's ferroxidase activity makes a major contribution to extracellular antioxidant protection against iron-driven lipid peroxidation and Fenton chemistry (Gutteridge and Stocks, 1981). Caeruloplasmin rapidly removes ferrous ions from solution and simultaneously reduces oxygen to water, with the transfer of 4 electrons at the enzymes active centre; a good example of protection by non-release of reactive forms of oxygen into the aqueous milieu.

In summary, iron is an important signal molecule both inside and outside cells. Inside cells it can safely exist because intracellular protein antioxidants remove reduced intermediates of oxygen. Extracellularly, however, iron is removed or inactivated by proteins in order to allow limited survival of $O_2^{\cdot}$, H_2O_2, LOOH, HOCl and $\dot{N}O$ which can act as humoral signal molecules. Lipid peroxidation of membranes may be consequential upon cell damage and death but nevertheless provides numerous active messenger molecules for transferring information on cell growth and control.

Acknowledgements

JMCG thanks the British Oxygen Group, the British Lung Foundation and the British Heart Foundation for their generous support.

References

Esterbauer H, Schaur RJ, Zollner H (1991). Chemistry and biochemistry of 4-hydroxynonenal, malonaldehyde and related aldehydes. Free Rad. Biol. Med. 11; 81-128.

Esterbauer H, Zollner H and Schaur RJ (1988). Hydroxyl alkenals: Cytotoxic products of lipid peroxidation. ISI Atlas of Sci. 1; 311-317.

Fridovich I (1974). Superoxide dismutases. Adv. Enzymol. 41; 35-48.

Gutteridge JMC (1982). Fate of oxygen radicals in extracellular fluids. Biochem. Soc. Trans. 10; 72-73.

Gutteridge JMC (1986). Iron promoters of the Fenton reaction and lipid peroxidation can be released from haemoglobin by peroxides. FEBS Lett. 201; 291-295.

Gutteridge JMC (1987). The antioxidant activity of haptoglobin towards haemoglobin stimulated lipid peroxidation. Biochim. Biophys. Acta. 917; 219-223.

Gutteridge JMC and Halliwell B (1988). The antioxidant proteins of extracellular fluids. In. Cellular Antioxidant Defence Mechanisms (Chow CK, ed). CRC Press, Boca Raton. pp 1-23.

Gutteridge JMC, Paterson SK, Segal AW and Halliwell B (1981). Inhibition of lipid peroxidation by the iron binding protein lactoferrin. Biochem. J. 199; 259-261.

Gutteridge JMC and Smith A (1988). Antioxidant protection by hemopexin of haem-stimulated lipid peroxidation. Biochem. J. 256; 861-865.

Gutteridge JMC and Stocks J (1976). Peroxidation of cell lipids. J. Med. Lab. Sci. 53; 281-285.

Gutteridge JMC and Stocks H (1981). Caeruloplasmin: physiological and pathological perspectives. CRC Crit. Rev. Clin. Lab. Sci. 14; 257-329.

Haile DJ, Rouault TA, Harford JB et al (1992a). Cellular regulation of the iron-responsive element binding protein: disassembly of the cubane iron-sulfur cluster results in high affinity RNA binding. Proc. Natl. Aca. Sci. USA. 89: 11735-11739.

Haile DJ, Rouault TA, Tang CK et al (1992b) Reciprocal control of RNA-binding and aconitase activity in the regulation of the iron-responsive element binding protein: role of the iron-sulfur cluster. Proc. Natl. Acad. Sci. USA. 89: 7536-7540.

Halliwell B and Gutteridge JMC (1984). Lipid peroxidation, oxygen radicals, cell damage and antioxidant therapy. Lancet. 1; 1396-1397.

Halliwell B and Gutteridge JMC (1986). Oxygen free radicals and iron in relation to biology and medicine: some problems and concepts. Arch. Biochem. Biophys. 246: 501-514.

Halliwell B, Gutteridge JMC (1989). Free Radicals in Biology and Medicine. Oxford University Press; Oxford.

Halliwell B and Gutteridge JMC (1990). The antioxidants of human extracellular fluids. Arch. Biochem. Biophys. 280: 1-8.

Kaplan J, Jordan I and Sturrock A (1991). Regulation of the transferrin-independent iron transport system in cultured cells. J. Biol. Chem, 266: 2997-3004.

Kennedy MC, Mende-Mueller L, Blondin GA and Beinert H (1992). Purification and characterisation of cytosolic aconitase from beef liver and its relationship to the iron-responsive element binding protein (IRE-BP). Proc. Natl. Acad. Sci. USA. 89 : 11730-11734.

Kalusner RD, Rouault TA and Harford JB (1993). Regulating the fate of mRNA: The control of cellular iron metabolism. Cell . 72: 19-28.

Marklund SL, Holme E and Hellner L (1982). Superoxide dismutase in extracellular fluids. Clin. Chim. Acta. 126; 41-51.

Natarajan V, Scribner WM and Taher MM (1993). 4-Hydroxynonenal, a metabolite of lipid peroxidation, activates phospholipase D in vascular endothelial cells. Free Rad. Biol. Med. 15: 365-375.

Saran M and Bors W (1989). Oxygen radicals acting as chemical messengers: A hypothesis. Free Rad. Res. Commun. 7: 213-220.

Saran M and Bors W (1994). Signalling by $0_2^{\cdot}$ and $NO^{\cdot}$: how far can either radical, or any specific reaction product, transmit a message under in vivo conditions? Chem. Biol. Interact. 90: 35-45.

Schaur RJ, Dussing G, Kink E et al (1994). The lipid peroxidation product 4-hydroxynonenal is formed by - and is able to attract-rat neutrophils in vivo. Free Rad. Res. 20: 365-373.

Vile GF and Tyrrell RM (1993). Oxidative stress resulting from ultraviolet A irradiation of human skin fibroblasts leads to a heme oxygenase-dependent increase in ferritin. J. Biol. Chem. 268: 14678-14681.

STRATEGIES OF ANTIOXIDANT DEFENSE: RELATIONS TO OXIDATIVE STRESS

Helmut Sies
Institut für Physiologische Chemie I
Heinrich-Heine-Universität Düsseldorf
Postfach 101007
D-40001-Düsseldorf
Germany

SUMMARY

Oxidants and antioxidants have attracted widespread interest in diverse scientific disciplines, ranging from free radical chemistry to biochemistry, nutrition research, biology and medicine. Life on this planet utilizes oxygen and oxygen metabolites in energy conversion, and it has become clear that constant generation of pro-oxidants, including oxygen free radicals, is an essential attribute of aerobic life. This challenge is met by a system of anti-oxidants which help to maintain the steady state of the living organism.

A disturbance in the prooxidant/antioxidant system has been defined as *'Oxidative Stress'*. Operationally, a useful definition would also include that the disbalance in favor of the prooxidants is associated with potential damage to the biological system. Damage products as indicators of oxidative stress include damaged DNA bases, protein oxidation products and products of lipid peroxidation. A loss of antioxidant capacity may concern enzymatic defense, e.g. superoxide dismutase, glutathione peroxidases, or catalase, or a weakening of nonenzymatic defense, notably the loss of micronutrients such as vitamins C, E, carotenoids and selenium.

An interesting aspect resides in the modulation of gene expression by oxidative stress. In consequence, this leads to new therapeutic concepts of employing compounds active as antioxidants. One example is the GSH peroxidase mimic, ebselen, a selenoorganic compound.

NATO ASI Series, Vol. H 92
Signalling Mechanisms – from Transcription Factors to Oxidative Stress
Edited by L. Packer, K. Wirtz

INTRODUCTION

Hydroperoxide metabolism in mammalian organs (Chance et al., 1979) and the biochemistry of oxidative stress (Sies, 1986) have been a focus of research since some time. The nature of various biological oxidants covers large ranges in biological lifetime, in concentration, and in occurrence in cells and organs. Experimental studies revealed that cells and organisms require defense against oxidants, without which survival under aerobic conditions is jeopardized. In view of the variety in oxidants, also called prooxidants, it is not surprising that nature has evolved a battery of different types of antioxidant.

This article examines the strategies of antioxidant defense in biological systems, it is based on a recent review (Sies, 1993a). Emphasis will be more on identifying the types or principles of defense rather than on completeness of coverage of the available literature.

OXIDATIVE STRESS AND THE PRINCIPLES OF PROTECTION

Aerobic metabolism entails the production of reactive oxygen species, even under basal conditions, hence a continuous requirement for their inactivation. This steady-state of prooxidants and antioxidants may be disrupted. A disbalance in favor of the prooxidants and in disfavor of the antioxidants, potentially leading to damage, has been called 'oxidative stress' (Sies, 1991). Such damage may afflict all types of biological molecules, including DNA, lipids, proteins, and carbohydrates. Thus, oxidative stress may be involved in processes such as mutagenicity, carcinogenicity, membrane damage, lipid peroxidation, protein oxidation and fragmentation, and carbohydrate damage.

In principle, the protection against such deleterious effects can be at three levels: prevention, interception, and repair. All three of these levels of protection are realized in biology, and examples will be given below. In order to lay out the variety of problems of protective measures, the nature of the prooxidants and the antioxidants will first be presented briefly.

NATURE AND DIVERSITY OF PROOXIDANTS

Molecular oxygen can be reduced to water. The intermediate steps of oxygen reduction are the superoxide anion radical, hydrogen peroxide, and the hydroxyl radical, corresponding to the steps of reduction by one, two and three electrons. Further, ground state molecular (triplet) oxygen as a diradical can be electronically excited to singlet molecular oxygen. Oxygen radical functions in combination with other atoms or in larger molecules can occur as $RO^{\bullet}$ or $ROO^{\bullet}$, alkyl or peroxyl radicals, e.g. in lipids. Also, there is nitric oxide, $NO^{\bullet}$, as one of the gaseous radicals of biological interest.

Oxidant functions are carried by different types of radiation, with X-irradiation generating the hydroxyl radical, and UV-irradiation generating electronically excited states with subsequent radical formation, encompassing the field of photochemistry. Ultrasound and microwave radiation can also generate reactive oxygen species. Even shear stress, e.g. as in homogenisation, is known to generate radicals.

As shown in Table 1, the half-lives of the major reactive oxygen species are vastly different, underscoring the necessity for different types of defense strategy. Highest rate constants for the reaction with target molecules are found for the hydroxyl radical; its reactions are diffusion-limited, i.e. they take place practically at the site of its generation. On the other hand, some peroxyl radicals are relatively stable, with half-lives into the range of seconds. Such molecules may diffuse away from their site of generation and thus transport the radical or oxidant function to other target sites.

Table 1. Estimate of the half-lives of reactive oxygen species.
Modified from (Pryor, 1986; Sies et al., 1992).

Reactive oxygen species	Half-life
	s
HO•, hydroxyl radical	**10^{-9}**
RO•, alkoxyl radical	**10^{-6}**
ROO•, peroxyl radical	**7**
H_2O_2, hydrogen peroxide	**-(enzymic)**
$O_2^{\bullet -}$, superoxide anion radical	**-(enzymic)**
$^1O_2$, singlet oxygen	**10^{-5}**
Q•, semiquinone radical	**days**
NO•, nitric oxide radical	**1 - 10**
$ONOO^-$, peroxynitrite	0.05-1

In cell metabolism, clandestine oxidant functions may exist and be transported to distant target sites to exert oxidant activity at that location. This would include compounds or enzyme activities that are innocuous in one environment but activatable to generate oxidants in other conditions.

The diet contains many compounds of oxidant and antioxidant nature (Ames, 1983). In the present context, it is important to note that there are dietary compounds acting as potential oxidants, including a variety of quinones capable of redox cycling (Kappus & Sies, 1981), and substrates for enzyme systems which generate oxidants.

NATURE AND DIVERSITY OF ANTIOXIDANTS

In their definition of the term 'antioxidant', Halliwell and Gutteridge (1989) formulate: 'any substance that, when present at low concentrations compared to that of an oxidizable substrate, significantly delays or inhibits oxidation of that substrate'. This definition would comprise compounds of nonenzymatic as well as of enzymatic nature. Table 2 gives an overview on some of the antioxidants of biological interest.

Table 2. Antioxidant defense in biological systems. Condensed list of antioxidant compounds and enzymes. Modified from (Sies, 1985).

System	Remarks
Non-enzymic	
α-tocopherol (vitamin E)	radical chain-breaking
β-carotene	singlet oxygen quencher
lycopene	singlet oxygen quencher
ubiquinol-10	radical scavenger
ascorbate (vitamin C)	diverse antioxidant functions
glutathione (GSH)	diverse antioxidant functions
urate	radical scavenger
bilirubin	plasma antioxidant
flavonoids	plant antioxidants (rutin, etc.)
plasma proteins	metal binding, e.g. coeruloplasmin
chemical	food additives, drugs (see text)
Enzymic (direct)	
superoxide dismutases	CuZn enzyme, Mn enzyme, Fe enzyme
GSH peroxidases	see enzymes (GPx, PHGPx)
ebselen	as enzyme mimic
catalase	heme protein, peroxisomes
Enzymatic (ancillary enzymes)	
conjugation enzymes	glutathione-S-transferases UDP-glucuronosyl-transferases
NADPH-quinone oxidoreductase	two-electron reduction
GSSG reductase	maintaining GSH levels
NADPH supply	NADPH for GSSG reductase
transport systems	GSSG export thioether (S-conjugate) export
repair systems	DNA repair systems oxidized protein turnover oxidized phospholipid turnover

Clearly, there is a diversity of antioxidants which matches that of the prooxidants. In the following, some of the principles underlying the antioxidant functions will be discussed.

PREVENTION

A first line of defense against reactive oxygen species is, of course, protection against their formation, i.e. prevention. There are numerous strategies in biology designed to evade oxidative stress, ranging from the plankton that descends from the surface of the seawater to lower levels under conditions of solar irradation, to the packaging of DNA in chromatin to shield the genetic material by providing alternate targets. Microbes have developed specialized strategies to prevent oxygen-dependent killing by phagocytes (Haas & Goebel, 1992). This extends to the concept that antioxidant defense in multicellular organisms can be more efficient than that in unicellular organisms, based on the observation that catalase in *Escherichia coli* can defend colonial, but not individual cells against hydrogen peroxide (Ma & Eaton, 1992). Thus, group protection by the activity of enzymes that mitigate oxidative stress was proposed as a driving force in the evolution of multicellular organisms. Numerous further biological, physiological and morphological examples could be given. However, here we focus on the biochemical mechanisms of prevention.

Regarding radical formation, first it should be mentioned that some of the enzymes prone to generate free radical species are ingeniously designed. Cytochrome oxidase, carrying the predominant cellular share of oxygen reduction, does not release superoxide or other radicals, even though it contains iron and copper ions. Likewise, the three-dimensional structure of the enzyme, ribonucleotide reductase, keeps the radical character of the tyrosyl function in subunit B from spreading to the environment by forming an appropriate 'cage' (Reichard & Ehrenberg, 1983).

Further, the prevention of initiation of chain reactions includes the binding of metal ions, in particular iron and copper ions. Metal chelation is a major means of controlling lipid peroxidation and DNA fragmentation. Thus, the metal-binding proteins, ferritin, transferrin, coeruloplasmin and others, e.g. metallothionein, are of central importance in the control of potential radical-generating reactions. Another strategy to increase the resistance to metal-ion-dependent oxidation is to modify the potential target site. An example is the stable modification of low-density lipoprotein by dehydroascorbate or decomposition products thereof to impart increased resistance to metal-ion-induced oxidation (Retsky et al., 1993).

Prevention of cells against incident radiation may occur through specialized pigments, e.g. the melanins for ultraviolet radiation or the carotenoids for electronically excited states such as singlet oxygen. Alternatively, compounds acting as photosensitizers may be reductively detoxified, for example as a mechanism of fungal resistance to cercosporin (Daub et al., 1992). However, these and other strategies are not completely preventative, because they operate by decreasing the yield of a given challenging agent with less than 100% efficiency (see below).

In this regard, there are many enzymic systems in cells and body fluids which control the level of reactive species which otherwise might generate a cascade of products which, in turn, lead to attacking oxidants. One important group of such enzymes is the glutathione S-transferases. This family of enzymes catalyzes the reaction of the major low-molecular-mass thiol, glutathione, with reactive electrophiles to form thioethers, called S-conjugates (Mannervik, 1985; Sies & Ketterer, 1988). Biologically reactive electrophilic intermediates can be formed in a variety of metabolic pathways, notably those involving cytochrome P-450, and are of interest in toxicology and pharmacology (Witmer et al., 1991).

A strategy of preventative antioxidation could therefore be formulated as prevention by diversion, i.e. by channeling an attacking species into a less harmful product, hence lowering the risk of further damage. In the extreme, this could involve whole cells, one example being the intestinal mucosal cells. These cells are exposed to a variety of reactive intermediates and xenobiotics, and the rate of accumulation of oxidative-damage products in these cells is high. The turnover and elimination of whole cells prevents further spread of the challenging species. Again, this type of prevention overlaps in part with the concept of interception.

Interception

Nonenzymic

This is the domain of the antioxidants as defined in a more narrow sense. The basic problem is to intercept a damaging species, once formed, from further activity. This is the process of deactivation. For radical compounds, the final deactivation consists of the formation of nonradical end products. Due to the nature of the free radicals, there is a tendency towards chain reaction, i.e. a compound carrying an unpaired electron will react with another compound to generate an unpaired electron in that compound ('radicals beget radicals').

A second objective of biological importance is to transfer the radical function away from more sensitive target sites to compartments of the cell in which an oxidative challenge would be less deleterious. In general, this means transferring the oxidizing equivalents from the hydrophobic phases into the aqueous phases, e.g. from the membrane to the cytosol or from lipoproteins in the blood plasma to the aqueous phase of the plasma. Biologically, the most efficient intercepting antioxidants combine optimal properties in both these objectives; firstly, they react with initial free radicals such as lipid peroxyl radicals at suitable rates, and, secondly, they are capable of interacting with water-soluble compounds for their own regeneration. This then transfers the radical function away from further potential targets. In biological membranes, where a high-efficiency back-up system is present, there may be the need for only 1-3 antioxidant molecules/1000 potential target molecules.

Such intercepting chain-breaking antioxidants are often phenolic compounds. (R,R,R)-alpha-Tocopherol probably is the most efficient compound in the lipid phase (Burton et al.,

1983). This biological antioxidant (Tappel, 1962) contains shielding methyl groups in the vicinity of the phenolic hydroxyl group of the chromane moiety, and it is optimally positioned in the membrane by its phytyl sidechain.

The maintenance of a steady-state rate of peroxyl radical reduction by tocopherol in the membrane is dependent on the reduction of the tocopheroxyl radical, once formed, by external reductants. These include ascorbate and thiols (Niki, 1987; Wefers & Sies, 1988). Whether the reaction occurs directly or through intermediate steps is still debated (see [Sies & Murphy, 1991] for review) but, *in vitro*, the reaction has been clearly demonstrated by pulse-radiolysis experiments (Packer et al., 1979) and occurs in membrane systems (Niki, 1987; Wefers & Sies, 1988).

A prerequisite for efficient interception by the phenolic antioxidants resides in the life-time of the radical to be intercepted. This predisposes the peroxyl radicals as major reaction partners, since their life-time extends into the range of seconds (Table 1). In contrast, the hydroxyl radical, with its high reactivity and extremely short life-time, cannot be intercepted with reasonable efficiency. It has been shown that up to 100 mM of an intercepting compound would be required for 90% efficiency (Czapski, 1984), eliminating interception as a useful strategy for defense against the hydroxyl radical, if only for osmotic reasons.

Interception of oxidants by cholesterol has also been proposed (Smith, 1991). The B-ring oxidized oxysterols of human blood were considered to represent past interception in vivo by cholesterol. The oxysterols are efficiently metabolized and excreted by the liver. Another example would be the function of plasmalogens, suited for the reaction with singlet molecular oxygen (Morand et al., 1988); the oxidation products would then be replaced by intact plasmalogen molecules, and the effect would be to avoid alternate targets and decrease the biological yield of the attacking species.

Highly efficient biological polyene quenchers for singlet molecular oxygen (Foote & Denny, 1968), notably carotenoids and oxycarotenoids, provide a suitable defense system against this oxygen species, in spite of its reactivity and short life-time. The localized concentrations of the carotenoids are decisive in determining the efficiency of the quenching of singlet oxygen and other electronically excited states (Krinsky, 1989; Sies et al., 1992).

Enzymic

All cells in eukaryotic organism contain powerful antioxidant enzymes (for review, see [Chance et al., 1979]). The three major antioxidant enzymes are the superoxide dismutases (McCord & Fridovich, 1969), catalase and glutathione (GSH) peroxidases. In addition, there are numerous specialized antioxidant enzymes reacting with and, in general, detoxifying oxidant compounds (Table 2). Indirect antioxidant funcions carried by enzymes are (a) the backup function, e.g. the replenishment of GSH from glutathione disulfide (GSSG) by the flavoprotein GSSG reductase, and (b) the transport and elimination of reactive compounds, e.g. the glutathione S-transferases and the transport systems for the glutathione S-conjugates.

For the present discussion, it is of interest to consider the fact that different subcellular sites and different cell types contain varying amounts of the antioxidant enzymes (see [Chance et al., 1979]).

Repair

Protection from the effects of oxidants can also occur by repair of damage once it has occurred. As prevention and interception processes are not completely effective, damage products are continuously formed in low yields and hence may accumulate. This refers to DNA damage, occurring as damaged bases or in the form of single-strand or double-strand breaks, to membrane damage, occurring as a variety of phospholipid oxidation products, and to proteins and other compounds as well. Correspondingly, there are multiple enzyme systems involved in DNA repair and lipolytic as well as proteolytic enzymes capable of serving the functions of restitution or replenishment. Many supportive strategies are operative, for example, in the surveillance of the building blocks for DNA synthesis, the dGTP pool is enzymically cleared from the contaminant oxidized base, 8-oxo-dG (Maki & Sekiguchi, 1992; Mo et al., 1992). This very extensive field of repair is not reviewed here in detail.

Adaption: adaptive responses

Prokaryotes

The control of antioxidant enzyme levels in cells is of key importance for survival in an aerobic environment (Harris, 1992; Remacle et al., 1992). While little is known about constitutive expression of antioxidant enzymes, the adaption of cells to oxidative stress has been a topic of active research, particularly with prokaryotes such as *Salmonella typhimurium* and *E. coli*. Bacteria adapt to the lethal effects of oxidants by inducing the expression of protective stress genes under the control of regulons, e.g. oxyR (Christman et al., 1985) and soxR (Greenberg et al., 1985). The oxyR gene product is redox sensitive and, in its oxidized form, activates gene expression (Storz et al., 1990; Tartaglia et al., 1991). It is suggested that oxidation of the oxyR protein brings about a conformational state that transduces the oxidative-stress signal to selectively activate DNA transcription.

Bacterial strains carrying deletions in oxyR exhibit significantly increased frequencies of mutagenesis (Storz et al., 1987; Greenber & Demple, 1988), which are pronounced under aerobic conditions. The high frequency of mutagenesis in oxyR deletion strains was suppressed by multicopy plasmids expressing high levels of catalase (katG gene), alkylhydroperoxide reductase (ahpCF gene) or superoxide dismutase (sodA gene) activities ([Storz et al., 1987]; see

Table 3). These observations provide evidence that the oxyR regulon plays an important role in protecting against oxidative DNA damage that would otherwise cause mutations.

Table 3. oxyR deletion strains have increased frequencies of spontaneous mutagenesis. The frequency of mutagenesis in the S. typhimurium oxyR mutant strains was assayed by tbe reversion of His-auxotrophy to His+ prototrophy (taken from [Storz et al., 1987]). pKM101 encodes mucA and mucB (analogues of the E. coli umuC and umuD genes) that make strains more susceptible to mutagenesis by a number of mutagens.

	Number of mutants/plate
$oxyR^+$ (wild type)	6
oxyR Δ2 (oxyR deletion)	76
$oxyR^+$/pKM101[b]	57
oxyR Δ2/pKM101	3102
oxyR Δ2/pKM101/pACYC184 (vector)	946
oxyR Δ2/pKM101/pAQ5 (oxyR)	33
oxyR Δ2/pKM101/pAQ6 (sodA)	196
oxyR Δ2/pKM101/pAQ7 (katG)	47
oxyR Δ2/pKM101/pAQ8 (ahp)	31

Mammalian cells

Adaptive responses to several types of challenge, including heat shock and oxidative stress, have also been identified in human cells. For example, heme oxygenase was found to be a major stress protein produced in responses to oxidative challenge (Keyse & Tyrrell, 1989). Reactive oxygen species activate NF-kappaB, a transcriptional regulator of genes involved in inflammatory and acute-phase responses (Schreck et al., 1991; 1992). Modulation of NF-kappaB-binding activity by oxidation/reduction has been demonstrated *in vitro* (Toledano & Leonard, 1991). Recently, expression of a human gene encoding a protein-tyrosine phosphatase was found to be greatly induced by oxidative stress and heat shock in skin cells (Keyse & Emslie, 1992) linking redox signaling to protein phosphorylation. Thus, there is a relationship between redox changes and regulation of receptor activity, cellular proliferation and the cell cycle. A variety of oxidative-stress models have been shown to lead to increased expression of proto-oncogenes, including c-fos, c-jun and c-myc (Cerutti et al., 1988).

Adaptive responses at the level of gene regulation were studied in the rat glutathione S-transferase Ya-subunit gene and the NAD(P)H:quinone reductase gene by mutation and deletion analyses (Rushmore et al., 1991; Nguyen & Pickett, 1992). An antioxidant responsive element (ARE) was identified in the 5'-flanking region of both genes.
The sequences,

5'-RGTGACNNNGC-3'

3'-YCACTGNNNCG-5',

where N is any nucleotide, represent the core sequence of the ARE required for transcriptional activation by phenolic antioxidants and metabolizable planar aromatic compounds. The observation that the ARE contains a recognition motif highly similar to the consensus binding sequence for the c-Jun/c-Fos heterodimer suggested a possible involvement of c-Jun in the ARE-regulatory-protein complex (Nguyen & Pickett, 1992). Induction of c-Jun expression in response to hydrogen peroxide has been demonstrated (Devary et al., 1991). In addition, a redox mechanism may regulate Jun-Fos DNA-binding activity (Abate et al., 1990). Jun-D and c-Fos were identified as two members of the ARE-protein complex in studies on the regulation of the human NAD(P):quinone oxidoreductase gene (Li & Jaiswal, 1992).

A nuclear protein, Ref-1, has been described, that stimulates DNA binding of Fos and Jun heterodimers, identifying it as a redox factor capable of regulating the function of transcription factors (Xanthoudakis & Curran, 1991). The activity occurs through a conserved cysteine residue in the DNA-binding domain of Fos and Jun (Xanthoudakis et al., 1991). The oxidation state of the cysteine has not yet been identified, but it does not involve the formation of a disulfide bond (Abate et al., 1990). The Ref-1 system probably constitutes a major switch function with regard to redox signaling.

Dietary constituents are capable of modifying the metabolism of carcinogens by the induction of antioxidant enzymes of detoxication, particularly the so-called phase-II enzymes, notably quinone reductase (DT diaphorase) and glutathione transferases (Prochaska & Talalay, 1991). Numerous epidemiological studies suggest that high consumption of yellow and green vegetables reduces the risk of cancer development. This could be directly due to protection by the antioxidant compounds contained in these vegetables. However, alternatively, inducing effects are exerted by compounds contained in these vegetables (Prochaska et al., 1991). The induction of quinone reductase was studied in particular with regard to the compound sulforaphane, an isothiocyanate derivative present in broccoli (Zhang et al., 1992). It appears possible that these dietary inducers of quinone reductase act through ARE.

Control of prooxidant enzyme activities: NADPH oxidase and nitric oxide synthase

The cellular production of reactive oxygen species by phagocytes is a well-studied phenomenon, forming the basis of an important sector of host defense. Recently, the control of the major enzymes involved in this host defense, NADPH oxidase and NO synthase, has been intensely studied. It is important to exert subtle control over the activity of these and other enzymes, because an overproduction of superoxide or nitric oxide might be harmful to the cell and the organism as a whole. Thus, the on/off switches are crucial.

NADPH oxidase

NADPH oxidase is the superoxide-forming enzyme of phagocytes and B-lymphocytes and is composed of cytosolic and membrane-associated components. The cytosolic components form a 240 kDa complex consisting of the p47phox-encoded and p67phox-encoded subunits, as well as a small GTP-binding protein, p21rac2. Subunits translocate from the cytosol to the plasma membrane where the oxidase is activated (Segal, 1989; Knaus et al., 1992; Park et al., 1992). This signaling system provides for tight control by mediators.

It has also been shown that non-phagocytes, e.g. fibroblasts, can generate superoxide under the control of signal molecules, e.g. interleukin-1 and tumor necrosis factor (Meier et al., 1989; 1990). Interestingly, the Mn-superoxide dismutase (SOD) has been found to be induced by interleukin-1 and tumor necrosis factor and to protect against subsequent oxidant injury (Wong & Goeddel, 1988; Wong et al., 1989), and likewise human Mn-SOD in pulmonary epithelial cells of transgenic mice conferred protection (Wispe et al., 1992).

There are several further consequences of the presence of superoxide in cells and in extracellular fluids. For example, *in vitro* there is superoxide-dependent stimulation of leukocyte adhesion by oxidatively modified LDL (Lehr et al., 1992), underlining the importance of control of superoxide production. In this regard, it is noteworthy that an adhesion protein has been found to inhibit superoxide release by human neutrophils (Wong et al., 1991), and that this adhesion protein may be considered as an antiinflammatory molecule preventing the inappropriate activation of neutrophils in the circulation (Wong et al., 1991).

Nitric oxide synthase

This family of enzymes has attracted considerable interest in biochemistry, physiology and pharmacology (Noack & Murphy, 1991; Nathan, 1992). Nitric oxide synthase is a catalytically self-sufficient cytochrome P-450 enzyme, containing both a reductase and a heme domain (White & Marletta, 1992). Whereas the enzyme in macrophages and several other cell types is only expressed following exposure of the cells to activating cytokines or microbial products and produces NO independently of added calcium and calmodulin (Stuehr, 1991), the brain enzyme is expressed constitutively and generates NO in response to calcium and calmodulin (Bredt & Snyder, 1990; Busse & Mülsch, 1990). The genes of the inducible (Xie et al.,

1992) and the constitutively expressed forms (Bredt et al., 1991) of nitric oxide synthase have been cloned and characterized. Regulatory sites of the latter were identified as phosphorylation sites and included different serines as substrates for cyclic-AMP-dependent protein kinase, protein kinase C and calcium/calmodulin protein kinase (Bredt et al., 1992). This complex regulation provides for multiple means of regulating NO levels and for cross-talk between different secondary-messenger systems. In particular, down-regulation of nitric oxide synthase activity by more than 66% was obtained by activation of protein kinase C by 50 nM phorbolester (Bredt et al., 1992).

Superoxide dismutase can catalyze the reversible interconversion of nitric oxide and the nitroxyl anion, so that the redox state of the copper in superoxide dismutase can influence the metabolic fate of the generated nitric oxide (Murphy & Sies, 1991).

Synthetic antioxidants

The strategies of antioxidant defense pursued with synthetic antioxidants basically overlap those employed by biological systems. Applications of synthetic antioxidants are similar to those of biological antioxidants but, in addition, they are of potential use in chemistry, in food industry and in medicine.

Nonenzymic

Phenolic antioxidants. There are a number of phenolic antioxidants, butylated hydroxytoluene and butylated hydroxyanisole being prominent examples. These compounds have been widely used as food antioxidants, but, because of their metabolism to potentially reactive intermediates, applications have been restricted recently (Kahl, 1991).

Probucol, a compound containing two phenoxyl moieties, has been particularly useful in studies on the protection of low-density lipoprotein against peroxidation (Carew et al.,1987).

Modified tocopherol ascorbate and carotenoids. Natural antioxidants have been modified to generate synthetic compounds exhibiting novel properties. For example, alpha-tocopherol has been modified to a water-soluble derivative, trolox, by exchanging the phytyl side chain for a carboxylate group. Conversely, ascorbate has been esterified with fatty acids such as palmitate to generate a more hydrophobic derivative. Synthetic carotenoids retaining the polyene structure have been examined for their ability to quench singlet oxygen (Devasagayam et al., 1992). The rationale for these and other types of derivative is to exploit activities exerted at different localizations in cells or fluids due to changes in solubility properties, retaining the functional end of the antioxidant molecule. This may open new sites for protection, employing the antioxidant principle of the natural parent compound.

Thiols. Since glutathione, as the major low-molecular-mass thiol in cells, does not enter most types of cells, glutathione ethylester has been synthesized as a precursor penetrating into cells to then be hydrolyzed to glutathione (Anderson & Meister, 1989). Alternatively, thiazolidine derivatives or N-acetyl cysteine have been employed as precursors for cysteine supporting GSH biosynthesis by the substrate supply, but also acting as antioxidants by themselves.

Therapeutic use of synthetic racemic lipoate (thioctic acid) is based, in part, on the antioxidant function of the dihydrolipoate/lipoate system shown by its protecion against microsomal lipid peroxidation (Bast & Haenen, 1988; Scholich et al., 1989), against DNA damage by singlet oxygen (Devasagayam et al., 1993), and against a decrease in membrane fluidity in hypoxia/reoxygenation (Scheer & Zimmer, 1993). Like N-acetyl cysteine (Staal et al., 1990), lipoate inhibits NF-*k*B activation in human T-cells (Suzuki et al., 1992).

Numerous other thiol compounds, notably those of aminothiol structure, were examined as radioprotectors.

Metal chelators. An important strategy of prevention is to bind metal ions (Weglicki & Mak, 1992). Desferrioxamine and many related metal chelators were designed to bind iron or copper ions (Halliwell, 1989).

Miscellaneous. The targeting of compounds to membrane sites in cells may have been the strategy involved in generating compounds known as lazaroids, with the steroid ring system as a basic building block (Hall et al, 1988).

Chemical modification of structural features of flavonoids has generated a multitude of synthetic antioxidant compounds.

Enzyme mimics

Low-molecular-mass compounds exhibiting catalytic activity, i.e. operating as enzyme mimics, have been used as antioxidants. Copper diisopropyl salicylate and other copper complexes were shown to mimic superoxide dismutase activity (Oberley et al., 1984). A selenoorganic compound, ebselen, was shown to mimic the GSH peroxidase reaction (Müller et al., 1984; Wendel et al., 1984). As discussed recently (Sies, 1993b), the kinetic mechanism of ebselen closely resembles that of the phospholipid hydroperoxide GSH peroxidase (Maiorino et al., 1988) and GSH peroxidase enzymes.

Enzymic

Synthetic enzymic antioxidants have many future perspectives. One route is to generate chimeric proteins that allow for targeting. The Hb-SOD, which binds to endothelial cells and was shown to positively affect elevated blood pressure in experimental animals (Nakazono et al., 1991), is one example. The engineering of SOD molecules with higher catalytic rates, employing the principle of electrostatic guidance, led to recombinant enzyme preparations more active that native SOD (Getzoff et al., 1992).

The site-specific mutation of a crucial oxidizable methionine residue to a nonoxidizable amino acid in the elastase inhibitor provides an interesting example of a preventative strategy (Rosenberg et al., 1984). Elastase activation, as a consequence of oxidative stress, is considerably diminished with the oxidation-resistant inhibitor.

Only a few examples of the strategies employed in recent years to generate a multitude of potential drug antioxidants synthetically (Emerit et al., 1990) have been presented here. A delicate balance exists between prooxidants and antioxidants in cells and relationships exist between the redox state and cellular gene expression, as described briefly above. Therefore, pharmacological applications of highly efficient antioxidant compounds or enzymes may potentially interfere with important cellular functions, including changes in the enzyme activity, enzyme patterns, membrane fluidity and responses to stimuli. While this aspect deserves attention in each case, it should be mentioned that, overall, the use of antioxidant compounds has largely been devoid of side effects and in the applications studied thoroughly, has proved predominantly beneficial.

ACKNOWLEDGEMENT

Studies from the author's laboratory were generously supported by the National Foundation for Cancer Research, Bethesda, by the *Bundesministerium für Forschung und Technologie*, Bonn, by the *Ministerium für Wissenschaft und Forschung*, NRW, and by the *Jung-Stiftung für Wisssenschaft und Forschung*, Hamburg.

REFERENCES

Abate C, Patel L, Rauscher, FJ & Curran, T (1990) Science 249, 1157-1161.
Ames, BN (1983) Science 221, 1256-1263.
Anderson, ME & Meister, A (1989) Anal. Biochem. 183, 16-20.
Bast, A & Haenen, GRMM (1988) Biochim. Biophys. Acta 963, 558-561.
Bredt, DS & Snyder, SH (1990) Proc. Natl. Acad. Sci. 87, 682-685.
Bredt, DS, Ferris, CD & Snyder, SH (1992) J. Biol. Chem. 267, 10976-10981.
Bredt, DS, Hwang, PM, Glatt, CE, Lowenstein, C, Reed RR &

Snyder, SH (1991) Nature 351, 714-718.
Burton, GW, Joyce, A & Ingold, KU (1983) Arch. Biochem. Biophys. 221, 281-290.
Busse, R & Mülsch, A (1990) FEBS Lett. 265, 133-136.
Carew, TE, Schwenke, DC & Steinberg, D (1987) Proc. Natl. Acad. Sci. USA 84, 7725-7729.
Cerutti, P, Larsson, R, Krupitzka, D, Muehlematter, D, Crawford, D & Amstad, P (1988) In Oxy-radicals in Molecular Biology and Pathology (Cerutti, PA, Fridovich, I & McCord, JM, eds.) pp 493-507, A. Liss, New York.
Chance, B, Sies, H & Boveris, A (1979) Physiol. Revs. 59, 527-605.
Christman, MF, Morgan, RW, Jacobson, FS & Ames, BN (1985) Cell 41, 753-762.
Czapski, G (1984) Isr. J. Chem. 24, 29-32.
Daub, ME, Leisman, GB, Clark, RA & Bowden, EF (1992) Proc. Natl. Acad. Sci. USA 89, 9588-9592.
Devary, Y, Gottlieb, RA, Lau, LF & Karin, M (1991) Mol. Cell. Biol. 11, 2804-2811.
Devasagayam, TPA, Subramanian, M, Pradhan, DS & Sies, H (1993) Chem.-Biol. Interact. 86, 79-92.
Devasagayam, TPA, Werner, T, Ippendorf, H, Martin H-D & Sies, H (1992) Photochem. Photobiol. 55, 511-514.
Emerit, I, Packer, L & Auclair, C eds. (1990) Antioxidants in Therapy and Preventive Medicine, Plenum Press, New York.
Foote, CS & Denny, RW (1968) J. Am. Chem. Soc. 90, 6233-6235.
Getzoff, ED, Cabelli, DE, Fisher, CL, Parge, HE, Viezzoli, M S, Banci, L & Hallewell, RA (1992) Nature 358, 347-351.
Greenberg, JT & Demple, B (1988) EMBO J. 7, 2611-2617.
Greenberg, JT, Monach, PA, Chou, JH, Josephy, PD & Demple, B (1990) Proc. Natl. Acad. Sci. 87, 6181-6185.
Haas, A & Goebel, W (1992) Free Radical Research Commun. 16, 137-157.
Hall, EO, Yonkers, PA, McCall, JM, Braughler, JM (1988) J. Neurosurg. 68, 456-461.
Halliwell, B & Gutteridge, JMC (1989) Free Radicals in Biology and Medicine (2nd edn.) Clarendon Press, Oxford.
Halliwell, B (1989) Free Radical Biol. Med. 7, 645-651.
Harris, ED (1992) FASEB J. 6, 2675-2683.
Kahl, R (1991) In Oxidative Stress: Oxidants and Antioxidants, (Sies, H, ed.) pp 245-273, Academic Press, London.
Kappus, H & Sies, H (1981) Experientia 37, 1233-1241.
Keyse, SM & Emslie, EA (1992) Nature 359, 644-647.
Keyse, SM & Tyrrell, RM (1989) Proc. Natl. Acad. Sci. USA 86, 99-103.
Knaus, UG, Heyworth, PG, Kinsella, T, Curnutte, JT & Bokoch, GM (1992) J. Biol. Chem. 267, 23575-23582.
Krinsky, NI (1989) Free Radical Biol. Med. 7, 617-635.
Lehr, HA, Becker, M, Marklund, SL, Hubner, C, Arfors, KE, Kohlschatter, A & Messmer, K (1992) Arterioscl. Thromb. 12, 824-829.
Li, Y & Jaiswal, A (1992) J. Biol. Chem. 267, 15097-15104.
Ma, M & Eaton, JW (1992) Proc. Natl. Acad. Sci. USA 89, 7924-7928.
Maiorino, M, Roveri, A, Coassin, M & Ursini, F (1988) Biochem. Pharmacol. 37, 2267-2271.

Maki, H & Sekiguchi, M (1992) Nature 355, 273-275.
Mannervik, B (1985) Adv. Enzymol. 57, 357-417.
McCord, JM & Fridovich, I (1969) J. Biol. Chem. 244,6049-6055.
Meier, B, Radeke, HH, Selle, S, Habermehl, GG, Resch,K & Sies, H (1990) Biol. Chem. Hoppe-Seyler 371, 1021-1025.
Meier, B, Radeke, HH, Selle, S, Younes, M, Sies, H, Resch, K & Habermehl, GG (1989) Biochem. J. 263, 539-545.
Mo, J-Y, Maki, H & Sekiguchi, M (1992) Proc. Natl. Acad. Sci. USA 89, 11021-11025.
Morand, OH, Zoeller, RA & Raetz, CRH (1988) J. Biol. Chem. 263, 11597-11606.
Müller, A, Cadenas, E, Graf, P & Sies, H (1984) Biochem. Pharmacol. 33, 3235-3240.
Murphy, ME & Sies, H (1991) Proc. Natl. Acad. Sci. USA 88, 10860- 10864.
Nakazono, K, Watanabe, N, Matsuno, K, Sasaki, J, Sato, T & Inoue, M (1991) Proc. Natl. Acad. Sci. USA 88, 10045-10048.
Nathan, C (1992) FASEB J. 6, 3051-3064.
Nguyen, T & Pickett, CB (1992) J. Biol. Chem. 267, 13535-13539.
Niki, E (1987) Chem. Phys. Lipids 44, 227-253.
Noack, E & Murphy, M (1991) In Oxidative Stress: Oxidants and Antioxidants (Sies, H, ed.) pp 445-489, Academic Press, London.
Oberley, LW, Leuthauser, SWC, Pasternack, RF, Oberley, TD, Schutt, L & Sorenson, JR (1984) Agents and Actions 15, 536-538.
Packer, JE, Slater, TF & Willson, RL (1979) Nature 278, 737-738.
Park, JW, Ma, M, Ruedi, JM, Smith, RM & Babior, BM (1992) J. Biol. Chem. 267, 17327-17332.
Prochaska, H & Talalay, P (1991) In Oxidative Stress: Oxidants and Antioxidants (Sies, H, ed.) pp 195-211, Academic Press, London.
Prochaska, H, Santamaria, AB & Talalay, P (1992) Proc. Natl. Acad. Sci. USA 89, 2394-2398.
Pryor, WA (1986) Annu. Rev. Physiol. 48, 657-667.
Reichard, P & Ehrenberg, A (1983) Science 221, 514-519.
Remacle, J, Lambert, D, Raes, M, Pigeolet, E, Michiels, C & Toussaint, O (1992) Biochem. J. 286, 41-46.
Retsky, KL, Freeman, MW & Frei, B (1993) J. Biol. Chem. 268, 1304-1309.
Rosenberg, S, Barr, PJ, Najarian, R & Hallewell, R. (1984) Nature 312, 77-79.
Rushmore, TH, Morton, MR & Pickett, CB (1991) J. Biol. Chem. 266, 11632-11639.
Scheer, B & Zimmer, G (1993) Arch. Biochem. Biophys. 302, 385-390.
Scholich, H, Murphy, ME & Sies, H (1989) Biochim. Biophys. Acta 1001, 256-261.
Schreck, R, Albermann, K & Baeuerle, PA (1992) Free Rad. Res. Comm. 17, 221-238.
Schreck, R, Rieber, P & Baeuerle, PA (1991) EMBO J. 10, 2247-2258.
Segal, AW (1989) J. Clin. Invest. 83, 1785-1793.
Sies, H (1986) Angew. Chem. Int. Ed. 25, 1058-1071.

Sies, H (1991) In: Oxidative Stress: Oxidants and Antioxidants (Sies, H, ed.)pp xv-xxii, Academic Press, London.
Sies, H (1993a) Eur. J. Biochem. 215, 213-219.
Sies, H (1993b) Free Radicals Biol. Med. 14, 313-323.
Sies, H & Ketterer, B eds. (1988) Glutathione S-Conjugation: Mechanisms and Biological Significance, Academic Press, London.
Sies, H & Murphy, ME (1991) J. Photochem. Photobiol. B: Biol. 8, 211-218.
Sies, H, Stahl, W & Sundquist, RA (1992) Ann. NY Acad. Sci. 669, 7-20.
Smith, LL (1991) Free Radical Biol. Med. 11, 47-61.
Staal, FJT, Roederer, M & Herzenberg, LA (1990) Proc. Natl. Acad. Sci. USA 87, 9943-9947.
Storz, G, Christman, MF, Sies, H & Ames, BN (1987) Proc. Natl. Acad. Sci. USA 84, 8917-8921.
Storz, G, Tartaglia, LA & Ames, BN (1990) Science 248,189-194.
Stuehr, DJ, Cho, HJ, Kwon, NS, Weise, MF & Nathan, CF (1991) Proc. Natl. Acad. Sci. USA 88, 7773-7777.
Suzuki, YJ, Aggarwal, BA & Packer, L (1992) Biochem. Biophys. Res. Commun. 189, 1709-1715.
Tappel, AL (1962) Vitam. Horm. 20, 493-510.
Tartaglia, LA, Storz, G, Farr, SB & Ames, BN (1991) In Oxidative Stress: Oxidants and Antioxidants (Sies, H, ed.) pp 155-169, Academic Press, London.
Toledano, MB & Leonard, WJ (1991) Proc. Natl. Acad. Sci. USA 88, 4328-4332.
Wefers, H & Sies, H (1988) Eur. J. Biochem. 174, 353-357.
Weglicki, WB, Mak, IT (1992) Mol. Cell. Biochem. 118, 105-111.
Wendel, A, Fausel, M, Safaghi, H, Tiegs, G & Otter, R (1984) Biochem. Pharmacol. 33, 3241-3245.
White, KA & Marletta, MA (1992) Biochemistry 31, 6627-6631.
Wispe, JR, Warner, BB, Clark, JC, Dey, CR, Neumann, J, Glasser, SW, Crapo, JD, Chang, LY & Whitsett, JA (1992) J. Biol. Chem. 267, 23937-23941.
Witmer, CM, Snyder, RR, Jollow, DJ, Kalf, GF, Kocsis, JJ & Sipes, IG eds. (1991) Biological Reactive Intermediates IV, Plenum Press, New York.
Wong, CS, Gamble, JR, Skinner, MP, Lucas, CM, Berndt, MC & Vadas, MA (1991) Proc. Natl. Acad. Sci.USA 88, 2397-2401.
Wong, GHW & Goeddel, DV (1988) Science 242, 941-944.
Wong, GH, Elwell, JH, Oberley, LW & Goeddel, DV (1989) Cell 58, 923-931.
Xanthoudakis, S & Curran, T (1991) EMBO J. 11, 653-665.
Xanthoudakis, S, Miao, G, Wang, F, Pan, Y-C & Curran, T (1991) EMBO J. 11, 3323-3335.
Xie, Q-W, Cho, HJ, Calaycay, J, Mumford, RA, Swiderek KM, Lee, TD, Ding, A, Troso, T & Nathan, C (1992) Science 256, 225-228.
Zhang, Y, Talalay, P, Cho, C-G & Posner, GH (1992) Proc. Natl. Acad. Sci. USA 89, 2399-2403.

Antioxidants: Molecular Interactions Between Vitamins E And C, Biothiols and Carotenes

Lester Packer
Department of Molecular and Cell Biology
University of California,
Berkeley, CA 94720-3200, USA

Antioxidants are coming under increasing scrutiny as preventive and ameliorative agents in a number of chronic degenerative conditions that have an oxidative component. A great deal of epidemiological and clinical evidence suggests that vitamin E and beta-carotene, in particular, may be beneficial.

The dietary intake of foods rich in carotenoids has been associated with a reduced risk of some types of cancer (Zieger, 1989) and cardiovascular disease (Riemersma et al., 1991). An inverse correlation between plasma vitamin E levels and mortality from ischemic heart disease in a cross cultural epidemiological study has also been reported (Gey et al. 1991). Evidence is now emerging that supplementation with these antioxidants may be even more effective than their dietary consumption. A continuing study of 22,000 male physicians in the US yielded an interim finding that those taking a supplement of beta-carotene reduced the occurrence of major cardiovascular events almost 50% (Hennekens and Eberlin, 1985). In a study of over 80,000 women, those who took vitamin E supplements for more than two years had over a 50% decrease in risk of heart attack compared to those whose intake was low[5].(Stampfer et al. 1993)

Because of such results, it is important to understand the molecular mechanisms of antioxidant actions of these compounds. Vitamin E is actually a group of eight compounds, and the carotenoids are an even more diverse group of molecules. In light of their potential in disease prevention, studies of the actions and bioavailability of various forms of these two classes of antioxidants are of considerable interest.

BIOCHEMISTRY OF VITAMIN E - TOCOPHEROLS AND TOCOTRIENOLS

Vitamin E is the collective name for a group of naturally-occurring tocopherols and tocotrienols found abundantly in plants, especially in plant oils (Sheppard et al. 1993). Both tocopherols and tocotrienols have a chromanol head group and a phytyl side chain. Differing

NATO ASI Series, Vol. H 92
Signalling Mechanisms – from Transcription Factors to Oxidative Stress
Edited by L. Packer, K. Wirtz

methyl substitutions around the aromatic ring of the head group determine whether a tocopherol or a tocotrienol is designated alpha, beta, gamma, or delta (α, β, γ or δ) (Packer, 1994). The tocotrienols and tocopherols thus designated have identical structures in the chromanol nucleus but the hydrophobic tail, which anchors vitamin E molecules into membranes or in lipoproteins, differs. In tocotrienols, there are three unsaturated linkages in the tail, whereas in tocopherols it is fully saturated. It is not surprising that some of the biological actions of tocotrienols and tocopherols differ because of this structural difference.

Antioxidant properties of vitamin E

The function of vitamin E was uncertain for quite a while after its discovery in 1922 by Herbert Evans at the University of California, Berkeley (Evans and Bishop, 1922). After its molecular structure was determined, it became evident that the phenolic hydroxyl group located at the C6 position on the aromatic ring was important to its antioxidant properties (Fig 1) (Burton and Ingold, 1981).

α-TOCOPHEROL

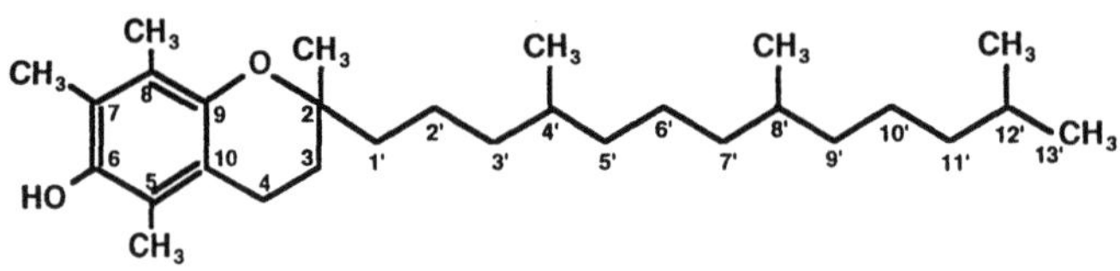

α-TOCOTRIENOL

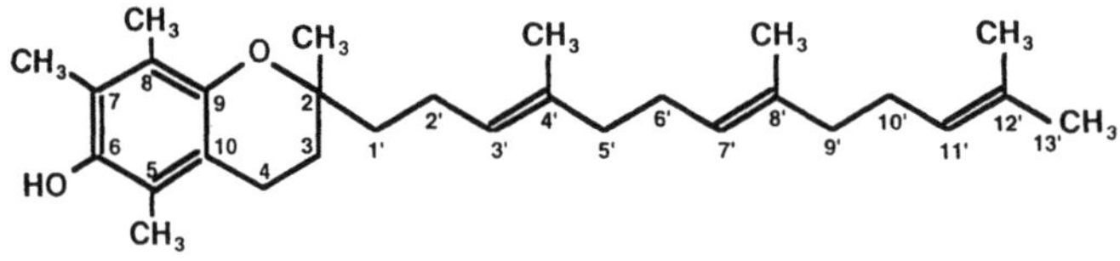

Fig. 1: Molecular structures of D-alpha-tocopherol and D-alpha-tocotrienol

Oxidation of polyunsaturated fatty acids of phospholipids in lipoproteins or in membranes can lead to lipid peroxidation. A polyunsaturated fatty acid will become a lipid radical after hydrogen abstraction by a strong oxidant (such as hydroxyl radical) or interaction with another free radical molecule, like a lipid radical (L•) . The L• reacts rapidly with oxygen to form a peroxyl radical (LOO•). This initiates a chain reaction in which different lipid radicals react with each other to form lipid hydroperoxides (LOOH) or lipid alkoxy radicals (LO•) and the chain of propagation of free radical reactions continues (Halliwell and Gutteridge, 1989). This process destroys lipids and neighboring molecules, including proteins and nucleic acids, with which radicals react. It is generally recognized that lipid peroxidation is an important factor in the progression of many chronic and degenerative diseases of aging (Halliwell and Chirico, 1993).

Vitamin E can break this chain reaction during the propagation phase of these free radical reactions (Burton and Ingold, 1989). However, in the process, vitamin E itself becomes a free radical. It may asked— what has been gained when vitamin E becomes a free radical? The answer is that when vitamin E becomes a free radical, the unpaired electron is delocalized around the aromatic ring. Hence, it is not as reactive a radical as the radical which it quenched. Because of the longer lived, persistent nature of vitamin E chromanoxyl radicals it is possible to regenerate the tocopherol or tocotrienol form by biochemical mechanisms.

Biochemical activities of α-tocopherol and α-tocotrienol

Although tocopherols and tocotrienols are closely related chemically, they have widely varying degrees of biological effectiveness. The potency of alpha-tocotrienol, evaluated by traditional gestation-resorption assays, has been shown to be 32% of the potency of alpha-tocopherol (Machlin, 1984). However, recent studies have challenged this traditional view, and show that tocotrienols may have more potent antioxidant effects than tocopherols, as well as having unique cholesterol-lowering properties.

Indeed, there is indirect evidence of higher antioxidant activity of tocotrienols in comparison with tocopherols. Tatsuta reported in her studies on hemolysis that alpha-tocotrienol showed higher efficiency in protecting red blood cells than alpha-tocopherol in vitro, the opposite of the result observed in vivo (Tatsuta, 1971). Tocotrienols have been reported to possess higher protective activity against cardiotoxicity of the anti-tumor redox cycling drug adriamycin. The cardiotoxicity is believed to be caused by free radicals generated by adriamycin (Komiyama et al., 1989). It was also found that alpha-tocotrienol showed higher inhibitory effect on lipid peroxidation induced by adriamycin in rat liver and murine microsomes than alpha-tocopherol (Kato et al 1985). Comparison of the antioxidative properties of different tocopherols and tocotrienols in prevention the oxidation of lard showed that tocotrienols were more active than the corresponding tocopherols (Seher and Ivanov 1973). The antioxidative efficiency of tocotrienol isomers measured at 110ºC in the dark was in the following order: alpha>beta>gamma>delta tocotrienol.

However, in liposomes prepared from microsomal phospholipids or from dipalmitoylphosphatidylcholine, tocopherol and tocotrienol were equally efficient in inhibiting iron-induced cholesterol 5-alpha-hydroperoxide decomposition (Nakano et al 1980). To evaluate the antioxidant activity of different molecular forms of vitamin E, we have developed a number of in vitro and in vivo assays which assesses their relative reactivity in radical quenching of peroxyl radicals formed by several means.

Antioxidant properties in hexane solution In vitro, we have used parinaric acid as a reporter group for evaluating the ability of vitamin E to prevent free radical reactions (Kuypers et al, 1987). Parinaric acid is a fluorescent lipid molecule whose fluorescence is destroyed when it reacts with free radicals. A constant stream of free radicals can be produced by 2, 2'-azobis (2,4-dimethylvaleronitrile) (AMVN), a lipophilic azo-initiator of peroxyl radicals which, at a given temperature, produces peroxyl radicals at a known rate. If parinaric acid in hexane solution is exposed to AMVN, one can observe, as shown in figure 2, that the fluorescence of parinaric acid is rapidly lost (Fig 2) (Suzuki et al. 1993).

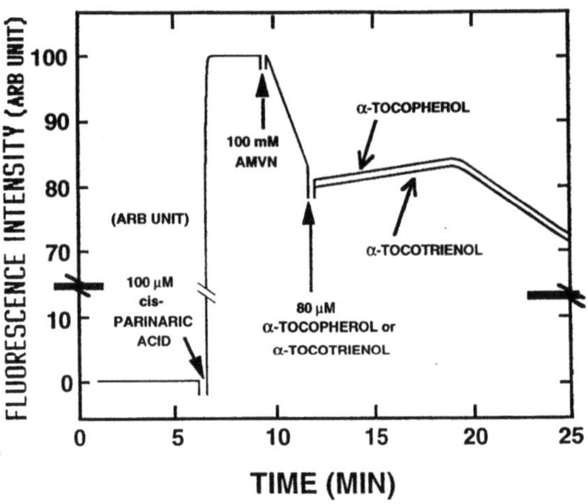

Fig 2: Comparison of the antioxidant activity between α-tocopherol and α-tocotrienol in hexane, using cis-parinaric acid as a reporter molecule. At the arrows, reactants were added to hexane, to the indicated final concentrations. When a-tocopherol or a-tocotrienol was added, the time delay in return of fluorescence decay corresponds to the time when these substances were quenching peroxyl radicals. Because the concentrations of antioxidants and rate of radical production from AMVN under these conditions are known, the stoichiometry of radicals quenched per antioxidant molecule can be calculated. Temp = 40°C.

Upon addition of either alpha-tocopherol or alpha-tocotrienol in a small but equal amounts, a temporary delay in the fluorescence decay rate is observed (until the amount of vitamin E added is exhausted -- whereupon loss of fluorescence commences once again). It is clear from this experiment that the antioxidant potency in quenching peroxyl radicals by d-alpha-tocopherol or d-alpha-tocotrienol in a pure chemical system is identical, and this is in agreement with other studies that show that the length or unsaturation of the hydrocarbon chain of forms of vitamin E does not affect their antioxidant potency in homogenous solution (Burton et al. 1985; Burkalova et al. 1980). It can be calculated that two peroxyl radicals are quenched per mole of either α-tocopherol or α-tocotrienol.

Antioxidant properties in model membranes In a membrane the phytyl side chains may well contribute to a difference in antioxidant activities (Niki et al. 1986; Burton and Ingold, 1986). To investigate this possibility, we have used two different systems. In the first system, cis-parinaric acid was incorporated in dipalmitoleylphosphatidyl choline (DPPC liposomes) AMVN initiated the loss of fluorescence; then either alpha-tocopherol or alpha-tocotrienol was added. Subsequently, the rate of fluorescence loss due to generation of peroxyl radicals was less for a given quantity of alpha-tocotrienol compared with an equal amount of alpha-tocopherol. (Fig. 3)

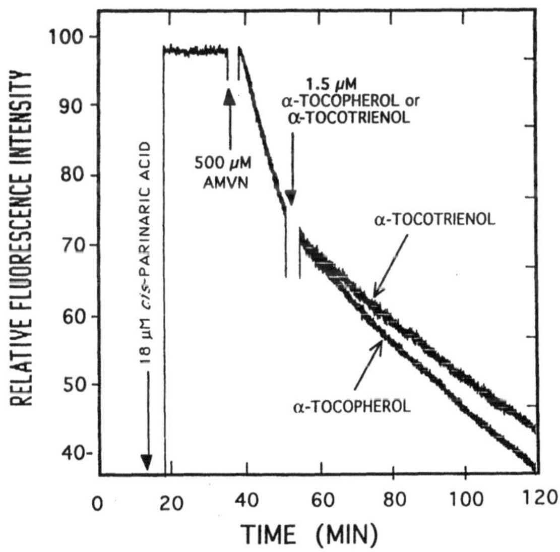

Fig. 3: Comparison of the antioxidant activity between α-tocopherol and α-tocotrienol in dipalmitylphosphatidyl choline liposomes, using cis-parinaric acid as a reporter molecule. Genration of peroxyl radicals by AMVN was detected by monitoring decay of fluorescence of cis-parinaric acid (λ_{excit} 328 nm, λ_{emiss} 415 nm). The reaction mixture (2 ml) contained AMVN (500uM) and cis-parinaric acid (18uM) in 20mM Tris-HCl (pH 7.4). α-Tocopherol or α-tocotrienol (1.5 uM) dissolved in ethanol was incorporated into the membrane by further sonicating the DPPC liposomes, which were initially prepared by sonication with AMVN under nitrogen. Temperature = 40 °C.

One cannot determine the stoichiometry of radical quenching in the membrane system owing to the absence of a clear discontinuity in the fluorescence loss (Suzuki et al. 1993). Therefore, in the second system, dipalmitoleylphosphatidyl choline (DPPC) or dioleylphosphatidyl choline (DOPC) liposomes containing various amounts of alpha-tocopherol or alpha-tocotrienol were used. However, in these studies, generation of peroxyl radicals, induced by AMVN, was detected by monitoring the chemiluminescence of luminol at various temperatures. It was found that both alpha-tocopherol and alpha-tocotrienol quenched the AMVN induced luminol-enhanced chemiluminescence in DPPC liposomes with a half quenching concentration of 50 and 15 nanomolar for alpha-tocopherol and alpha-tocotrienol, respectively. Similar data were also obtained using DOPC liposomes. Such findings indicate that alpha-tocotrienol is a more efficient scavenger of peroxyl radicals, than alpha-tocopherol in these model membrane systems. (Fig 4)

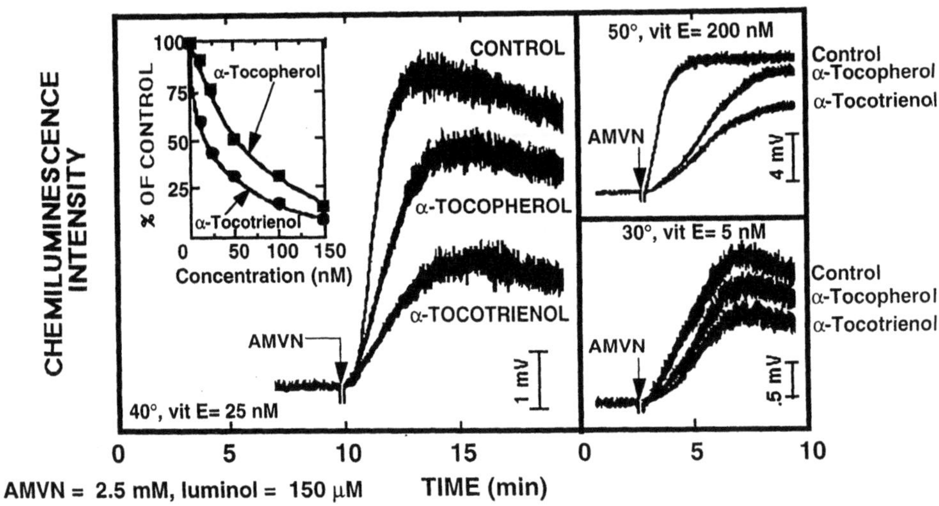

Fig. 4: Chemiluminescence assays for the comparison of the antioxidant activity of alpha-tocopherol and alpha-tocotrienol in DPPC liposomes. Generation of peroxyl radicals by AMVN (2.5 mM) was detected by monitoring the chemiluminescence of luminol (150 uM) at various temperatures, as noted in the Figure. At 40 °C, [vitamin E] = 25 nM; at 50 °C, [vitamin E] = 200 nM; at 30 °C, [vitamin E] = 5 nM.

It is clear from these experiments that at all concentrations of the vitamin E forms used, alpha-tocotrienol is always more effective in quenching chemiluminescence than alpha-tocopherol.

Antioxidant properties in microsomes The antioxidant activity of alpha-tocopherol and alpha-tocotrienol was compared previously by Serbinova, et al (1991) who observed, using several systems for inducing lipid peroxidation (ascorbate + Fe^{+2}, or NADPH + Fe^{+2}) in isolated microsomal suspensions, that lipid peroxidation was inhibited to a much greater extent by alpha-tocotrienol compared with alpha-tocopherol. Quantitative comparison of the antioxidant potencies of alpha-tocopherol and alpha-tocotrienol in a more physiological system (Fe(II)+ascorbate or Fe(II)+NADPH induced lipid peroxidation in rat liver microsomes) has shown that alpha-tocotrienol exerts much higher antioxidant activity than alpha-tocopherol in these systems. The concentrations of alpha-tocopherol producing 50% inhibition (K_{50}) are 40- and 60-times higher than those for alpha-tocotrienol for (Fe^{2+} + NADPH)- and (Fe^{2+} + ascorbate)-dependent lipid peroxidation, respectively (Serbinova et al. 1991). Similar results were obtained for rat heart mitochondria and microsomes (**Table 1**)

TABLE 1

CONSTANT FOR 50% INHIBITION OF (Fe(II) + Ascorbate) INDUCED LIPID PEROXIDATION BY ALPHA-TOCOPHEROL AND ALPHA-TOCOTRIENOL IN RAT HEART MICROSOMES AND MITOCHONDRIA*

	CONSTANTS, M	
	microsomes	mitochondria
Alpha-tocopherol	7.1×10^{-6}	2.8×10^{-6}
Alpha-tocotrienol	0.4×10^{-6}	0.3×10^{-6}

Time of lipid peroxidation - 5 min. Incubation medium contained 0.5 mg protein/ml, 10μM $FeSO_4 \times 7H_2O$, ascorbate 0.5 mM in Na,K - phosphate buffer, pH 7.4 at 37ºC. N=6

*Adapted from: Methods in Enzymology, 234D: 331-345.

Feeding studies Because of the greater antioxidant potency of tocotrienols in vitro, it is of interest to determine the extent of their biological assimilation, compared to tocopherols. For this reason, we studied the distribution of pure alpha-tocopherol and alpha-tocotrienol in rat tissues obtained after feeding these forms of vitamin E to vitamin E-deficient animals.

Sprague-Dawley rats were first fed a vitamin-E deficient diet until the level of vitamin E in tissues was 0.05-0.1 nmols/mg protein (Serbinova and Packer, 1994). They were then fed alpha-tocotrienol or alpha-tocopherol- supplemented diet (3 g/kg diet each) for 8 weeks.. In the liver alpha-tocotrienol was observed after 2 weeks of supplementation and reached a maximum of 1.1 nmoles/mg protein after 5 weeks. In the heart, after 2 weeks the level of alpha-tocotrienol was 0.1 nmoles/mg protein and after 7 weeks reached a maximal level of 0.9 nmoles/mg protein. The maximal levels of tocotrienol accumulated in the heart was about 2-3 times lower than the level of alpha-tocopherol and its accumulation was slower than that

of alpha-tocopherol. When susceptibility to induced peroxidation was compared for hearts with similar concentrations of alpha-tocopherol and alpha-tocotrienol, tocotrienol was found to confer more protection than tocopherol (Fig. 5)
No detectable amounts of tocotrienols were found in brain, kidney and skin.

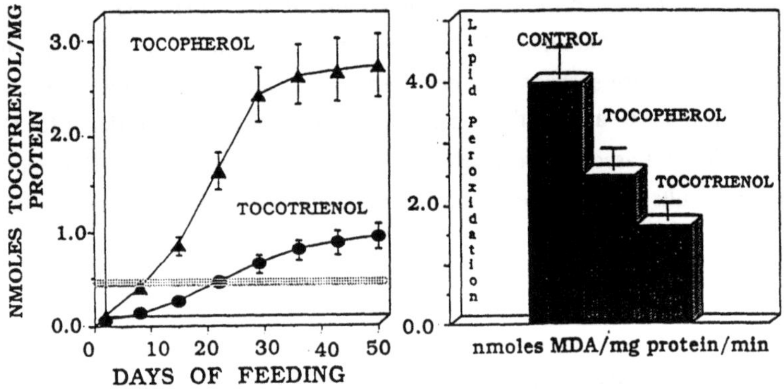

Fig. 5: Accumulation of alpha-tocopherol and alpha-tocotrienol in rat heartsand their susceptibility to *in vitro* (Fe (II) + Ascorbate)-induced lipid peroxidation. Incubation medium contained 0.5 mg protein/ml, 10µM $FeSO_4 \times 7H_2O$, ascorbate 0.5 mM in Na,K -phosphate buffer, pH 7.4 at 37ºC. N=6

Because of its in vitro antioxidant potency, and because it does appear to be assimilated from the diet in significant amounts, alpha-tocotrienol would be predicted to afford even more protection than alpha-tocopherol against conditions involving an oxidative component. One such condition is ischemia-reperfusion injury of the heart, widely believed to be primarily caused by reactive oxygen species.

We studied the protection afforded by palm oil vitamin E (POE), a mixture of 55% tocotrienols and 45% tocopherols, in Langendorff perfused rat hearts subjected to 40 minutes of global ischemia (Serbinova et al. 1992). After 40 min of ischemia and 20 min of reperfusion, the mechanical recovery rate of hearts from the POE-supplemented group (7g POE/kg diet) was 90% as compared to the 24% recovery of the hearts obtained from rats fed the normal diet (30 IU alpha-tocopherol/kg) and to the 80% recovery of the hearts obtained from rats kept on alpha-tocopherol (10 000 IU/kg) supplemented diet (Serbinova et al. 1992).

These results demonstrated that palm oil vitamin E was more efficient in the protection of the isolated Langendorff heart against ischemia-reperfusion injury than tocopherol, as measured by mechanical recovery. Palm oil vitamin E also completely suppressed LDH enzyme leakage from ischemic hearts, prevented the decrease in ATP and creatine phosphate levels and inhibited the formation of endogenous lipid peroxidation products (Serbinova et al. 1992). Hence, tocotrienols from a natural source such as palm oil may have significantly more protective effect in ischemia-reperfusion injury than do tocopherols.

Vitamin E recycling Vitamin E is usually present in low molar ratios in natural membranes, perhaps one molecule per 2000-3000 phospholipids (Esterbauer et al. 1992). Lipid peroxyl radicals can be generated in membranes at the rate of 1 to 5 nmol per mg of membrane protein per minute, yet destructive oxidation of membrane lipids does not normally occur, nor is vitamin E rapidly depleted. Furthermore, vitamin E deficiency is difficult to induce in adults animals. In response to this seeming paradox, models of vitamin E recycling, whereby the intact molecule is regenerated from its radical formed, have been proposed, and there is a great deal of experimental evidence to support such models. In these models (**Fig. 6**), the vitamin E radical can be converted back to vitamin E by ascorbate or by ubiquinols, or by enzymatic pathways.

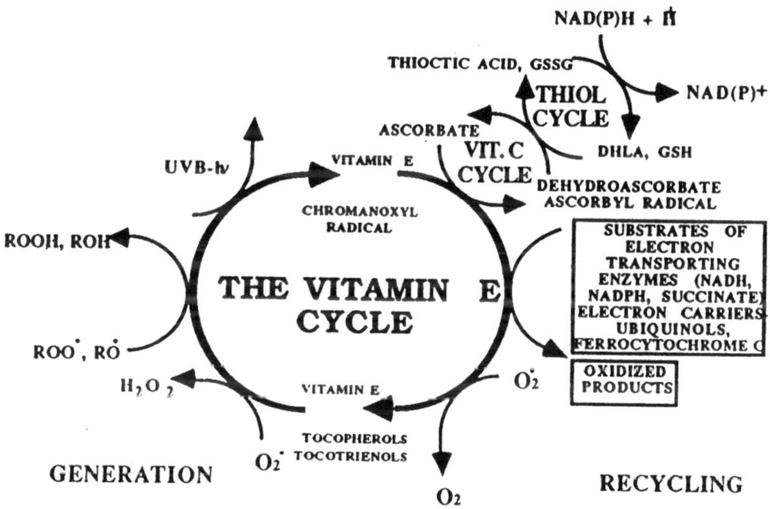

Fig. 6: Pathways for generation of chromanoxyl radicals and their recycling to vitamin E

The rate of regeneration, or recycling, of the vitamin E radicals that form during its antioxidant action may affect both its efficiency in antioxidant action and its lifetime in biological systems. To study the recycling and regeneration of vitamin E during its antioxidant action, we have carried out studies using human low density lipoprotein suspensions (LDL) and in membranes from animal tissues that had been enriched with vitamin E by feeding studies.

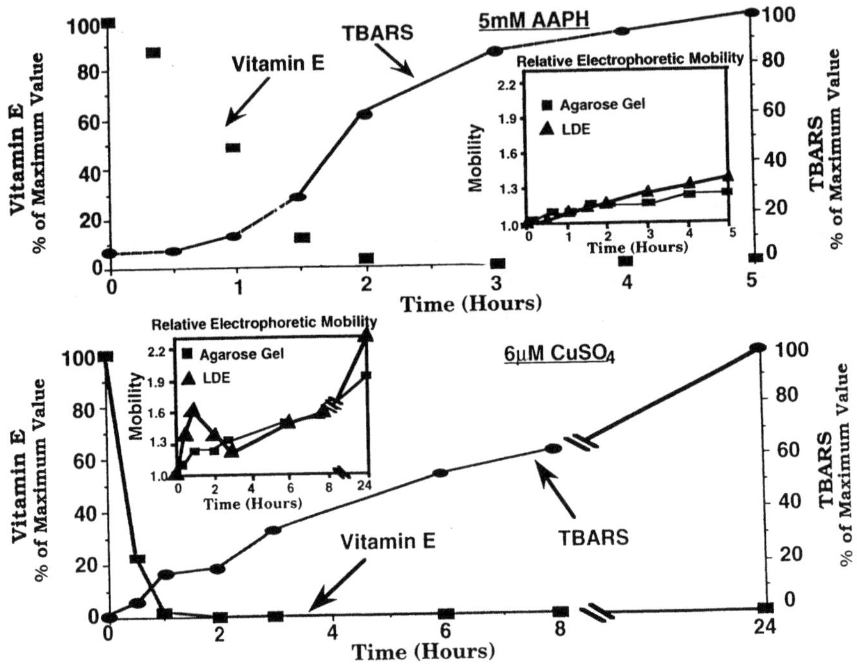

Fig. 7. Kinetics of vitamin E disappearance and TBARS accumulation during AAPH and $CuSO_4$ induced oxidation of human LDL. LDL concentration = 1mg protein (apo B)/ml, AAPH and $CuSO_4$ were at the concentrations indicated. Insets: change in negative charge on LDL measured on agarose gels or by laser doppler electrophoresis (LDE).

We have compared the ability of d-alpha-tocopherol, d-alpha-tocotrienol and a short chain homologue of alpha-tocopherol, C6 chromanol (shorter phytyl tail), to be recycled after a given amount of vitamin C is added to the suspension (Kagan et al. 1992b). That is, in the presence of vitamin C the vitamin E radical signal is suppressed, but when the vitamin C is consumed, the vitamin E radical appears because the vitamin C is no longer present to regenerate tocopherol from the tocopheroxyl radical. We term this process "recycling". Figure 9 shows the kinetics of vitamin E radicals, with and without vitamin C, during the time course of oxidation of LDL.

In such experiments, we have found that alpha-tocotrienol is more efficiently recycled than alpha-tocopherol (note the longer lag time before the reappearance of the ESR signal); the short chain homologue of vitamin E displays even greater capacity for recycling.

LDL is the transport vehicle for vitamin E to the peripheral tissues (Traber and Kayden 1984). In human lipoproteins which contain perhaps 2000 - 2500 molecules, the molar ratio of vitamin E to phospholipids is between 0.5 and 1 mole percent. This is much higher than the ratio in membranes and makes LDL an ideal preparation in which to stuydy vitmain E oxidation, using ESR detection (vitamin E concentrations in membranes are too low to allow this). In a typical human LDL preparation, on average there are about 7 molecules vitamin E per LDL particle, or about 88% of all the lipophilic antioxidants in the particle (Esterbauer et al. 1992). The other antioxidants in LDL are the carotenoids; perhaps up to eight different carotenoids may be observed. On the average, they constitute about 10% of the antioxidants. Also, there are trace amounts of ubiquinol 10, usually about one percent of the total antioxidants.

LDL suspensions can be oxidatively stressed. A commonly used method is copper-induced oxidation, and we also use water soluble azo-initiators such as 2, 2'-azobis (2-amidinopropane) hydrochloride (AAPH) for generation of peroxyl radicals. This causes the loss of aqueous antioxidants, like Vitamin C and others. When these defenses are diminished, the vitamin E content of the LDL suspension begins to be drastically decreased. **Figure 7** shows the kinetics of vitamin E disappearance and accumulation of thio-barbituric acid reactive substances (TBARS, markers of lipid peroxidation) during AAPH and copper sulfate induced oxidation of human LDL. As the levels of vitamin E are depleted, accumulation of lipid peroxidation products (TBARS) are observed. Also, changes in the physiochemical properties of the surface of LDL preparation can be seen by agarose gel electrophoresis, or more sophisticated methods such as laser doppler electrophoresis (inset, Fig. 7), which shows that the surface of LDL become more net negatively charged as oxidation process (Arrio et al. 1993).

As vitamin E is depleted during the oxidation of LDL, significant amounts of chromanoxyl radicals are present, which can be observed directly by ESR spectroscopy. This is because the molar ratio of vitamin E is high enough without supplementation in human LDL to directly observe such signals. Note this is not true for natural biological membranes, because the molar ratio of vitamin E is too low for detection by ESR. If during this time, radical-radical reactions of vitamin E with itself or other lipid radicals occur, vitamin E is degraded and slowly lost from the system. If however, regeneration of vitamin E is accomplished by adding ascorbic acid, the steady state signal of the vitamin E radicals is lowered drastically, such that the vitamin E radical signal cannot even be detected by ESR (Fig. 8).

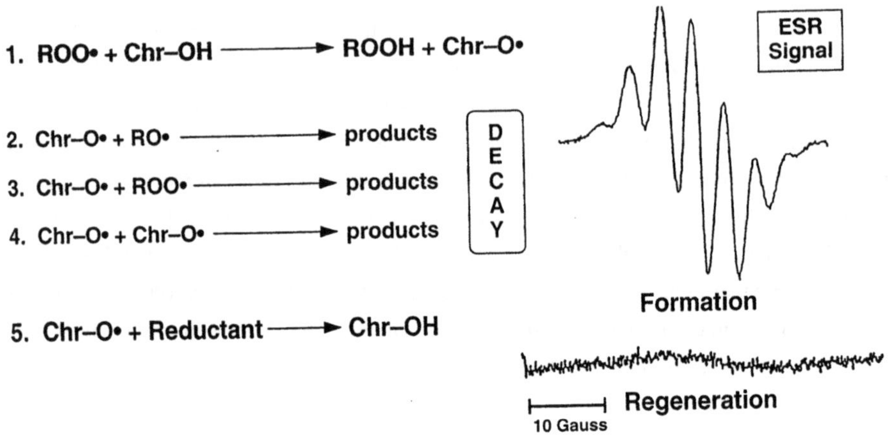

Fig. 8: Some reactions of chromanoxyl radicals in the course of lipid peroxidation

Biophysical basis of the differences between tocopherol and tocotrienol To what can we attribute this greater antioxidant activity of tocotrienol? ESR studies provide clues.

Slow motions of the 5-doxylstearic acid (5-DSA) spin label were studied in DPPC

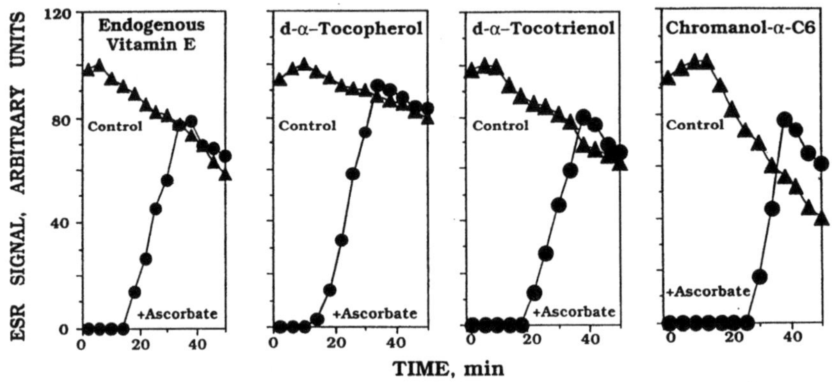

Fig. 9: Time-course of chromanoxyl radical ESR signals generated by the lipoxygenase + linolenic acid oxidation system in LDL from endogenous vitamin E or exoogenously added α-tocopherol, α-tocotrienol, or chromanol-α-C6. Effect of ascorbate. In each case ascorbate (1.5mM) was added before the recording of the first ESR spectrum. The concentrations of endogenous vitamin E were 6.2 nmol/mg protein in the sample to which no exogenous chromanols was added and 2.5 nmol/mg protein in the samples to which chromanols (80 nmol/mg protein) were added. All values are given as a percentage of the maximal magnitude obtained.

liposomes, using saturation transfer ESR studies, which provides information about the reorientation dynamics. Typical saturation transfer spectra are shown in figure 8. The positions of the central field [C/C'] and low field [L/L"] parameters are indicated in the control sample at 5 °C. (Fig 10)

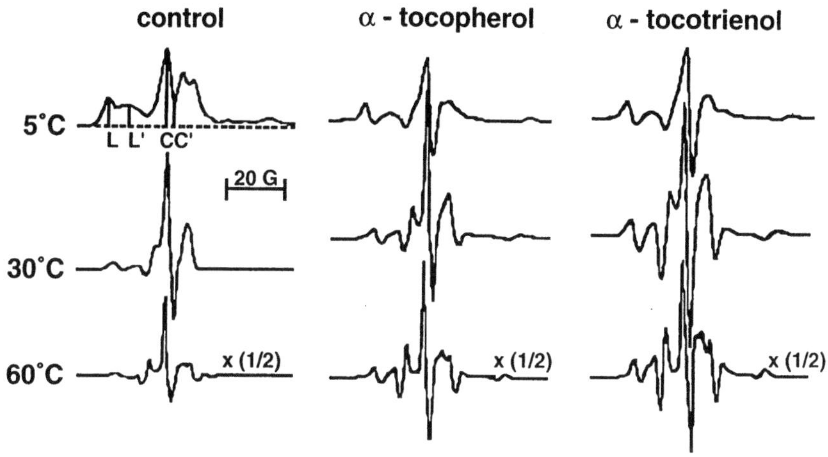

[DPPC] = 68 mM; [5 - doxylstearic acid (5-DSA)] = 0.68 mM;
[α - tocopherol] = 13.6 mM; [α - tocotrienol] = 13.6 mM

Fig. 10: Saturation transfer ESR spectra of 5-doxylstearic acid, at 5°, 30° and 60° C made in DPPC liposomes in the absence and presence of either alpha-tocopherol or alpha-tocotrienol. [DPPC] = 68 mM; [5-DSA] = 0.68 mM; [α-tocopherol] = 13.6 mM ; [α-tocotrienol] = 13.6 mM

Temperature dependence of the central field [C/C'] and low field [L/L"] parameters were calculated for each of these spectra and are plotted in figure 11.

Using the data shown in figure 9, the anisotropic motion at various temperatures was computed by the method of Marsh (1980), and is shown in figure 12..

Higher anisotropic motion was found at all temperatures in the presence of vitamin E. However, tocotrienol had a greater anisotropic effect than did tocopherol at all temperatures. Although introduction of alpha-tocopherol increased the anisotropy of motion in the gel phase, as also shown by Severcan and Cannistraro (1990) our data established that the motion anisotropy is greater with alpha-tocotrienol. Thus, a clear distinction between the effects of molecular organization between alpha-tocopherol and alpha-tocotrienol were identified at the molecular level in these membranes. Alpha-tocotrienol causes a higher degree of membrane disorganization, which allows it an increased mobility as compared with alpha-tocopherol. Further, this increased mobility should allow alpha-tocotrienol a greater accessibility to radicals.

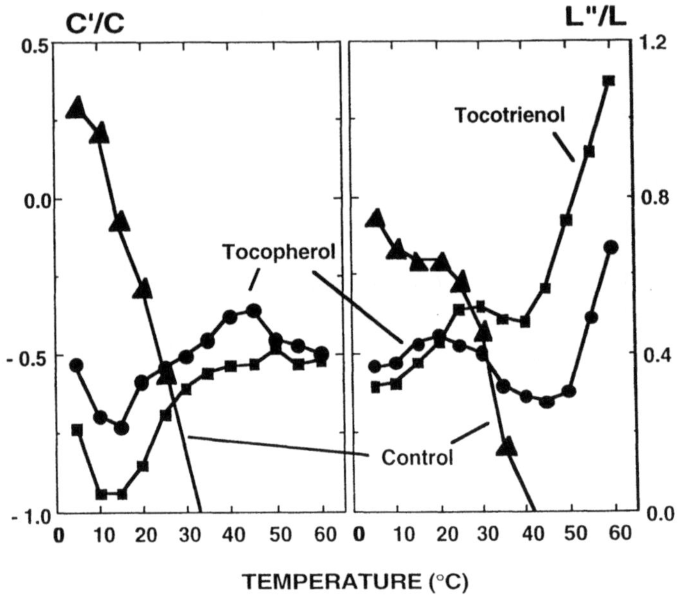

Fig. 11: Temperature dependence of parameters C'/C and L''/L from saturation transfer ESR spectra of 5-doxylstearic acid. [DPPC] = 68 mM; [5-DSA] = 0.68 mM; [α-tocopherol] = 13.6 mM ; [α-tocotrienol] = 13.6 mM

We speculate that the differences in membrane reorientation dynamics and ordering that have been identified correlate with the greater antioxidant activity of δ-alpha-tocotrienol as compared with δ-alpha-tocopherol.

Conclusions From these experiments, we have concluded that the greater antioxidant activity seen in isolated membranes and lipoproteins in vitro with alpha-tocotrienol as compared with alpha-tocopherol is due to a number of different factors: its more uniform distribution in the membrane bilayer, which has been observed in fluorescence self-quenching studies (Serbinova et al. 1991), the more effective collision of radicals with vitamin E as determined by its behavior in model membranes (described above), and the greater recycling activity of chromanoxyl radicals by Vitamin C, shown above for LDL (Kagan et al. 1992b) and also occurring in membranes (Serbinova et al. 1991). We have found that the greater recycling activity correlates with the increased inhibition of lipid peroxidation. The different molecular structures of tocotrienols and tocopherols described here appear to account in vitro for its more effective antioxidant activity in membranes.

Action of vitamin E as a biological response modifier

Besides its antioxidant action, vitamin E has been observed to have effects on transcellular signaling mechanisms and gene expression. It is not entirely clear whether the effects of vitamin E observed in these systems are due to its antioxidant activity or to other properties of vitamin E. Some examples of its action in this regard are given below.

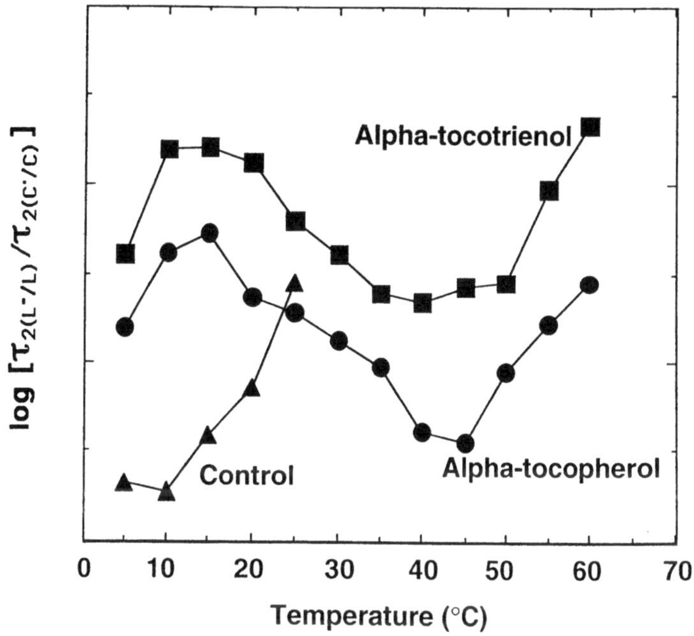

Fig. 12: Temperature dependence of log of the ratio of correlation times derived from L''/L and C'/C DPPC liposomes in the presence of 5-doxylstearic acid and either alpha-tocopherol or alpha-tocotrienol. [DPPC] = 68 mM; [5-DSA] = 0.68 mM; [α-tocopherol] = 13.6 mM ; [α-tocotrienol] = 13.6 mM

Together with A. Pentland's laboratory (Pentland et al. 1992), we have shown that various structural homologues of vitamin E with alterations of its sidechain have effects on pathways involving arachidonic acid metabolism. Suppression of ionophore stimulated prostaglandin synthesis in human keratinocytes by the short chain homologue of vitamin E, PMC, the alpha C6 and alpha C11 homologues were readily observed. These effects appear to be directly attributed to inhibition of phospholipase A2 activity, since the isolated enzyme was inhibited by PMC. The short chain homologues are more effective than the natural form of vitamin E, which we have attributed to the poorer cellular permeability of the natural form.

Another transcellular signaling mechanism, protein kinase C (PKC) has been studied in Angelo Azzi's laboratory (Chatelain et al. 1993). They demonstrated that various tocopherols and tocotrienols can down-regulate protein kinase C activity in a variety of cell types, e.g., smooth muscle cells in vitro. In this regard, alpha-tocopherol exhibits a much greater inhibition that beta-tocopherol (which is almost without action) whereas alpha-tocotrienol has only a small effect. These highly specific effects of different tocopherol and tocotrienol derivatives on PKC activity have been shown to be linked with specific times in the cell growth cycle. Investigations in Azzi's laboratory suggest that this may not be due to the antioxidant properties of vitamin E.

Vitamin E and its homologues have also been shown to regulate the activation of oxidatively induced nuclear transcription factors, such as NFκB. When NFκB is activated by tumor necrosis factor or phorbol esters, its activation can be prevented by forms of vitamin E which are cell permeable. These include vitamin E acetate and vitamin E succinate and the short chain homologue of vitamin E, PMC. In addition, it has been observed (Suzuki and Packer, 1993) that vitamin E succinate has a specific effect, not shown by PMC or vitamin E acetate, in inhibiting the binding of NFkB to DNA. The DNA binding inhibition is not demonstrated by these other homologues. It is difficult to explain these effects purely based on the antioxidant properties of vitamin E.

The actions of vitamin E as a biological response modifier are summarized in Fig. 13.

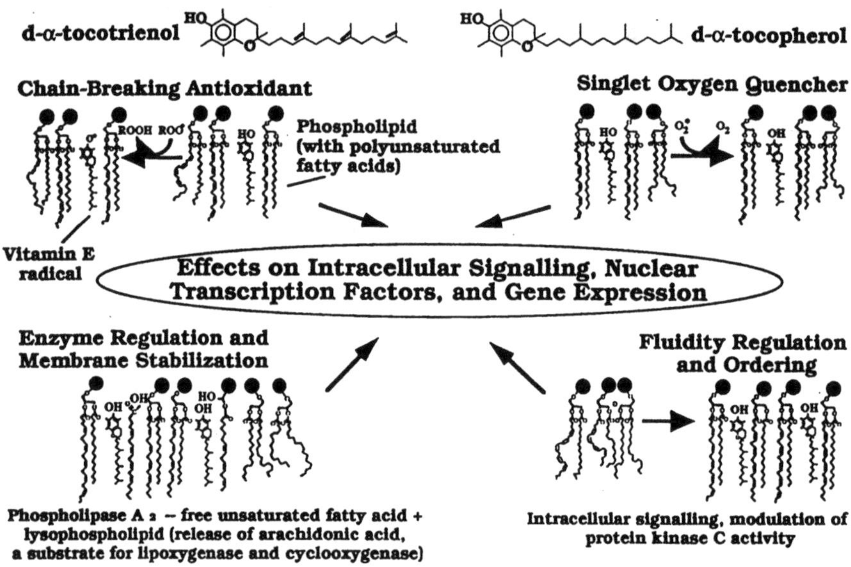

Fig. 13: Biological roles of vitamin E

In this diagram, alpha-tocopherol and alpha-tocotrienol are depicted to be active as a chain breaking antioxidants, as quenchers of singlet oxygen in the lipophilic domain, and also to exhibit effects on transcellular signaling, nuclear transcription and gene expression.

Thus vitamin E exhibits actions in biological systems from the membrane to the gene. All of these need to be factored together in evaluating its biological effects in aging, chronic and degenerative diseases, and in acute clinical conditions where vitamin E exerts actions in health and disease.

CAROTENOIDS

Carotenoids are primarily symmetrical, C-40 polyisoprenoid structures with an extensive conjugated double bond system **(Fig 14)**. They are widely distributed in nature and are synthesized by photosynthetic organisms (Olson, 1992).

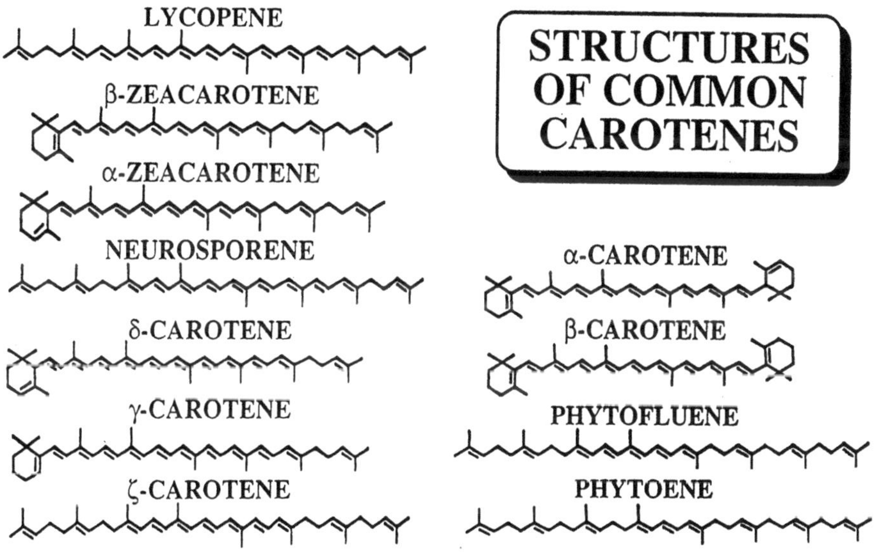

Fig. 14: Structures of common carotenes

Carotenoids have aroused considerable interest as antioxidants in the past decade. Epidemiological studies indicate that they may play a protective role against a variety of chronic degenerative diseases in which reactive oxygen species are thought to play a role (Gey, 1993). In our laboratory we have taken a multi-faceted approach to the study of these compounds, as well as retinoids, investigating their antioxidant capacity in a variety of *in vitro* systems, in LDL, and measuring the uptake of carotenoids from dietary sources. These experiments make it clear that the system in which the activity of these compounds is investigated greatly influences their antioxidant capacity, and conclusions about antioxidant potency can only be drawn after studies in a number of *in vitro* systems, as well as *in vivo* experiments, are conducted.

In vitro assessement of antioxidant capacity

In assessing *in vitro* antioxidant activity, several assays should be used. In this way, factors such as solubility or steric hindrance, which may be of overiding importance in one environment but not another, can be varied and the antioxidant capacity of a substance in a variety of milieus may be evaluated. For these reasons, we have used four different types of assays to measure *in vitro* antioxidant activity of carotenoids: interaction with the stable free radical 1,1-diphenyl-2-picrylhydrazyl radical (DPPH), interaction with the water-soluble fluorescent reporter molecule phycoerythrin, interaction with the lipid-soluble reporter molecule parinaric acid (in hexane and in liposomes), and chemiluminescence of DOPC liposomes.

Retinoids and DPPH radicals DPPH is a stable free radical whose signal is easily followed by electron spin resonance (ESR); antioxidants can quench the signal (Hiramatsu and Packer 1990). The concentration dependence of this quenching is the basis for the assay. In the experiments described here, 30 ul of DPPH (200 uM) and 30 ul of sample in ethanol were mixed (30 ul of ethanol only for controls) and 50 ul of this solution was placed in a capillary tube, which was placed in the ESR cavity. Manganese oxide was used as an internal standard; its peak height was assumed to be constant and the peak height of the DPPH control was taken as 100%. **Table 2** shows the results of this assay for 11 retinoids. Only Ro 22-1318 and Ro 4-3870 (13-*cis*-retinoic acid), at a concentration of 2.5 mM, had any quenching effect on the DPPH signal. For comparison, glutathione, vitamin C and vitamin E completely quenched the signal at this concentration; hence, in this assay the antioxidant action of Ro 4-3870 and Ro 22-1318 are small in comparison to vitamins C and E, and glutathione.

Phycoerythrin and parinaric acid assays Phycoerythrin is a water-soluble fluorescent compound whose fluorescence is quenched upon oxidation. Thus, by measuring the effect of an antioxidant on the rate of decay of fluorescence in the presence of an oxidizer, one may evaluate the antioxidant potency of the substance (Glazer 1990). In the case of phycoerythrin, sparingly water-soluble substances such as retinoids and carotenoids may be used as long as the concentration is kept below the critical micellar concentration. The antioxidant efficiency of a compound may be evaluated in terms of number of moles of peroxy radical quenched per mole of antioxidant added; **Fig. 15** illustrates the effect of adding Trolox, a water-soluble analog of vitamin E, to a phycoerythrin system in which free radicals were generated by the decay of the temperature-sensitive compound 2, 2'-azobis(2-amidino-propane)-HCl (AAPH). In Fig. 15A the Trolox was added before radical generation was initiated, whereas in Fig. 15B it was added during the course of the reaction.

TABLE 2
EFFECT OF RETINOIDS ON THE ELECTRON SPIN RESONANCE SIGNAL OF 1,1-DIPHENYL-2-PICRYLHYDRAZYL (DPPH) RADICAL[a]

Common name	Ro number	Inhibition (% of control)
Furyl analog of retinoic acid	22-1318	77.3
13-*cis*-Retinoic acid	04-3870	81.7
Acitretin	10-1670	95.6
Etretin	10-9359	97.5
all-*trans*-Retinoic acid	01-5488	99.7
Arotinoid acid	13-7410	100.1
Theinyl analog of retinoic ester	21-6583	101.5
Arotinoid sulfone ester	15-1570	111.7
Motretinide	11-1430	112.3
Termarotene	15-0778	115.1
Arotinoid ester	13-6298	118.8

[a] Retinoids at 2.5 mM were added to a 100 uM solution of DPPH. Each value is the mean of two or three determinations.

Trolox is a highly efficient antioxidant, quenching 2 moles of free radicals per mole of Trolox. Table 3 shows the antioxidant efficiency for four retinoids and for Trolox as evaluated by this assay. Two retinoids which showed no activity in the DPPH assay (all-

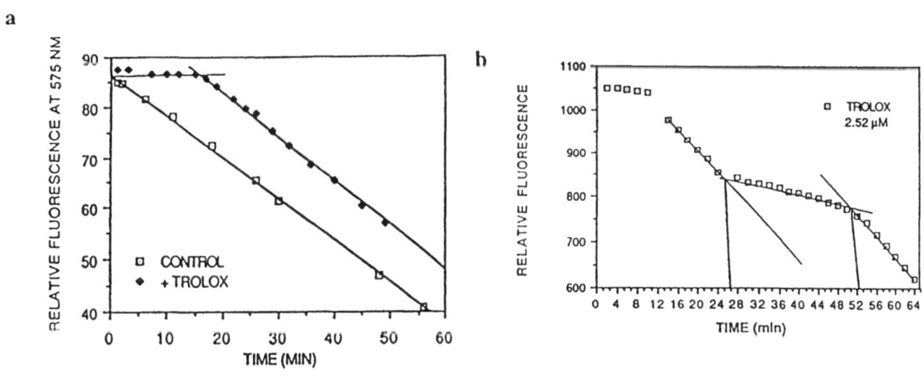

Fig. 15. Fluorescence decay assay of Trolox (a vitamin E analog) antioxidant activity

trans-Retinoic acid and Acitretin) are quite effective in the paraneric acid assay; in fact, acitretin had the highest efficiency in quenching free radicals of the retinoids tested. By this assay, all four retinoids had high antioxidant efficiency (between 1 and 1.6), though not as great as Trolox (antioxidant efficiency of 2).

TABLE 3
REACTION STOICHIOMETRY OF ANTIOXIDANTS, USING PHYCOBILIPROTEIN FLUORESCENCE DECAY ASSAY

Substance	Amount added (umol)	Reaction stoichiometry (Mol peroxy radicals consumed/mol)
13-*cis*-Retinoic acid	13.40	1.10
	2.52	1.04
all-*trans*-Retinoic acid	7.92	1.06
	2.52	1.05
Acitretin	2.50	1.60
Furyl analog	2.50	1.52
Trolox	2.52	2.00

The lipid-soluble molecule *cis*-parinaric acid is a fluorescent molecule whose fluorescence is quenched on oxidation, similar to phycoerythrin. Assays of antioxidant efficiency may be carried out in hexane solution, using the lipid-soluble thermo-labile free radical initiator 2,2'-azobis(2,4-dimethylvaleronitrile) (AMVN) (Tsuchiya et al. 1993). In addition, the generation of peroxyl radicals by AMVN and interaction of peroxyl radicals with retinods and carotenoids in DOPC liposomes may be followed by chemiluminescence produced in the presence of luminol and detected with a luminometer (Tsuchiya et al. 1993). The antioxidant is incorporated into the liposomes.

In the hexane system, addition of carotenoids but not retinoids delayed the AMVN-induced *cis*-parinaric acid fluorescence decay (**Fig. 16**). This indicates that the carotenoids have a high scavenging capacity for hydrophobic radicals but retinoids are not as efficient. The values for the number of moles of peroxyl radicals scavenged per mole of antioxidant may be calculated and are shown in Table 4 for a number of retinoids, carotenoids, and, for comparison, α–tocopherol and Trolox . Trolox has the same chromanol nucleus as α–tocopherol, with similar antioxidant activity, yet it had no antioxidant activity in this system, owing to its hydrophilic character, compared to an efficiency of 2 for α–tocopherol, as expected. This illustrates the necessity of testing antioxidants in a variety of aqueous and lipid environments to adequately reflect the variety of cellular and extracellular environments in which they may operate. The contrast between the high efficiencies of the carotenoids and the lack of efficiency of the retinoids in this system may be partially explained by the number of conjugated double bonds in the two classes of compounds. Retinoids have 5 conjugated

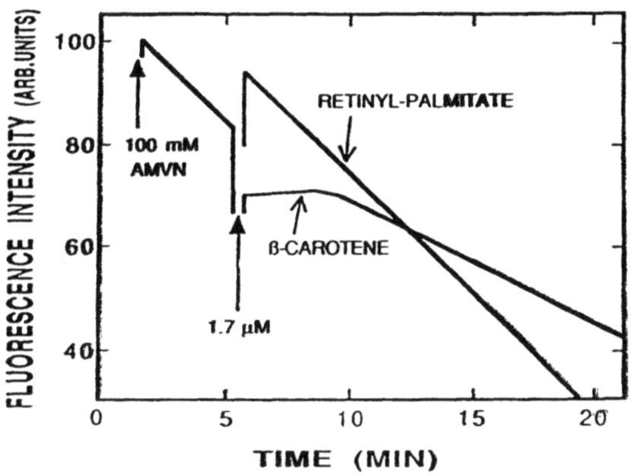

Fig. 16. Effects of β–carotene and retinyl palmitate in the cis-parinaric acid fluorescence decay assay.

TABLE 4
REACTION STOICHIOMETRY OF CAROTENOIDS AND RETINOIDS WITH PEROXYL RADICALS IN HEXANE AS COMPARED TO ALPHA-TOCOPHEROL

Substance	Reaction stoichiometry (mol peroxyl radicals consumed/mol, mean ± SE)	No. of conjugated double bonds
β-Carotene	30.8 ± 0.8	11
Crypoxanthin	29.4 ± 1.2	11
Lutein	29.6 ±1.6	11
Canthaxanthin	30.2 ± 0.8	11\
Lycopene	27.2 ± 1.0	11
Zeaxanthin	31.4 ± 0.8	11
Retinyl-palmitate	0.0	5
Retinoic acid	0.0	5
Retinol	0.0	5
α-Tocopherol	2.0 ± 0.0	—
Trolox	0.0	—

double bonds while carotenoids have 11; thus, carotenoids may compete highly efficiently with *cis*-parinaric acid (which has only 4 conjugated double bonds) for AMVN-derived peroxyl radicals while retinoids cannot compete so overwhelmingly. Still, there must be some other factor or factors at work to explain the efficiency for the retinoids. There must also be some other factor at work to explain the extremely high efficiency of the carotenoids, around 30 moles of radical quenched per mole of carotenoid. Possibly, oxidation products of carotenoids may possess enough quenching potential to still compete successful for peroxyl radicals with *cis*-parinaric acid. In this system carotenoids were about fifteen times as efficient as α–tocopherol in radical quenching.

In the chemiluminescent DOPC liposome system, retinoids and carotenoids inhibited the chemiluminescence caused by radical formation, as did α–tocopherol. The inhibition was concentration-dependent in all cases (**Fig. 17**).

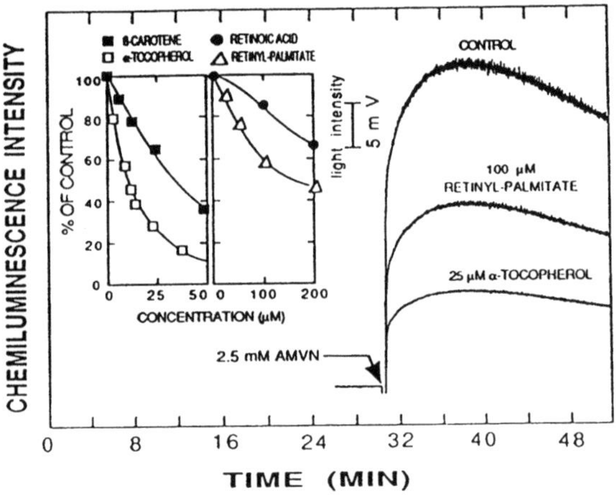

Fig. 17. Comparison of the effects of α–tocopherol, β–carotene, retinyl-palmitate, and retinoic acid on luminol-amplified chemiluminescence in DOPC liposomes. Inset shows the concentration dependence of chemiluminescence quenching by the compounds tested.

The concentrations for half-inhibition were 10, 37, 160 and over 200 uM for α–tocopherol, β–carotene, retinyl-palmitate, and retinoic acid, respectively. β–carotene, which was a strong scavenger than α–tocopherol in hexane, was weaker in the liposomal system, and retinoids, which had no radical-scavenging activity in the hexane, showed marked activity in the liposomal system. These results indicate that it is not simple chemical reactivity (as measured in the hexane solution) that determines antioxidant capacity in a membrane system; other factors, such as the interaction of antioxidants with membrane components, are also involved. Mobility of scavenging molecules in membranes is one of the possible factors. Note that carotenoids were still about five times as efficient as retinoids in the membrane system, again indicating that the number of double bonds is involved in the radical-scavenging efficiency of these compounds.

A final lipid system is a hybrid of the hexane-parinaric acid system and the DOPC liposome system. In this model, parinaric acid is incoroporated into DOPC liposomes and acts as a reporter molecule in the same way as in the simpler hexane system (Tsuchiya et al 1992; Kuypers et al. 1987). When the protective effects of α–tocopherol, ubiquinol, and β–carotene were compared in this system, α–tocopherol was clearly superior (**Fig 18**), though both β–carotene and ubiquinol possessed

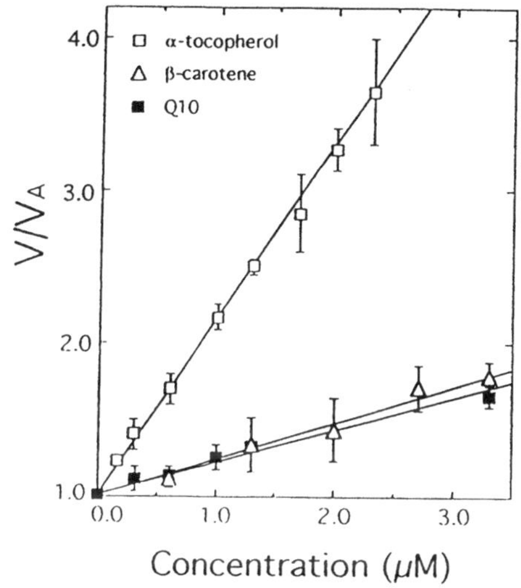

Fig. 18. Relationship between V/V_A and antioxidant concentration in *cis*-parinaric acid-incorporated liposomes. V = the rate of initial fluorescence decay of *cis*-parinaric acid in the absence of antioxidants; V_A = the rate of initial fluorescence decay of *cis*-parinaric acid in the presence of the antioxidants indicated in the figure.

antioxidant capability. One can also compare the antioxidant efficiency of liposomes containing α–tocopherol alone to that of liposomes containing α–tocopherol plus one other antioxidant, either in the liposome or in the surrounding medium. If the effects are additive, it indicates little or no interaction between the antioxidants. But if the effects are synergistic, there is probably interaction between the antioxidants. **Fig 19** illustrates this for ascorbate and β–carotene. The effect of ascorbate is synergistic with that of α–tocopherol, consistent with the well-known ability of ascorbate to regenerate tocopherol from the tocopheroxyl radical (Packer et al. 1979; Niki 1987). On the other hand, the effect of β–carotene appears to be merely additive, indicating little or no interaction between the antioxidants in this system.

Carotenoid Animal Feeding Studies Little quantitative data are available on the absorption of carotenoids from biological sources. Therefore, we examined the absorption of various carotenoids from Palm Oil Carotene enriched fraction (POC) incorporated into the diet of

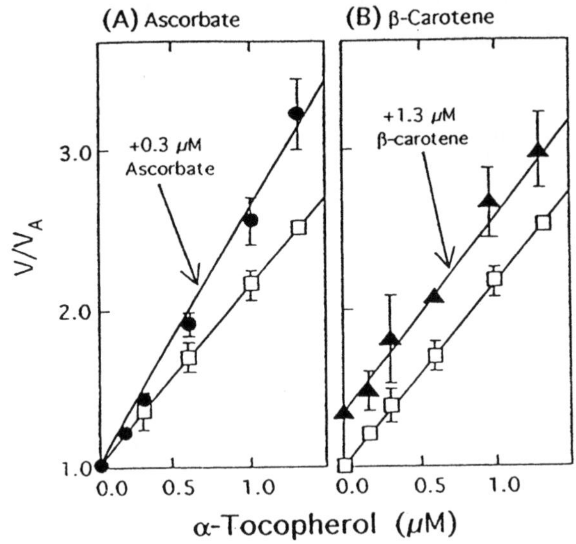

Fig. 19 Effect of ascorbate and β–carotene on the plot of V/VA against α–tocopherol concentration. A. Ascorbate B. β–Carotene

rats, and also examined the resistance of various tissues to lipid peroxidation in supplemented and unsupplemented animals. The carotenoid content of POC is given in Table 5. It was incorporated

into the diet of male Sprague-Dawley rats at a concentration of 7.5 g POC/kg diet, and rats were fed this diet for periods up to 10 weeks. Rats were sacrificed at various times, and their tissue concentrations of lycopene, β–carotene, α–carotene, and total carotenoids were measured by HPLC. In addition, lipid peroxidation was induced in liver homogenates by the lipophilic azo-initiator AMVN. Lipid peroxidation products were measured by TBARS assay.

At no time were carotenoids detectable in adipose tissue or skin. Heart and skeletal muscle initially had no detectable carotenoids, but by 10 weeks levels of β–carotene in these tissues were 17 ± 4 ng/g and 6 ± 1 ng/g wet tissue, respectively. The greatest increase in β–carotene concentration was in liver, where levels increased from 7.3 to 145 ng/g wet tissue over 10 weeks (Fig. 20). When liver homogenates were exposed to AMVN the correlation between TBARS and carotenoid content was found to be in the order α–carotene> β–

TABLE 5
THE CAROTENE COMPOSITION OF PALM OIL CAROTENE ENRICHED FRACTION (POC)

Carotene	mg/kg POC
phytoene	1200
phytofluene	240
β-carotene	40,000
α-carotene	26,640
ξ-carotene	trace
γ-carotene	trace
δ-carotene	480
neusporene	trace
β-zeacarotene	trace
α-zeacarotene	trace
lycopene	2720
cis-β-carotene	720
cis-α-carotene	4400

carotene > lycopene (r = 0.673 > 0.487 > 0.401).

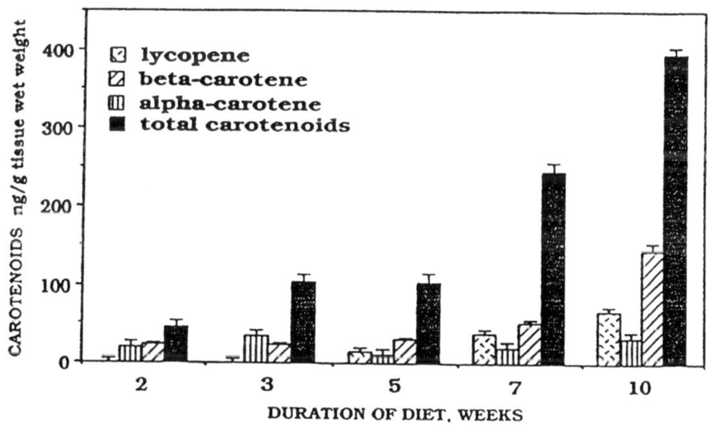

Fig. 20. Accumulation of carotenoids in liver of animals fed a palm oil carotene enriched fraction. N = 3.

These results indicate that high amounts of not only β–carotene, but also lycopene and α–carotene, can be absorbed by the liver, and can afford significant antioxidant protection to that organ. In agreement with other researchers (Lakshman et al. 1989; Mathews-Roth et al. 1990) we confirmed that the rat, unlike humans, does not accumulate carotenoids in adipose tissue.

Carotenoids as protective agents in LDL Low-density lipoproteins (LDL) are an attractive system in which to study the interactions of antioxidants. They contain the major lipid-soluble antioxidants, including vitamin E, carotenoids, and ubiquinols (Esterbauer et al. 1990a, 1990b, 1991; Dieber-Rotheneder et al. 1991). They are easily isolated and represent an important component of the plasma, especially in terms of cardiovascular disease (Goldstein and Brown 1977). Many methods have been developed for inducing oxidation in LDL and for studying the effects (Bonnefont-Rousselot et al. 1992; Steinbrecher 1987) For these reasons we decided to examine the effects of oxidation on carotenoids in LDL.

When LDL are exposed to UV light of a wavelength around 295 nm, the vitamin E is selectively destroyed. By doing this, one may examine the interaction of other antioxidants with vitamin E, since antioxidants which are depleted under this condition must interact with vitamin E in some way. When this experiment is carried out, one finds that not only is E destroyed, but β–carotene as well (see Table 6). Thus, there must be some interaction between the

TABLE 6
EFFECT OF UV IRRADIATION, DIHYDROLIPOIC ACID, AND ASCORBATE ON BETA-CAROTENE CONTENT OF HUMAN LOW-DENSITY LIPOPROTEIN

Treatment	β-carotene pmol/mg protein	% control	% spared
Control	320	100.0	—
+UV	80	25.0	0.0
+UV and DHLA	120	37.5	12.5
+UV and ascorbate	175	54.7	29.7
+UV, DHLA, and ascorbate	300	93.75	68.7

vitamin E radical formed during UVB radiation and β–carotene in LDL. These results imply that the components of the LDL are arranged in such a way that the two antioxidants can interact, even though some investigators believe that they are located in different parts of the LDL particle. Furthermore, when antioxidants such as dihydrolipoic acid or ascorbate are added to the LDL preparation, they exert some protective effect on the β–carotene, and together they have a synergistic protective effect (Table 6). These antioxidants can reduce oxidized compounds and "recycle" antioxidants such as vitamin E (Kagan et al. 1992b) and they exhibit synergism in this recycling. But β–carotene is chemically destroyed when it reacts with radicals, and cannot be regenerated by oxidation-reduction reactions (Palozza and Krinsky 1992). These compounds thus probably protect β–carotene indirectly, by reducing the vitamin E radical as it is formed, back to vitamin E, and keeping the radical's steady-state concentration low.

We have, in collaboration with Bernard Arrio and Dominique Bonnefort Rousselot, investigated the ability of carotenoid supplementation to protect LDL from oxidation. We have used a new technique, laser doppler electrophoresis (LDE), to measure the changes in mobility in human LDL caused by oxidation. LDE measures the electrophoretic mobility of particles, such as LDL, whose size can range between 0.01 and 30 um, in liquid suspensions. The particles are analyzed by independent laser Doppler measurements at four different angles simultaneously with 256-channel resolution each. Comparisons of simultaneous laser Doppler spectra from four angles allow the detection of very small particles as well as the separation of effects due to electrophoretic inhomogeneity, thermal diffusion, and flow inhomogeneity (Rivier et al. 1988; Bloomfield and Lim 1978). For these experiments, the 25.6° angle was used to obtain data for graphs.

In one experiment, LDL from a normolipidemic subject was isolated. The LDL was either left untreated (control) or was supplemented with an extract from Dunaleiella salina which initially consists of equal amounts of all-*trans* and 9-*cis* β–carotene (Betatene, a gift of Henkel Corp., La Grange, Ill.). It was first dissolved in hexane at 60 °C and thereafter diluted in ethanol. LDL suspensions were pre-treated with 12.5 uM betatene 24 hours before inducing oxidation. 10 ul of appropriate stock solution was added to 1 ml of LDL preparation. In all cases control LDL preparations were treated with the same volume of ethanol alone for the same length of time as the β–carotene treated samples. Incorporation of β–carotene into the LDL was confirmed by scanning the absorption spectrum of the β–carotene-treated samples vs. control samples (all samples were scanned after dialysis): the treated samples absorbed strongly at 465 nm (λmax for β–carotene), while the control samples showed only weak absorbance at this wavelength. Both control and β–carotene supplemented samples were then subjected to oxidation by three methods: Cu^+ oxidation, AAPH oxidation, and γ–radiolysis.

In LDL treated with Cu^+, there is a pronounced shift toward a more negative mobilty as well as a greater heterogeneity of mobilities. The betatene treated LDL has a less pronounced shift toward a negative mobility and displays greater homogeneity of populations; it thus appears, at least by LDE, that β–carotene is protective against oxidative stress in LDL. Biochemical analyses give further indication that betatene pretreatment was protective for LDL in all three different methods of oxidation. In the betatene-treated LDL, compared to control, there was 1-23% less loss of vitamin E, especially when oxidation was initiated by γ–radiolysis, and 10-50% less lipid peroxidation, as measured by TBARS, the degree of protection depending on the type of oxidation used.

LDL was also obtained from a vitamin E-deficient subject. This LDL contained 15% the vitamin E concentration of LDL from normolipidemic subjects (Kohlschütter et al 1988). Using this LDL allowed us to determine whether supplementation with β–carotene could compensate for the loss of vitamin E in this LDL. Control samples and betatene-treated samples were oxidized by treatment with AAPH and compared by LDE. In this case the population shifts toward a more negative mobility, and heterogeneity appears. Fig 21 shows betatene-pretreated E-deficient

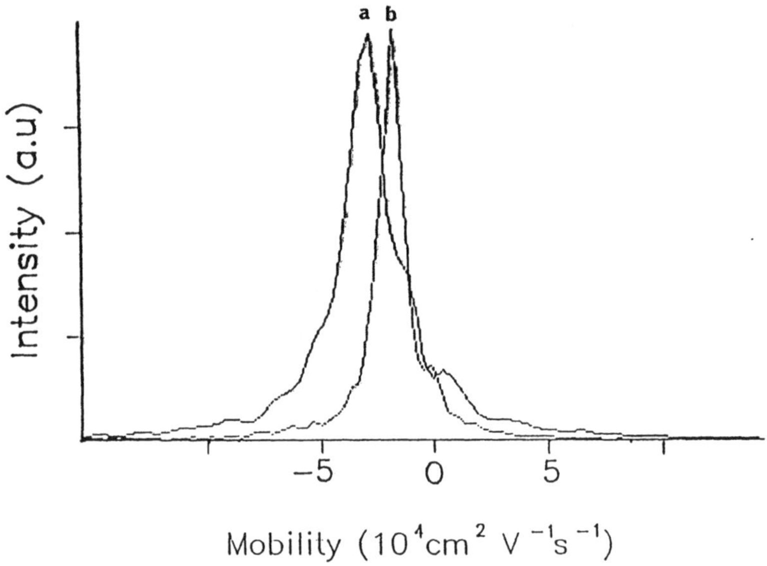

Fig. 21. LDE spectrum illustrating the protecitve effect of β–carotene against AAPH oxidation.a. Vitamin E-deficient LDL; b. Same LDL sample treated with β–carotene. Both samples oxidized by treatment with AAPH.

DL oxidized with AAPH compared to control E-deficient LDL oxidized with AAPH. Betatene prevents some of the shift toward a more negative mobility and there is no loss of homogeneity on oxidation. This indicates that β–carotene can at least partially compensate for loss of vitamin E in LDL.

Although the concentration of carotenoids in normal LDL is far lower than that of vitamin E (Esterbauer et al. 1992), these results indicate that β–carotene can play a significant protective role against oxidation in LDL. LDL samples prepared from the vitamin E deficient subject, which contained almost undetectable endogenous levels of vitamin E, could be readily oxidized, but could be afforded protection by pre-treatment with β–carotene prior to oxidant (AAPH) exposure. Our results are in agreement with β–carotene supplementation of LDL from normolipidemic subjects, both by us in the present studies, and by others, where supplementation was found to prevent increased negative charge changes as measured by agarose gel electrophoresis (Jialal et al. 1991). Our findings provide direct evidence for the importance of β-carotene in protection of LDL against oxidation, even in the presence of low exogenous vitamin E. Carotenoids and vitamin E may act independently to protect LDL, since β–carotene antioxidant action is not based on redox reactions as is the case with vitamin E. However, it is not yet known if a synergistic protective antioxidant effect is afforded by the presence of both vitamin E and carotenoids in the same LDL particle.

REFERENCES

Arrio, B., Bonnefort-Rousselot, D., Catudioc, J. and Packer, L. (1993) *Biochem. Mol. Biol. Int.* 30: 1101-1114.

Bloomfield, V.A., & T.K. Lim. (1978). *Meth. Enzymol.* 48: 415-494.

Bonnefont-Rousselot, D., M. Gardes-Albert, S. Lepage, J. Delattre, & C. Ferradini. (1992). *Radiat. Res.* 132: 228-236.

Burkalova, Ye. B., Kuchtina, Te. E., Ol'khovskhaya, I.P., Sarycheva, I.K., Sinkiona, Ye. B., Khrapova, N.G. (1980) *Biofizika* 24: 989-993.

Burton, G.W. and Ingold, K.U. (1981) *J. Amer. Chem. Soc.* 103: 6472-6477.

Burton, G.W. and Ingold, K.U. (1989) *Ann. N.Y. Acad. Sci.* 570: 7-22.

Burton, G.W., and Ingold, K.U. (1986) *Acc. Chem. Res.* 19: 194-201.

Burton, G.W., Doba, T., Gabe, E.J., Hughes, L., Lee, F.L., Prasad, L., Ingold, K.U. (1985) *J. Am. Chem. Soc.* 107: 7053-7065.

Chatelain, E., Boscoboinik, D.O., Bartoli, G.-M., Kagan, V.E., Gey, F., Packer, L. and Azzi, A. (1993) *Biochim. Biophys. Acta* 1176: 83-89.

Dieber-Rotheneder, M., H. Puhl, G. Waeg, G. Striegl, & H. Esterbauer. (1991). *J. Lipid Res.* 32: 1325-1332.

Esterbauer, H. M. Dieber-Rotheneder, G.Waeg, H. Puhl, & F. Tatzber. (1990b). *Biochem Soc. Transact.* 18: 1059-1061.

Esterbauer, H., Gebicki, J., Puhl, H. and Jurgens, G. (1992) *Free Rad. Biol. Med.* 13: 341-90.

Esterbauer, H., H. Puhl, M. Dieber-Rotheneder, G. Waeg, & H. Rabl. (1991). *Annals Med.* 23: 573-581.

Esterbauer, H., J. Gebicki, H. Puhl, & G. Jürgens. (1992). *Free Rad. Biol. Med.* 13: 341-390.

Esterbauer, H., M. Dieber-Rotheneder, G. Waeg, G. Striegl, & Jurgens, G. (1990a). *Chem. Res. Toxicol.* 3: 77-92.

Evans, H.M. and Bishop, K.S. (1922) *Science* 56: 650-651.

Gey K., Purka P. Jordan P. and Moser H (1991) *Am. J. Clin. Nutr.* Suppl. 53, 326S-334S

Gey, KF (1993). *Brit. Med. Bull.* 49: in press.

Glazer, A.N. (1990). In *Meth. Enzymol.*, L. Packer & A.N. Glazer, Eds., Vol. 186: 161-168. Academic Press, New York, NY.

Goldstein, J.L., & M.S. Brown. (1977). *Ann. Rev. Biochem.* 46: 897-930.

Halliwell, B. and Chirico, S. (1993) *Am. J. Clin. Nutr.* 57: 715S-725S.

Halliwell, B. and Gutteridge, J.M.C. (1989) *Free Radicals in Biology and Medicine.*, Clarendon Press, Oxford.

Hennekens C. and Eberlin K (1985) *Prev. Med* 14, 165-168

Hiramatsu, M., R. & L. Packer. (1990). In *Meth. Enzymol.*, L. Packer, Ed. Vol. 190: 273-280, Academic Press, New York, NY.

Jialal, I., E.P. Norkus, L. Cristol., & S.M. Grundy. (1991). *Biochim. Biophys. Acta.* 1086: 134-138.

Kagan, V.E., E.A. Serbinova, T. Forte, G. Scita, & L. Packer. (1992b). *J. Lipid Res.* 33: 385-397.
Kagan, V.E., Serbinova, E.A., Forte, T., Scita, G. and Packer, L. (1992b) *J. Lipid Res.* 33: 385-397.
Kato A., Yamaoka M. and Tanaka A. (1985) *Abura Kagaku (Jpn.)*, 34, 373-376
Kohlschütter, A., C. Hubner, W. Jansen, & S.G. Lindner. (1988). *J. Inherited Met. Dis.* 11 Suppl. 2: 149-152.
Komiyama K., Iizuka K. Yamaoka M., Watanabe H. Tsuchiya M. and Ukezawa I. (1989) *Chem Pharm Bull (Tokyo)*, 37, 1369-71
Kuypers, F.A., van den Berg, J.J., Schalkwijk, C., Roelofsen, B. and Op den Kamp, J.A. (1987) *Biochim. Biophys. Acta* 921: 266-274.
Lakshman, M., K. Asher, M. Attlesey, S. Satchithanandam, I Mychkovsky, & P. Coutakis. (1989). *J. Lipid Res.* 30: 1545-1550.
Machlin, L.J., Ed. (1984) *Handbook of Vitamins: Nutritional, Biochemical and Clinical Aspects*, New York: Marcel Dekker; .
Marsh, D. (1980) *Biochem.* 19: 1632-1637.
Mathews-Roth, M., S. Welankiwar, P. Sehgal, C. Lausen, M. Russett, And N. Krinsky. (1990). *J. Nutr.* 120: 1205-1209.
Nakano M., Siguoka K., Nakamura T and Oki T (1980) *Biochim Biophys Acta*, 619-274-286
Niki, E. (1987). *Chem. Phys. Lipids* 44: 227-253.
Niki, E., Kawakami, A., Saito, T., Yamamoto, Y., Tsuchiya, J., Kamiya, Y. (1986) *J. Biol. Chem.* 260: 2191-2196.
Olson J., (1992) In: *Lipid soluble antioxidants: Biochemistry and Clinical Applications*, A. Ong and L. Packer, eds., Birkhauser Verlag, Basel, Boston, Berlin, , 178-192
Packer, J.E., T.R. Slater, & R.L. Willson. (1979). *Nature* 278: 737-738.
Packer, L. (1994) *Sci. Am. Sci. Med.* 1: 54-63.
Palozza, P. & N.I Krinsky. (1992). In: *Meth. Enzymol.*, L. Packer, Ed. Vol. 213: 403-420. Academic Press, New York, NY.
Pentland, A.P., Morrison, A.R., Jacobs, S.C., Hruza, L.L., Hebert, J.S. and Packer, L. (1992) *J. Biol. Chem.* 267: 15578-15584.
Riemersma, R.A., Wood, D., Macintyre, C., Elton, R., Gey, K., and Oliver, M., (1991), *Lancet*, 337, 1-5
Riviere, M.-E., B. Johannin, D. Gamet, V. Molitor, G.A. Peschek, & B. Arrio. (1988) In: Meth. Enzymol. 167: 691-700.
Seher V and Ivanov S (1973) *Fette Seifen Anstri*, 75, 606-609
Serbinova E. and Packer L. (1994) *Am. J. Nutr.*, (in press)
Serbinova E., Khwaja S., Catudioc J., Torres Z., Kagan V. and Packer L. (1992) *Nutr. Res. (Suppl.)*,12, S203-S215
Serbinova, E., Kagan, V., Han, D. and Packer, L. (1991) *Free Rad. Biol. Med.* 10: 263-275.
Severcan, F. and Cannistraro, S. (1990) *Chem. Phys. Lipids* 53: 17-26.
Sheppard, A.J., Pennington, J.A.T. and Weihrauch, J.L. (1993) *Vitamin E in Health and Disease* Ed. L. Packer and J. Fuchs. New York, NY, Marcel Dekker, Inc. pp. 9-31.
Stampfer, M.J., Hennekens, C.H., Manson, J.E., Colditz, G.A., Rosner, B., and Willett, W.C. *N. Engl. J. Med.* 328: 1444-1449, 1993.
Steinbrecher, V.P. (1987). *J. Biol. Chem.* 262: 3603-3608.
Suzuki, Y.J. and Packer, L. (1993) *Biochem. Biophys. Res. Comm.* 93: 277-283.
Suzuki, Y.J., Tsuchiya, M., Wassall, S.R., Choo, Y.M., Govil, G., Kagan, V.E. and Packer, L. (1993) *Biochemistry* 32: 10692-10699.
Tatsuta T. *Vitamins (Jpn)* (1971), 44, 185-170
Traber, M.G. and Kayden, H.J. (1984) *Am. J. Clin. Nutr.* 40: 747-751.
Tsuchiya, M, G. Scita, H.J. Freisleben, V.E. Kagan, & L. Packer. (1992). In *Meth. Enzymol.*, L. Packer, Ed. Vol. 213: 460-472. Academic Press, New York, NY.
Tsuchiya, M., G. Scita, D.F.T. Thompson, L. Packer, V.E. Kagan, & M.A. Livrea. (1993). In Retinoids: Progress in Research and Clinical Applications, M.A. Livrea and L. Packer, Eds. pp. 525-536. Marcel Dekker, Inc. New York, NY.
Zieger, R.G. (1989), *J. Nutrition*, 119, 116-122

BIOKINETICS OF VITAMIN E IN LIPOPROTEINS AND CELLS

Maret G. Traber, Ph.D.
Department of Molecular and Cell Biology
University of California
Berkeley, CA 94720-3200
United States of America

The paper will describe 1) the various naturally occurring and synthetic forms of vitamin E, 2) lipoprotein metabolism as it relates to plasma vitamin E transport, 3) the mechanism for the preferential enrichment of plasma and tissues with only one form of vitamin E, namely *RRR*-α-tocopherol, 4) kinetic estimates of the turnover of vitamin E in human plasma and 5) determinants of vitamin E tissue concentrations.

I. VITAMIN E FORMS

Dietary vitamin E occurs in 8 different forms. The number and position of methyl groups on the chromanol ring is denoted by α– (3 methyl groups; positions 5,7 and 8; see Figure 1), β, γ (2 methyl groups; positions 5, 8 or 7,8, respectively) and δ (1 methyl group, position 8). Tocopherols have a phytyl tail with 3 chiral centers at positions 2, 4' and 8', while tocotrienols have an unsaturated tail.

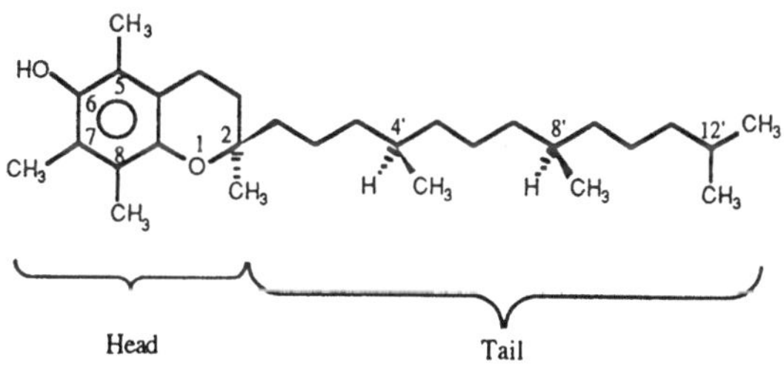

Figure 1. α-tocopherol

The biological activity of these forms of vitamin E has been assessed using a rat fetal resorption assay (Bunyan 1961). α–Tocopherol has the highest activity; that is, the smallest amount is necessary for addition to the diet of a vitamin E deficient animal to prevent fetal resorption.

Differences in biologic activity are also present between different stereoisomers of α–tocopherol (Weiser 1986). *All racemic* α–tocopherol, synthesized for use as vitamin E supplements, contains eight stereoisomers, arising from the two possible orientations for each

NATO ASI Series, Vol. H 92
Signalling Mechanisms – from Transcription Factors
to Oxidative Stress
Edited by L. Packer, K. Wirtz

chiral center. These are named *RRR*, *RSR*, *RRS*, *RSS*, *SSS*, *SRS*, *SRR*, *SSR*. Only 12.5% of synthetic vitamin E (*all rac a*–tocopherol) is the naturally occurring form, *RRR*-α–tocopherol.

It is likely that differences in the biological activities are a result of differences in lipoprotein transport of vitamin E, as discussed below.

II. VITAMIN E TRANSPORT IN PLASMA LIPOPROTEINS

Vitamin E transport in plasma lipoproteins was first reported about 40 years ago (Lewis 1954; McCormick 1960). Other than lipoprotein transport, there are no known plasma carrier proteins for vitamin E (Granot 1988). In general, vitamin E is distributed in all of the plasma lipoproteins, with the highest proportion of vitamin E in the lipoprotein with the greatest abundance (Ogihara 1988). Studies using deuterated tocopherols have been instrumental in understanding how vitamin E is incorporated and transported in human lipoproteins and these will be reviewed briefly here.

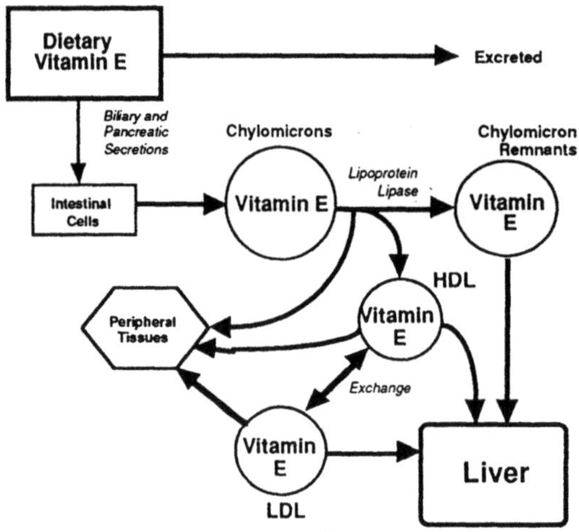

Figure 2. Vitamin E is absorbed from the intestinal lumen in the presence of bile acids and pancreatic secretions into enterocytes. These intestinal cells then secrete chylomicrons containing various forms of vitamin E into the lymph. The chylomicrons are catabolized in the circulation by lipoprotein lipase, which delipidates the chylomicrons and transfers fatty acids and various forms of vitamin E to peripheral tissues. During this process, excess chylomicron surface is formed, which is transferred to HDL. Vitamin E is also transferred to HDL, which can transfer tocopherols to other circulating lipoproteins, such as LDL and VLDL. Thus, during chylomicron catabolism, some vitamin E is delivered to peripheral tissues, and the remainder is delivered to the liver in the form of chylomicron remnants.

A. Absorption and chylomicron secretion

Following uptake of dietary fats by enterocytes in the small intestine, these fats are incorporated into chylomicrons, which are secreted by the intestine into the lymph (Cohn 1988a). Chylomicrons similarly transport newly absorbed dietary vitamin E (Gallo-Torres 1970; Traber 1990a; Traber 1988; Traber 1986; Traber 1990c).

Plasma concentrations of the various forms of vitamin E are not determined by differences in the degree of intestinal absorption. Studies in humans have shown that various forms of vitamin E are absorbed and secreted in chylomicrons similarly.

Specifically, no differences in the efficiencies of absorption and chylomicron secretion between α– and γ–tocopherols (Meydani 1989; Traber 1992a; Traber 1989), or between *RRR*- and *SRR*-α–tocopherols were observed (Traber 1992a; Traber 1990a).

Chylomicrons are catabolized in the circulation by lipoprotein lipase hydrolyzing triglycerides and releasing free fatty acids. During delipidation, the size of the chylomicron triglyceride core is reduced and excess surface is created, which pinches off, forming high density lipoproteins (HDL) (Havel 1994). Some vitamin E is also transferred to HDL, and from HDL to other lipoproteins (Bjornson 1975; Granot 1988; Kayden 1972; Massey 1984; Traber 1992b). This distribution of vitamin E to all of the circulating lipoproteins is readily observed during the first few hours following a dose of deuterated tocopherols to humans (Traber 1992a; Traber 1990a; Traber 1993b).

To test if post-absorptive phenomenon are responsible for discrimination between tocopherols, intestinal absorption was circumvented by administering an intravenous infusion of a lipid emulsion resembling chylomicrons to normal human volunteers (Traber 1993a). The soybean oil-derived lipid emulsion contained primarily γ–tocopherol. During the 6 h infusion, plasma γ–tocopherol concentrations increased about 8-fold (to 25 nmol/ml). Cessation of the infusion resulted in a rapid decrease in plasma γ–tocopherol concentrations, which by 24 h were within pre-infusion concentrations (5 nmol/ml). This study emphasizes that despite raising emulsion γ–tocopherol to concentrations observed for α–tocopherol, γ–tocopherols were not maintained at these high concentrations. Thus, discrimination between tocopherols is a metabolic process, and is not a result of differences in intestinal absorption.

B. Very low density lipoprotein secretion and catabolism

Once chylomicron remnants reach the liver, the dietary fats are repackaged and secreted into the plasma in very low density lipoproteins (VLDL) (Havel 1994). Like chylomicrons, VLDL have a triglyceride-rich core, but they have apolipoprotein B–100 (apoB 100), instead of apoB 48 present in chylomicrons, as the major apolipoprotein (Havel 1994). In the circulation, VLDL are delipidated forming LDL, which retain apoB 100, allowing the LDL to be directed to LDL receptors for catabolism.

Unlike other lipid soluble vitamins, such as vitamin A or D, that have specific plasma transport proteins, vitamin E is secreted from the liver in VLDL (Bjørneboe 1987; Cohn 1988b). Remarkably, only one form of vitamin E, *RRR*-α–tocopherol, is preferentially secreted by the liver, as illustrated in the figure. Nascent VLDL, isolated from perfusates of livers from cynomolgus monkeys fed 24 hours previously with various deuterated tocopherols, contained about 80% *RRR*-α–tocopherol of the total deuterated tocopherols (Traber 1990b). Thus, the liver, not the intestine, discriminates between tocopherols. These data suggest that the liver contains a mechanism, likely a protein, which selects *RRR*-α–tocopherol and inserts it into nascent VLDL. A possible candidate for this process is the tocopherol binding protein, which is discussed further below.

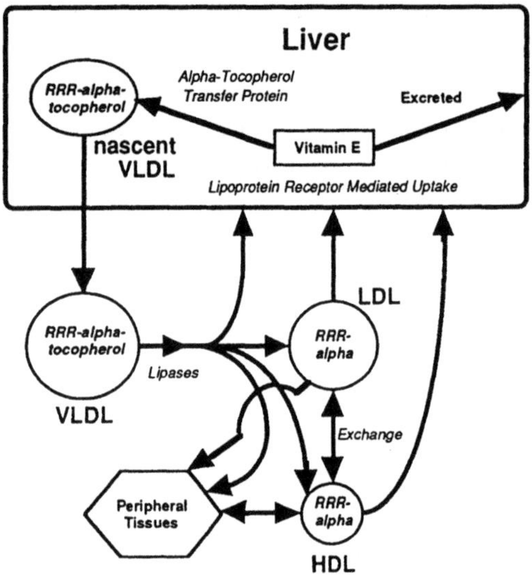

Figure 3. Of the various forms of vitamin E delivered to the liver by chylomicrons and other circulating lipoproteins, only RRR-alpha-tocopherol is preferentially secreted in VLDL, likely as a result of the tocopherol transfer protein. Excess vitamin E is probably excreted in the bile. Following VLDL secretion into the plasma, both lipoprotein lipase and hepatic triglyceride lipase delipidate VLDL to LDL. Only about half of the VLDL is converted to LDL; the remainder is taken up by the liver. During triglyceride hydrolysis by lipases, tocopherol is also transferred to HDL, as well as to tissues. Tissues with LDL receptors can also take up LDL containing alpha tocopherol. Peripheral tissues apparently acquire most of their vitamin E as a result of the processes shown because most tissues contain primarily alpha tocopherol, not other forms of vitamin E.

As a result of VLDL catabolism, the plasma becomes preferentially enriched in *RRR*-α–tocopherol (Traber 1992a; Traber 1990a; Traber 1988; Traber 1990c). Similar to the process during chylomicron catabolism, the conversion of VLDL to LDL results in the equilibration of *RRR*-α–tocopherol between LDL and HDL. The distribution of α–tocopherol between LDL and HDL is dependent upon plasma concentrations of these two lipoproteins (Traber 1992b).

From the foregoing discussion it is apparent that the liver, not the intestine, discriminates between forms of vitamin E. Data supporting the role of the liver in this process comes from studies using deuterated tocopherols in rats (Burton 1988; Cheng 1987; Ingold 1987), guinea pigs (Burton 1990b), monkeys (Traber 1990b), and humans (Traber 1990a), especially patients with genetic abnormalities of lipoprotein metabolism (Traber 1992a) and patients with genetic abnormalities of vitamin E transport (Traber 1990c; Traber 1993b), who are further discussed below.

III. PREFERENTIAL ENRICHMENT OF THE PLASMA WITH *RRR*-α–TOCOPHEROL

The mechanism for the preferential incorporation of *RRR*-α–tocopherol is dependent upon the secretion of VLDL enriched in *RRR*-α–tocopherol from the liver. The putative mechanism for the process involves the hepatic tocopherol binding protein.

A. Hepatic Tocopherol Binding Protein

An hepatic tocopherol binding protein (30-35 kDa) has been purified to homogeneity from rat liver (Sato 1991; Yoshida 1992) and the complete amino acid sequence of the protein has been reported (Sato 1993). It is present only in the liver, not other tissues; of the various liver cells; it has only been identified in hepatocytes (Yoshida 1992). A similar tocopherol binding protein has been isolated from human liver (Kuhlenkamp 1993).

The purified protein transfers α–tocopherol between liposomes and microsomes (Sato 1991). Both α– and β–tocopherols are effective competitors, γ–tocopherol is about half as effective and δ–tocopherol is about 1/3 as effective; α–tocopheryl acetate, tocopherol quinone and cholesterol were ineffective competitors (Sato 1991).

Likely the tocopherol binding protein participates in the preferential incorporation of *RRR*-α–tocopherol into nascent VLDL for secretion by from the liver. It, however, has not yet been demonstrated that the purified protein is involved in VLDL assembly. Credence for this hypothesis comes from patients with familial isolated vitamin E deficiency (FIVE deficiency), who appear to lack or have a defective form of this protein.

B. Patients with isolated vitamin E deficiency

Patients with familial isolated vitamin E (FIVE) deficiency have a rare genetic disorder, which results in neurologic abnormalities similar to those in patients with vitamin E deficiency secondary to lipid malabsorption, yet FIVE deficiency patients have no abnormalities known to cause vitamin E deficiency (Burck 1981; Harding 1985; Kohlschütter 1988; Krendel 1987; Laplante 1984; Sokol 1988; Stumpf 1987; Traber 1990c; Traber 1993b; Traber 1987; Trabert 1989; Yokota 1987). In contrast to patients with lipid malabsorption, where oral vitamin E supplementation is ineffective, if FIVE deficiency patients consume vitamin E supplements (400-1200 IU per day), their plasma α–tocopherol concentrations increase from less than 2 nmol/ml to within the normal range (10-30 nmol/ml). However, if supplementation is stopped, their plasma α–tocopherol concentrations fall to deficient levels within days (Sokol 1988).

To investigate the transport of vitamin E in patients with FIVE deficiency, studies using deuterated *RRR*-α–tocopheryl acetate were carried out (Traber 1990c). The patients absorbed and transported vitamin E in chylomicrons normally, but secretion and transport in VLDL was defective. Apparently, these patients are unable to maintain normal plasma vitamin E concentrations because they either lack, or have a defective, hepatic tocopherol binding protein (Traber 1990c). Since the tocopherol binding protein is thought to be involved in discrimination between tocopherols, the ability of patients with FIVE deficiency to perform this function was also assessed (Traber 1993b). When FIVE deficiency patients consumed a dose containing both *RRR*- and *SRR*-α–tocopherols labeled with different amounts of deuterium, the patients segregated into two groups. Four of the 8 patients did not discriminate between the two tocopherols (non-discriminators) and 4 were able to discriminate (discriminators).

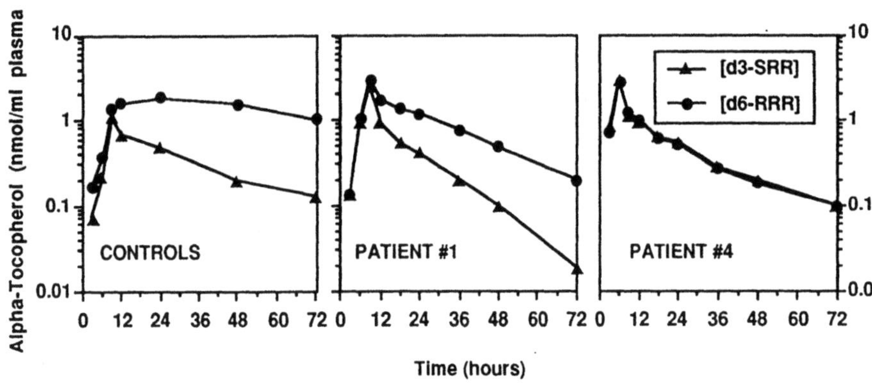

FIGURE 4. Subjects were given an oral dose containing 20 mg each of 2R, 4'R, 8'R-alpha-(5,7-(C2H3)2) tocopheryl acetate (d6-RRR-alpha-tocopheryl acetate), 2S, 4'R, 8'R-alpha-(5-(C2H3) tocopheryl acetate (d3-SRR-alpha-tocopheryl acetate), and blood samples were obtained at the indicted intervals. Plasma concentrations of d6-RRR- and d3-SRR-alpha-tocopherols in controls (mean of seven subjects), in a patient with familial isolated vitamin E deficiency, who discriminated between these tocopherols (patient #1, discriminator), and a patient who did not (patient #1, non-discriminator). From (Traber 1993b) Copyright © 1993 Lipid Research, Inc.

The figure shows plasma concentrations of both labeled tocopherols in controls, and two representative patients. In non-discriminators, both *SRR*- and *RRR*–α–tocopherols disappeared rapidly from the plasma. Patient # 1 is an example of a discriminator, while patient #4 is an examples of a non-discriminator Apparently, non-discriminator patients have a genetic defect that results in a complete lack of functionality of the tocopherol binding protein, either due to its complete absence, or its inability to insert *RRR*–α–tocopherol into nascent VLDL. In discriminators, plasma *RRR*-α–tocopherol concentrations decreased more slowly than did *SRR*-α–tocopherol, but more rapidly than *RRR*–α–tocopherol in controls. Because the discriminators retain some ability to distinguish between tocopherols, they may have a defect that impairs the insertion of *RRR*-α–tocopherol into nascent VLDL. The identification of the actual genetic defects awaits isolation and cloning of the human tocopherol binding protein.

C. Isolated Vitamin E Deficiency Mimics Friedreich's Ataxia

In addition to patients with FIVE deficiency cited above, there appears to be other patients who have a similar disorder. Ben Hamida et al.(Ben Hamida 1993a; Ben Hamida 1993b) reported that they have identified an unusual variant of Friedreich's Ataxia in large, consanguineous Tunisian families. These patients have an autosomal recessive neurodegenerative disorder, but do not have the characteristic linkage of Friedreich's ataxia to chromosome 9. Instead, the patients were found to have remarkably low serum vitamin E concentrations; less than 1/100 of normal. This selective vitamin E deficiency phenotype was localized to chromosome 8q by homozygosity mapping.

The genetic abnormality in these Tunisian patients is likely the same as the genetic abnormality in the FIVE deficiency patients because both groups develop vitamin E deficiency without abnormalities of lipid metabolism. It is, therefore, also probable that the gene for the tocopherol binding protein is therefore located on chromosome 8.

IV. KINETIC ESTIMATES OF PLASMA *RRR*-α–TOCOPHEROL

A kinetic model of vitamin E transport in patients with FIVE deficiency and in normal humans has been developed using data from studies with deuterium labeled stereoisomers of α–tocopherol (*RRR*- and SRR-) (Traber 1994). The mathematical model was designed assuming 1) the absorption and secretion of the two deuterated tocopherols (*RRR*- and SRR-α–tocopherols) into chylomicrons were similar, 2) initial inputs into the plasma occurred simultaneously for the two labels.

In three non-discriminator patients with familial isolated vitamin E deficiency (FIVE deficiency), the fractional disappearance rates of deuterium labeled *RRR*- and *SRR*-α–tocopherols in plasma were 1.4 ± 0.6 and 1.3 ± 0.3 pools per day, respectively, with a half-life of approximately 13 h for both α–tocopherols In controls, the fractional disappearance rates of deuterium labeled *RRR*-α–tocopherol (0.4 ± 0.1 pools per day) were significantly ($P< 0.01$)

slower than for SRR- (1.2 ± 0.6) with an apparent half-life of *RRR*-α–tocopherol of approximately 48 h, consistent with the "slow" disappearance of α–tocopherol from the plasma (Traber 1994).

The similarity in the fractional disappearance rates for the two stereoisomers of α–tocopherol in the non-discriminator patients, along with the similarity in *SRR*-α–tocopherol fractional catabolic rates between patients and controls, support the idea that *SRR*-α–tocopherol can be used in normal subjects to trace the irreversible loss of vitamin E from the plasma. The differences (0.8 ± 0.6 pools/day) between the rates in controls estimate the rate that *RRR*-α–tocopherol, which had left the plasma, was returned to the plasma. Although plasma labeled *RRR*-α–tocopherol concentrations in controls appear to change slowly, this study shows that both *RRR*- and *SRR*-α–tocopherols leave the plasma rapidly, but only *RRR*-α–tocopherol is returned to the plasma, likely in nascent VLDL. This recycling of *RRR*-α–tocopherol accounts for nearly 1 pool of α–tocopherol per day.

V. VITAMIN E TISSUE CONCENTRATIONS

Relatively little is known about the vitamin E concentrations of human tissues with the exception of those that can readily be obtained, such as blood cells, needle biopsies of adipose tissue, biopsies of nerve tissue or heart muscle, and scrapings of buccal mucosal cells. Insights into vitamin E concentrations of other tissues, therefore, must come from studies carried out in animals.

A. Delivery of vitamin E to tissues.

Likely plasma vitamin E concentrations dictate the forms of vitamin E that are delivered to the tissues. Because plasma α–tocopherol concentrations increase in response to α–tocopherol supplementation and γ–tocopherol concentrations decrease (Baker 1986; Handelman 1985), it is not surprising to find the same changes in adipose tissue (Handelman 1994).

Vitamin E is likely distributed to tissues as well as lipoproteins during chylomicron catabolism by a mechanism involving lipoprotein lipase. In vitro this enzyme can deliver tocopherols to cells by functioning as a transfer protein for both fatty acids and tocopherols (Traber 1985).

One other important mechanism for the delivery of tocopherols to tissues is the LDL receptor mechanism. LDL-vitamin E was taken up more effectively by fibroblasts with functional

LDL receptor activity compared with fibroblasts lacking receptors (Traber 1984) . More recently, Cohn et al (Cohn 1992) have demonstrated that *in vivo* both LDL receptor pathways and non-receptor dependent mechanisms are important for tissue uptake of tocopherols.

B. Tissues with slow versus fast turning over vitamin E pools.

Deuterated tocopherols have been used to assess the kinetics and distribution of tocopherols into various tissues both in rats and in guinea pigs (Burton 1990b; Ingold 1987). From these studies it is apparent that a group of tissues is in rapid equilibrium with the plasma tocopherol pool. Thus, tissues such as erythrocytes, liver, spleen quickly replace "old" with "new" tocopherol (Burton 1990a). In contrast, other tissues such as heart, muscle and spinal cord have slower tocopherol turnover times. By far, the tissue with the slowest tocopherol turnover times appears to be the brain.

Recent studies in adult beagle dogs demonstrated that the peripheral nerve is the most responsive of the nervous system to vitamin E concentrations in the diet (Pillai 1993). In dogs fed a vitamin E deficient diet for nearly two years the peripheral nerve concentration was lower than any of the another nervous tissues and was as low as most of the non-nervous tissues. By contrast, in dogs fed a vitamin E supplemented diet, peripheral nerve concentrations were higher than other nervous tissues. Thus, it is not surprising that in humans the peripheral nerve (Sokol 1993), albeit the sensory neurons, are the most susceptible to vitamin E deficiency (Traber 1987). Furthermore, in vitamin E deficient humans, the vitamin E concentrations of sural nerve decreases prior to histologic or functional defects (Traber 1987).

VI. CONCLUSIONS

Recent advances in our knowledge about the plasma transport of vitamin E in lipoproteins and its delivery to tissues have suggested that the hepatic tocopherol transfer protein is important in the regulation of plasma concentrations of vitamin E. Apparently, the discrimination between forms of vitamin E and the preference for *RRR*-α–tocopherol is the result of the hepatic tocopherol binding protein because patients with a genetic defect in vitamin E plasma transport both are unable to maintain plasma vitamin E concentrations and cannot discriminate between forms of vitamin E. Although some tissues have been reported to have tocopherol binding proteins which could regulate tissue concentrations of vitamin E (Nalecz 1992), the question of what determines tissue concentrations of vitamin E remains open.

VII. REFERENCES

Baker, H, Handelman, GJ, Short, S, Machlin, LJ, Bhagavan, HN, Dratz, EA, Frank, O (1986) Comparison of plasma α- and γ-tocopherol levels following chronic oral administration of either all-rac-α-tocopheryl acetate or RRR-α-tocopheryl acetate in normal adult male subjects. Am. J. Clin. Nutr. 43: 382-387.

Ben Hamida, C, Doerflinger, N, Belal, S, Linder, C, Reutenauer, L, Dib, C, Gyapay, G, Bignal, A, Le Paslier, D, Cohen, D, Pandolfo, M, Mokini, V, Novelli, G, Hentati, F, Ben Hamida, M, Mandel, JL, Koenig, M (1993a) Localization of Friedreich ataxia phenotype with selective vitamin E deficiency to chromosome 8q by homozygosity mapping. Nature Genetics 5: 195-200.

Ben Hamida, M, Belal, S, Sirugo, G (1993b) Friedreich's ataxia phenotype not linked to chromosome 9 and associated with selective autosomal recessive vitamin E deficiency in two inbred Tunisian families. Neurology 2179-2183.

Bjørneboe, A, Bjørneboe, G-EA, Hagen, BF, Nossen, JO, Drevon, CA (1987) Secretion of α-tocopherol from cultured rat hepatocytes. Biochim. Biophys. Acta 922: 199-205.

Bjornson, LK, Gniewkowski, C, Kayden, HJ (1975) Comparison of exchange of α-tocopherol and free cholesterol between rat plasma lipoproteins and erythrocytes. J. Lipid Res. 16: 39-53.

Bunyan, J, McHale, D, Green, J, Marcinkiewicz, S (1961) Biological potencies of ε- and ζ_1-tocopherol and 5-methyltocol. Br. J. Nutr. 15: 253-257.

Burck, U, Goebel, HH, Kuhlendahl, HD, Meier, C, Goebel, KM (1981) Neuromyopathy and vitamin E deficiency in man. Neuropediatrics 12: 267-278.

Burton, GW, Ingold, KU, Foster, DO, Cheng, SC, Webb, A, Hughes, L, Lusztyk, E (1988) Comparison of free α-tocopherol and α-tocopheryl acetate as sources of vitamin E in rats and humans. Lipids 23: 834-840.

Burton, GW, Traber, MG (1990a) Vitamin E: antioxidant activity, biokinetics and bioavailability. Annu. Rev. Nutr. 10: 357-382.

Burton, GW, Wronska, U, Stone, L, Foster, DO, Ingold, KU (1990b) Biokinetics of dietary *RRR*-α-tocopherol in the male guinea pig at three dietary levels of vitamin C and two levels of vitamin E. Evidence that vitamin C does not "spare" vitamin E in vivo. Lipids 25: 199-210.

Cheng, SC, Burton, GW, Ingold, KU, Foster, DO (1987) Chiral discrimination in the exchange of α-tocopherol stereoisomers between plasma and red blood cells. Lipids 22: 469-473.

Cohn, JS, McNamara, JR, Cohn, SD, Ordovas, JM, Schaefer, EJ (1988a) Plasma apolipoprotein changes in the triglyceride-rich lipoprotein fraction of human subjects fed a fat-rich meal. J. Lipid Res. 29: 925-936.

Cohn, W, Goss-Sampson, M, Grun, H (1992) Plasma clearance and net uptake of alpha-tocopherol and low-density lipoprotein by tissues in WHHL and control rabbits. Biochem J. 287: 247-254.

Cohn, W, Loechleiter, F, Weber, F (1988b) α-Tocopherol is secreted from rat liver in very low density lipoproteins. J. Lipid Res. 29: 1359-1366.

Gallo-Torres, H (1970) Obligatory role of bile for the intestinal absorption of vitamin E. Lipids 5: 379-384.

Granot, E, Tamir, I, Deckelbaum, RJ (1988) Neutral lipid transfer protein does not regulate α-tocopherol transfer between human plasma lipoproteins. Lipids 23: 17-21.

Handelman, GJ, Epstein, WL, Peerson, J, Spiegelman, D, Machlin, LJ (1994) Human adipose α-tocopherol and γ-tocopherol kinetics during after 1 y of α-tocopherol supplementation. Am. J. Clin. Nutr. 59: 1025-1032.

Handelman, GJ, Machlin, LJ, Fitch, K, Weiter, JJ, Dratz, EA (1985) Oral α-tocopherol supplements decrease plasma γ-tocopherol levels in humans. J. Nutr. 115: 807-813.

Harding, AE, Matthews, S, Jones, S, Ellis, CJK, Booth, IW, Muller, DPR (1985) Spinocerebellar degeneration associated with a selective defect of vitamin E absorption. N. Engl. J. Med. 313: 32-35.

Havel, R (1994) McCollum Award Lecture, 1993: Triglyceride-rich lipoproteins and atherosclerosis--new perspectives. Am J. Clin. Nutr. 59: 795-799.

Ingold, KU, Burton, GW, Foster, DO, Hughes, L, Lindsay, DA, Webb, A (1987) Biokinetics of and discrimination between dietary *RRR*- and *SRR*-α-tocopherols in the male rat. Lipids 22: 163-172.

Kayden, HJ, Bjornson, LK (1972) The dynamics of vitamin E transport in the human erythrocyte. Ann. N.Y. Acad. Sci. 203: 127-140.

Kohlschütter, A, Hubner, C, Jansen, W, Lindner, SG (1988) A treatable familial neuromyopathy with vitamin E deficiency, normal absorption, and evidence of increased consumption of vitamin E. J. Inher. Metab. Dis. 11: 149-152.

Krendel, DA, Gilchrest, JM, Johnson, AO, Bossen, EH (1987) Isolated deficiency of vitamin E with progressive neurologic deterioration. Neurology 37: 538-540.

Kuhlenkamp, J, Ronk, M, Yusin, M, Stolz, A, Kaplowitz, N (1993) Identification and purification of a human liver cytosolic tocopherol binding protein. Prot. Exp. Purific. 4: 382-389.

Laplante, P, Vanasse, M, Michaud, J, Geoffroy, G, Brochu, P (1984) A progressive neurological syndrome associated with an isolated vitamin E deficiency. Can. J. Neurol. Sci. 11: 561-564.

Lewis, LA, Quaife, ML, Page, IH (1954) Lipoproteins of serum, carriers of tocopherol. Am. J. Physiol. 178: 221-222.

Massey, JB (1984) Kinetics of transfer of α–tocopherol between model and native plasma lipoproteins. Biochim. Biophys. Acta 793: 387-392.

McCormick, EC, Cornwell, DG, Brown, JB (1960) Studies on the distribution of tocopherol in human serum lipoproteins. J. Lipid Res. 1: 221-228.

Meydani, M, Cohn, JS, Macauley, JB, McNamara, JR, Blumberg, JB, Schaefer, EJ (1989) Postprandial changes in the plasma concentration of α- and γ-tocopherol in human subjects fed a fat-rich meal supplemented with fat-soluble vitamins. J. Nutr. 119: 1252-58.

Nalecz, K, Nalecz, M, Azzi, A (1992) Isolation of tocopherol-binding proteins from the cytosol of smooth muscle A7R5 cells. Eur. J. Biochem. 209: 37-42.

Ogihara, T, Miki, M, Kitagawa, M, Mino, M (1988) Distribution of tocopherol among human plasma lipoproteins. Clin. Chim. Acta 174: 299-306.

Pillai, SR, Traber, MG, Steiss, JE, Kayden, HJ, Cox, NR (1993) α-Tocopherol concentrations of the nervous system and selected tissues of dogs fed three levels of vitamin E. Lipids 28: 1101-1105.

Sato, Y, Arai, H, Miyata, A, Tokita, S, Yamamoto, K, Tanabe, T, Inoue, K (1993) Primary structure of alpha-tocopherol transfer protein from rat liver. Homology with cellular retinaldehyde-binding protein. J Biol Chem 268: 17705-10.

Sato, Y, Hagiwara, K, Arai, H, Inoue, K (1991) Purification and characterization of the α-tocopherol transfer protein from rat liver. FEBS Lett 288: 41-45.

Sokol, RJ (1993). Vitamin E deficiency and neurological disorders. Vitamin E in Health and Disease, Packer, L, Fuchs, J, eds. New York, NY, Marcel Dekker, Inc. 815-849.

Sokol, RJ, Kayden, HJ, Bettis, DB, Traber, MG, Neville, H, Ringel, S, Wilson, WB, Stumpf, DA (1988) Isolated vitamin E deficiency in the absence of fat malabsorption - familial and sporadic cases: Characterization and investigation of causes. J. Lab. Clin. Med. 111: 548-559.

Stumpf, DA, Sokol, R, Bettis, D, Neville, H, Ringel, S, Angelini, C, Bell, R (1987) Friedreich's disease: V. Variant form with vitamin E deficiency and normal fat absorption. Neurology 37: 68-74.

Traber, MG, Burton, GW, Hughes, L, Ingold, KU, Hidaka, H, Malloy, M, Kane, J, Hyams, J, Kayden, HJ (1992a) Discrimination between forms of vitamin E by humans with and without genetic abnormalities of lipoprotein metabolism. J. Lipid Res. 33: 1171-1182.

Traber, MG, Burton, GW, Ingold, KU, Kayden, HJ (1990a) *RRR*- and *SRR*-α-tocopherols are secreted without discrimination in human chylomicrons, but *RRR*-α-tocopherol is preferentially secreted in very low density lipoproteins. J. Lipid Res. 31: 675-685.
Traber, MG, Carpentier, YA, Kayden, HJ, Richelle, M, Galeano, N, Deckelbaum, RJ (1993a) Alterations in plasma α- and γ- tocopherol concentrations in response to intravenous infusion of lipid emulsions in humans. Metabolism 42: 701-709.
Traber, MG, Ingold, KU, Burton, GW, Kayden, HJ (1988) Absorption and transport of deuterium-substituted 2*R*,4'*R*,8'*R*-α-tocopherol in human lipoproteins. Lipids 23: 791-797.
Traber, MG, Kayden, HJ (1984) Vitamin E is delivered to cells via the high affinity receptor for low density lipoprotein. Am. J. Clin. Nutr. 40: 747-751.
Traber, MG, Kayden, HJ (1989) Preferential incorporation of α-tocopherol vs γ-tocopherol in human lipoproteins. Am. J. Clin. Nutr. 49: 517-526.
Traber, MG, Kayden, HJ, Green, JB, Green, MH (1986) Absorption of water miscible forms of vitamin E in a patient with cholestasis and in rats. Am. J. Clin. Nutr. 44: 914-923.
Traber, MG, Lane, JC, Lagmay, N, Kayden, HJ (1992b) Studies on the transfer of tocopherol between lipoproteins. Lipids 27: 657-663.
Traber, MG, Olivecrona, T, Kayden, HJ (1985) Bovine milk lipoprotein lipase transfers tocopherol to human fibroblasts during triglyceride hydrolysis in vitro. J. Clin. Invest. 75: 1729-1734.
Traber, MG, Ramakrishnan, R, Kayden, HJ (1994) Human plasma vitamin E kinetics demonstrate rapid recycling of plasma *RRR*-α-tocopherol. Proc. Natl. Acad. Sci. USA In Press.
Traber, MG, Rudel, LL, Burton, GW, Hughes, L, Ingold, KU, Kayden, HJ (1990b) Nascent VLDL from liver perfusions of cynomolgus monkeys are preferentially enriched in *RRR*- compared with *SRR*-α tocopherol: studies using deuterated tocopherols. J. Lipid Res. 31: 687-694.
Traber, MG, Sokol, RJ, Burton, GW, Ingold, KU, Papas, AM, Huffaker, JE, Kayden, HJ (1990c) Impaired ability of patients with familial isolated vitamin E deficiency to incorporate α-tocopherol into lipoproteins secreted by the liver. J. Clin. Invest. 85: 397-407.
Traber, MG, Sokol, RJ, Kohlschütter, A, Yokota, T, Muller, DPR, Dufour, R, Kayden, HJ (1993b) Impaired discrimination between stereoisomers of α-tocopherol in patients with familial isolated vitamin E deficiency. J. Lipid Res. 34: 201-210.
Traber, MG, Sokol, RJ, Ringel, SP, Neville, HE, Thellman, CA, Kayden, HJ (1987) Lack of tocopherol in peripheral nerves of vitamin E-deficient patients with peripheral neuropathy. N. Engl. J. Med. 317: 262-265.
Trabert, W, Stober, T, Mielke, V, Siu Heck, F, Schimrigk, K (1989) Isolierter Vitamin-E-Mangel. Fortschr. Neurol. Psychiat. 57: 495-501.
Weiser, H, Vecchi, M, Schlachter, M (1986) Stereoisomers of α-tocopheryl acetate. IV. USP units and α-tocopherol equivalents of *all-rac-*, *2-ambo-* and *RRR-α*-tocopherol evaluated by simultaneous determination of resorption-gestation, myopathy and liver storage capacity in rats. Int. J. Vit. Nutr. Res. 56: 45-56.
Yokota, T, Wada, Y, Furukawa, T, Tsukagoshi, H, Uchihara, T, Watabiki, S (1987) Adult-onset spinocerebellar syndrome with idiopathic vitamin E deficiency. Ann. Neurol 22: 84-87.
Yoshida, H, Yusin, M, Ren, I, Kuhlenkamp, J, Hirano, T, Stolz, A, Kaplowitz, N (1992) Identification, purification and immunochemical characterization of a tocopherol-binding protein in rat liver cytosol. J. Lipid Res. 33: 343-350.

ROLE OF MITOCHONDRIA IN RADICAL GENERATION AND TOXICITY OF τ-BUTYLHYDROPEROXIDE TOWARDS CULTURED HUMAN ENDOTHELIAL CELLS.

Valerie B. O'Donnell[1], Mark J. Burkitt[2] & Jonathan D. Wood[3].

Valerie B. O'Donnell,
Rowett Research Institute,
Greenburn Road,
Bucksburn,
Aberdeen, AB2 9SB
Scotland.

The structural integrity of the endothelium is of critical importance in the maintenance of normal vascular function. Damage to the endothelial layer can result in loss of several critical functions, including maintenance of vascular tone, control of clotting and production of essential mediators. It has been proposed that oxidant-induced injury to the endothelium is an important component of the damage which leads to atherosclerotic plaque formation (Hefner & Repine, 1989). Oxidatively modified lipids and lipoproteins can alter a variety of endothelial functions including pinocytosis and PDGF production (Borsum *et al*, 1985; Fox *et al*, 1987). Oxidised low density lipoprotein (LDL) can also enhance monocyte adhesion to endothelial cells (Berliner *et al*, 1990), and stimulate production of prostacyclins (Triau *et al*, 1988), colony-stimlation factors (Rajavashisth *et al*, 1990) and tissue factor (Drake *et al*, 1991). Lipid hydroperoxides are a major component of oxidised LDL and have many cytotoxic properties in their own right. Linoleic acid hydroperoxide can induce metalloproteinases (Sasaguri *et al*, 1993) and activate protein kinase C (Taher *et al*, 1993) in cultured endothelial cells, while injection into laboratory animals leads to injury of aortic endothelium (Yagi *et al*, 1981). The mechanism of induction of endothelial damage by lipid hydroperoxides is largely unknown, but has been suggested to involve production of reactive oxygen species that are capable of rapid reaction with important biomolecules (Yagi *et al*, 1991; Yagi *et al*, 1993).

---

[1]Institute of Biochemistry and Molecular Biology, Buhlstrasse 28, CH-3012 Bern, Switzerland.
[2]Rowett Research Institute, Greenburn Rd., Bucksburn, Aberdeen AB2 9SB, Scotland.
[3]Department of Biochemistry, University of Bristol, Bristol BS8 1TD, England

NATO ASI Series, Vol. H 92
Signalling Mechanisms – from Transcription Factors to Oxidative Stress
Edited by L. Packer, K. Wirtz

Organic hydroperoxides, such as τ-butyl hydroperoxide (τ-BOOH), have been used to study the effects of oxidant stress in many cell types. Enzymatically, peroxides can be removed by either selenium-dependent glutathione (GSH) peroxidases (GSHPx) (Flohe, 1982) or selenium-independent glutathione-S-transferases (GST) (Prohaska, 1980). Two selenoperoxidases have been identified; classical GSH peroxidase (cytosolic, mitochondrial matrix) (Mills, 1958) and phospholipid hydroperoxide GSH peroxidase (cytosolic, membrane bound) (Ursini *et al*, 1991). The net reaction catalysed by all of these involves a 2-electron reduction of the peroxide to its corresponding alcohol and simultaneous oxidation of 2GSH to glutathione disulfide (GSSG). Regeneration of GSH is accomplished by the activity of glutathione disulfide-reductase (GRD) which utilises NADPH as a 2-electron reductant.

In red blood cells, peroxide administration causes membrane lipid peroxidation (Trotta *et al*, 1981), increased cation permeability (Van de Zee *et al*, 1985), glutathione depletion (Srivastava *et al*, 1974), protein crosslinking (Corry *et al*, 1980) and oxidative denaturation of hemoglobin (Trotta *et al*, 1981). The mechanism of organic hydroperoxide toxicity in isolated hepatocytes may involve at least two mechanisms. Studies on liver mitochondria isolated from selenium deficient rats have paradoxically suggested a major role for the GSHPx/GRD system in the toxicity of τ-BOOH (Sies & Moss, 1978; Lotscher *et al*, 1979). Following depletion of mitochondrial NAD(P)H, alterations in calcium homeostasis occur which ultimately lead to membrane damage and loss of mitochondrial integrity by free radical-independent mechanisms (Bellomo *et al*, 1984). However, in whole hepatocyte studies, increases in intracellular calcium could be prevented with the iron chelator desferrioxamine and the reductant, dithiothreitol (Nicotera *et al*, 1988). In addition, lipid peroxidation occured independently of intracellular alterations of glutathione or calcium homeostasis (Masaki *et al*, 1989). This suggests that in whole hepatocytes, alterations in intracellular calcium homeostasis may follow peroxidative damage to cell membranes.

Free radical generation by τ-BOOH mitochondria from both liver and heart has been demonstrated (Kennedy *et al*, 1986; Kennedy *et al*, 1992; Radi *et al*, 1993). The observation that lipid peroxidation in submitochondrial particles can prevented by depletion of cytochrome *c* suggests a role for this component in electron donation to/from the peroxide (Radi *et al*, 1993). Indeed, free radical reactions between cytochrome *c* (reduced or oxidised) and τ-BOOH *in vitro* giving rise to alkoxyl, methyl and peroxyl radicals have been shown using electron spin resonance techniques (Kennedy *et al*, 1992; Davies, 1988). Inactivation of catalytic activity of a variety of heme-containing systems other than cytochrome *c* by organic hydroperoxides has been described (Davies, 1988; Pichorner *et al*, 1993; Kadlubar *et al*, 1973; Tajima *et al*, 1993; Yao *et al*, 1993; Maples *et al*, 1990). Formation of both peroxyl and alkoxyl radicals was seen following addition of τ-BOOH to cytochrome P450, metmyoglobin, oxyhemoglobin, methemoglobin, catalase, horseradish peroxidase or hematin (Davies, 1988).

In endothelial cells, major changes induced by organic hydroperoxides include lipid peroxidation (Elliot *et al*, 1993), alterations in ion transporter activities (Elliot *et al*, 1992a), and lipid peroxidation-independent changes in calcium homeostasis (Elliot *et al*, 1992b). Previously, using spin trapping techniques, we demonstrated mitochondrial 1-electron reduction of τ-BOOH to free radicals in human endothelial cells (Figure 1) (O'Donnell & Burkitt, 1994). The ESR spectrum shown in Fig. 1a consists of signals from two free radical adducts to the spin trap 5,5-dimethyl-1-pyrroline-N-oxide (DMPO). The methyl radical is formed via a β-scission reaction of τ-butoxyl radicals , which are generated *via* 1-electron reduction of τ-BOOH. The weaker 4-line signal is believed to be from the DMPO hydroxyl radical adduct, generated as an artefact *via* radical-induced nucleophilic addition of water to DMPO (O'Donnell & Burkitt, 1994).

These signals were unaffected by the inclusion of the metal chelator desferrioxamine indicating a lack of involvement of free metal ions in radical generation. Using mitochondrial inhibitors, the site of reduction was demonstrated to be distal to cytochrome *b* of complex 3 (i.e. cytochromes *c*, c_1, a/a_3). These results are in agreement with those of Kennedy *et al* (1992) who demonstrated alkoxyl and methyl radical formation following 1-electron reduction of τ-BOOH by actively respiring isolated rat liver mitochondria. Interestingly, in the absence of respiratory substrates, they also observed formation of peroxyl radicals from 1-electron oxidation of τ-BOOH. This was not observed in our system, most likely as our intracellular mitochondria would be expected to be actively respiring (37°C, 5mM glucose) during the assay period. Previous studies implicating GSHPx/GRD in the damaging effects of τ-BOOH in isolated mitochondria have shown NADH depletion (within 2 min) to proceed that of NADPH (greater than 10 min) (Bellomo *et al*, 1984). As GRD does not utilize NADH as reducing cofactor, depletion of this pyridine nucleotide was suggested to occur as a result of transhydrogenase activity. In the light of our results, it is equally possible that NADH depletion occurs as a result of hydroperoxide reduction *via* the electron transfer chain.

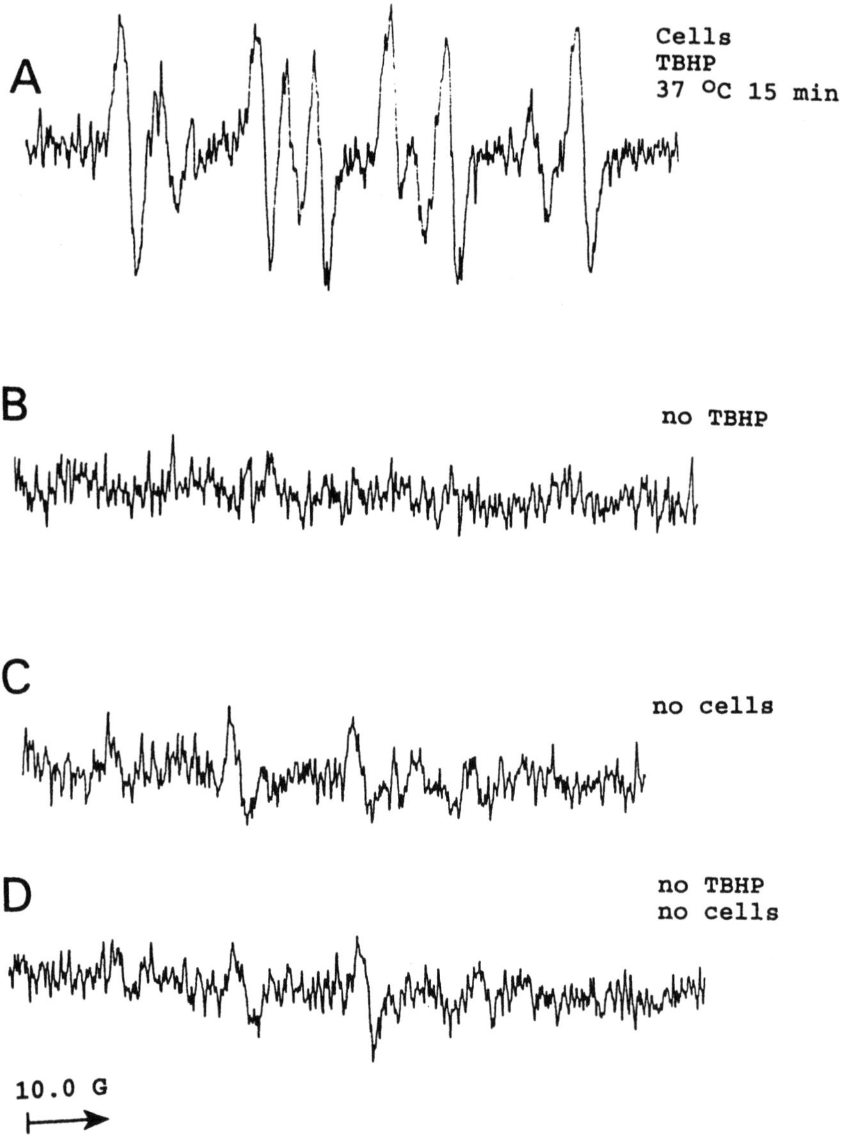

Figure 1. ESR spectra of radical adducts detected following incubation of endothelial cells with τ-BOOH. Cells (5 x 10^5) were preincubated, with stirring at 37°C in PBS/glucose containing 80 mM DMPO for 5 min before the addition of 200μM τ-BOOH. After a further 15 min incubation, cells were transferred to a quartz flat cell and spectra were recorded on a Varian E104 spectrometer using the following instrument settings: modulation frequency, 100 kHz; sweep width, 100 G; scan time, 8 min; time constant, 0.5 sec.; modulation amplitude, 2.5 G; gain, 5 x 10^4. As indicated, incubations were also carried out in the absence of τ-BOOH, cells or both.

To examine whether radical generation plays a role in the cytotoxicity of τ-BOOH, we examined the effects of the spin trap DMPO, on τ-BOOH-mediated release of the cytosolic enzyme, lactate dehydrogenase (LDH). As can be seen in Figure 2, inclusion of DMPO led to complete dose-dependent protection against peroxide toxicity.

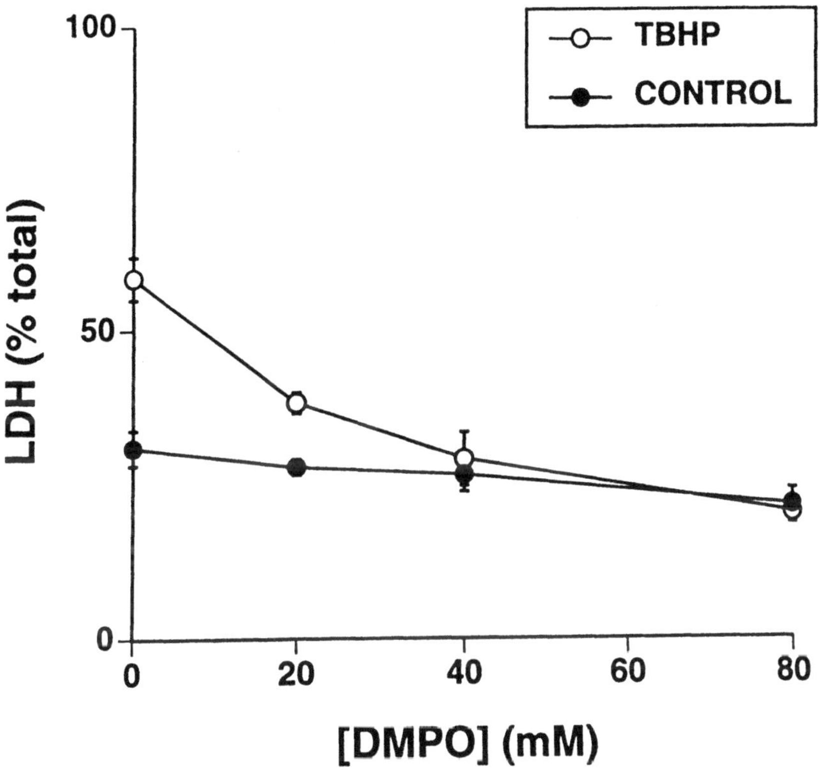

Figure 2. Effect of DMPO on τ-BOOH-induced LDH leakage from endothelial cells. Cells were incubated in the presence or absence of 200μM τ-BOOH in Kreb's Ringer buffer in 24-well plates. DMPO (0-80 mM) was added to control and treated samples immediately before the peroxide. Plates were then incubated for 6 hours at 37 °C/5% CO_2 before determination of cell death by LDH release.

Unexpectedly, inclusion of the metal chelator desferrioxamine also protected against killing (Figure 3). This may indicate participation of iron in propagation of cytotoxic reactions following mitochondrial radical generation.

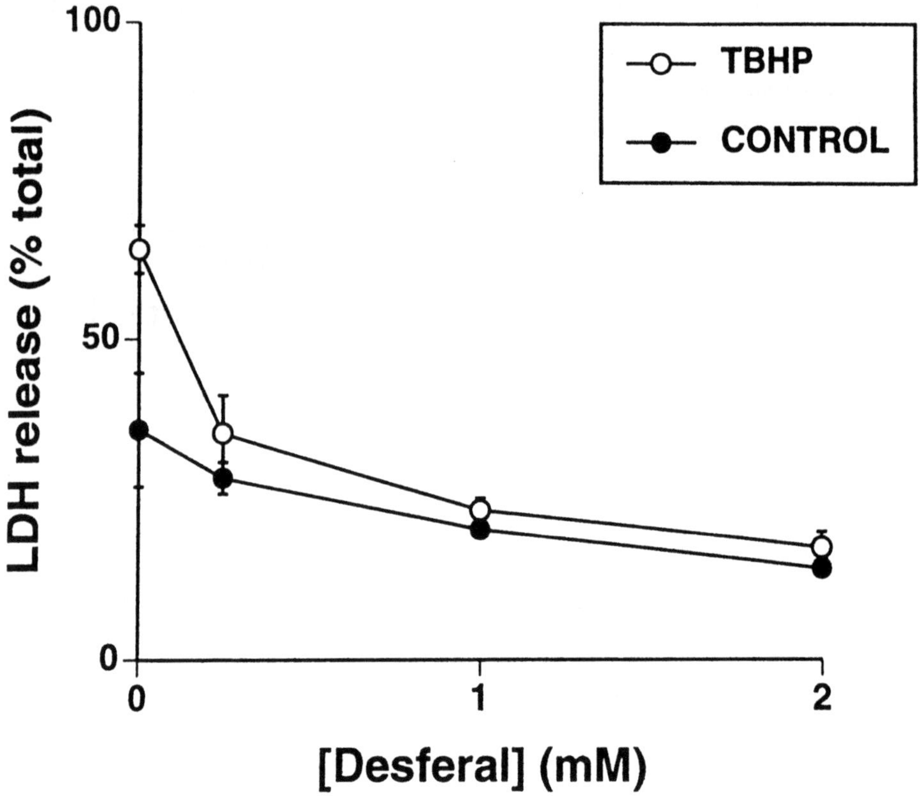

Figure 3. Effect of desferrioxamine on τ-BOOH-induced LDH leakage from endothelial cells. Cells were incubated in the presence or absence of 200μM τ-BOOH in Kreb's Ringer buffer in 24-well plates. Desferrioxamine (0-2 mM) was added to control and treated samples immediately before the peroxide. Plates were then incubated for 6 hours at 37°C/5% CO_2 before determination of cell death by LDH release.

We next sought to examine whether the GSHPx/GRD detoxification system played a role in modulation of toxicity, either in protection of the cells by peroxide removal, or in potentiation of damage by depletion of reduced pyridine nucleotides, as previously described for isolated rat liver mitochondria (Sies & Moss, 1978; Lotscher *et al*, 1979; Bellomo *et al*, 1984). Using the γ-glutamyl cysteine synthase inhibitor, L-buthionine-(S-R)-sulfoximine (BSO), we could deplete intracellular GSH by 97.2%, without effecting viability (Figure 4).

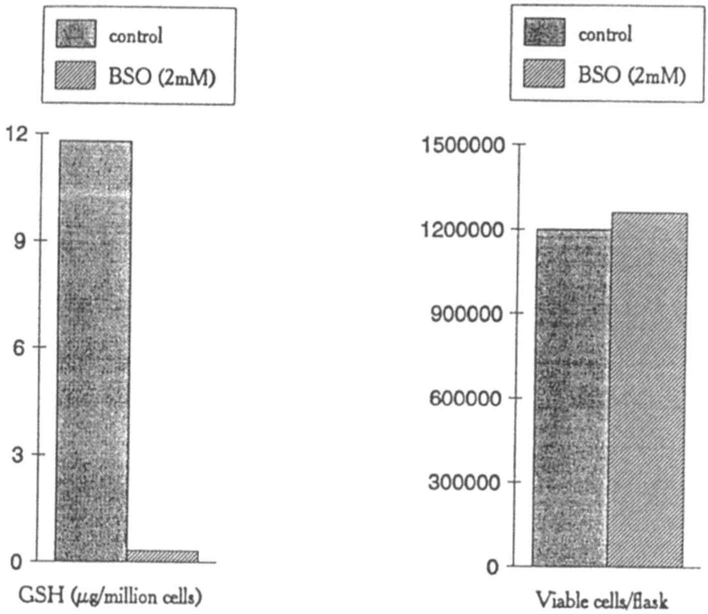

Figure 4. Effect of BSO treatment on HUVEC glutathione content and viability. HUVECs (3.5 x 10^6 total) were incubated for 19 hours with/without BSO (2 mM) at 37°C. Total and oxidised (which was undetectable) glutathione content was determined by the method of Allen *et al* (1988). Cell viability was determined by trypan blue exclusion after trypsinisation of the cells.

Unexpectedly, following this treatment cells became significantly resistant to subsequent hydroperoxide challenge (Figure 5). It is unlikely that this resistance to peroxide toxicity is due to prevention of (free radical-independent) NAD(P)H depletion and subsequent alterations in calcium homeostasis, as removal of free radicals by scavenging with DMPO completely prevented toxicity in these cells. Therefore, we sought an alternative explanation.

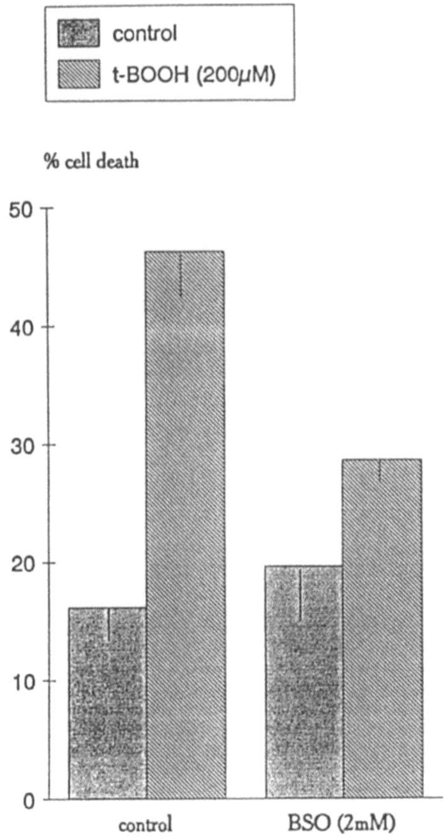

Figure 5. Effect of BSO treatment on sensitivity of HUVECs to τ-BOOH. HUVECs (3.4 x 10^4 per well) were seeded into 24-well plates and grown for 19 hours with/without BSO (2 mM). Following this, they were washed with PBS and incubated with/without τ-BOOH (200μM) in Kreb's Ringer buffer for 6 hours at 37°C. Cell death was determined by LDH release (n = 3 for each bar).

Regulation of gene expression by redox processes is well known in several systems. In eukaryotic cells, the DNA binding activities of the Fos-Jun heterodimer (Abate *et al*, 1990) and nuclear factor/κB (Staal *et al*, 1990) are highly sensitive to sulfhydryl modifying agents. Recently it has been reported that modulation of intracellular glutathione levels can lead to alterations in expression of the heme oxygenase 1 (HO1) gene (Lautier *et al*, 1992). Induction of this enzyme in tandem with ferritin is believed to mediate protection against oxidative stress by removal of potentially prooxidant heme moieties and subsequent chelation of redox active iron (Vile & Tyrrell, 1993).

Following an overnight exposure to BSO (2mM), a 12-fold induction of mRNA (as judged by densitometry scanning, normalised to α-actin) coding for heme oxygenase 1 could be clearly seen in our cells (Figure 6). In addition, a 2.4-fold induction of GSHPx was found. However, in the absence of intracellular GSH, it is unclear if this enzyme could play any significant protective role.

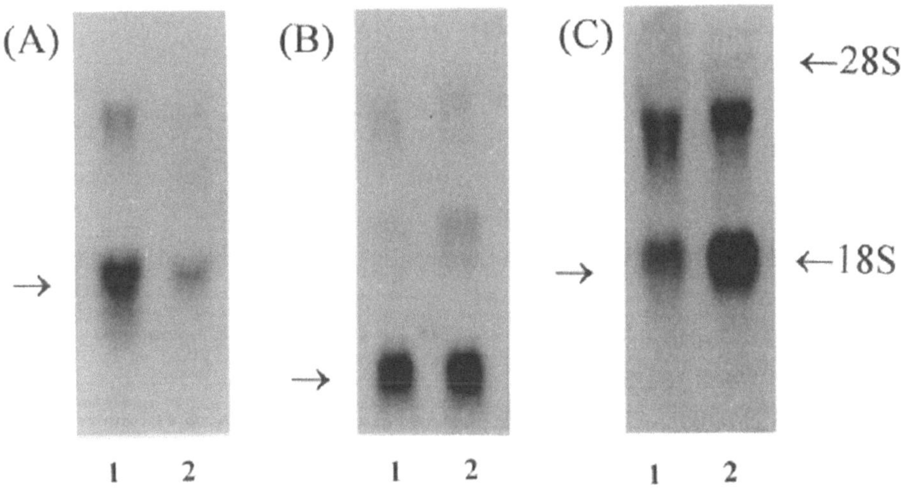

Figure 6. Induction of heme oxygenase 1 (HO1) and glutathione peroxidase by BSO treatment of HUVECs. Northern blot was probed for (a) HO1, (b) GSHPx and (c) α-actin. Cells were grown with (lane 1) or without (lane 2) BSO (2 mM) for 19 hours before mRNA extraction.

Previously, Jurnot & Junod (Jurnot & Junod, 1993) examined for induction of antioxidant enzymes in human endothelial cells in response to a variety of oxidant challenges, including glutathione depletion by BSO. Although they found induction of HO1 (1.8 fold) by BSO, they report no protection of cells against subsequent oxidant challenge. Our findings are in agreement with theirs, although we have much greater induction of HO1 (12-fold v's 1.8-fold) and protection against τ-BOOH toxicity. This is most probably due to differences in incubation time (19 h v's 16 h) and concentration of BSO (2mM v's 200μM).

In conclusion, our results demonstrate a central role for mitochondrial-derived free radicals in hydroperoxide toxicity towards cultured endothelial cells. In addition, it was not possible to show a protective role for the GSHPx/GRD detoxification pathway as BSO treatment led to induction of protective enzyme(s) which resulted in cells which displayed resistance to peroxide challenge.

Acknowlegements:
We would like to thank Dr. R.M. Tyrrell for HO1 cDNA and Dr. P. Harrison for GSHPx cDNA. We would also like to thank S. Jungi for photography of figures. This research was supported by SOAFD.

References:

Abate, C., Patel, L., Rauscher, F.J. & Curran, T. (1990) *Science Wash. DC* **249**, 1157-1161

Allen, K.G.D., Arthur, J.R., Morrice, P.C., Nicol, F. & Mills, C.F. (1988) *Proc. Soc. Exp. Biol. Med.* **187**, 38-43

Bellomo, G., Martino, A., Richelmi, P., Moore, GA., Jewell, S.A. & Orrenius, S. (1984) *Eur. J. Biochem.* **140**, 1-6

Berliner, J.A., Territo, M.C., Sevanian, A., Ramin, S., Kim, J.A., Bamshad, B., Esterson, M. & Fogelman, A.M. (1990) *J. Clin. Invest.* **85**, 1260-1266

Borsum, T., Henriksen, T. & Reisvaag, A. (1985) *Atherosclerosis*, **58**, 81-96

Corry, W.D., Meiselman, H.J. & Hochstein, P. (1980) *Biochim. Biophys. Acta.***597**, 224-234

Davies, M.J. (1988) *Biochim. Biophys. Acta.* **964**, 28-35

Drake, T.A., Hannani, K., Fei, H., Lavi, S. & Berliner, J.A. (1991) *Am. J. Pathol.* **138**, 601-607

Elliot, S.J., Meszaros, J.G. & Schilling, W.P. (1992a) *Free Rad. Biol. Med.* **13**, 635-650

Elliot, S.J. & Schilling, W.P. (1992b) *Am. J. Physiol.* **263**, H96-H102

Elliot, S.J., Doan, T.N., & Schilling, W.P. (1993) *J. Pharmacol. Exp. Ther.* **264**, 1063-1070

Flohe, L. (1982) In *Free Radicals in Biology*, Vol.5. Prior, W.A. Ed., Academic Press, New York, 223-254

Fox, P.L., Chisolm, G.M. & DiCorleto, P.E. (1987) *J. Biol. Chem.* **262**, 6046-6054

Heffner, J.C. & Repine, J.E. (1989) *Am. Rev. Respir. Dis.* **140**, 531-554

Jurnot, L & Junod, A.F. (1993) *Am. J. Physiol.* L482-L489

Kadlubar, F.F., Morton, K.C. & Ziegler, D.M. (1973) *Biochem. Biophys. Res. Commun* **54**, 1255-1261

Kennedy, C.H., Church, D.F., Winston, G.W. & Prior, W.A. (1992) *Free Rad. Biol. Med.* **12**, 381-387

Kennedy, C.H., Prior, W.A., Winston, G.W. & Church, D.F. (1986) *Biochem. Biophys. Res. Commun.* **141**, 1123-1129

Lautier, D., Luscher, P. & Tyrrell, R.M. (1992) *Carcinogenesis*, **13**, 227-232

Lotscher, H.R., Winterhalter, K.H., Carafoli, E. & Richter, C. (1979) *Proc. Natl. Acad. Sci. U.S.A.* **76**, 4340-4344

Maples, K.R., Kennedy, C.H., Jordan, S.J. & Mason, R.P. (1990) *Arch. Biochem. Biophys.* **277**, 402-409

Masaki, N., Kyle, M.E. & Farber, J.L. (1989) *Arch. Biochem. Biophys.* **269**, 390-399

Mills, G.C. (1958) *J. Biol. Chem.* **229**, 189-197

Nicotera, P., McConkey, D., Svensson, S.A., Bellomo, G & Orrenius, S. (1988) *Toxicology*, **52**, 55-63

O'Donnell, V.B. & Burkitt, M.J. (1994) *Biochem. J.* In press.

Pichorner, H., Jessner, G. & Ebermann, R. (1993) *Arch. Biochem. Biophys.* **300**, 258-264

Prohaska, J.R. (1980) *Biochim. Biophys. Acta.* **611**,87-98

Radi, R., Bush, K.M. & Freeman, B.A. (1993) *Arch. Biochem. Biophys.* **300**, 409-415

Rajavashisth, T.B., Andalabi, A., Territo, M.C., Berliner, J.A., Navab, M., Fogelman, A.M. & Lusis, A.J. (1990) *Nature (Lond).* **344**, 254-257

Sasaguri, Y., Kakita, N., Murahashi, N., Kato, S., Hiraoka, K., Morimatsu, M. & Yagi, K.(1993) *Atherosclerosis*, **100**, 189-196

Sies, H. & Moss, K.M. (1978) *Eur. J. Biochem.* **84**, 377-383

Srivastava, S.K., Awasthi, Y.C. & Beutler, E. (1974) *Biochem. J.* **139**, 289-295

Staal, F.J.T., Roederer, M. & Herzenberg, L.A. (1990) *Proc. Natl. Sci. USA.* **87**, 9943-9947

Taher, M.M., Garcia, J.G.N., & Viswanathan, N. (1993) *Arch. Biochem. Biophys.* **303**, 260-266

Tajima, K., Edo, T., Ishizu, K., Imaoka, S., Funae, Y., Oka, S. & Sakurai, H. (1993) *Biochem. Biophys. Res. Commun* **191**, 157-164

Triau, J.E., Meydani, S.N. & Schaefer, E.J. (1988) *Arteriosclerisis*, **8**, 810-818

Trotta, R.J., Sullivan, S.G & Stern, A. (1981) *Biochim. Biophys. Acta.* **679**, 230-237

Ursini, F., Maiorino, M. & Sevanian, A. (1991) In *Oxidative stress: Oxidants and antioxidants.* Academic Press, New York, 319-336

Van de Zee, J., Dubbleman, T.M.A.R. & Van Steveninck, J. (1985) *Biochim. Biophys. Acta.* **818**, 38-44

Vile, G.F. & Tyrrell, R.M. (1993) *J. Biol. Chem.* **268**, 14678-14681

Yagi, K., Ishida, N., Komura, S., Ohishi, N., Kusai, M. & Kohno, M. (1991) *Biochem. Biophys. Res. Commun.* **183**, 945-951.

Yagi, K., Komura, S., Ishida, N, Nagata, N., Kohno, M. & Ohishi, N. (1993) Biochem. Biophys. Res. Commun. **190**, 386-390

Yagi, K., Ohkawa, H., Ohishi, N., Yamashita, M. & Nakashima, T. (1981) *J. Appl. Biochem.* **3**, 58-65

Yao, K., Falick, A.M., Patel, N. & Correia, M.A. (1993) *J. Biol. Chem.* **268**,59-65

Hypoxia and Endothelial Cell Adhesiveness

Anneke Pietersma, Netty de Jong, Johan F. Koster and Wim Sluiter
Department of Biochemistry
Cardiovascular Research Institute (COEUR)
Erasmus University Rotterdam
P.O. Box 1738
3000 DR Rotterdam
The Netherlands

Introduction

Ischemia is a common clinical event leading to local and remote tissue injury. Reperfusion of the ischemic tissues is associated with a local activation and infiltration of granulocytes (Lucchesi *et al.*, Werns *et al.*). These cells form a potential threat to the function and integrity of the endothelial barrier (Wedmore *et al.*), since they can mediate tissue damage by the release of cytotoxic agents, e.g., oxygen derived free radicals and lysosomal enzymes.

Normally, the adherence of granulocytes to the vascular wall is minimized through the production of anti-adhesive factors, e.g., nitric oxide and prostacyclin (Kubes *et al.*, Prescott *et al.*), and a low to moderate expression of the adhesion molecules intercellular adhesion molecule (ICAM)-1 and ICAM-2 by the endothelial cells (Springer, Sluiter *et al.*). In response to tissue injury various inflammatory mediators are generated that have been shown to stimulate the expression of different adhesion molecules on endothelial cells and granulocytes. These adhesion molecules include granular membrane protein-140 (GMP-140 or P-selectin), endothelial leucocyte adhesion molecule-1 (E-selectin) and ICAM-1 on endothelial cells (Sluiter *et al.*). Their expression follows a distinct course in

NATO ASI Series, Vol. H 92
Signalling Mechanisms – from Transcription Factors to Oxidative Stress
Edited by L. Packer, K. Wirtz

time and depends upon the nature of the stimulus. In general, P-selectin is rapidly and transiently expressed by endothelial cells, while the expression of E-selectin and ICAM-1 requires *de novo* protein synthesis. Granulocytes can be stimulated to an increased expression (Miller *et al.*), e.g., of CD11b/CD18 the receptor for the opsonic fragment of the third component of complement iC3b, and/or an increased affinity, e.g., of the granulocyte lymphocyte function associated antigen (CD11a/CD18) for its ligand ICAM-1 (Springer). The increased expression and activation of adhesion molecules results in an increased adherence of granulocytes to endothelial cells. Granulocytes ultimately pass between endothelial cell junctions and penetrate the basement membrane to access the underlying tissues.

The accumulation of granulocytes in the ischemic tissues could very well be a secondary effect of ischemic tissue damage mediated by soluble factors like activated complement and cytokines (Ginis *et al.*). However, since the increased margination of granulocytes is an early event in experimental myocardial ischemia hypoxia may also directly trigger the local acccumulation of these cells. This notion has led a number of investigators to study the effect of hypoxia on the adhesiveness of the endothelial cells (Yoshida *et al.*, Palluy *et al.*, Arnould *et al.*, Ginis *et al.*, Pietersma *et al.*). These studies, in which only human umbilical vein endothelial cells were used, are summarized in Table I. These studies all agree that hypoxia can modulate the adherence of granulocytes to the endothelial cells. The majority of the studies show an *increase* in the adherence of granulocytes following exposure of the endothelial cells to hypoxia. We have found, however, that after 2 h of hypoxia the adherence of granulocytes *decreased* to 50% of the normoxic control. This decreased adherence could not be attributed to a decreased expression of ICAM-1 or ICAM-2, nor to the increased generation of the anti-adhesive factors NO, adenosine or prostacyclin. We therefore concluded that hypoxia cannot directly stimulate the adhesiveness of the endothelial cells (Pietersma *et al.*).

Could these divergent results be attributed to differences in the experimental protocols used? The summary in Table I suggests that the most striking differences are 1) the severity of the hypoxia, 2) the presence or absence of reoxygenation and 3) the number of granulocytes used in the adherence assay.

Table I. Comparison of the experimental protocols used in studying the effect of hypoxia on endothelial cell adhesiveness.

author	HUVEC, passage	ratio cm^2/V	hypoxia gasphase $N_2/CO_2/H_2$	hypoxia time	PMN/cm^2 of EC (*1000)	reoxygenation	adherence
N. Yoshida *et al.*	1-3	9.5[a]	95/3/5	30'	260[a]	yes	increased
O. Palluy *et al.*	3-4	6.7	95/5/-	5 h	250	yes	increased
T. Arnould *et al.*	1	13.7	100/-/-	90'	500	yes	increased
I. Ginis *et al.*	4-6	3.3	95/5/-	3 h	1000	yes	increased
A. Pietersma *et al.*	3-6	4.0	95/5/-	2 h	250	no	decreased

[a]: From reference Inauen *et al.* 1990.

Table II. Comparison of the adhesion assays used in studying the effect of hypoxia on endothelial cell adhesiveness.

author	duration of hypoxia	adherence time	reoxygenation time (adh.time included)	ligands EC	ligands PMN
N. Yoshida *et al.*	30'	1 h	10'-1 h	PAF/ICAM-1	CD11a/CD11b/CD18
O. Palluy *et al.*	5 h	15'	20'-24 h	GMP-140/ELAM-1	not studied
T. Arnould *et al.*	30'-2 h	5'	5'	PAF/ICAM-1/GMP-140	CD11b/CD18
I.Ginis *et al.*	3 h	1 h	1 h	unknown ligand	CD11a/CD18

Hypoxia

The severity of the hypoxic exposure depends on the pO_2 of the gasphase and the rate at which oxygen diffuses from the supernatant medium. (The oxygen consumed by the endothelial cells is in the nanomolar range and is therefore considered to be negligble (Mertens *et al.*)). The rate at which oxygen diffuses from the supernatant medium depends on the area of the supernatant surface and the volume of the supernatant medium. In other words the greater the area the quicker the decrease in medium pO_2, and the greater the supernatant volume the slower the depletion of oxygen. Thus the severity of the hypoxic condition can be expressed as the ratio of the supernatant surface area and the supernatant volume. Figure 1

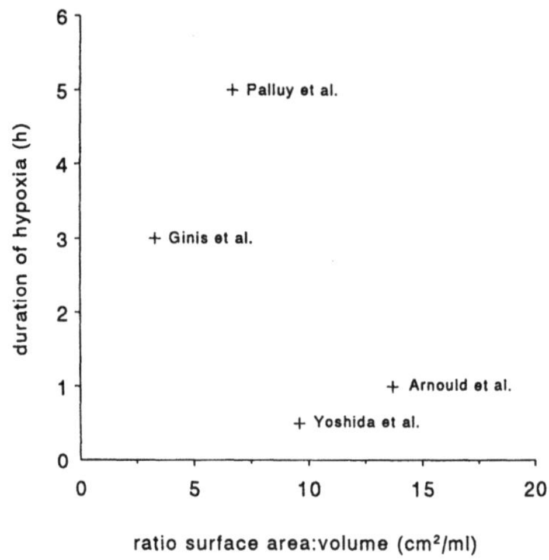

Fig 1. Severity of hypoxia vs. time of exposure, before the earliest reported effect on endothelial cell adhesiveness occurred.

indicates that the earliest effect of hypoxia on the endothelial cell adhesiveness is observed after a shorter incubation period if the hypoxia is more severe. This would implicate that the relative mild hypoxia we have applied in our experimental protocol could only induce increased endothelial cell adhesiveness after a prolonged exposure, i.e., more than 2 hours. (It is important to already note here that we did not reoxygenate the endothelial cells in our experimental protocol as will be discussed in the next section). Milhoan *et al.* have exposed endothelial cells from porcine aorta, coronary microvessel and human umbilical vein to a pO_2 of 2%, a condition which is less severe than ours, and did observe an increased adherence of granulocytes already after 2 hours. Therefore we have to conclude that the severity of the hypoxic condition cannot satisfactorily explain our divergent results.

Reoxygenation

Table I suggests that reoxygenation after the hypoxic incubation could be an important factor in the increased adherence of granulocytes to the endothelial cells. In order to prevent reoxygenation of the endothelial cells during the adherence assay we have used an hypoxic incubator in which cells can be handled under ambient conditions and do not find an increased adherence of granulocytes to endothelial cells under hypoxic conditions. Yoshida *et al.* reported that the adherence of granulocytes increases with increasing duration of reoxygenation of the endothelium. While Palluy *et al.* observed a biphasic response with an elevated adherence at 20 min and at 4 h *after* reoxygenation, fixation of the endothelial cells immediately after hypoxia prevented this effect. Arnould *et al.* reported that reoxygenation could not have been a major factor in the increased adherence they observed. They found an increased adherence after 60 min to 90 min of hypoxia and 5 min of reoxygenation, which could be completely inhibited by the PAF antagonist WEB2086. Since the production of PAF by the endothelial cells started already *during* hypoxia, reoxygenation was not the trigger for the increased production of this stimulator of granulocyte adhesion. In a similar experimental protocol Yoshida *et al.* could completely inhibit the increased adherence either by the PAF antagonist WEB2086 or catalase. This indicates that PAF and/ or hydrogen peroxide are part of the same adherence stimulating mechanism in reoxygenated endothelial cells. The source of the reactive oxygen species (ROS) superoxide radical and or hydrogen peroxide is not known. It could very well be that they originate from the endothelial cells since the increased adherence observed by Palluy *et al.* could be inhibited by superoxide dismutase, catalase and oxypurinol. *In vivo* granulocytes that are located in the ischemic area could be an important source of ROS as well.

Granulocytes

Table I shows that the number of granulocytes plated onto the endothelial cells in the adherence assays varied considerably. Ginis *et al.* showed that the effect of hypoxia on the adherence of granulocytes, administered in a dose range of $3.0x10^5$ to $40.0x10^5$ per cm^2 of endothelial monolayer, was more prominent when granulocytes were added at a number sufficient to saturate the endothelial ligands. They did not observe an increased adherence at the lowest concentration of granulocytes used. We have also found no increase in granulocyte adherence using a similar dose of granulocytes ($2.5x10^5$ granulocytes per cm^2 of endothelial monolayer). We disagree, however, that this lack of effect could be attributed to the administration of an insufficient number of granulocytes since under normoxic conditions only 10% of the granulocytes are bound by the endothelial cells. Furthermore, at higher numbers of granulocytes interaction of these cells with the endothelial cells cannot be excluded, and may actively contribute to the adhesive mechanism. To rule out such an effect fixed granulocytes should be used in the adherence assay.

It is well established that a firm adherence requires the active contribution of both the endothelial cell and the granulocyte. In all five studies granulocytes were isolated from whole human blood. The activation status of circulating granulocytes is reflected by their expression of specific membrane receptors, which can be modulated by microbial products (e.g., CD11b/CD18) and cytokines (e.g., CD64) (Allen *et al.*). Circulating granulocytes are therefore thought to reflect the *in vivo* state of the immune response (Allen *et al.*). Increased levels of IFN-gamma for example might induce a so called primed state of circulating granulocytes in an individual. The activation status of granulocytes can also be affected by preparative procedures. Granulocytes are purified by gradient centrifugation techniques and lysis of contaminating erythrocytes. Especially the temperature at which the purification is performed can induce changes in granulocyte function (Forsyth *et al.*). Cooling granulocytes to 4°C and rewarming the cells to 37°C during 30 min induces an increased expression of all the leukocyte integrins (Forsyth *et al.*). Therefore it needs to be studied whether the activation status of the granulocytes may have contributed to the observed effects.

Ligands

Table II suggests that the increased adherence of granulocytes to endothelial cells after hypoxia and reoxygenation involves multiple adhesive mechanisms. There seems to be an early - less than 1.5 h of hypoxia followed by reoxygenation - and a late response - more that 2 h of hypoxia followed by reoxygenation. A comparison of the results of Yoshida *et al.* and Arnould *et al.*, who studied the early response, shows that there is a great similarity in the adhesive mechanism they have identified. Arnould *et al.* have shown that PAF is synthesized already during hypoxia. In order to stimulate the adherence of granulocytes, PAF would have to be secreted in the supernatant medium and bind the granulocytic PAF-receptor. That PAF indeed is released has been shown by several investigators (Camussi *et al.*, Yoshida *et al.*, Caplan *et al.*). PAF could then trigger the increased expression of CD11b/CD18 (Yoshida *et al.*), and could stimulate the affinity of CD11a/CD18 for its ligand ICAM-1. Another way in which this adhesion stimulating mechanism may operate is through the co-operative actions of P-selectin and PAF. Binding of P-selectin would potentiate the CD11/CD18 dependent adhesive response to PAF (Lorant *et al.*). Arnould et al. show that P-selectin is indeed involved. The results by Palluy *et al.* strongly indicate that the increased expression of P-selectin requires the reoxygenation of the endothelial cells. However, this group submitted the endothelial cells to 5 h of hypoxia followed by reoxygenation and have therefore studied late effects of hypoxia. In contrast to the early effects of hypoxia the late effects do not seem to require the presence of soluble mediators like PAF. The late response involves a direct stimulation of the expression of adhesion molecules by the endothelial cells (Ginis *et al.*, Palluy *et al.*). In addition to P-selectin , E-selectin is expressed. Ginis *et al.* presented evidence for a novel ligand for CD11a/CD18 expressed by endothelial cell after longer periods of hypoxia and subsequent reoxygenation.

Conclusions

In this report we have discussed five studies that investigate the effect of hypoxia on the adhesiveness of endothelial cells. In four studies an increased adherence *after* hypoxia was found. Whereas we observed the opposite. Although our own results appeared contradictory at first glance we have identified two major differences in the experimental protocols used, that could explain the divergent results. In the first place, reoxygenation seems to be required for the increased adhesiveness. Secondly, the number of granulocytes used in the adherence assay may influence the final outcome. Further research is needed to resolve these issues.

Literature references

Allen RC, Stevens DL (1992) The circulating phagocyte reflects the in vivo state of immune defense. Curr Opinion Infect Dis 5:389-398

Arnould T, Michiels C, Remacle J (1993) Increased PMN adherence on endothelial cell after hypoxia: involvement of PAF, CD18/CD11b, and ICAM-1. Am J Physiol (Cell Physiol 33) 264:C1102-C1110

Camussi G, Aglietta M, Malavasi F, Tetta C, Piacibello W, Sanavio F, Bussolino F

(1983) The release of platelet-activating factor from human endothelial cells in culture. J Immunol 131:2397-2403

Caplan MS, Adler L, Kelly A, Hsueh W (1992) Hypoxia increases stimulus-induced PAF production and release from human umbilical vein endothelial cells. Biochim Biophys Acta 1128:205-210

Forsyth KD, Levinsky RJ (1990) Preparative procedures of cooling and re-warming increase leukocyte integrin expression and function on neutrophils. J Immunol Methods 128:159-163

Ginis I, Mentzer SJ, Faller D (1993) Oxygen tension regulates neutrophil adhesion to human endothelial cells via an LFA-1 dependent mechanism. J Cell Physiol 157:569-578

Inauen W, Granger DN, Meininger CJ, Schelling ME, Granger HJ, Kvietys PR (1990) Anoxia-reoxygenation-induced, neutrophil mediated endothelial cell injury: role of elastase. Am J Physiol (Heart Circ Physiol 28) 259:H925-H931

Kubes P, M Suzuki, DN Granger (1991) Nitric oxide: an endogenous modulator of leukocyte adhesion. Proc Natl Acad Sci USA 88:4651-4655

Lorant DE, Patel KD, McIntyre TM, McEver RP, Prescott SM, Zimmerman GA (1991) Coexpression of GMP-140 and PAF by endothelium stimulated by histamine or thrombine: a juxtacrine system for adhesion and activation of neutrophils. J Cell Biol 115:223-234

Lucchesi BR, KM Mullane (1986) Leukocytes and ischemia-induced myocardial injury. Ann Rev Pharmacol Toxicol 26:201-224

Mertens S, Noll T, Spahr A, Krützfeldt A, Piper HM (1990) Energetic response of coronary endothelial cells to hypoxia. Am J Physiol (Heart Circ Physiol 27) 258:H689-H694

Milhoan KA, Lane TA, Bloor CM (1992) Hypoxia induces endothelial cells to increase their adherence for neutrophils: role of PAF. Am J Physiol (Heart Circ Physiol 32) 263:H956-H962

Miller LJ, DF Bainton, N Borregaard, TA Springer (1987) Stimulated mobilization of monocyte MAC-1 and p150.95 adhesion proteins from an intracellular vesicular compartment to the cell surface. J Clin Invest 80:535-544

Palluy O, Morliere L, Gris JC, Bonne C, Modat G (1992) Hypoxia/reoxygenation stimulates endothelium to promote neutrophil adhesion. Free Rad Biol Med 13:21-30

Pietersma A, N de Jong, JF Koster, W Sluiter (1994) Effect of hypoxia on the adherence of granulocytes to endothelial cells in vitro. Am J Physiol (Heart Circ Physiol 36) 267:H874-H879

Prescott SM, McIntyre TM, GA Zimmerman (1990) The role of platelet-activating factor in endothelial cells. Thromb Haemost 64:99-103

Sluiter W, A Pietersma, JMJ Lamers, JF Koster (1993) Leukocyte adhesion molecules on the vascular endothelium: their role in the pathogenesis of cardiovascular disease and the mechanism underlying their expression. J Cardiovasc Pharmacol 22(Suppl 4):S37-S44

Springer TA (1990) Adhesion receptors of the immune system. Nature 346:425-434
Werns SW, and BR Lucchesi (1988) Oxygen Radicals in the Pathophysiology of Heart Disease. Kluwer Academic Publishers Dordrecht 1988:123-140

Wedmore CV, TJ Williams (1981) Control of vascular permeability by polymorphonuclear leukocytes in inflammation. Nature 289:647-650

Yoshida N, Granger DN, Anderson DC, Rothlein R, Lane C, Kvietys PR (1992) Anoxia and reoxygenation-induced neutrophil adherence to cultured endothelial cells. Am J Physiol (Heart Circ Physiol) 262:H1891-H1898

Hypoxia activates endothelial cells to release inflammatory mediators and growth factors

Carine Michiels, Thierry Arnould, Khalid Bajou, Isabelle Géron, José Remacle
Laboratoire de Biochimie Cellulaire
Facultés Universitaires Notre Dame de la Paix
61 rue de Bruxelles
5000 Namur
Belgium

SUMMARY

Ischemia is a common situation involved in several pathologies. Besides the reperfusion injury which is now well established, ischemia by itself also induces damages. However, the biochemical mechanisms and the cell types involved in these damages are still relatively unknown. We focused our attention on this problem using an *in vitro* model where human umbilical vein endothelial cells were submitted to a severe hypoxia. We found that before affecting viability, hypoxia is able to strongly activate the endothelial cells. We observed hypoxia induces an increase in the cytosolic calcium concentration, which is then responsible for the activation of phospholipase A_2. Phospholipase A_2 activity releases arachidonic acid which is transformed in endothelial cells into prostaglandins and lyso-PAF which leads to PAF (platelet-activating factor). The synthesis of both PAF and prostaglandins is actually induced by hypoxia.

The physiological consequences of this endothelial cell activation are numerous. First, we found that hypoxia-activated endothelial cells have an increased adhesiveness for neutrophils leading to their activation. Secondly, they release mitogenic molecules for vascular smooth muscle cells identified as $PGF_{2\alpha}$ and bFGF. These results explain how ischemia by itself can to lead to a local inflammation and induces changes within the vascular wall like its thickening observed in some pathological situations.

NATO ASI Series, Vol. H 92
Signalling Mechanisms – from Transcription Factors to Oxidative Stress
Edited by L. Packer, K. Wirtz

INTRODUCTION

The advances in endothelial research over the past decade have revolutionized ideas about the endothelium. It is no longer seen just as a physical barrier but it is now recognized as a tissue fulfilling important physiological functions (Malik *et al.*, 1989). As a hematocompatible container, the endothelium controls coagulation and platelet functions by releasing antithrombotic molecules such as PGI_2 (prostacyclin). It is very active in the synthesis and secretion of various metabolites like extracellular matrix components and it metabolizes lipoproteins and angiotensin (Jaffe, 1985). It modulates smooth muscle cell contraction and relaxation by releasing vasoactive molecules such as EDRF (endothelium-derived relaxing factor), PGI_2 or endothelin (Henderson, 1991; Henrich, 1991) and their proliferation by synthesizing growth factors like ECDRF (endothelium cell-derived growth factor), PDGF (platelet-derived growth factor) and bFGF (basic fibroblast growth factor) (Schwatrz *et al.*, 1981). Finally, endothelial cells closely interact with leukocytes : these interactions are specific for each type of leukocytes and are precisely regulated by the induction of the expression of different adhesion molecules (Springer, 1990) as well as by the secretion of various chemotactic agents like PAF (platelet activating factor), IL-8 (interleukin-8) and CSFs (colony stimulating factors) (Borsum, 1991).

Ischemia is a common feature of various vascular diseases : impairment of blood circulation is involved in thrombosis, myocardial infarction or cerebral ischemia. A deficit in blood transport leads to a decrease in oxygen and nutriment supply to the tissue and to dramatic changes in the metabolism and functions of this tissue but also to the development of an inflammatory situation with increased vascular permeability and neutrophil recruitment (Entman *et al.*, 1991; Leff & Repine, 1990). The mechanism which links ischemia and tissue damages and which triggers this inflammatory situation is still poorly understood. The site of the endothelium as a barrier between blood where ischemia takes place and the tissue which undergoes the damages makes it necessarily a key actor in this process. It is already known that the endothelium is very sensitive to oxygen depletion during ischemia (Tsao *et al.*, 1990) but as yet the consequences of the impairment of endothelial functions on the whole tissue have not been deeply investigated. We focused our work on this problem. For this purpose, we used an *in vitro* model which mimics the *in vivo* ischemia. We will summarize the main results obtained on the activation of the endothelial cells by hypoxia and see how it affects their interactions with neutrophils and smooth muscle cells.

EXPERIMENTAL MODEL

Human umbilical vein endothelial cells (HUVEC) are incubated in a saline solution containing 5 mM glucose under 100 % N_2, leading to a PO_2 in the medium below 10 mm Hg. In these conditions, the viability of the cells remains higher than 95 % up to 120 min incubation (Michiels *et al.*, 1992).

During hypoxia incubation, the first modification observed in HUVEC is a decrease in the ATP content of the cells : ATP content is already lowered after 30 min hypoxia and a 43 % decrease is observed after 120 min hypoxia. This decrease can be inhibited if ß-hydroxybutyrate, which is an energy-supplier molecule, is added during the hypoxia incubation (figure 1A).

Hypoxia also induces an increase in the cytosolic calcium concentration : this concentration is about 22 nM in normoxic conditions and rises up to 300-400 nM after 120 min hypoxia. This modification is reversible : the calcium homeostasis is recovered in 45 min when cells are replaced in normoxic conditions. Both calcium influx and calcium release from intra-

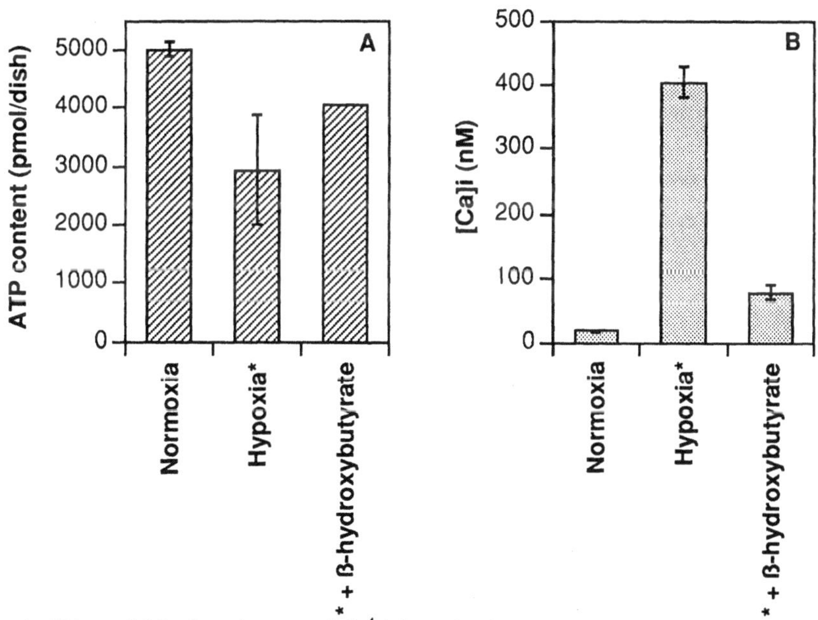

Figure 1 : Effect of ß-hydroxybutyrate (10^{-4} M) on the decrease in ATP content (expressed as in pmol/dish) and the increase in intracellular calcium concentration (expressed in nM) induced in HUVEC by 120 min of hypoxia incubation.

cellular stores account for this increase respectively for 60 and 40 % (Arnould *et al.*, 1992). As for the decrease in ATP content, ß-hydroxybutyrate inhibits the hypoxia-induced increase in cytosolic calcium concentration (figure 1B), suggesting that both events are linked.

Calcium is well known to act as second messenger for the activation of various enzymes. Amongst them, phospholipase A_2 (PLA_2) is very important in endothelial cells. We have shown that hypoxia activates PLA_2 activity in HUVEC already after 30 min, this activation is specific for PLA_2 since only arachidonic acid is released from membrane phospholipids but not fatty acids in the *sn*-1 position of these phospholipids (Michiels *et al.*, 1993). In HUVEC, arachidonic acid is mainly transformed by cyclooxygenase into different prostaglandins (PG). In normal conditions, HUVEC synthesize four different PGs : 26.9, 7.7, 1.4 and 1.0 ng PG/mg of proteins respectively for 6-keto-$PGF_{1\alpha}$ (a stable metabolite of PGI_2), $PGF_{2\alpha}$, PGE_2 and PGD_2. When HUVEC are incubated under hypoxia for 120 min, there is a strong induction of the synthesis of all four PGs, up to 6-fold leading respectively to the synthesis of 139.2, 34.8, 12.0 and 9.5 ng PG/mg of proteins (Michiels *et al.*, 1993).

PLA_2 activity also releases lyso-PAF from membrane phospholipids. In normal conditions, HUVEC do not synthesize PAF but PAF synthesis is rapidly induced in hypoxic

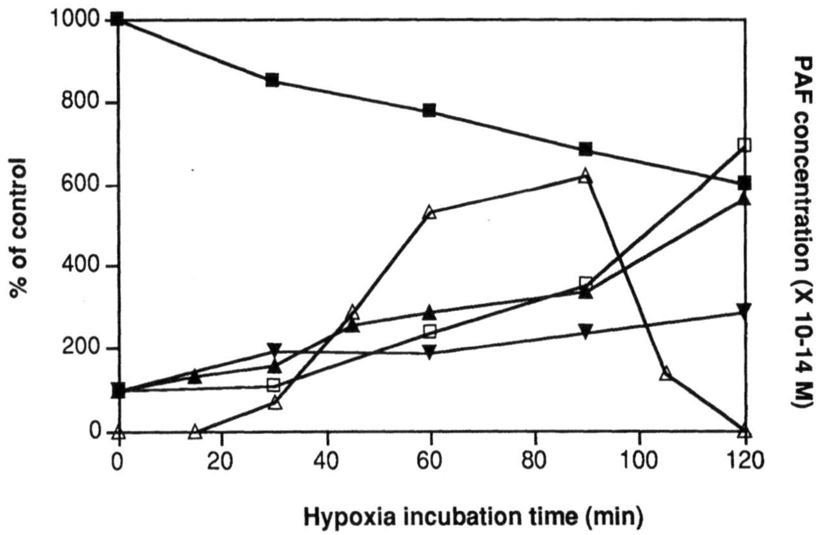

Figure 2 : Time curve of the decrease in ATP content (■, expressed in ‰ of control), the increase in cytosolic calcium concentration (□, expressed in % of control), in phospholipase A_2 activity (▼, expressed in % of control), in prostaglandin synthesis (▲, expressed in % of control) and in PAF synthesis (Δ, expressed in 10^{-14} M) induced in HUVEC by hypoxia.

conditions leading to a maximal concentration of 0.24 10^{-12} mol/10^{6} cells after 90 min (Arnould *et al.*, 1993). Afterwards, PAF synthesis decreases and PAF could no longer be detected after 120 min hypoxia, probably due to a rapid hydrolysis. PAF is not released into the medium but remains associated to the cells.

The time curves of the different processes induced in HUVEC by hypoxia are illustrated in figure 2. The first major change which is observed is the decrease in ATP content, then the cytosolic calcium concentration and PLA_2 rise in parallel as well as the induction of the synthesis of both PGs and PAF.

WHICH TRANSDUCTION SIGNAL IS INDUCED BY HYPOXIA IN HUVEC ?

These different results show that hypoxia is able to activate the endothelial cells to release different inflammatory mediators. This represents a strong stimulation when compared for example to the activation induced by a classical stimulus like thrombin (Michiels *et al.*, 1993; Arnould *et al.*, 1993) but this activation is in one way very different and unusual : it is not mediated by a receptor. The important question raised by these observations is what could be the transduction signal that hypoxia induces and which leads to the release of inflammatory mediators such as PGs and PAF.

The results presented in figure 1 suggest that there is a biochemical link between the decrease in ATP content and the rise in the cytosolic calcium concentration. It is well accepted that the oxidative phosphorylations of the mitochondrial respiration slow down during hypoxia due to the oxygen lack leading to a decrease in ATP regeneration. As a compensatory mechanism glycolysis is induced. We have shown that it is also the case in HUVEC exposed to hypoxia : there is an increase in glucose incorporation and in lactate production (Janssens *et al.*, submitted). This induction is however transient and glycolysis stops after 120 min hypoxia probably due to an accumulation of NADH.

We are currently investigating the hypothesis that due to the glycolysis activation, the intracellular pH decreases. This could induce the Na^{+}/H^{+} exchanger to maintain the cytosolic proton concentration close to physiological values and the influx of sodium ion would then induce the Na^{+}/Ca^{++} exchanger leading to an influx of calcium ion. That calcium influx could be responsible for the increased cytosolic calcium concentration in HUVEC observed during hypoxia incubation. This calcium will then act as a second messenger as it does in other situations. This hypothesis which has been evidenced on cardiomyocytes (Karmazyn & Moffat, 1993) but which still needs experimental confirmation in endothelial cells is presented in figure 5.

CONSEQUENCES OF THE RELEASE OF INFLAMMATORY MEDIATORS

Hypoxia which mimics the *in vivo* ischemia activates endothelial cells to release inflammatory mediators. *In vivo*, the endothelium closely interacts with the circulating blood cells and mainly leukocytes. Amongst them, polymorphonuclear neutrophils (PMN) are the most abundant ones. We thus wanted to know whether hypoxia could modulate the interactions between endothelial cells and PMN.

In normal conditions, the adherence of PMN to HUVEC is very low but increases if either PMN or endothelial cells are activated. When HUVEC are incubated under hypoxia, their adhesiveness for PMN gradually increases to reach 20 to 25 % after 120 min. This adherence is mediated by PAF after 90 min of hypoxia as indicated by the inhibition obtained with PAF receptor antagonists and with PAF synthesis inhibitor. When tested on HUVEC incubated for 120 min under hypoxia, PAF antagonists could not inhibit the PMN adherence, whereas inhibition of PAF synthesis during hypoxia could block the process, suggesting a role of PAF acting as a second messenger. In addition, the inhibitory effects obtained using monoclonal antibodies indicates that this increased adherence is also mediated by ICAM-1 (intercellular adhesion molecule -1) on HUVEC and by CD18/CD11b on PMN. P-selectin seems also to be involved after 90 min hypoxia but not after 120 min, which correlates well with the presence of

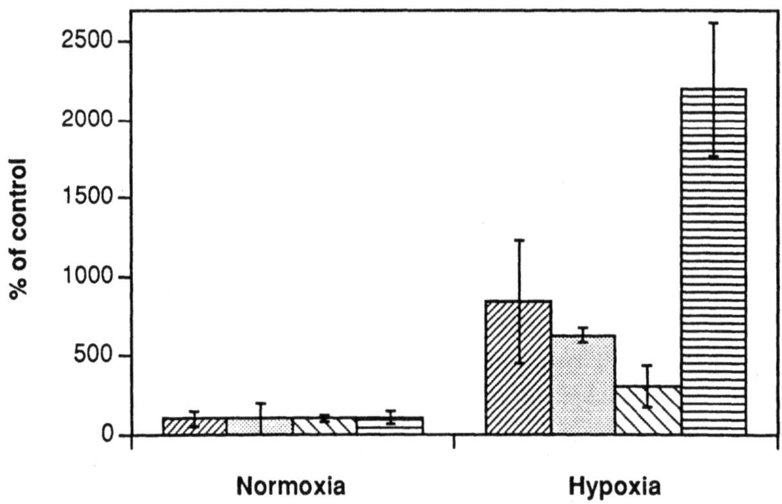

Figure 3 : Co-incubation of unstimulated human PMN for 5 min with HUVEC which were incubated 120 min in normoxic or hypoxic conditions : the adherence of PMN to HUVEC (▨), their cytosolic calcium concentration (░), their production of superoxide anion (▧) and their production of leukotriene B4 (☰) were measured. The results are expressed in % of the corresponding control.

PAF. However, the precise mechanism which triggers PMN adherence after 120 min hypoxia is still unclear but is not mediated by HETE (hydroxyeicosatetraenoic acid) or IL-8 and does not depend on *de novo* protein synthesis (Arnould *et al.*, 1993).

Hypoxia is thus able to activate the endothelial cells to become adhesive for neutrophils and this depends at least in part on their PAF synthesis. Not only hypoxia does induce PMN adherence to HUVEC but there is a strong activation of the PMN which adhere to the hypoxic HUVEC as evidenced by an increase in intracellular calcium concentration as well as by superoxide anion release and leukotriene B_4 synthesis by these PMN (figure 3). On the other hand, conditioned medium from hypoxic HUVEC failed to activate PMN and when PMN are in the presence of hypoxic HUVEC but if the adherence process is blocked for example by inhibiting PAF synthesis, no activation of PMN can be observed. These observations indicate that the activation of PMN is probably induced by the adherence process to the hypoxic HUVEC (Arnould *et al.*, 1994).

These different results indicate that hypoxia is able to modulate the interactions between endothelial cells and neutrophils and induces an inflammatory situation where neutrophils become adherent to the endothelial cells and are activated during this process. This inflammatory situation is initiated by the activation of the endothelial cells by hypoxia. The figure 5 presents a schematic illustration of these interactions.

HYPOXIA ALSO ACTIVATES ENDOTHELIAL CELLS TO RELEASE GROWTH FACTORS

In addition to modulate leukocyte functions, endothelial cells also regulate smooth muscle cell (SMC) contraction and proliferation by releasing vasoactive molecules and growth factors. In several pathological situations, the delicate balance between growth inhibitor and mitogen synthesis by the endothelium is disturbed and SMC migrate into the intima where they proliferate, leading to a thickened vessel wall. Whether ischemic situations could play a role in this process is still unknown.

To answer this question, HUVEC-conditioned media were added to SMC and the proliferation of these cells was measured. We observed pro-proliferative activity for SMC of the hypoxic HUVEC-conditioned medium but not of the normoxic HUVEC one. This activity is evidenced after 3 day incubation by an increased cell number and an increased DNA synthesis (Michiels *et al.*, 1994). It can be inhibited if the activation of HUVEC by hypoxia is blocked by adding ß-hydroxybutyrate as well as if the synthesis of PGs by HUVEC during hypoxia incubation is inhibited by indomethacin (figure 4). $PGF_{2\alpha}$ at the concentration found in the hypoxic HUVEC-conditioned medium was demonstrated to have a mitogenic effect on SMC. However, when tested on fibroblast which do not respond to $PGF_{2\alpha}$, hypoxic HUVEC-conditioned medium also has a pro-proliferative activity. In addition, cycloheximide does not inhibit the release of the mitogens by HUVEC during hypoxia incubation and anti-bFGF antibodies but not anti-PDGF block the mitogenic activity of this conditioned medium on SMC. Finally, the mitogenic effects of $PGF_{2\alpha}$ and bFGF on SMC are additive (Michiels *et al.*, 1994). These results indicate that $PGF_{2\alpha}$ could act in synergy with bFGF and both molecules are probably the molecules responsible for the pro-proliferative activity observed in hypoxic HUVEC-conditioned medium. These findings could be of great importance since they provide an explanation for the excessive growth of SMC in blood vessels following chronic ischemic situations.

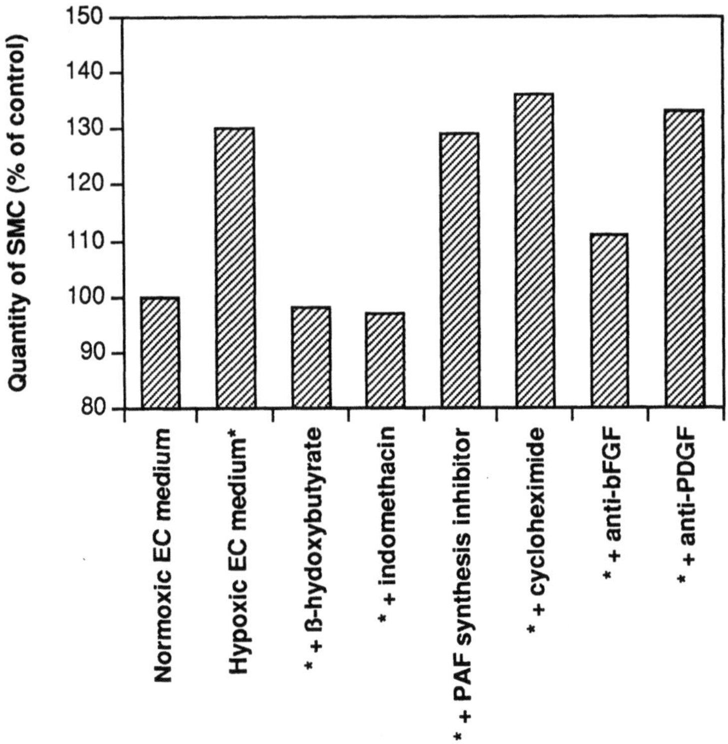

Figure 4 : Effect of conditioned medium from HUVEC incubated 120 min under normoxic or hypoxic conditions on the proliferation of SMC. HUVEC were also incubated under hypoxia in the presence of 10^{-4} M ß-hydroxybutyrate, 10^{-5} M indomethacin, 10^{-7} M oleic acid (= PAF synthesis inhibitor) or preincubated for 4 hours with 5 10^{-5} M cycloheximide before the hypoxic incubation or their conditioned media were preincubated 90 min with 50 µg/ml anti-bFGF or anti-PDGF. Results are expressed in % of the corresponding control.

CONCLUSION

Ischemic events occur in various pathological situations. Before inducing cytotoxic effects, ischemia can trigger other effects which can also be of importance for the recovery of the organ. We have shown that hypoxia *in vitro* can activate endothelial cells to release inflammatory mediators and growth factors. The first series of molecules is responsible for the recruitment of neutrophils leading to their adherence to the endothelial cells and their activation. *In vivo,* this would lead to their infiltration into the underlying tissue where they can induce damages. In addition, hypoxic endothelial cells release growth factors which can trigger the proliferation of SMC. Infiltrated activated neutrophils and proliferating SMC would induced profound alterations in the ischemic tissue and lead to a pathological situation. All these events originate from the activation of endothelial cells by hypoxia which is thus a key event warranting further investigations for example to precisely determine the transduction signal involved in this process.

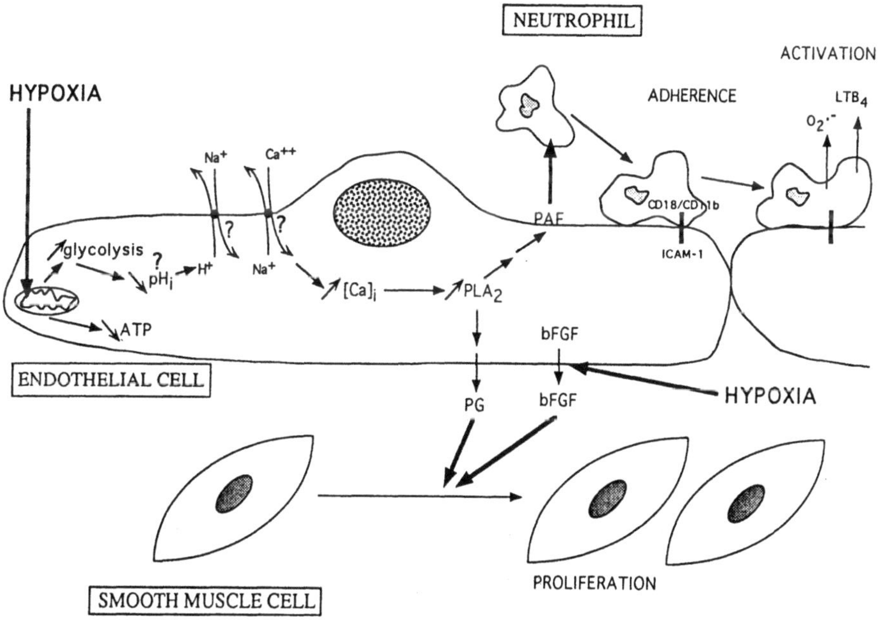

Figure 5 : Schematic representation of the effects of hypoxia on the metabolism of endothelial cells and on their interactions with neutrophils and smooth muscle cells.
? = hypothetic pathways.

Acknowledgment : CM is Senior Research Assistant of the FNRS (Fonds National de la Recherche Scientifique, Belgium) and TA has a grant from FNRS. We would like to acknowledge the support of SSTC and FRRC (Belgium) and the nurses of the Clinique Sainte Elisabeth in Namur for providing the umbilical cords.

REFERENCES

Arnould T, Michiels C, Alexandre I, Remacle J (1992) Effect of hypoxia upon intracellular calcium concentration of human endothelial cells. J. Cell. Physiol. 152:215-221

Arnould T, Michiels C, Remacle J (1994) Hypoxic human umbilical vein endothelial cells induce activation of adherent polymorphonuclear leukocytes. Blood 83:3705-3716

Arnould T, Michiels C, Remacle J (1993) Increased PMN adherence on endothelial cells after hypoxia: involvement of PAF, CD18/CD11b and ICAM-1. Am. J. Physiol. 264:C1102-C1110

Borsum T (1991) Biochemical properties of vascular endothelial cells. Virchows Arch. B Cell Pathol. 60:279-286

Entman ML, Michael L, Rossen RD, Dreyer WJ, Anderson DC, Taylor AA, Smith CW (1991) Inflammation in the course of early myocardial ischemia. FASEB J. 5:2529-2537

Henderson AH (1991) Endothelium in control. Br. Heart J. 65:116-125

Henrich WL (1991) The endothelium: a key regulator of vascular tone. Am. J. Med. Sci. 302:319-328

Jaffe EA (1985) Physiological functions of normal endothelial cells. Ann. NY Acad. Sci. 454:279-291

Janssens D, Michiels C, Delaive E, Eliaers F, Drieu K, Remacle J (submitted) Protection of hypoxia-induced ATP decrease in endothelial cells by *Ginkgo biloba* extract and bilobalide. Biochem. Pharmacol.

Karmazyn M, Moffat MP (1993) Role of Na^+/H^+ exhange in cardiac physiology and pathophysiology: mediation of myocardial reperfusion injury by the pH paradox. Cardiovasc. Res. 27:915-924

Leff JA, Repine JE (1990) Blood cells and ischemia-reperfusion injury. Blood Cells 16:183-192

Malik AB, Lynch JJ, Cooper JA (1989) Endothelial barrier function. J. Invest. Dermatol. 93:62S-67S

Michiels C, Arnould T, Houbion A, Remacle J (1992) Human endothelial cells submitted to hypoxia-reoxygenation: implication of free radicals, xanthine oxidase and energy deficiency. J. Cell. Physiol. 153:53-61

Michiels C, Arnould T, Knott I, Dieu M, Remacle J (1993) Stimulation of prostaglandin synthesis by human endothelial cells exposed to hypoxia. Am. J. Physiol. 264:C866-C874

Michiels C, Arnould T, Remacle J (1993) Hypoxia-induced activation of endothelial cells as a possible cause of venous diseases: hypothesis. Angiology 44:639-646

Michiels C, De Leener F, Arnould T, Dieu M, Remacle J (1994) Hypoxia stimulates human endothelial cells to release smooth muscle cell mitogens: role of prostaglandins and bFGF. Exp. Cell Res. 213:43-54

Schwartz SM, Gajdusek CM, Selden SC (1981) Vascular wall growth control: the role of the endothelium. Arteriosclerosis 1:107-161

Springer TA (1990) Adhesion receptors of the immune system. Nature 346:425-434

Tsao PS, Aoki N, Lefer DJ, Johnson G, Lefer AM (1990) Time course of endothelial dysfunction and myocardial injury during myocardial ischemia and reperfusion in the cat. Circulation 82:1402-1412

Oxidative Stress Induced by γ-interferon and Lipopolysaccharide in Rat Hepatocyte Cultures. Relationship with Nitric Oxide Production

Odile SERGENT, Isabelle MOREL, Martine CHEVANNE, Pierre CILLARD and Josiane CILLARD.

Laboratory of Cellular Biology, INSERM U49
Faculty of Pharmacy
2, av. Pr. Léon Bernard
35043 Rennes Cedex
FRANCE

Introduction

γ-interferon (IFN), a proinflammatory polypeptide, and lipopolysaccharide (LPS), a bacterial endotoxin, are known to enhance the antimicrobial activity of macrophages and to increase their capacity to secrete reactive oxygen species such as hydrogen peroxide and superoxide anion (Bautista AP et al, Clark IA et al, Ding AH et al, Nathan CF et al).

In some inflammatory circumstances like systemic infection, IFN which is normally not present in the tissues, appears rapidly (Billiau AB et al). Recently, it was shown that rat hepatocytes and even human hepatocytes can be stimulated to produce nitric oxide (NO) by a mixture of LPS and cytokines as IFN, interleukin 1 or tumor necrosis factor; but LPS alone or IFN alone cannot lead the NO production in rat hepatocytes (Geller DA et al, Nussler AK et al (a)). An inducible NO synthase was found following exposure of hepatocytes to a combination of LPS and cytokines. While NO has been implicated as mediator of the cytotoxic action of activated macrophages (Hibbs JB et al, Jiang X et al, Keller R), the real toxicity of NO production in hepatocytes is incompletely understood. *In vitro* NO can inhibit mitochondrial respiration and protein synthesis in hepatocytes (Curran RD et al, Stadler J et al). Furthermore NO has been described to react with superoxide anion to form the powerful oxidant, peroxynitrite anion (Beckman et al) or with hydrogen peroxide to produce singlet oxygen, another oxidant (Noronha-Dutra et al).

These observations suggested that a combination of IFN and LPS would be able to generate reactive oxygen species and NO simultaneously in hepatocytes. The purpose of this work was to investigate whether NO could be implicated in an oxidative stress induced by IFN

NATO ASI Series, Vol. H 92
Signalling Mechanisms – from Transcription Factors to Oxidative Stress
Edited by L. Packer, K. Wirtz

and LPS in rat hepatocytes. In this paper oxidative stress has been studied by estimating lipid peroxidation and superoxide scavenging capacity of the cells.

Materials and methods

Materials

IFN, LPS obtained from Escherichia coli serotype 055:B5 and l-N^G-monomethylarginine acetate (lNMMA) were purchased from Sigma (Saint Quentin Fallavier, France).

Cell isolation and culture

According to a method previously described (Guguen-Guillouzo C et al), adult rat hepatocytes were isolated and cultured. Usually 20 x 10^6 hepatocytes, plated in 175 cm^2 Nunclon® flasks, were supplemented with IFN (500 U/ml) and LPS (20 μg/ml). The cultures were then incubated at 37°C for 6, 12, 15, 18 and 27 hours. Some control cultures were incubated without IFN and LPS.

For another experiment, 500 mM l-N^G-monomethylarginine (lNMMA) was added to the cultures with IFN and LPS or not.

Evaluation of lipid peroxidation

The extent of lipid peroxidation was measured using two markers as previously described (Sergent O et al) : extracellular free malondialdehyde (MDA) estimated on the ultrafiltrate of culture medium by size exclusion chromatography and conjugated dienes evaluated in the cells by the second derivative ultraviolet spectroscopy of the lipid extract.

Estimation of superoxide anion scavenging capacity of hepatocytes

Superoxide anion was generated *in vitro* by the xanthine/xanthine oxidase reaction in presence of the spin trap 5,5-dimethyl-1-pyrroline N-oxide (DMPO). Superoxide anion content was estimated by the formation of DMPO-OOH adduct detected by Electron Paramagnetic Resonance (EPR). EPR spectra were obtained at ambient temperature with the EPR parameters as following : 0.804 G modulation amplitude, 100 kHz modulation frequency, 9.71 GHz frequency and 20 mW microwave power. To the reaction mixture (xanthine oxidase (0.8 UI/ml) : 30 µl; xanthine (10 mM) : 30 µl; DMPO (800 mM) : 20 µl; absolute ethanol : 100 µl) was added 20 µl of the methanol supernatant of the cellular extract obtained for the conjugated dienes estimation (Sergent O et al). The DMPO-OOH signal intensity was evaluated by double-integration of both lines of lowest field.

Measurement of NO production

Two methods were simultaneously used.

First, the signal of iron-nitrosyl complex formed by the binding of NO on iron containing protein was observed directly in the intact cells using EPR. Briefly, after each incubation time, culture media were removed and hepatocytes were scraped up, washed, resuspended in 400 µl of buffer composed with 50 mM hepes and 250 mM sucrose at pH 7.42. Then, the hepatocytes were transferred to quartz EPR tubes and frozen. EPR examination was performed at 100 K using a Bruker 106 spectrometer at 10 G modulation amplitude, 100 kHz modulation frequency, 9.34 GHz frequency and 10 mW microwave power. The iron-nitrosyl complex intensity was estimated by double-integration of both lines and then corrected for the protein concentration.

The second method for NO production analysis was to evaluate nitrite concentration by the Griess colour reaction in the culture medium. Briefly 100 µl of aliquots were removed from the culture medium and incubated with 100 µl of the Griess reagent (1 % sulfanilamide, 0.1 % naphthylethylenediamine dihydrochloride in 5 % phosphoric acid) at room temperature for 10 minutes. Then the absorbance at 541 nm was measured and nitrite concentration was determined using a curve calibrated with sodium nitrite standards.

Protein assay

Protein concentration was measured by the Bradford method using Bio-Rad assay solution with bovine serum albumin as standard.

Results

Lipid peroxidation

During the incubation time with IFN and LPS, lipid peroxidation extent remained close to that observed in controls, except at 18 hours (Figure 1). For this incubation time both conjugated dienes and malondialdehyde increased above control values ($p < 0.01$ for both indexes).

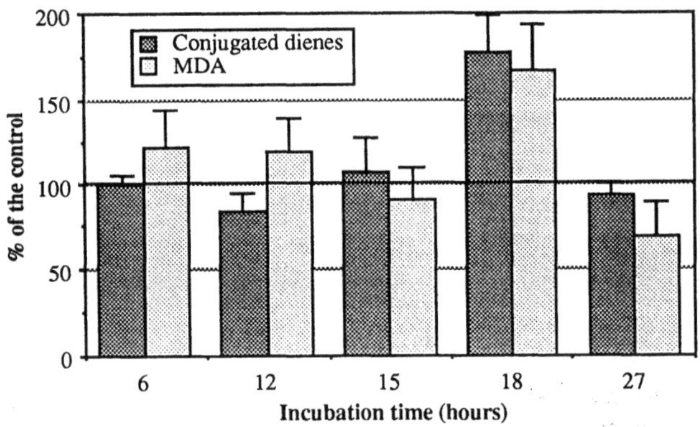

Figure 1:
Lipid peroxidation induced in rat hepatocyte cultures (20 x 10^6 cells/flask) incubated with IFN (500 U/ml) and LPS (20 µg/ml). 100 % corresponded to free MDA or conjugated diene levels evaluated in the control hepatocyte cultures.

Superoxide scavenging capacity of the hepatocytes

Superoxide scavenging capacity of hepatocytes showed a gradual decrease during the incubation with IFN and LPS (Figure 2). At the early stage of incubation (6 hours), the superoxide scavenging capacity was above that of the control ($p < 0.01$). For further incubation times, the scavenging activity was close to the controls, whereas on and after 18 hours, it was significantly below (18 hours : $p< 0.05$; 27 hours : $p < 0.01$).

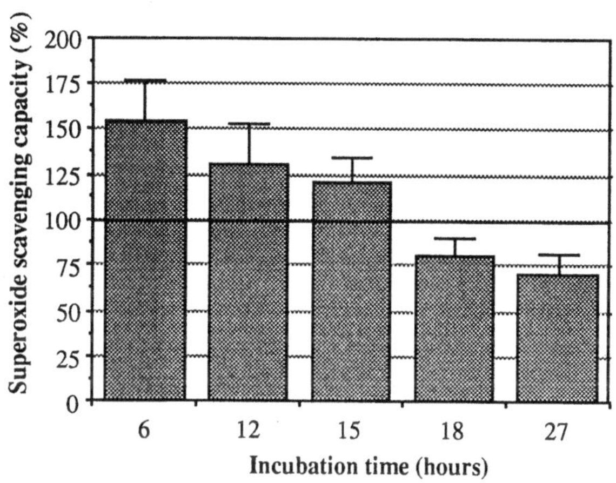

Figure 2 :
Superoxide scavenging capacity of rat hepatocytes (20 x 10^6 cells/flask) supplemented with IFN (500 U/ml) and LPS (20 µg/ml). 100 % corresponded to the DMPO-OOH signal observed by EPR in the control hepatocytes.

NO production

On and after 12 hours of incubation with IFN and LPS, iron-nitrosyl complex signal showed an elevation when compared to the controls ($p < 0.01$ at 12 and 15 hours) (Figure 3). The maximal increase was observed at 18 hours. The first significant increase of nitrite level was also detected at 12 hours ($p < 0.001$). Nevertheless, the maximal value obtained for nitrite concentration in the culture media was not at 18 hours as for iron-nitrosyl complex, but at 27 hours.

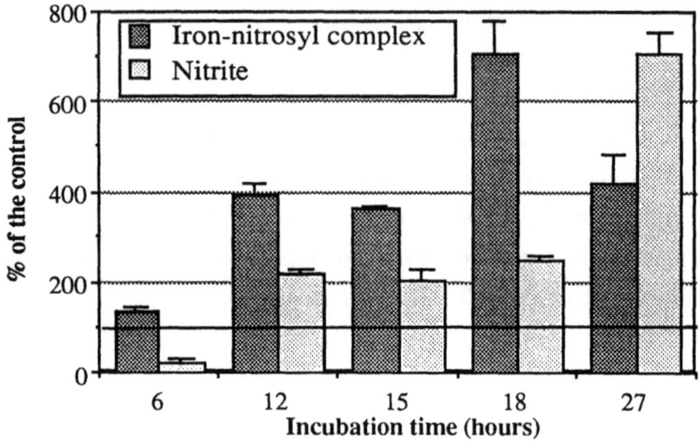

Figure 3 :
Nitric oxide production in rat hepatocyte cultures (20 x 10^6 cells/flask) supplemented with IFN (500 U/ml) and LPS (20 µg/ml). 100 % corresponded to the iron-nitrosyl complex signal observed by EPR in the control hepatocytes or to the nitrite concentration in the media of the control cultures.

Effect of a NO synthase inhibitor, l-N^G-monomethylarginine

When hepatocyte cultures were supplemented with 500 mM lNMMA and incubated with IFN and LPS for 18 hours, a significant inhibition of NO production was observed (Figure 4). Simultaneously, lipid peroxidation, estimated by MDA release in the culture media and conjugated diene content in the cells was inhibited. The superoxide scavenging capacity was increased and even reached a value superior to that of the control ($p < 0.05$).

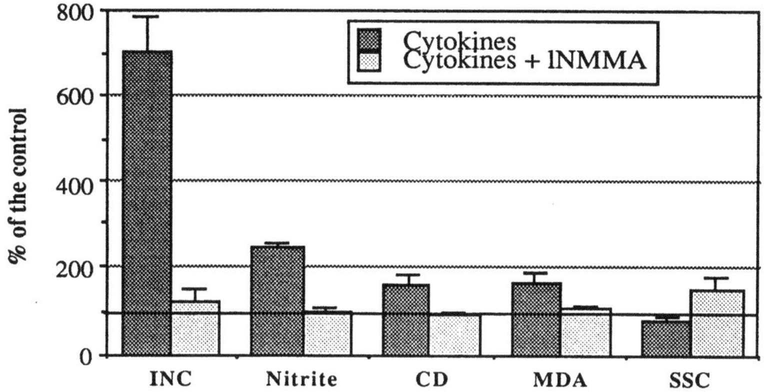

Figure 4 :
Effect of l-N^G-monomethylarginine (lNMMA) on oxidative stress (conjugated dienes : CD; malondialdehyde : MDA; superoxide scavenging capacity : SSC) and NO production (iron-nitrosyl complex : INC and nitrite) induced by IFN (500 U/ml) and LPS (20 µg/ml) in rat hepatocytes incubated for 18 hours. 100 % corresponded to the values observed in the control cultures.

Discussion

In rat hepatocyte cultures, IFN and LPS added for 18 hours induced an oxidative stress characterized by a lipid peroxidation and the loss of a part of the cellular antioxidant capacity. This result was in good agreement with Sugino K et al, who found a maximal MDA level 16 hours after the administration of LPS to mice. In our experiment, the oxidative stress was related to NO production since lNMMA, an NO synthase inhibitor, totally inhibited lipid peroxidation and prevent the decrease of superoxide scavenging capacity. Nevertheless some authors studiing LPS hepatic toxicity in Corynebacterium parvum-treated mice observed that lNMMA administration increased the liver enzyme release, supporting the idea that NO would have a protective function in sepsis (Harbrecht BG et al). Moreover they showed that NO protected from reactive oxygen species such as superoxide anion or hydroxyl radical. But in their model, the amount of circulating nitrogen oxides was extremely high (nearly 45 times the control value) while in our model of hepatocyte cultures supplemented with IFN and LPS, the levels of NO were found at the most 7 times greater than those of the control cultures. This observation could suggest opposite effects of NO in function of its concentration. Besides nitric oxide levels observed in our experiment was in accordance with previous results obtained by Geller (Geller K et al) and Nussler (Nussler AK et al (b)) who showed that the LPS and IFN combination did not induce an important production of NO synthase ARNm and nitrogen oxides. A discrepancy between iron-nitrosyl signal intensity and nitrite content seemed to occur. In fact, a delay was observed between the maximal level of nitrite (27 hours) compare to that of iron-nitrosyl complex (18 hours). This could be explained by the time required for the diffusion of NO and/or nitrite through the cellular membranes.

A part of the mechanism of lipid peroxidation induction could be the decrease of superoxide scavenging capacity which occured at the same time than lipid peroxidation. This decrease which was related to the NO production would lead to an elevation of superoxide anion concentration. Furthermore it has been shown that IFN was able to induce an increase of liver xanthine oxidase activity especially the oxidized form, which produces superoxide anion (Ghezzi P et al, Deloria L et al). Therefore in the rat hepatocytes at 18 hours of incubation with LPS and IFN, superoxide anion concentration would be increased. Superoxide anion is well-known to produce hydroxyl radical, a powerful oxidant, via Haber-Weiss reaction. Moreover the simultaneous elevation of NO and superoxide anion levels could generate peroxynitrite, another powerful oxidant (Beckman JS et al). At last, superoxide anion is dismutated into H_2O_2 either spontaneously or catalytically by superoxide dismutase; then H_2O_2 could react with NO to form singlet oxygen, a cytotoxic molecule (Noronha-Dutra AAet al). Obviously further experiment will be required to conclusively demonstrate this hypothesis.

References

Bautista AP, Spitzer JJ (1990) Superoxide anion generation by in situ perfused rat liver : effect of in vivo endotoxin. Am J Physiol 259:G907-G912

Beckman JS, Beckman TW, Chen J, Marshall PA, Freeman BA (1990) Apparent hydroxyl radical production by peroxynitrite : implications for endothelial injury from nitric oxide and superoxide Proc Natl Acad Sci USA 87:1620-1624

Billiau TR, Vanderkerckhove F (1991) Cytokines and their interactions with other inflammatory mediators in the pathogenesis of sepsis and septic shock. Eur J Clin 21:559-573

Clark IA, Hunt NH, Butcher GA, Cowden WB (1987) Inhibition of murine malaria (Plasmodium chabaudi) in vivo by recombinant interferon-γ or tumor necrosis factor, and its enhancement by butylated hydroxyanisole J Immunol 139:3493-3496

Curran RD, Ferrari FK, Kispert PH, Stadler J, Stuehr DJ, Simmons RL, Billiar TR (1991) Nitric oxide and nitric oxide-generating compounds inhibit hepatocyte protein synthesis FASEB J 5:2085-2092
Deloria L, Abbott V, Gooderham N, Mannering GJ (1985) Induction of xanthine oxidase and depression of cytochrome P-450 by interferon inducers : genetic difference in the response of mice Biochem Biophys Res Comm 131:109-114
Ding AH, Nathan CF, Stuehr DJ (1988) Release of reactive nitrogen intermediates and reactive oxygen intermediates from mouse peritoneal macrophages J Immunol 141:2407-2412
Geller DA, Nussler AK, Di Silvio M, Lowenstein CJ, Shapiro RA, Wang SC, Simmons RL, Billiar TR (1993) Cytokines, endotoxin, and glucocorticoids regulate the expression of inducible nitric oxide synthase in hepatocytes Proc Natl Acad Sci USA 90:522-526
Ghezzi P, Bianchi M, Gianera L, Landolfo S, Salmona M (1985) Role of reactive oxygen intermediates in the interferon-mediated depression of hepatic drug metabolism and protective effect of N-acetylcysteine in mice Cancer Res 45:3444-3447
Guguen-Guillouzo C, Guillouzo A, Boisnard M, Le Cam A, Bourel M (1975) Etude structurale de monocouches d'hépatocytes de rat adulte cultivés en présence d'hémisuccinate d'hydrocortisone Biol Gastroenterol 8:223-231
Harbrecht BG, Billiar TR, Stadler J, Demetris AJ, Ochoa J, Curran RD, Simmons RL (1992) Inhibition of nitric oxide synthesis during endotoxemia promotes intrahepatic thrombosis and an oxygen radical-mediated hepatic injury J Leuk Biol 52:390-394
Hibbs JB, Taintor RR, Vavrin Z, Rachlin EM (1988) Nitric oxide : a cytotoxic activated macrophage effector molecule Biochem Biophys Res Comm 157:87-94
Jiang X, Leonard B, Benson R, Baldwin CL (1993) Macrophage control of brucella abortus : role of reactive oxygen intermediates and nitric oxide Cell Immunol 151:309-319
Keller R (1990) Induction of tumoricidal activity and generation of reactive nitrogen intermediates in macrophages : a comparative analysis in vitro and in vivo in Moncada S, Higgs EA (eds) Nitric oxide from l-arginine : a bioregulatory system Elsevier Science Publisher Amsterdam 409-413
Nathan CF, Murray HW, Weibe ME, Rubin BY (1983) Identification of interferon-gamma as the lymphokine that activates human macrophage oxidative metabolism and antimicrobial activity. J Exp Med 158:670-689
Noronha-Dutra AA, Epperlein MM, Woolf N (1993) Reaction of nitric oxide with hydrogen peroxide to produce potentially cytotoxic singlet oxygen as a model for nitric oxide-mediated killing FEBS lett 321:59-62
Nussler AK (a), Di Silvio M, Billiar TR, Hoffman RA, Geller DA, Selby R, Madariaga J, Simmons RL (1992) Stimulation of the nitric oxide synthase pathway in human hepatocytes by cytokines and endotoxin. J Exp Med 176:261-264
Nussler AK (b), Billiar TR (1993) Inflammation, immmunoregulation and inducible nitric oxide synthase J Leuk Biol 54:171-178
Sergent O, Morel I, Cogrel P, Chevanne M, Pasdeloup N, Brissot P, Lescoat G, Cillard P, Cillard J (1993) Simultaneous measurements of conjugated dienes and free malondialdehyde used as a micromethod for the evaluation of lipid peroxidation in rat hepatocyte cultures Chem Phys Lipids 65:133-139
Stadler J, Billiar TR, Curran RD, Stuehr DJ, Ochoa JB, Simmons RL (1991) Effect of exogenous and endogenous nitric oxide on mitochondrial respiration of rat hepatocytes Am J Physiol 260:C910-C916
Sugino K, Dohi K, Yamada K, Kawasaki T (1987) The role of lipid peroxidation in endotoxin-induced hepatic damage and the protective effect of antioxidants Surgery 101:746-752

Effects of Nitric Oxide on Morris Water Maze Performance in Rats: Correlation with cGMP Levels

Serdar Demirgören and Şakire Pöğün
Ege University School of Medicine
Department of Physiology
Bornova, Izmir 35100
Turkey

The relationship between cGMP (guanosine 3',5' cyclic monophosphate) and CNS amino acid neurotransmission systems have been known since two decades and glutamate has been shown to increase cGMP levels in mouse striatal slices (Ferandelli JA et al., 1973). Activation of glutamate receptors triggers biochemical events stimulating calcium influx (Wroblewski JT et al., 1985) and cGMP accumulation (Novelli A et al., 1987). NO (nitric oxide) is a novel biological messenger molecule in the CNS which is enzymatically formed from L-Arginine by nitric oxide synthase (NOS) ,a Ca^{2+}-Calmodulin requiring enzyme; citrulline is the co-product (Garthweite J 1991). Glutamate and N-methyl-D-aspartate (NMDA) receptor activation stimulate arginine-citrulline formation and increase cGMP levels in the cerebellum; L-N^G-monomethyl-arginine (L-NMMA) inhibits NOS and prevents the stimulation of cGMP formation (Bredt DS and Snyder SH 1989, Garthweite J et al 1989). NMDA receptor activation has thus been linked to NO.

Long-term potentiation (LTP) is considered a cellular analogue of learning and memory in the mammalian nervous system (Haley JE et al., 1992). LTP in the hippocampus, which may underlie memory formation, involves NMDA receptor activation (Böhme GA et al., 1991). There is compelling evidence that NO is the early retrograde messenger in long-term potentiation in hippocampal slices (Fazeli MS 1992, O'Dell TJ et al., 1991). Endogenous nitric oxide release has been shown to be essential for synaptic plasticity in the cerebellum (Shibuki K and Okada D 1991) and in the hippocampus (East SJ and Garthweite J 1991, Haley et al., 1992, Schuman EM and Madison DV 1991). Long-term depression (LTD) in the cerebellum is implicated as the cellular mechanism for cerebellar motor learning and is due to

NATO ASI Series, Vol. H 92
Signalling Mechanisms – from Transcription Factors to Oxidative Stress
Edited by L. Packer, K. Wirtz

desensitization of quisqualate-specific glutamate receptors (Ito M 1989). Glutamate receptor desensitization in the cerebellum contrasts to the sensitization of glutamate receptors that underlie LTP in the hippocampus (Lynch G and Baudry M 1984).

Several theories acknowledge the critical role of the hippocampus for spatial learning (Schmajuk, NA 1984, Schwegler H et al., 1988). The Morris Water Maze (MWM) is a widely accepted test of spatial learning in rodents in which the animal has to learn to find the hidden platform submerged several centimeters below water level in a circular pool filled with opaque water (Brandeis R et al., 1989, Morris RGM 1981). Infusion of an NMDA receptor antagonist, AP5, into the lateral ventricle or the hippocampus impairs spatial learning and blocks LTP in rats (Halliwell RF and Morris RGM, 1986, Morris RGM 1981).

The present study was undertaken to study the effects of in-vivo administration of sodium nitroprusside (SNP: sodium nitroferricyanide) , a NO generator, and L-NMMA on spatial learning using the MWM and ex-vivo determination of cGMP levels in the hippocampus and cerebellum in rats.

The findings presented in this study have appeared in abstract form (Demirgören S and Pöğün Ş 1993).

METHODS

Experimental Animals

Male Sprague Dawley rats (3 months old) that were housed in large cages in groups of 6 or 7 which were kept in a 12 hour dark-light cycle and were on an *ad-lib* diet were used in the experiments. All the treatments and behavioral testing were done between 9:00-11:00, alternating the order of the experimental groups.

Morris Water Maze Experiments

The MWM was a circular pool with 110 cm diameter filled with opaque water. The platform was placed 3 cm below water level; the quadrant where the animals were placed in the beginning of the experiments each day was alternated and the animals were given three trials each day not to exceed 2 minutes each; after each trial, the rats were allowed to stay on the platform for 10 seconds.

Following the initial handling period with the guide, animals were trained in the MWM for 5 consecutive days; the platform was removed on the last day and there was only one trial session of 2 minutes: probe trial. Learning performance was based on the time spent in the quadrant where the platform had been.

Treatment

Rats were injected i.p. with either 4 mg/kg SNP (Sigma) or 5 mg/kg L-NMMA (Sigma) 2 minutes before learning experiments. The control group received only the vehicle, saline. The injection volume was 0.250 ml. The animals received a total of six injections: one before the initial handling period with the guide and five before each testing session. The number of animals in each group are given in Table 1.

cGMP Assay

On the 5th day of the experiments, approximately 10 minutes after the injections and 2 minutes after the termination of the MWM learning experiments, the rats were decapitated, hippocampi and cerebella quickly dissected on ice, frozen in liquid nitrogen and stored at -70 C^0. cGMP levels were determined using cGMP [^{125}I] scintillation proximity assay (Amersham, UK).

RESULTS

All the groups studied, spent more time in the quadrant where the platform had been than in the other quadrants during the final day of the experiments; if they had swam in the maze randomly, we would expect to see approximately 30 seconds for each quadrant. The SNP treated group spent most time in the correct quadrant searching for the platform, followed by the control and the L-NMMA treated groups (Table 1).One way ANOVA of the data reveal a significance of only $p<0.06$ ($F=4.03$) between the groups, implying a trend which did not reach statistical significance but the Duncan's multiple range test show that the SNP and the L-NMMA groups are significantly different from each other ($p<0.05$).

The cGMP levels in the hippocampus were significantly different between the groups, the SNP treated rats had the highest levels, followed by the control and the L-NMMA treated groups (ANOVA, $F=5.45$, $p<0.03$) (Table 1). Duncan's multiple range test show that the SNP and the L-NMMA groups ($p<0.01$) and the control and SNP groups ($p<0.05$) are significantly different from each other.

The cGMP levels were also different in the cerebellum but in a different direction: The control group had the highest values followed by the L-NMMA and SNP treated groups (ANOVA, $F=7.81$, $p<0.02$) (Table 1). Duncan's multiple range test show that the control and SNP groups are significantly different from each other ($p<0.05$).

Table 1: Effects of SNP or L-NMMA on learning performance and cGMP levels

Treatment	Learning perf.[a] (sec)	cGMP Levels (fmol/10 mg tissue) Hippocampal	Cerebellar
Saline (n=6)	46.33 ± 5.09	27.68 ± 2.86*	251.16 ± 20.32*
SNP (n=5)	54.80 ± 2.91	47.84 ± 11.68	178.05 ± 22.16
L-NMMA (n=6)	39.80 ± 5.65*	20.72 ± 3.99**	192.61 ± 13.15
ANOVA	$F=4.03$, $p<0.06$	$F=5.45$, $p<0.032$	$F=7.81$, $p<0.017$

[a]Learning performance shows the time spent in the target quadrant in 2 minutes on the last day of MWM testing.
Figures represent mean ± SEM
Duncan's multiple range tests: Different from SNP *$p<0.05$, **$p<0.01$

There was a positive correlation ($p<0.05$)between the cGMP levels in the hippocampus and the learning performance when data from all the animals were handled together (Figure 1).

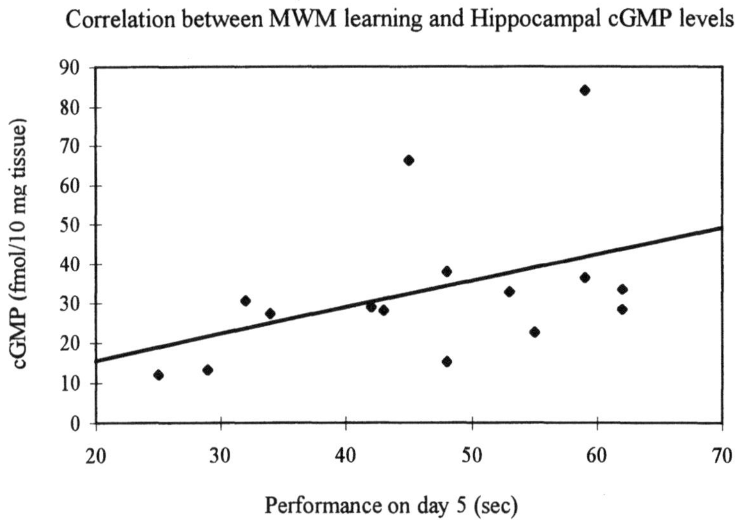

Figure 1: Correlation between hippocampal cGMP levels and learning performance measured as the time spent in the target quadrant in 2 minutes on the last day of MWM testing. ($p<0.05$)

DISCUSSION

Our results show that SNP facilitates and that NOS inhibition impairs spatial learning in rats. The observed effects are correlated with the cGMP levels in the hippocampus. Ingram, D.K. et al. (Ingram DK et al., 1992) has also shown the impairment of learning of male F-344 rats in a 14-unit T-maze by NOS inhibition through i.p. injections of N^{ω}-nitro arginine (NNA). Even though NNA has a higher potency in inhibiting brain NOS compared to L-NMMA, the inhibition is irreversible in-vitro and in-vivo (Dwyer MA et al., 1991). In our behavioral studies, we did not want to interfere with the integrity of the biological system and therefore chose to inhibit NOS transiently using a reversible inhibitor of the enzyme NOS at moderately low concentrations. The SNP injected animals did show some motor disability mainly due to the cardiovascular effects of the drug, especially during the initial days, but were able to complete the experiments.

The observed effects of NO-related compounds on cGMP levels in the cerebellum were contrasting those in the hippocampus. Our results support the view of the desensitization of the glutamatergic system in the cerebellum that is observed in LTD (Ito M 1989, Shibuki K and Okada D 1991). Even though NO may mediate in the enhancement of LTD initially, continued exposure, together with behavioral testing may be leading to desensitization and resulting in lower cGMP levels in the SNP treated group.

The positive correlation observed between hippocampal cGMP levels and learning performance provides further support for the involvement of NO in mediating in learning and memory, possibly through the retrograde activation of NMDA-type glutamate receptors with simultaneous increases in cGMP levels. It would be interesting to show the involvement of the NMDA-type glutamate receptors in-vivo, using receptor antagonists, under similar experimental conditions, in conjunction with NO generators and NOS inhibitors.

Our study, to the best of our knowledge, is the first linking the literature findings obtained in tissue slices pointing to the involvement of NO mediation of the glutamatergic system in obtaining LTP in the hippocampus and behavioral testing in intact animals with resulting elevations in cGMP levels.

ACKNOWLEDGEMENTS

This study was supported by Grant no. 91/024 from Ege University Research funds. The authors thank Dr. Tolga Uz for his assistance in MWM testing. The authors are also grateful to Drs M.J.Kuhar and E.D. London for their careful evaluation of the manuscript.

REFERENCES

Böhme GA, Bon C, Stutzmann JM, Doble A and Blanchard JC (1991) Possible invol-vement of nitric oxide in long-term potentiation. Eur J Paharmacol. 199: 379-381

Brandeis R., BrandysY, and Yehuda S (1989) The use of the Morris Water Maze in the study of memory and learning. Intern. J. Neurosci. 48:29-69

Bredt DS and Snyder SH (1989) Nitric oxide mediates glutamate-linked enhancement of cGMP levels in the cerebellum. Proc Natl Acad Sci USA 86:9030-9033

Demirgören S, and Poğün Ş (1993) In-vivo effects of nitric oxide on spatial learning in rats. Soc for Neurosci Abstracts 19: 310.3

Dwyer MA, Bredt DS, and Snyder SH (1991) Nitric Oxide Synthase: Irreversible inhibition by L-N^{ω}-nitroarginine in brain in-vitro and in-vivo. Biochem Biophys Res Com 176(3):1136-1141

East SJ, and Garthweite J (1991) NMDA receptor activation in rat hippocampus induces cyclic GMP formation through the L-arginine-nitric oxide pathway. Neuroscience letters 123:17-19

Fazeli MS (1992) Synaptic plasticity: on the trail of the retrograde messenger. TINS 15(4):115-117

Ferandelli JA, Chang MM, and Kinscherf DA. (1973) Elevation of cyclic GMP levels in central nervous system by excitatory and inhibitory amino acids. J Neurochem 22: 535-540

Garthweite J (1991) Glutamate, nitric oxide and cell-cell signalling in the nervous system. TINS, 14(2):60-67

Garthweite J, Garthweite G, Palmer RMJ, and Moncada S (1989) NMDA receptor activation induces nitric oxide synthesis from arginine in rat brain slices. Eur J Pharmacol 172:413-416

Haley JE, Wilcox GL. and Chapman PF (1992) The role of nitric oxide in hippocampal long-term potentiation. Neuron 8:211-216

Halliwell RF, and Morris RGM (1986) Intrahippocampal microinfusion of an N-methyl-D-aspartate antagonist (AP5) blocks LTP in-vivo and impairs spatial learning in rats. Soc for Neurosci Abstr 12: 519

Ingram DK, Spangler E, Roberts D, Iijima S, and London ED (1992) Nitric oxide synthase inhibition by N^{G}-nitro L-arginine impairs learning of rats in a 14-unit T-maze. Soc.for Neurosci. Abstracts. 18: 509.4

Ito M (1989) Long-term depression. Ann Rev Neurosci, 12:85-102

Lynch, G. and Baudry, M. (1984) The biochemistry of memory: A new and specific hypothesis. Science 224:1057-1063

Morris RGM (1981) Spatial localization does not require the presence of local cues. Learn Motiv 12: 239-249

Morris RGM, Anderson E, Lynch GS, and Baudry M (1986) Selective impairment of learning and blockade of long-term potentiation by an N-methyl-D-aspartate antagonist, (AP5). Nature 297:681-683

Novelli A, Nicoletti F, Wroblewski JT, Alho H, Costa E, and Guidotti A (1987) Excitatory amino acid receptors coupled with guanylate cyclase in primary cultures of cerebellar granule cells. J Neurosci 7:40-47

O'Dell TJ, Hawkins RD, Kandel ER, and Arancio O (1991) Tests of the roles of two diffusable substances in long-term potentiation: Evidence for nitric oxide as a posible early retrograde messenger. Proc Natl Acad Sci USA 88:11285-11289

Shibuki K, and Okada D (1991) Endogenous nitric oxide release required for long-term synaptic depression in the cerebellum. Nature 349:326-328

Schmajuk NA (1984) Psychological theories for hippocampal function. Physiol Psychol. 12: 166-183

Schuman EM, and Madison DV (1991) A requirement for the intracellular messenger nitric oxide in long-term potentiation. Science 254: 1503-1506

Schwegler H, Crusio WE, Lipp HP, and Meimrich B (1988) Water maze learning in the mouse correlates with variations in hippocampal morphology. Behav Genetics, 18:153-165

Wroblewski JT, Nicoletti F, and Costa E (1985) Different coupling of excitatory amino acid receptors with Ca^{2+} channels in primary cultures of cerebellar granule cells. Neuropharmacology, 24: 919-921

Signal Transduction from the Cytoplasm to the Cell Nucleus by NF-κB/Rel Transcription Factors

M. Lienhard Schmitz and Patrick A. Baeuerle
Institute of Biochemistry
Albert-Ludwigs-University
Hermann-Herder-Str. 7
D-79104 Freiburg
Germany

An eucaryotic transcription factor system is reviewed that is specialized in the transduction of primarily pathogenic signals from the cytoplasm into the cell's nucleus. This system, called NF-κB/Rel, comprises a family of five distinct DNA-binding subunits and five regulatory proteins with inhibitory function, called IκB proteins. By interaction of IκB proteins with dimers of the DNA binding subunits, nuclear transport and DNA binding are suppressed, leading to the cytoplasmic accumulation and sequestration of the NF-κB/Rel transcription factors. Following extracellular stimulation, IκB proteins are rapidly degraded and the released factors are able to migrate into the nucleus where they initiate transcription upon DNA-binding to enhancer elements. The activation of NF-κB is controlled indirectly and directly by protein kinases. We review the molecular biology, biochemistry and physiology of this unique signal-transducing system, which is activated by the inducible proteolysis of inhibitory subunits.

Introduction

Eucaryotic gene expression is primarily regulated at the level of transcription (for recent reviews, see Cell Vol. 77, 1994). RNA polymerase II-dependent promoters contain indispensable core elements, such as the TATA box or, alternatively, an initiator element. The TATA box is directly bound by the basic transcription factor TFIID, a multiprotein complex which contains the TATA-binding protein (TBP) and various associated proteins, called TBP-associated factors (TAFs). At least seven more general transcription factors are required to assemble a preinitiation complex in the

NATO ASI Series, Vol. H 92
Signalling Mechanisms – from Transcription Factors
to Oxidative Stress
Edited by L. Packer, K. Wirtz

promoter region immediately upstream from the transcription start site. However, a strong increase in the rate of transcription initiation of a particular gene requires additional cis-regulatory elements, the so-called enhancer sequences. These are located in variable position and distance with respect to the promoter and may be several kilobases away from the transcription start site. Very frequently, enhancer elements are located upstream of the promoter elements, but they are also found in introns and downstream of the gene´s coding sequence. Cis-acting enhancer elements are bound by transcription factors, which can either stimulate or repress the transcriptional activity of a promoter. A variety of such transcription factors has been shown to directly interact with the basal transcription factors constituting the initiation complex at the promoter region. The DNA which spaces the enhancer element and the promoter region must be looped out in order to allow a physical interaction between two DNA-bound proteins. In this way, a given transcription factor can influence transcription over variable distances of DNA. Apart from DNA binding domains, a hallmark of transcription factor proteins are domains required to recruit coactivator molecules or basal transcription factors. These so-called transactivating domains help to stabilize the preinitiation complex and facilitate the elongation of RNA synthesis by RNA polymerase II. Typical promoters and enhancers contain several sequences binding a variety of distinct transcription factors. The interplay of the factors, for instance, their synergistic activation or mutual repression, allows a highly controlled regulation of gene expression integrating information from the cell´s environment and developmental status.

The transcription factor NF-κB was first described as a tissue-specific factor that bound to and activated the immunoglobulin κ light chain intronic enhancer in mature B cells but not in pre-B cells (Sen and Baltimore, 1986). Now, it is known that NF-κB is a ubiquitous multisubunit factor that is present in most cell types in an inducible, pre-existing form. NF-κB binds to a decameric DNA sequence element (consensus: 5´-GGGPuNNPyPyCC-3´) which displays an imperfect dyad symmetry in its two half sites. High-affinity binding sites for NF-κB are found in the upstream promoter regions of many cellular and viral genes and in many instances it has been shown that mutational alteration of this sequence, such that NF-κB-binding to DNA is impaired, interferes with the transcriptional activation of these genes. The DNA-binding activity of NF-κB is induced by an extreme variety of conditions, including bacterial lipopolysaccharide (LPS), phorbol ester (PMA), tumor necrosis factor-α (TNF-α), interleukin-1 and by many viruses (Grilli et al., 1993). The induction of NF-κB´s DNA-binding activity does not require de novo protein synthesis; on the contrary, costimulation with the protein synthesis inhibitor cycloheximide causes a superinduction of NF-κB. The identification and purification of an inhibitory protein, termed IκB, was possible because NF-κB could be activated in a cell-free system by treatment of cytoplasmic fractions with the detergent desoxycholate (Baeuerle and Baltimore, 1988a; 1988b; Zabel and Baeuerle, 1990).

The NF-κB/Rel transcription factor family

Structural and functional characteristics

The DNA-binding subunits of NF-κB were purified to homogeneity from human cell cultures (Baeuerle and Baltimore, 1989; Kawakami et al., 1988), human placenta (Zabel and Baeuerle, 1990) and rabbit lung (Ghosh et al., 1990) using sequence-specific DNA affinity chromatography. In each case, two proteins with molecular weights of approximately 50 and 65 kDa were purified, referred to as p50 and p65. These proteins were shown to contact their cognate decameric DNA as heterodimers (Urban et al., 1991). UV-crosslinking experiments revealed that p50 contacts the first half site of the κB site 5′-GGGAAATTCC-3′ while the p65 protein showed a preference for the second half site, which is usually more degenerated when sites from different genes are compared.

Molecular cloning of the cDNA for the p50 DNA-binding subunit (Kieran et al., 1990; Ghosh et al., 1990; Meyer et al., 1991) revealed that the N-terminal 300 amino acids of the protein were highly homologous to the morphogen dorsal from *Drosophila melanogaster* (Steward, 1987) and the oncogene *v-rel* from the turkey retrovirus REV-T (Stephens et al., 1983), as well as its cellular homologue *c-rel* (Wilhelmsen et al., 1984). After cloning of the p65 subunit of NF-κB (Nolan et al., 1991; Ruben et al., 1991) it became apparent that p65 also contains the rel-homology region in its N-terminal half. Low-stringency hybridization and PCR cloning extended the NF-κB/Rel-family by two additional members: p52 (Schmidt et al., 1991) and RelB (Ryseck et al., 1992; Ruben et al., 1992). The primary structures of the NF-κB/Rel/Dorsal (NRD) family members are schematically displayed in Figure 1. The NRD domain is sufficient for DNA-binding, dimerization, nuclear localization and IκB interaction. Accordingly, all members of the NF-κB/Rel family were found to be conditionally nuclear, to dimerize, to bind with similar specificity to DNA and to control gene expression.

A short stretch of 4 or 5 basic amino acids at the C-terminal end of the NRD domain functions as nuclear location signal (NLS) sequence as shown first in a study on v-Rel (Gilmore and Temin, 1988). Mutational studies of the related NLS sequences in c-Rel, p50 and p65 all proved the functional importance of this domain in controlling nuclear uptake (Beg et al., 1992; Ganchi et al., 1992; Henkel et al., 1992; Zabel et al., 1993). A study employing trans-dominant negative mutants mapped the region responsible for homo- and heterotypic dimerization to the more C-terminal half of the NRD domain (Logeat et al., 1991). A mutant of p50 lacking the region between amino acids 110 and 201 was unable to bind to DNA, but interferred with the DNA-binding activity of other NRD domain proteins by forming inactive heterodimers. Evidence for the importance of the more N-terminal part of the NRD domain in DNA-binding came from the finding that alkylation or mutation of Cys62 in p50 prevents DNA-binding but not dimerization of p50. Replacement of the N-terminal region of p65 with the corresponding region from p50 switched its fine DNA-binding specificity towards that of p50 (Toledano et al., 1993).

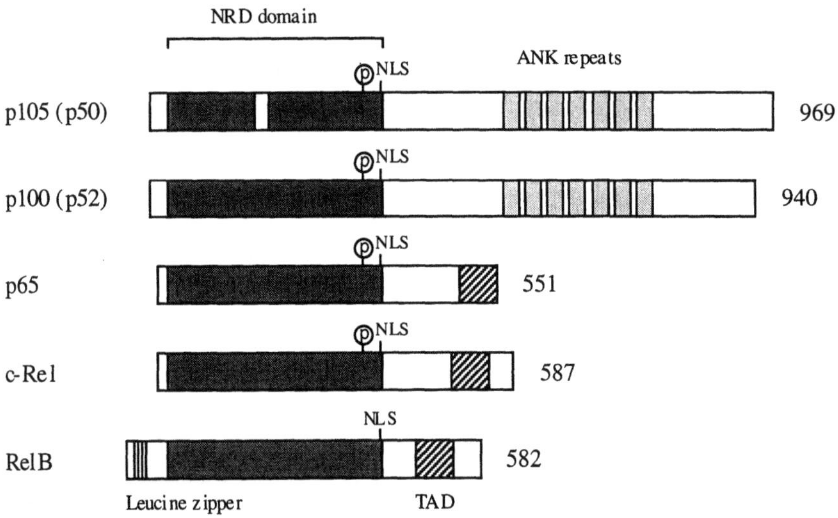

Figure 1. Members of the NRD family. The NRD domain is indicated by dark shading, the ankyrin repeats by light shading. A putative PKA phosphorylation site is indicated by a circled P and is located next to the nuclear location signal (NLS) close to the carboxyl end of the NRD domain. The positions of the transactivation domains (TADs) are indicated by the hatched boxes, the potential leucine zipper in RelB is indicated by the vertical lines. The total number of amino acids for each human protein is given at the right.

A potential phosphorylation site for protein kinase A (Arg-Arg-X-Ser) lies about 25 residues N-terminal from the NLS in all members of the NRD domain family except for RelB. Constitutive phosphorylation of this site was mimicked by replacing the serine by glutamic or aspartic acid in v-rel and c-rel (Mosialos et al., 1991). The mutations impaired the cytoplasmic retention of the c-Rel protein and the transcriptional repression activities of the v-Rel protein. However, the physiological regulation of this phosphorylation still remains to be investigated in intact cells and for other family members. Some of the NRD family members display unique sequence elements in their NRD domains, which may be involved in protein/protein interactions. The p50 molecule contains a stretch of about 30 unique amino acids in the middle of its NRD domain and the relB protein carries a putative leucine zipper motif in its first 50 amino acids.

p50 and p52 are not synthesized as active DNA-binding proteins but as large precursor molecules of 105 and 100 kDa molecular weight, respectively. Both precursor proteins carry repeated sequence motifs in their C-termini. This sequence motif with a length of 30-33 amino acids was originally discovered in the SWI6 protein of *Saccharomyces cerevisiae* (Breeden and Nasmyth, 1987) and is called SWI6/ankyrin (ANK) repeat. The ANK repeat is reiterated 7 times in both p105 and in p100. The so far unidentified natural cleavage sites in p105 and p100 are close to a glycine-rich region that might serve as a 'hinge' region between both halves of p105. Both precursor proteins are unable to bind to DNA. The protease responsible for p105 processing has been proposed to be an ATP-dependent enzyme (Fan and Maniatis, 1991). In vitro experiments demonstrated that purified HIV-1 protease specifically cleaved the p105 protein generating a smaller version of the DNA-binding p50 subunit (Rivière et al., 1991). Northern blot analysis provided no evidence that p50 is generated by alternative splicing, an alternative mechanism to proteolysis that may give rise to p52 (Schmidt et al., 1991). The C-terminal domains of p105 and p100 serve several functions, which will be described in the forthcoming section.

The recent cloning of NF-ATp and NF-ATc, two protein components of the nuclear factor of activated T cells (NF-AT), revealed that they contain a region with a low degree of homology to the DNA-binding NRD domain (reviewed in Nolan, 1994). This homology region is located in the middle part of both molecules, but the maximal cumulative identity of these proteins to the NRD family members does not exceed 20 %.

Transcription activation by members of the NRD protein family and Bcl-3

The NRD family proteins can be functionally grouped into two classes, the first one consisting of the transactivating subunits c-Rel, RelB and p65, and the second one consisting of p50 and p52, which have none or little transactivating potential. Interestingly, the same classification is apparent with respect to their phylogenetic relationship (Schmid et al., 1991). The C-terminal sequences of c-Rel, p65 and RelB contain potent transcription activation domains, which are enriched in serine, acidic and hydrophobic amino acid residues (Bull et al., 1989; Schmitz and Baeuerle, 1991; Ryseck et al., 1992). Only RelB contains an additional activation domain in its N-terminus with the putative leucine zipper.

The role of p50 as a transcriptional activator is a matter of debate. Cell-free transcription experiments showed that dimeric p50 could stimulate transcription from reporter constructs with the HIV promoter and the palindromic motif from the MHC class I enhancer but not with the motif from the β-interferon enhancer (Fujita et al., 1992; Kretzschmar et al., 1992). Further studies revealed that p50 can induce transcription of a reporter gene in yeast cells, however, only to a minor extent when compared with the transcriptional activity of p65 (Moore et al., 1993). In contrast, in transient transfection studies using mammalian cells the p50 subunit consistently failed to activate expression of κB-dependent reporter genes. Coexpression of p50 even suppressed the transactivation by p65 most probably by competitive DNA binding of transcriptionally inactive p50 homodimers (Schmitz and Baeuerle, 1991). The repressing property of p50 is also observed in intact cells under physiological conditions (Brown et al., 1994). The promoter of the MHC class II-associated invariant chain I_i gene contains two binding sites for NF-κB. The activation of this promoter in B and T cell lines corresponds well with the presence of c-Rel and p65 subunits. Likewise, the absence of promoter activity in myelomonocytic and glial cell lines is correlated with the predominance of p50 and p52 dimers. The IL-2 promoter was found to be complexed in resting $CD4^+$ T-cells with transcriptionally inactive p50 homodimers. Antigenic stimulation of the T cells triggered the IL-2 expression and resulted in the replacement of the p50 homodimers by newly activated p50/p65 heterodimers (Kang et al., 1992). The absence of reporter gene activation by p50 homodimers was also seen in experiments employing transgenic mice (Lernbecher et al., 1993). Tissues containing nuclear p50 homodimer showed no activation of the transgene which was controlled by three κB sites. In the light of these findings, two different functions can be assigned to the p50 subunit. Firstly, as part of a heterodimer it serves as a helper subunit by increasing the affinity of a transactivating subunit for a specific DNA target. Secondly, excess p50 can downregulate gene expression by occupying NF-κB binding sites with transcriptionally inactive p50 homodimers.

An alternative mechanism of κB-site specific transcriptional activation is brought about by the Bcl-3 protein, which belongs to the family of IκB proteins. The Bcl-3 protein, which itself is unable to bind to DNA, was found to stimulate κB-dependent gene expression with the help of both p52 and p50 (Franzoso et al., 1992; Bours et al., 1993; Fujita et al., 1993). This occurs via two fundamentally distinct molecular mechanisms. Firstly, since Bcl-3 can inhibit the DNA-binding of the p50 homodimer but not of the p50/p65 heterodimer (Nolan et al., 1993), Bcl-3 can remove excess inhibiting p50 homodimers from the DNA thereby allowing the heterodimer to bind to DNA and activate transcription. Secondly, Bcl-3 can selectively associate with p52 dimers on the DNA and form a ternary complex with strong transactivating potential (Bours et al., 1993). Consistent with this model, Bcl-3 is predominantly located in the nucleus. Some aspects of Bcl-3 function are a matter of debate. Since Bcl-3 is highly phosphorylated in the cell and dephosphorylation impairs its activity (Nolan et al., 1993) it should be taken into account whether the Bcl-3 protein studied was of eukaryotic or bacterial origin and whether studies were performed using saturating or finely titrated amounts of Bcl-3.

IκB proteins

Structural characteristics of IκB proteins

In the majority of cell types, NF-κB was found to exist in an inducible form requiring distinct stimuli in order to appear as a DNA-binding factor in the nucleus. The inducible form is stabilized in the cytoplasm by IκB proteins which can be released from NF-κB experimentally in cytosolic fractions from non-stimulated cells upon treatment with dissociating agents, such as formamide or desoxycholate (Baeuerle and Baltimore, 1988a; 1988b). The inhibitory subunit IκB was found to exert its function on NF-κB's DNA binding in a highly specific and reversible fashion (Baeuerle and Baltimore, 1988b). IκB was subsequently purified from human placenta and rabbit lung employing classical column chromatography. Two variants were isolated. One was a protein of 37 kDa, termed IκB-α; and a second form, IκB-β, had an apparent molecular weight of 43 kD (Zabel and Baeuerle, 1990; Ghosh et al., 1990; Link et al., 1992). Purified IκB proteins were found to actively dissociate NF-κB from its cognate DNA in vitro (Zabel and Baeuerle, 1990). Both variants of IκB had isoelectric points between 4.8 and 5. Phosphatase treatment of purified IκB-β but not of -α abolished its inhibitory activity (Link et al., 1992).

Cloning of human, rat, pig and avian IκB-α revealed that the protein also contains ANK repeats like the C-terminal parts of p105 and p100 (reviewed in Beg and Baldwin, 1993). Soon it became evident that also the IκB proteins constitute a gene family. The five members and their schematic primary structures are displayed in Figure 2. The C-terminus of p105 was found to be separately expressed as a splicing product in a variety of lymphoid cells. The resulting 70 kDa protein, called IκB-γ, as well as the entire precursor p105 were shown to complex and inactivate various NRD family proteins in the cytoplasm (Inoue et al., 1992a).

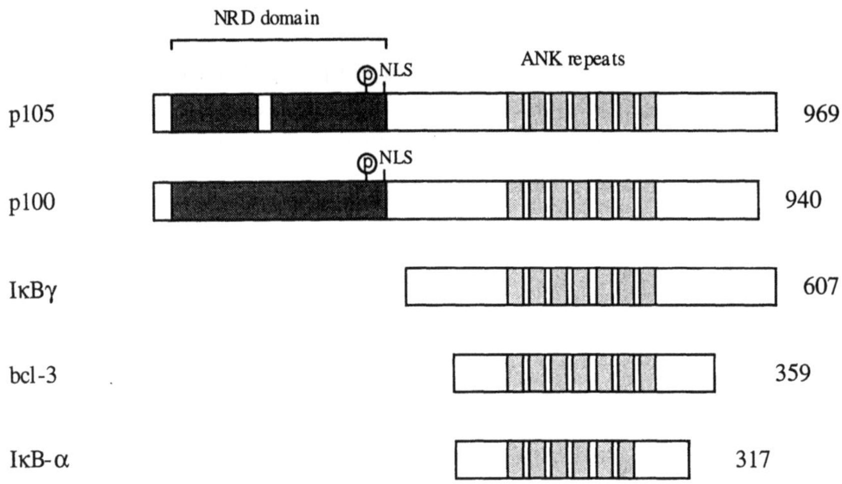

Figure 2. Members of the IκB protein family. The symbols used to highlight the functional domains are identical to those used in Figure 1. The total number of amino acids for each human protein is shown at the right.

Detailed studies revealed, that the different IκB proteins preferentially interact with various members of the NRD domain family (for a review, see Beg and Baldwin, 1993). IκB-α and IκB-β were found to specifically inhibit heterodimeric complexes containing p65 or c-Rel (Baeuerle and Baltimore, 1989; Schmitz et al., 1991; Beg et al., 1992) while Bcl-3 preferentially inhibits p50 homodimers but does not interfere with DNA-binding of c-Rel and p65 complexes (Nolan et al., 1993; Zhang et al., 1994). The 70 kDa form of IκB-γ inhibits p50 homodimers, p50-p65 and c-Rel (Inoue et al., 1992a). Alternative splicing of IκB-γ generates new isoforms with different inhibitory activities. The 63 kDa isoform IκB-γ-1 and the shorter 55 kDa isoform IκB-γ-2 were specific for p50 dimers (Grumont and Gerondakis, 1994). The p105 and p100 molecules displayed no remarkable degree of specificity for individual NF-κB subunits. They were found to be complexed with p50, p65, p52 and c-Rel (Rice et al., 1993; Mercurio et al., 1993; Scheinmann et al., 1993; Naumann et al., 1993). Future studies have to investigate to what extent NF-κB complexes containing Bcl-3, p100, p105 and splice variants of p105 confer activation in response to extracellular stimuli as seen with IκB-α and, presumably, IκB-β. It is conceivable that p100, p105 and splice variants serve as cytoplasmic buffer to inactivate transactivating complexes transiently released from inducibly regulated IκB proteins.

Functions of IκB proteins

A structural hallmark of IκB molecules is that they contain 5-7 complete ANK repeat motifs. These are important for protein-protein recognition (for a review, Michaly and Bennett; 1992) and, as an entity, are indispensable for the function of IκB-like molecules (Hatada et al., 1993; Inoue et al., 1992b). ANK repeats were found to be involved in interaction with the conserved NRD domain (Ganchi et al., 1992; for a review, see Beg and Baldwin, 1993).

With the exception of Bcl-3, all other IκB proteins retain their target molecules in the cytoplasm by masking their NLS sequence. An antibody specifically recognizing the NLS of p65 and p50 could bind to its epitope only in the absence of IκB (Beg et al., 1992; Henkel et al., 1992; Zabel et al., 1993). Masking of an NLS also accounts for the cytoplasmic localization of p105 but in this case this is likely to occur by an intramolecular mechanism (Henkel et al., 1992).

An important feature of IκB-like molecules seems to be their ability to dissociate a complex formed between NF-κB and its cognate DNA. This has been shown for IκB-α and -β (Zabel and Baeuerle, 1990), as well as for p105 (Rice et al., 1992). Accordingly, cell-free transcription by NF-κB was immediately abrogated when purified IκB-α was added to an in vitro transcription reaction (Kretzschmar et al.,

1992). The inhibiting function of IκB is also evident from transient transfection studies where overexpressed IκB attenuates κB-specific transcriptional activity (Beg et al., 1992; Hatada et al., 1992). Although IκB-α is predominantly cytoplasmic it can move to the nucleus under certain conditions, for instance, following serum stimulation (Cressman and Taub, 1993). Nuclear IκB is therefore likely to influence gene expression in the nucleus, although this has so far not been experimentally shown in intact cells.

Fusion of the C-terminus of p105 to the DNA-binding domain of the yeast Gal4 protein created a transcriptionally active protein when tested with a Gal4-dependent reporter gene in yeast cells (Morin and Gilmore, 1992). However, the physiological relevance of this finding has to await further analysis since p105, unlike IκB-α or Bcl-3, is not seen in the nucleus.

Target genes of NF-κB

NF-κB participates in the expression of numerous target genes (listed in Table 1).

Class	Target Gene
Immunoreceptors	Immunoglobulin κ light chain
	T cell receptor β
	T cell receptor α chain (human)
	Major histocompatibility complex class I (H-2K^b)
	Major histocompatibility complex class II (Eα^d)*
	β_2-Microglobulin
	Invariant chain I_i
	Tissue factor-1
Cell adhesion molecules	Endothelial leucocyte adhesion molecule 1 (ELAM-1)
	Vascular cell adhesion molecule 1 (VCAM-1)
	Intercellular cell adhesion molecule 1 (ICAM-1)*
Cytokines and hematopoetic growth factors	β-Interferon
	Granulocyte/macrophage colony-stimulating factor (GM-CSF)
	Granulocyte colony-stimulating factor (G-CSF)
	Macrophage colony-stimulating factor (M-CSF)
	Melanoma growth stimulating activity (*gro*α-γ/MGSA)
	Interleukin-2
	Interleukin-6

	Interleukin-8 TNF-α Lymphtoxin (TNF-β) Proenkephalin MPC-1/JE*
Acute phase proteins	Angiotensinogen Serum amyloid A precursor Complement factor B Complement factor C4 Urokinase-type plasminogen activator*
Transcription factors	c-Rel NF-κB precursor p100 NF-κB precursor p105 IκB-α Bcl-3 c-Myc Interferon regulatory factor 1*
Viruses	Human immunodeficiency virus 1 (HIV-1) Cytomegalovirus (CMV) Adenovirus Simian virus 40 (SV40)
Others	Vimentin NO-Synthase*

Table 1. List of genes, in whose expression NF-κB participates as an inducible transcriptional activator. For references, see Baeuerle and Henkel, 1994. The asterisks indicate genes requiring additional analysis of their NF-κB-sites for the evaluation of the significance of NF-κB binding.

The target genes for NF-κB have in common that they are rapidly induced in response to an extracellular stimulus. Ther products of most genes are directly or indirectly important for the cellular defense program to pathogens and can be grouped into several classes: immunoreceptors, including the κ light chain and IL-2 receptor α chain, cell adhesion molecules and acute phase proteins. The largest class are the cytokines and hematopoetic growth factors. In addition, transcription factor genes were found to be regulated by NF-κB, allowing an increase in NF-κB´s pleiotropic effects and regulatory crosstalks. Some viruses use NF-κB to stimulate their own expression. A well-studied example is the human immunodeficiency virus (HIV-1) which contains two identical NF-κB-binding sites in its long terminal repeat

(LTR). Mutational analysis with HIV-1-LTR-controlled reporter constructs and recombinant viruses showed that the activity of the HIV-LTR largely depends on NF-κB as well as on the transcription factor Sp1 (Nabel and Baltimore, 1987).

Various subunit combinations of NRD proteins display subtle differences in binding site preference and transcriptional activity, as will be discussed below. It is thus possible that target genes may use distinct subunit combinations for optimal activation. An example is the IL-8 gene which appears to be preferentially induced by p65 homodimers (Kunsch and Rosen, 1993).

Signal transduction by NF-κB

Numerous, seemingly unrelated stimuli lead to the activation of NF-κB. Table 2 gives a list of physiological inducers of NF-κB which all have in common that they are either primary pathogenic conditions that menace the cell or are secondary signals, which are endogenously produced in response to pathogens. The signals include very diverse stimuli such as bacterial lipopolysaccharide (LPS), inflammatory cytokines TNF and IL-1, viruses, T cell mitogens and physical stress factors such as UV light and γ-ray exposure (reviewed in Grilli et al., 1993, Meyer et al., 1994). Despite the fact that NF-κB inducers use different receptor systems and intracellular signal-transducing pathways they, ultimately, cause the same reaction in the cytosol, i.e., the release of IκB from NF-κB.

Physiological NF-κB-Inducing Conditions

Class	**Condition**
Bacterial Products	Lipopolysaccharide Exotoxin B Toxic shock syndrome toxin 1 Muramyl peptides
Viruses	Human immunodeficiency virus type 1 Human T cell leukemia virus type 1 (HTLV-1) Hepatitis B virus (HBV) Herpes simplex virus type 1 Human herpes virus-6 Newcastle disease virus Sendai virus

	Epstein-Barr virus (EBV) Influenza virus
Viral products	Double-stranded RNA intermediates Tax (from HTLV-1) HBx (from HBV) MHBs[t] (from HBV) EBNA2 (EBV) Latent membrane protein (EBV) Hemagglutinin (Influenza)
Eucaryotic Parasites	Theileria parva
Cytokines	Tumor necrosis factor-α Lymphotoxin Interleukin-1 Leukotriene B4 Leukemia inhibitory factor Macrophage colony-stimulating factor
Mitogens	Antigen (T and B cells) CD2/CD28 ligands (T cells) Platelet-derived growth factor Serum
Protein accumulation in the ER	Viral secretory proteins IgM heavy chain Toxins (tunicamycin, monensin, brefeldin A)
Apoptotic and necrotic stimuli	H_2O_2, growth factor deprivation, TNF (cell type-dep.)
Radiation	UV-A, -B and -C light γrays
Oxidants	Hydrogen peroxide Oxidized Low Density Lipoprotein

Table 2: List of physiological inducers of NF-κB activity. For references, see Baeuerle and Henkel, 1994.

The roles of phosphorylation

Since it was known that PMA, a direct activator of protein kinase C (PKC), induces NF-κB early experiments on the signal transduction focussed on the role of protein kinases. Further support came from the findings that activation by PMA was blocked by the kinase inhibitors H7 and staurosporine and that the phosphatase inhibitors okadaic acid and calyculin A lead to an activation of NF-κB binding activity (Thevenin et al., 1991; Menon et al., 1993).

Treatment of cytosolic extracts with PKC, protein kinase A (PKA) and heme-regulated eIF-2 (HRI) kinase led to an in vitro activation of the DNA-binding activity of NF-κB apparently by phosphorylating IκB (Shirakawa and Mizel, 1989; Ghosh and Baltimore, 1990). However, it still remains to be clarified whether these kinases also phosphorylate IκB-α in intact cells. PMA-sensitive PKC isotypes did not appear to play a role in the activation of NF-κB by TNF-α and IL-1 (Meichle et al., 1990). Signal transduction by IL-1 apparently necessitates the cytoplasmic domain of the IL-1 receptor and is dependent on tyrosine kinase activity (Leung et al., 1994; Joshi-Barve et al., 1993). Further evidence for the involvement of protein kinase pathways came from studies on the UV-induced transcriptional activition of the NF-κB-dependent HIV-1 LTR. Induction was impaired by negative alleles of v-src, Ha-ras and raf-1 kinase (Devary et al., 1993). The involvement of tyrosine kinases is supported by the finding that the specific tyrosine kinase inhibitor AG213 inhibited the UV response of NF-κB (Devary et al., 1993). UV stimulation of Hela cells rapidly activates the cytoplasmic Src family tyrosine kinases, followed by the activation of the small guanosine triphosphate (GTP)-binding protein Ha-ras and the cytoplasmic serine-threonine kinase raf-1.

The importance of ha-Ras and Raf-1 for NF-κB activation is emphasized by several other reports (Bruder et al., 1993; Devary et al., 1993; Finco and Baldwin, 1993). The mechanism, by which Raf-1 activates NF-κB is a matter of debate. Bruder et al. (1993) suggest that activation takes place by a mechanism independent from IκB-α inactivation while other reports showed IκB-α elimination induced by Raf-1 (Li and Sedivy, 1993; Finco and Baldwin, 1993). The importance of Raf-1 in directly phosphorylating IκB-α is challenged by a recent study directly comparing the in vitro IκB-phosphorylating activity of different protein kinases including raf-1 (Diaz-Meco et al., 1994). These studies suggested, that neither MAPK, MKK nor Raf-1 are directly phosphorylating IκB-α. Rather, a ζPKC-associated kinase was identified to have this property. In-gel-kinase assays predicted this IκB-kinase to have a molecular weight of approximately 50 kDa (Diaz-Meco et al., 1994). The functional importance of ζPKC as a member of the signalling cascade leading to the phosphorylation of IκB-α (and the activation of NF-κB) is supported by the finding that overexpression of the enzyme stimulates a permanent translocation of active NF-κB to the nucleus of NIH3T3 cells. Transfection of a dominant negative mutant of ζPKC inhibited TNF-induced transactivation of a co-transfected reporter gene, which was under the control of NF-κB binding sites (Diaz-Meco, 1993).

Phosphorylation of IκB-α is also seen in intact cells by the induction of a more slowly-migrating IκB variant in response to stimulation of cells with PMA, TNF, IL-1 and LPS (Brown et al., 1993; Beg et al., 1993). Phosphatase treatment led to the disappearance of the more slowly migrating form, suggesting that its decreased mobility was caused by phosphorylation. However, the appearance of this more slowly migrating IκB form did not coincide but preceded the appearance of active NF-κB. Coimmunoprecipitation assays from Hela cells metabolically labelled with ^{35}S-methionine using an anti IκB-α antibody demonstrated that newly phosphorylated IκB-α is still attached to the p65 subunit of NF-κB (B. Traenckner and P. Baeuerle, unpublished). This suggests that phosphorylation does not release IκB, as was observed in cell-free kinase reactions, but may target the inhibitor for a subsequent proteolytic degradation (Sun et al., 1993; Brown et al., 1993; Beg et al., 1993; Henkel et al., 1993). In fact, it is the decay of IκB-α which occurs simultaneously with the activation of NF-κB.

Phosphorylation events may also lead to an inhibition of NF-κB activation. Transformation of a pre B-cell line with the Abelson murine leukemia virus resulted in inactivation of NF-κB. In conditionally transformed pre-B cell lines, the v-viral transforming protein v-Abl, a tyrosine kinase, abrogated the activity of NF-κB, presumably by a mechanism resulting in an increased stability of IκB-α (Klug et al., 1994).

The activity of the IκB-like proteins IκB-β and Bcl-3 is modulated by phosphorylation in a way distinct from IκB-α. In contrast to IκB-α, purified IκB-β and Bcl 3 lost their inhibitory activity upon treatment with phosphatase (Link et al., 1992; Nolan et al., 1993). Phosphatase treatment could also release active NF-κB from a complex with IκB-β, suggesting that NF-κB complexed to IκB-β may be activated in intact cells by dephosphorylation rather than by phosphorylation (Link et al., 1992).

The DNA-binding subunits can also be modified by phosphorylation. The human c-Rel and p105 proteins were found to be constitutively phosphorylated on serines and threonines. PMA/PHA-treatment of Jurkat T cells induced an additional tyrosine phosphorylation (Neumann et al., 1992). Phosphate-labelling experiments showed that treatment of cells with TNF, double-stranded RNA and phorbol esters led to an increase of p105 phosphorylation, followed by an increased rate of p105 degradation (Mellits et al., 1993). The p65 kDa subunit was also found to be closely associated with a serine/threonine kinase activity (Ostrowski et al., 1991). Stimulation of cells with IL-1α and LPS led to an increased phosphorylation of p65. The activity of the kinase was found to correlate with the binding activity of NF-κB, as determined by electrophoretic mobility shift assays. An increased phosphorylation of the transcription activation domain 2 (TA_2) of p65 was observed after stimulation of cells with PMA. This increased phosphorylation is presumably responsible for the PMA-enhanced transcriptional activity of the p65 TA_2 domain (M. L. Schmitz et al., unpublished).

The role of proteolysis

Following stimulation of cells with PMA, the half-life of IκB-α is reduced from 2 h to only 1.5 min (Henkel et al., 1993). Polyclonal antibodies directed against IκB-α could not detect any breakdown products in Western blots, suggesting a rapid and complete proteolysis of IκB-α. The enzyme responsible for the proteolytic degradation of IκB-α might be a subunit of the ubiquitous multicatalytic protease. This is evident from the finding that a specific membrane-permeable peptide inhibitor of the chymotrypsin-like subunit of the proteasome can significantly increase the half-life of IκB-α after stimulation of cells with TNF (B. Traenckner and P. Baeuerle, unpublished). Moreover, the inhibitor stabilized the phosphorylated form of IκB-α.

A small fraction of not more than 10-20% of the cellular IκB-α can escape the degradation by the protease. This might be due to the absence of a modification necessary for degradation of IκB-α. Alternatively, it is possible that a differently compartimentalized fraction of IκB-α (e.g. in the nucleus) can escape the action of the IκB-degrading enzyme. Another plausible explanation would presume that the protease can only recognize modified IκB-α complexed to NF-κB. This model would explain the finding that overexpressed IκB-α in transfected cells is more or less resistant to proteolytical degradation.

Following the depletion of IκB, the inhibitor is rapidly resynthesized. This is possible because the IκB-α gene is under transcriptional control by NF-κB itself (reviewed in Beg and Baldwin, 1993).

The role of redox signals

The induction of NF-κB activity by TNF is mediated by the 55 kDa TNF receptor (Schütze et al., 1992). Binding of TNF-α to the receptor was found to activate a phosphatidylcholine-specific phospholipase C to produce 1,2-diacylglycerol (DAG). DAG stimulates an acidic (or neutral) sphingomyelinase (SMase) and leads to the release of ceramide from sphingomyelin. An increase in ceramide leads to the stimulation of a ceramide-activated protein kinase (Dressler et al., 1992) which may trigger a cellular signalling cascade finally removing IκB. Addition of TNF, phospholipase C, synthetic DAG analogues, purified acidic SMase or ceramide to permeabilized cells all induced the DNA-binding form of NF-κB (Schütze et al., 1992). Another report confirmed the stimulatory effects of TNF, sphingomyelinase and ceramide, but failed to show the stimulatory effects of a synthetic DAG analogue and phospholipase C (Yang et al., 1993). Whereas both TNF and IL-1 seem to induce NF-κB by the ceramide pathway, other inducers of NF-κB may use distinct pathways.

The first indications for the probable involvement of redox signals came from the observation that the antioxidative compound N-acetyl-L-cysteine (NAC) can suppress the activity of NF-κB (Staal et al., 1990) and that H_2O_2 can activate the transcription factor (Schreck et al., 1991). Also other antioxidants, such as dithiocarbamates, 2-mercaptoethanol, α-lipoic acid, butylated hydroxyanisol and chelators of iron and copper ions were found to interfere with the activation of NF-κB (reviewed in Meyer et al., 1994). These compounds are chemically very diverse, suggesting that not their particular structure but their common antioxidative potential was responsible for their inhibitory effect. In the meantime, it was also possible to demonstrate that overexpression of the antioxidative enzymes thioredoxin (Schenk et al., 1994) and catalase (K. Schmidt and P.A. Baeuerle, unpublished) suppress NF-κB activation.

Direct evidence for the activating role of reactive oxygen intermediates (ROIs) comes from the inducing effect of micromolar amounts of hydrogen peroxide in Hela and Jurkat cells (Schreck et al., 1991). Reagents leading to the generation of superoxide, nitrous oxide, singlett oxygen and hypochlorite failed to activate, suggesting that NF-κB is a peroxide-inducible transcription factor, similar to the bacterial oxyR protein (reviewed in Pahl and Baeuerle, 1994). Many of the different inducers of NF-κB were shown to increase the cellular production of hydrogen peroxide. These include TNF, IL-1, LPS, UV-A, γ-rays and okadaic acid, as measured by electron spin resonance techniques, the determination of lipid peroxidation products, the depletion of glutathione levels or chemoluminescent methods (reviewed in Schreck et al., 1992). The generation of hydrogen peroxide (or secondary ROIs) is common to most (if not all) inducers of NF-κB, because the inducing effects of TNF,

IL-1, PMA, four viral transactivator proteins, muramyl peptides, LPS, γ-rays, UV-A, leukotriene B4 and double-stranded RNA could be efficiently blocked by antioxidative compounds (reviewed in Meyer et al., 1994). The fact that all these different inducers are inhibited by antioxidants indicates that the generation of hydrogen peroxide may be a relatively late event in the signalling cascade activating NF-κB. It is currently not clear by which enzymes the reactive oxygen indermediates are produced. Candidates are plasma membrane-associated NADPH-oxidases, xanthine oxidase, glucose oxidase and enzymes involved in arachidonic acid metabolism. In the case of TNF it was shown that the cytotoxic and NF-κB-activating effect relies on the production of ROIs originating from the mitochondria (Schulze-Osthoff et al., 1993). The electron transport inhibitors rotenone and amytal blocked cell killing by TNF and prevented NF-κB activation. Cells depleted of mitochondria by ethidium bromide treatment were found to be insensitive to the cytotoxic and NF-κB-activating effect of TNF.

How can hydrogen peroxide activate NF-κB? Various kinases and phosphatases are known to be regulated in their activity by the redox status of the cell but is not known yet whether the distinct enzymes control phosphorylation and degradation of IκB. The different signal transduction pathways leading to the activation of NF-κB are schematically displayed in Figure 3.

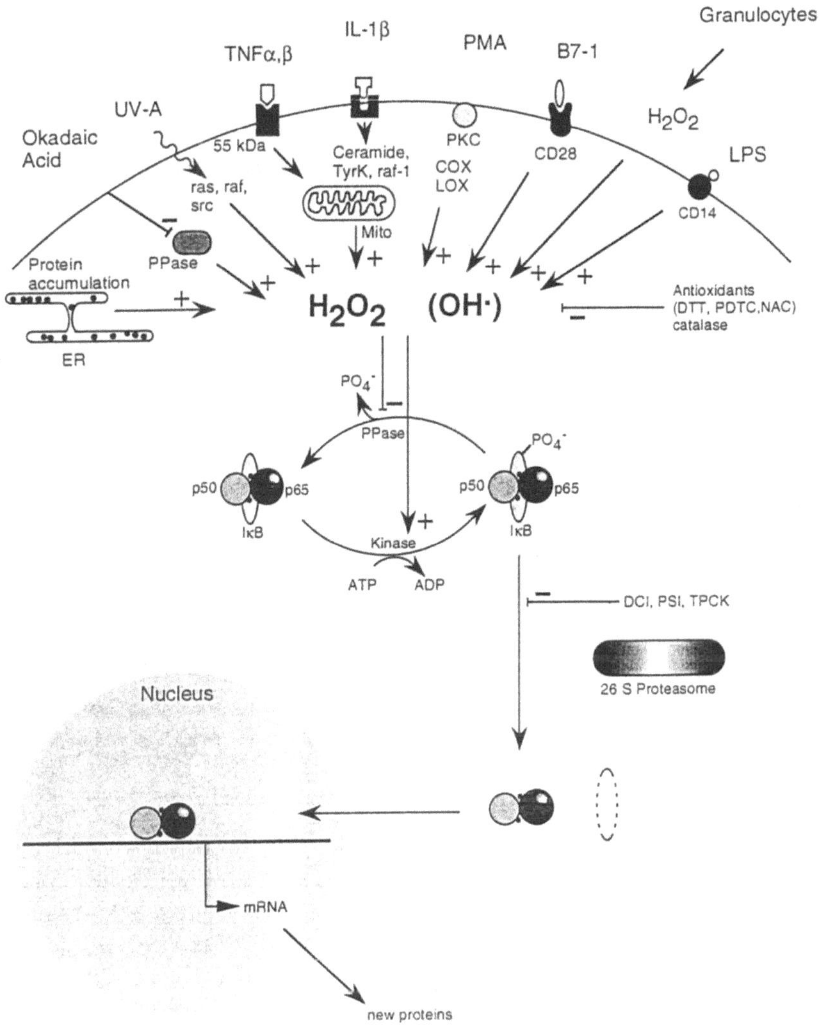

Fig. 3. Schematic representation of activation pathways for NF-κB. Abbreviations not used in the text are: B7-1: B-cell activation antigen 7-1; COX: cyclooxygenase; DCI: 3,4 dichloroisocoumarin; DTT: dithiothreitol ; ER: endoplasmatic reticulum; 55KDa: 55 kDa TNF receptor; LOX: lipoxygenase; Mito: mitochondrion; PDTC: pyrrolidinedithiocarbamate; PPase: phosphatase; PSI: Cbz-Ile-Glu(o-t-Bu)-Ala-Leucinal; TPCK: tosylphenylchloromethylketone; TyrK: tyrosine kinase. For further details, see text.

Conclusions

The regulation of the activity of NF-κB/Rel transcription factors displays several unique aspects of how gene expression can be inducibly controlled. Dimer combinations made up from five distinct DNA-binding subunits are sequestered in the cytoplasm by at least five different inhibitory proteins. Stimulation of cells with a great variety of pathogenic agents leads to the generation of ROIs, the proteolytical degradation of IκB molecules and, ultimately, the release of active transcription factor dimers from a subset of the cytoplasmically sequestered complexes. The loss of IκB unmasks the NLS sequences in the DNA-binding subunits which now can be recognized by general receptors delivering proteins to or into the nucleus. In the nucleus, NF-κB induces transcription of its target genes. Among them are genes encoding IκB proteins. The increased transcription and re-synthesis of IκB attenuates NF-κB-mediated induction of gene expression by means of an autoregulatory loop. The highly controlled and regulated shuttling of the NF-κB/Rel family members between the cytoplasm and the cell nucleus make NF-κB an ideal paradigm to study the point of intersection between signal transduction pathways and gene transcription.

Acknowledgements: We thank Dr. Heike L. Pahl for helpful discussions and Patricia Müller for her help in the preparation of the manuscript. This work was supported by grants from the Bundesministerium für Forschung und Technologie, the Deutsche Forschungsgemeinschaft and the European Community (Biotechnology Programme) awarded to P.A.B.

Literature

Baeuerle PA and Baltimore D (1988a) Activation of DNA-binding activity in an apparently cytoplasmic precursor of the NF-κB transcription factor. Cell 53:211-217

Baeuerle PA and Baltimore D (1988b) IκB: a specific inhibitor of the NF-κB transcription factor. Science 242:540-546

Baeuerle PA and Baltimore D (1989) A 65 kD subunit of active NF-κB is required for inhibition of NF-κB by IκB. Genes Dev 3:1689-1698

Baeuerle PA and Henkel T (1994) Function and activation of NF-κB in the immune system. Annu Rev Immunol 12:141-179

Beg AA, Ruben SM, Scheinman RI, Haskill S, Rosen CA and Baldwin Jr AS (1992) IκB interacts with the nuclear localization sequences of the subunits of NF-κB: a mechanism for cytoplasmic retention. Genes Dev 6:1899-1913

Beg, AA, Finco TS, Nantermet PV and Baldwin Jr AS (1993) Tumor Necrosis Factor and Interleukin-1 lead to phosphorylation and loss of IκBα: a mechanism for NF-κB activation. Mol Cell Biol 13:3301-3310

Beg AA and Baldwin Jr AS (1993) The IκB proteins: multifunctional regulators of Rel/NF-κB transcription factors. Genes Dev 7:2064-2070

Bours V, Franzoso G, Azarenko V, Park S, Kanno T, Brown K and Siebenlist U (1993) The oncoprotein bcl-3 directly transactivates through κB motifs via association with DNA-binding p50B homodimers. Cell 72:729-739

Breeden L and Nasmyth K (1987) Similarity between cell-cycle genes of budding yeast and fission yeast and the Notch gene of Drosophila. Nature 329:651-654

Brown AM, Linhoff MW, Stein B, Wright KL, Baldwin Jr AS, Basta PV and Ting PY (1994) Function of NF-κB/Rel Binding Sites in the Major Histocompatibility Complex Class II Invariant Chain Promoter is Dependent on Cell-Specific Binding of Different NF-κB/Rel Subunits. Mol Cell Biol 14:2926-2935

Brown K, Park S, Kanno T, Franzoso G and Siebenlist U (1993) Mutual regulation of the transcriptional activator NF-κB and its inhibitor, IκB-α. Proc Natl Acad Sci USA 90:2532-2536

Bruder JT, Heidecker G, Tan TH, Weske JC, Derse D and Rapp UR (1993) Oncogene activation of HIV-LTR-driven expression via the NF-κB binding sites. Nucl Acids Res 21:5229-5234

Bull P, Morley KL, Hoekstra MF, Hunter T and Verma IM (1990) The mouse c-rel protein has an N-terminal regulatory domain and a C-terminal transcriptional transactivation domain. Mol Cell Biol 10:5473-5485

Cressman DE and Taub R (1993) IκBα can localize in the nucleus but shows no direct transactivation potential. Oncogene 8:2567-2573

Devary Y, Rosette C, DiDonato JA and Karin M (1993) NF-κB activation by ultraviolet light not dependent on a nuclear signal. Science 261:1442-1445

Diaz-Meco MT, Berra E, Municio MM, Sanz L, Lozano J, Dominguez I, Diaz-Golpe V, Lain de Lera MT, Alcami J, Paya CV, Arenzana-Seisdedos F, Virelizier JL and Moscat J (1993) A dominant negative protein kinase C ζ subspecies blocks NF-κB activation. Mol Cell Biol 13:4770-4775

Diaz-Meco MT, Dominguez I, Sanz L, Dent P, Lozano J, Municio MM, Berra E. Hay RT, Sturgill TW and Moscat J (1994) ζPKC induces phosphorylation and inactivation of IκB-α *in vitro*. EMBO J 13:2842-2848

Dressler KA, Mathias S and Kolesnik RN (1992) Tumor necrosis factor-α activates the sphingomyelin signal transduction pathway in a cell-free system. Science 255:1715-1718

Fan CM and Maniatis T (1991) Generation of p50 subunit of NF-κB by processing of p105 through an ATP-dependent pathway. Nature (London) 354:395-398

Finco TS and Baldwin AS Jr (1993) κB site-dependent induction of gene expression by diverse inducers of Nuclear Factor κB requires Raf-1. J Biol Chem 268:17676-17679

Franzoso G, Bours V, Park S, Tomita-Yamaguchi M, Kelly K and Siebenlist U (1992) The candidate oncoprotein Bcl-3 is an antagonist of p50/NF-κB-mediated inhibition. Nature (London) 359:339-342

Fujita T, Nolan GP, Gosh S and Baltimore D (1992) Independent modes of transcriptional activation by the p50 and p65 subunits of NF-κB. Genes Dev 6: 775-787

Fujita T, Nolan GP, Liou HC, Scott, ML and Baltimore D (1993) The candidate proto-oncogene *bcl-3* encodes a transcriptional coactivator that activates through NF-κB p50 homodimers. Genes Dev 7:1354-1363

Ganchi PA, Sun SC, Greene WC and Ballard DW (1992) IκB/Mad-3 masks the nuclear localization signal of NF-κB p65 and requires the transactivation domain to inhibit NF-κB p65 DNA binding. Mol Cell Biol 3:1339-1352

Gilmore TD and Temin HM (1988) v-rel oncoproteins in the nucleus and in the cytoplasm transform chicken spleen cells. J Virol 62:703-714

Ghosh S, Gifford AM, Riviere LR, Tempst P, Nolan GP and Baltimore D (1990) Cloning of the p50 DNA binding subunit of NF-κB: homology to rel and dorsal. Cell 62:1019-1029

Ghosh S and Baltimore D (1990) Acivation in vitro of NF-κB by phosphorylation of its inhibitor IκB. Nature (London) 344:678-682

Grilli M, Chiu JJ and Lenardo MJ (1993) NF-κB and rel - participants in a multiform transcriptional regulatory system. Int Rev Cytol 143:1-62

Grumont RJ and Gerondakis S (1994) Alternative splicing of RNA transcripts encoded by the murine p105 NF-κB gene generates IκBγ isoforms with different inhibitory activities. Proc Natl Acad Sci 91:4367-4371

Hatada EN, Nieters A, Wulczyn, FG, Naumann M, Meyer R, Nucifora G, McKeithan TW and Scheidereit C (1992). The ankyrin repeat domains of the NF-κB precursor p105 and the proto-oncogene bcl-3 act as specific inhibitors of NF-κB DNA binding. Proc Natl Acad Sci USA 89:2489-2493

Henkel T, Zabel U, van Zee K, Müller JM, Fanning E and Baeuerle PA (1992) Intramolecular masking of the nuclear location signal and dimerization domain in the precursor for the p50 NF-κB subunit. Cell 68:1121-1133

Henkel T, Machleidt T, Alkalay I, Ben-Neriah Y, Krönke KM and Baeuerle PA (1993) Rapid proteolysis of IκB-α is necessary for activation of transcription factor NF-κB. Nature (London) 365:182-184

Inoue J, Kerr LD, Kakizuka A and Verma IM (1992a). IκB-γ, a 70 kD protein identical to the C-terminal half of p110 NF-κB: A new member of the IκB-family. Cell 68:1109-1120

Inoue J, Kerr LD, Rashid D, Davis N, Bose Jr HR and Verma IM (1992b) Direct association of pp40/IκB-β with rel/NF-κB transcription factors: Role of ankyrin repeats in the inhibition of DNA binding activity. Proc Natl Acad Sci USA 89:4333-4337

Joshi-Barve SS, Rangnekar VV, Sells SF and Rangnekar VM (1993) Interleukin-1-inducible Expression of gro-β via NF-κB Activation is dependent upon Tyrosine Kinase Signaling. J Biol Chem 268:18018-18029

Kang SM, Tran AC, Grilli M and Lenardo MJ (1992) NF-κB subunit regulation in nontransformed $CD4^+$ T lymphocytes. Science 256:1452-1456

Kawakami K, Scheidereit C, and Roeder RG (1988) Identification and purification of a human immunoglobulin-enhancer-binding protein (NF-κB) that activates transcription from a human immunodeficiency virus type 1 promoter in vitro. Proc Natl Acad Sci USA 85:4700-4704

Kieran M, Blank V, Logeat F, Vandekerckhove J, Lottspeich F, Le Bail O, Urban MB, Kourilsky P, Baeuerle PA and Israel A (1990) The DNA binding subunit of NF-κB is identical to factor KBF1 and homologous to the rel oncogene product. Cell 62:1007-1018

Klug CA, Gerety SJ, Shah PC, Chen YY, Rice NR, Rosenberg N and Singh H (1994) The v-*abl* tyrosine kinase negatively regulates NF-κB/Rel factors and blocks κ gene transcription in pre-B lymphocytes. Genes Dev 8:678-687

Kretzschmar M, Meisterernst M, Scheidereit C, Li G and Roeder RG (1992) Transcriptional regulation of the HIV-1 promoter by NF-κB in vitro. Genes Dev 6: 761-774

Kunsch C and Rosen CA (1993) NF-κB subunit-specific regulation of the Interleukin-8 promoter. Mol Cell Biol 13:6137-6146

Leung K, Betts JC, Xu L and Nabel GJ (1994) The Cytoplasmic Domain of the Interleukin-1 Receptor is required for Nuclear Factor-κB Signal Transduction. J Biol Chem 269:1579-1582

Lernbecher T, Müller U and Wirth T (1993) Distinct NF-κB/Rel transcription factors are responsible for tissue-specific and inducible gene activation. Nature 365:767-770

Li S and Sedivy JM (1993) Raf-1 protein kinase activates the NF-κB transcription factor by dissociating the cytoplasmic NF-κB-IκB complex. Proc Natl Acad Sci 90:9247-9251

Link E, Kerr LD, Schreck R, Zabel U, Verma I and Baeuerle PA (1992). Purified IκB-β is inactivated upon dephosphorylation. J Biol Chem 267:239-246

Logeat F, Israel N, Ten R, Blank V, LeBail O, Kourilsky P and Israel A (1991) Inhibition of transcription factors belonging to the rel/NF-κB family by a transdominant negative mutant. EMBO J 10:1827-1832

Meichle A, Schütze S, Hensel G, Brunsing D and Krönke M (1990) Protein kinase C-independent activation of nuclear factor κB by tumor necrosis factor. J Biol Chem 265:8339-8343

Mellits KH, Hay RT and Goodbourn S (1993) Proteolytic degradation of MAD3 (IκBα) and enhanced processing of the NF-κB precursor p105 are obligatory steps in the activation of NF-κB. Nucl Acids Res 21:5059-5066

Menon SD, Qin S, Guy GR and Tan YH (1993) Differential Induction of Nuclear NF-κB by Protein Phosphatase Inhibitors in Primary and Transformed Human Cells. J Biol Chem 268:26805-26812

Mercurio F, DiDonato JA, Rosette C and Karin M (1993) p105 and p98 precursor proteins play an active role in NF-κB-mediated signal transduction. Genes Dev 7:705-718

Meyer R, Hatada EN, Hohmann HP, Haiker M, Bartsch C, Rötlisberger U, Lahm HW, Schlaeger EJ, van Loon AP and Scheidereit C (1991) Cloning of the DNA-binding subunit of human nuclear factor κB: the level of its mRNA is strongly regulated by phorbol ester or tumor necrosis factor α. Proc Natl. Acad Sci USA 88: 966 - 970

Meyer M, Schreck R. Müller JM and Baeuerle PA (1994) Redox Control of Gene Expression by Eukaryotic Transcription Factors NF-κB, AP-1 and SRF/TCF. In Oxidative Stress, Cell Activation and Viral Infection (Pasquier C et al, eds) pp 217-235, Birkhäuser Verlag Basel/Schweiz

Michaely P and Bennett V (1992) The ANK repeat: a ubiquitous motif involved in macromolecular recognition. Trends Cell Biol 2:127-129

Moore PA, Ruben SM and Rosen CA (1993) Conservation of Transcriptional Activation Functions of the NF-κB p50 and p65 subunits in mammalian cells and Saccharomyces cerevisiae. Mol Cell Biol 13: 1666-1674

Mosialos G, Hamer P, Capobianco AJ, Laursen RA and Gilmore TD (1991) A protein kinase-A recognition sequence is structurally linked to transformation by p59v-rel and cytoplasmic retention of p68c-rel. Mol Cell Biol 11:5867-5877

Morin PJ and Gilmore TD (1992) The C terminus of the NF-κB p50 precursor and an IκB isoform contain transcription activation domains. Nucl Acids Res 20:2453-2458

Nabel G and Baltimore D (1987) An inducible transcription factor activates expression of human immunodeficiency virus in T cells. Nature (London) 326:711-713

Naumann M, Wulczyn FG and Scheidereit C (1993) The NF-κB precursor p105 and the protooncogene bcl-3 are IκB-molecules and control nuclear translocation of NF-κB. EMBO J 12:213-222

Neumann M, Tsapos K, Schoppler JA, Ross J and Franza BR (1992) Identification of complex formation between two intracellular tyrosine kinase substrates:human c-Rel and the p105 precursor of p50 NF-κB. Oncogene 7:2095-2104
Nolan GP, Ghosh S, Liou HC, Tempst P and Baltimore D (1991) DNA binding and IκB inhibition of the cloned p65 subunit of NF-κB, a rel-related polypeptide. Cell 64:961-969
Nolan GP, Fujita T, Bhatia K, Huppi C, Liou HC, Scott ML and Baltimore D (1993) The *bcl-3* oncogene encodes a nuclear IκB-like molecule that preferentially interacts with NF-κB p50 and p52 in a phosphorylation-dependent manner. Mol Cell Biol 13:3557-3566
Nolan GP (1994) NF-AT-AP-1 and Rel-bZIP: Hybrid Vigor and Binding under the Influence. Cell 77:795-798
Ostrowski J, Sims JE, Hopkins Sibley C, Valentine MA, Dower SK, Meier KE and Bomsztyk K (1991) A Serine/Threonine Kinase Activity is closely associated with a 65-kDa Phosphoprotein specifically recognized by the κB enhancer Element. J Biol Chem 266:12722-12733
Pahl HL and Baeuerle PA (1994) Oxygen and the Control of Gene Expression. BioEssays 16:497-502
Rice NR, MacKichan ML and Israël A (1992). The precursor of NF-κB p50 has IκB like functions. Cell 71:243-253
Rivière Y, Blank V, Kourilsky P and Israel A (1991). Processing of the precursor of NF-κB by the HIV-1 protease during acute infection. Nature (London) 350:625-626
Ruben SM, Dillon PJ, Schreck R, Henkel T, Chen CH, Maher M, Baeuerle PA and Rosen CA (1991) Isolation of a rel-related human cDNA that potentially encodes the 65-kD subunit of NF-κB. Science 251:1490-1493
Ruben SM, Klement JF, Coleman TA, Maher M, Chen CH and Rosen CA (1992) I-Rel, a novel rel-related protein that inhibits NF-κB transcriptional activity. Genes Dev 6:745-760
Ryseck RP, Bull P, Takamiya M, Bours V, Siebenlist U, Dobrzanski P and Bravo R (1992) RelB, a new Rel family transcription activator that can interact with p50-NF-κB. Mol Cell Biol 12:674-684
Scheinmann RI, Beg AA and Baldwin Jr AS (1993) NF-κB p100 (Lyt-10) is a Component of H2TF1 and can function as an IκB-like Molecule. Mol Cell Biol 13:6089-6101
Schenk H, Klein M, Erdbrügger W, Dröge W and Schulze-Osthoff K (1994) Distinct effects of thioredoxin and antioxidants on the activation of transcription factors NF-κB and AP-1. Proc Natl Acad Sci USA 91:1672-1676
Schmid RM, Perkins ND, Duckett CS, Andrews PC and Nabel GJ (1991) Cloning of an NF-κB subunit which stimulates HIV transcription in synergy with p65. Nature (London) 352:733-736
Schmitz ML and Baeuerle PA (1991) The p65 subunit is responsible for the strong transcription activation potential of NF-κB. EMBO J 10:3805-3817
Schmitz ML, Henkel T and Baeuerle, PA (1991) Proteins controlling the nuclear uptake of NF-κB, rel and dorsal. Trends Cell Biol 1:130-137
Schreck R, Rieber P and Baeuerle PA (1991) Reactive oxygen intermediates as apparently widely used messengers in the activation of the NF-κB transcription factor and HIV-l. EMBO J 10:2247-2258
Schreck R, Albermann K and Baeuerle PA (1992) Nuclear factor kappa B: an oxidative stress-responsive transcription factor of eucaryotic cells (a review). Free Rad Res Comms 17:221-237
Schütze S, Pothoff K, Machleidt T, Bercovic D, Wiegmann K and Krönke M (1992) TNF activates NF-κB by phosphatidylcholine-specific phospholipase C-induced "acidic" sphingomyelin breakdown. Cell 71:765-776

Schulze-Osthoff K, Beyaert R, Vandervoorde V, Haegeman G and Fiers W (1993) Depletion of the mitochondrial electron transport abrogates the cytotoxic and gene induction effects of TNF. EMBO J 12:3095-3104
Sen R and Baltimore D (1986) Multiple nuclear factors interact with the immunoglobulin enhancer sequences. Cell 46:705-716
Shirakawa F and Mizel S (1989) In vitro activation and nuclear translocation of NF-κB catalyzed by cyclic AMP-dependent protein kinase and protein kinase C. Mol Cell Biol 9:2424-2430
Staal FJ, Roederer M, Herzenberg LA and Herzenberg LA (1990) Intracellular thiols regulate activation of nuclear factor κB and transcription of human immunodefiency virus. Proc Natl Acad Sci USA 87:9943-9947
Stephens RM, Rice NR, Hiebsch RR, Bose Jr HR and Gilden RV (1983) Nucleotide sequence of v-rel: The oncogene of reticuloendotheliosis virus. Proc Natl Acad Sci 80:6229-6233
Steward R (1987) Dorsal, an embryonic polarity gene in Drosophila, is homologous to the vertebrate proto-oncogene c-rel. Science 238:692-694
Sun SC, Ganchi PA, Ballard DW and Greene WC (1993) NF-κB controls expression of inhibitor IκB-α: Evidence for an inducible autoregulatory pathway. Science 259:1912-1915
Thevenin C, Kim SJ and Kekol SJ (1991) Inhibition of protein phosphatases by okadaic acid induces AP-1 in human T-cells. J Biol Chem 166:9363-9366
Toledano MB, Ghosh D, Trinh F and Leonard WJ (1993) N-Terminal DNA-Binding Domains contribute to differential DNA-binding specificities of NF-κB p50 and p65. Mol Cell Biol 13:852-860
Urban MB, Schreck R and Baeuerle PA (1991) NF-κB contacts DNA by a heterodimer of the p50 and p65 subunit. EMBO J 10:1817-1825
Wilhelmsen KC, Eggleton K and Temin HM (1984) Nucleic acid sequences of the oncogene v-rel in reticuloendotheliosis virus strain T and its cellular homolog, the proto-oncogene c-rel. J Virol 52:172-182
Yang Z, Costanzo M, Golde DW and Kolesnick RN (1993) Tumor Necrosis Factor activation of the sphingomylein pathway signals Nuclear Factor κB translocation in intact HL-60 cells. J Biol Chem 268:20520-20523
Zabel, U. und Baeuerle, P. A. (1990). Purified human IκB can rapidly dissociate the complex of the NF-κB transcription factor with its cognate DNA. Cell 61:255-265.
Zabel U, Henkel T, dos Santos Silva M and Baeuerle PA (1993) Nuclear uptake control of NF-κB by MAD-3, an IκB protein present in the nucleus. EMBO J 12:201-211
Zhang Q, DiDonato JA, Karin M and McKeithan TW (1994) Bcl3 encodes a nuclear protein which can alter the subcellular location of NF-κB proteins. Mol Cell Biol 14:3915-3926

Functional Analysis of the Glucocorticoid Receptor

Per-Erik Strömstedt, Jan Carlstedt-Duke and Jan-Åke Gustafsson
Dept. of Medical Nutrition
Huddinge University Hospital,
Novum,
141 86 Huddinge
Sweden

THE STEROID/THYROID HORMONE RECEPTOR SUPERFAMILY

An increased knowledge of the mechanisms underlying signal transduction and regulation of gene expression is fundamental for our understanding of developmental biology, cell cycle regulation, endocrinology and carcinogenesis. In this context studies on the basic molecular biology and biochemistry of receptors and other transcription factors are of paramount interest. The steroid/thyroid hormone receptor superfamily of ligand inducible transcriptional activators includes receptors for steroid and thyroid hormones, vitamin D, retinoic acid, and a large group, still growing in number, of so called orphan receptors with unknown ligands and physiological functions (Evans, 1988; Beato, 1989; Carson-Jurica *et al.,* 1990; Wahli and Martinez, 1991; Power *et al.,*1992). The cloning and characterization of cDNAs encoding the various members of this superfamily has revealed a high level of molecular identity among the nuclear receptors. Common to all the nuclear receptors of this superfamily is their modular structural architecture (Fig. 1). They all harbor three major autonomous functional domains; an N-terminal domain with important functions for full transcriptional activity, a central DNA binding domain which shows the highest degree of structural homology among the members, and a C-terminal ligand-binding domain (Gustafsson *et al.,*1987; Carson-Jurica *et al.,* 1990; Wahli and Martinez, 1991; Gronemeyer, 1992). These three domains have been further subdivided into six regions, A to F (Krust *et al.,*1986), that are more or less homologous between the various members: A/B, F, modulating regions; C, DNA binding region; D, "hinge" region: E, ligand binding region (Fig. 1).

NATO ASI Series, Vol. H 92
Signalling Mechanisms – from Transcription Factors to Oxidative Stress
Edited by L. Packer, K. Wirtz

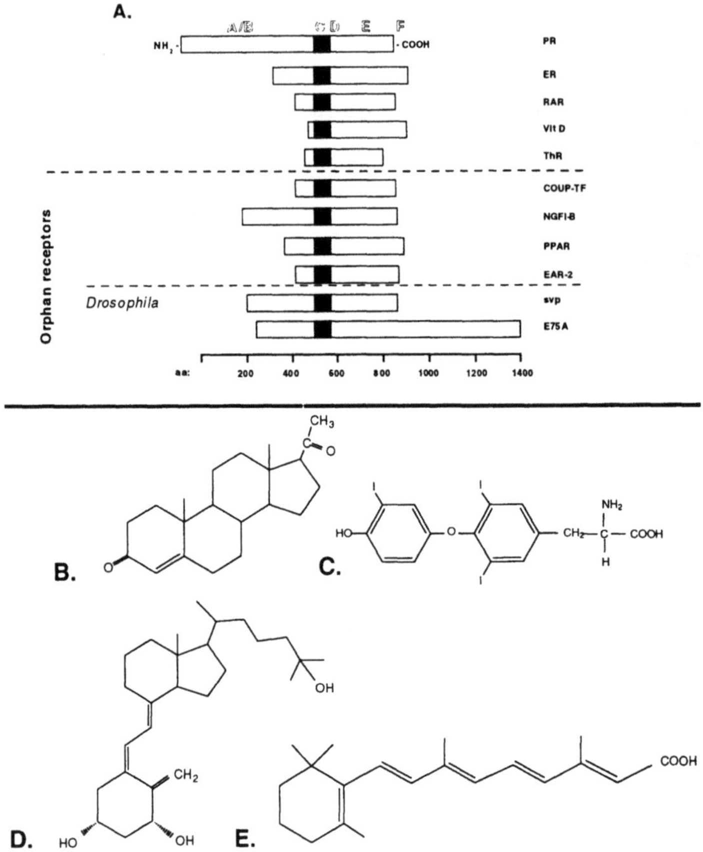

Fig. 1. Schematic comparison of primary structures between members of the nuclear receptor superfamily and the chemical structures of several known ligands. (A) schematic representations are shown for PR, thyroid hormone receptor α (ThRα), vitamin D receptor (VDR), retinoic acid receptor α (RARα), chicken ovalbumin upstream promoter transcription factor (COUP-TF), nerve growth factor inducible protein (NGFI-B), peroxisome proliferator activated receptor (PPAR), v-erbArelated receptors (EAR-2), seven-up (svp), and ecdysone inducible gene E75A. For references see text. The filled box represents the DNA binding domain. Ligands shown are (B) progesterone, (C) triiodothyronine, (D) calcitriol, and (E) retinoic acid.

The receptor subfamilies

The nuclear receptor superfamily can be grouped into at least three major subfamilies based on the similarity within the DNA binding domain: I. thyroid hormone and retinoic acid receptors, II. orphan receptors, and III. steroid receptors. Several of the orphan receptors have been cloned from *Drosophila* indicating evolutionary conservation of signal transduction pathways mediated by nuclear receptors (Segraves, 1991). Phylogenetic analysis of the superfamily has suggested that the nuclear receptors have evolved from a common ancestor gene (Amero *et al.*, 1992; Laudet *et al.*, 1992). Duplications and mutations of this gene could have resulted in the variety of receptors that exists today. Furthermore, this putative ancestor gene is believed to have been divergent from ancestor genes of other transcription factors, ligand-binding proteins, or zinc finger proteins (Amero *et al.*, 1992). Further break down into smaller subgroups within the three major subfamilies is possible, *e.g.* the glucocorticoid receptor-like subgroup consisting of receptors for progestins (PR), mineralocortcoids (MR), androgens (AR) and glucocorticoids (GR) (Evans, 1988; Wahli and Martinez, 1991; Laudet *et al.*,1992).

Ligands

The basic structural unity between the different members of the nuclear receptor superfamily stands in great contrast to the diversity of the chemical structures of their cognate ligands (Fig. 1). The three classes of ligands currently known are steroids, retinoids and thyroid hormone. Steroids and retinoids share a common biosynthetic pathway. Both are terpenes which are derived by assembly of isoprene units (Fig. 2). This common biosynthetic link raises interesting issues with regard to the evolutionary process of the receptor superfamily and the possible existence of receptors for other terpene derivatives including juvenile hormone and certain plant hormones (Moore, 1990). Thyroid hormone is the only currently known nonterpenoid ligand for members of the superfamily. However, there is evidence for groups of compounds other than the terpenoids as potential ligands for some orphan receptors. Binding of ligand is believed to be a prerequisite for activation of members of the nuclear receptor superfamily and great efforts are invested in order to identify putative ligands for the orphan receptors.

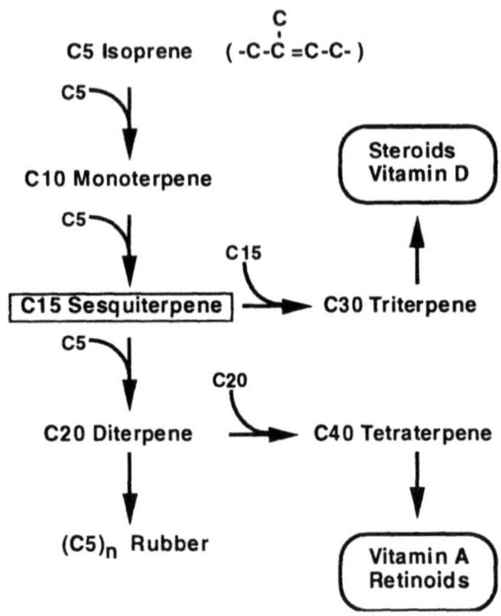

Fig. 2. Biosynthesis of steroids and retinoids. The isomeric C5 isopentenyl pyrophosphate and dimethylallyl pyrophosphate condense to form the C10 geranyl pyrophosphate. Addition of another isopentenyl pyrophosphate generates farnesyl pyrophosphate, the C15 precursor common to the steroid/vitamin D and retinoid/vitamin A families. Adapted from Moore (1990).

The peroxisome proliferator activated receptor (PPAR) is one example of an orphan receptor which can be activated by compounds other than terpenoids and thyroid hormones (Green, 1992; Poellinger *et al.*, 1992). As the name implies, potent peroxisome proliferators including hypolipidaemic drugs such as clofibrate and Wy-14,643, have been shown to activate PPAR. Fatty acids such as arachidonic, linoleic, or lauric acid have also, in a recent study, been demonstrated to activate PPAR (Göttlicher *et al.*, 1992). However, this diversity in putative ligands for PPAR has led to speculation whether they bind to PPAR directly or modulate its activity through some indirect mechanism. Importantly, it has hitherto not been possible to demonstrate direct binding by any of the activating compounds to PPAR. Moreover, there is accumulating evidence for alternative activation mechanisms besides direct ligand binding for other members of the nuclear receptor superfamily. The catecholaminergic neurotransmitter dopamine appears to initiate a signal transduction pathway from its membrane receptor that also, in a ligand-independent fashion, activates several orphan receptors (Power *et al.*, 1991*a*; Lydon *et al.*, 1992) and certain steroid hormone receptors (Power *et al.*, 1991*b*). This activation response is presumably mediated by inducing and/or increasing phosphorylation of target receptors or some other rate-limiting co-factor(s) (posttranslational modifications; Power *et al.*, 1992).

Smith et al. (1993) have shown that ligand-independent activation by dopamine of ER can be modulated by hormone and antihormone. The presence of an agonist (17β-estradiol) further stimulates the activation of ER by dopamine while a pure antagonist (ICI 164,384) inhibits activation by dopamine. Finally, expression of the orphan receptor NGFI-B, an immediate early protein, which stimulates transcription of the steroid 21-hydroxylase gene, seems to be regulated at a transcriptional level by adrenocorticotropic hormone (ACTH) (Wilson *et al.,* 1993). This apparent cross-talk between membrane-associated receptors and the intracellular nuclear receptors adds to the complexity and diversity of these systems. Further dissection of these signal transduction pathways will probably reveal many more common factors and regulatory mechanisms.

Classes of HREs

The nuclear receptors modulate transcription of target genes by their interaction with specific gene sequences known as hormone responsive or regulatory elements (HRE) (Lucas and Granner, 1992). These sequence elements are cis-acting and enhancer-like since they function in a manner relatively independent of position and orientation (Chandler *et al.,* 1983; Ponta *et al.,* 1985). Three general classes of HREs have been defined. These are shown in Fig. 3 and include glucocorticoid/progesterone response elements (GRE/PRE), estrogen response elements (ERE), and thyroid hormone/retinoic acid/vitamin D response elements (TRE/RRE/VDRE). Two different hexameric core sequences (half-sites) are conserved in all these elements and the half-sites are arranged in either palindromic or direct-repeat patterns. In addition to the consensus sequences the spacing between the half-sites (i.e. the conserved core motif) is also of importance for receptor specificity (de Luca, 1991). The GRE/PRE is also recognized by MR and AR (von der Ahe *et al.,* 1985). In the case of the ER-like members of the superfamily, including RAR, RXR, VDR, ThR, and PPAR, varying spacing between identical direct repeat sequences has been demonstrated to mediate different specificity for these receptors, suggesting that protein-protein interactions within homodimers/heterodimers confer different spatial arrangements of the DNA binding domains, thus resulting in different specificity (Umesono *et al.,* 1991; Yu *et al.,* 1991). This specific subgroup of the superfamily has been shown to require the presence of nuclear auxiliary proteins in order to achieve high affinity DNA binding to the specific response element *in vitro* (Murray and Towle, 1989; Burnside *et al.,*1990; Lazar and Berrodin, 1990; Darling *et al.,*1991). RXR seems under certain conditions to function as an auxiliary protein enabling high affinity DNA binding by PPAR, ThR, VDR, and RAR (Yu *et al.,*1991; Zhang etal., 1992; Kliewer *et al.,*1992a, 1992b; Leid *et al.,*1992; Marks *et al.,*1992; Gearing *et al.,*1993).

With regard to the HRE specificity, the subfamilies can be distinguished by the primary DNA-recognition sequence (P-box) (Fig. 10 compares the GR-like members with ER) within the first so called zinc finger of the DNA binding domain (DBD) (Wahli and Martinez, 1991).

GRE/PRE	$\mathbf{AG^{A}/_{G}ACA}$ NNN $\mathbf{TGT^{T}/_{C}C^{C}/_{T}}$
ERE	$\mathbf{^{N}/_{A}GGTCA}$ NNN $\mathbf{TGACC^{N}/_{T}}$
TRE/RRE/VDRE	**AGGTCA** $N_{0\text{-}6}$ **AGGTCA**

Fig. 3. Hormone response elements (consensus HREs) for a variety of nuclear receptors. Note the conserved core motif (bold letters)

The response elements may represent a family of closely related sequences (Klock *et al.*, 1987) and furthermore, the insect ecdysone-responsive element is a palindromic sequence very similar to the vertebrate ERE (Fig. 3). Thus, it is possible that invertebrate and vertebrate HREs originated from a common ancestral DNA motif (Martinez *et al.*, 1991). In view of all the common structural, functional, and evolutionary links between the members of this superfamily, and given that many of the functions and characteristics are shared among the members of this gene superfamily, more ligands, DNA binding sites and dimerization partners can be expected to be found.

The best studied subfamily of this large superfamily of transcription factors is the subfamily of steroid hormone receptors. This review will focus on the glucocorticoid receptor and its interaction with its cognate ligand.

THE GLUCOCORTICOID RECEPTOR

It has been known for more than three decades that steroid hormones act via intracellular receptor proteins and large efforts have since then been invested to elucidate their mechanism of action. In target cells, receptors are activated by hormonal ligands and thereafter modulate the expression of a network of specific target genes (Yamamoto and Alberts, 1976; Katzenellenbogen, 1980; Yamamoto, 1985). The first nuclear receptor to be cloned was the glucocorticoid receptor

(GR) including rat GR (rGR) (Miesfeld *et al.*, 1984, 1986), human GR (hGR) (Hollenberg, *et al.*, 1985), and mouse GR (mGR) (Danielsen *et al.*, 1986). The cloning of GR cDNAs and of cDNAs encoding other nuclear receptors belonging to this superfamily has been fundamental to reveal many aspects of receptor structure and function. The following sections will focus on certain key issues of glucocorticoid receptor action, and will attempt to present a balanced view of the conspicuous features of other receptors in the superfamily.

Glucocorticoid hormones

Glucocorticoids influence a wide variety of important biochemical, physiological, and pathological processes that involve virtually every organ system. In some cases the steroid is active but not sufficient by itself; it may, for instance, be required for the maximal action of some other hormones (Reshef and Shapiro, 1961; Boyet and Hofert, 1972; Granner, 1979). Biosynthesis of glucocorticoids takes place in the zona fasciculata and zona reticularis of the adrenal cortex and begins with cholesterol, as does the synthesis of all steroid hormones, and involves a series of enzymatic reactions mainly catalyzed by cytochrome P-450 isozymes (Gower, 1988; Miller, 1988; Simpson and Waterman, 1988). Loss of adrenal cortical function ultimately results in death unless replacement therapy is instituted. The major glucocorticoid secreted in humans is cortisol while in rodents it is corticosterone (Fig. 4).

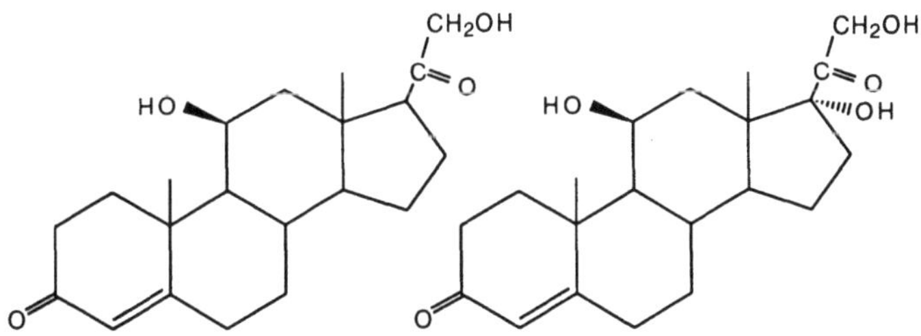

CORTICOSTERONE CORTISOL

Fig. 4. The major glucocorticoid produced by the adrenal cortex in rodents and human is corticosterone and cortisol, respectively.

The secretion of glucocorticoids is dependent on ACTH which in turn is regulated by corticotropin-releasing factor (CRF). Secretion of these hormones is regulated by a classic negative feedback loop mechanism (Fig. 5). The negative feedback control by adrenal steroids on pituitary-adrenal function has been subdivided into rapid, intermediate, and delayed feedback processes (Dallman *et al.,* 1987). Transient, rapid feedback actions are seen within minutes as steroid levels rise or fall, and these effects modulate pituitary-adrenal output. In contrast, delayed steroid effects on levels of CRF and ACTH peptides or their mRNA levels are usually manifested over several days, and reflect changes in the tonic control over pituitary-adrenal output. Excessive levels of free hormone inhibit CRF release, while abnormally low levels of free hormone activate the system by enhancing CRF release from the hypothalamus (Fig. 5).

ACTH release (and glucocorticoid secretion) is controlled by neural input from a number of sites within the central nervous system. There is an endogenous rhythm that controls the release of CRF and therefore ACTH. This circadian cycle is normally set to provide an increase of plasma glucocorticoids in the latter half of the sleep period. The plasma glucocorticoid level gradually increases during sleeping hours, peaks shortly after waking up, gradually falls over the next hours, and reaches the lowest point at about midnight. This general pattern is due to a series of episodic, pulsatile bursts of ACTH release (Jones and Gillham, 1988). Glucocorticoid secretion is also affected by physical and emotional stress, apprehension, fear, anxiety, and pain. These responses can override both the negative feedback system and the diurnal rythm.

Glucocorticoids circulate in plasma in protein-bound and free forms. The main plasma binding protein is called transcortin or corticosteroid-binding globulin (CBG), and binds most of the hormone when plasma cortisol levels are within normal range (Ballard, 1979; for reviews of CBG structure and function see Hammond, 1990; Rosner, 1990). A much smaller amount of cortisol is bound to albumin. The unbound or free fraction constitutes about 8 % of total plasma cortisol and represents the biologically active fraction of cortisol.

Metabolic effects of glucocorticoids

Glucocorticoid hormones are named for their ability to promote glucose production. The effects on intermediary metabolism are minimal in the fed state. However, during fasting or starvation, glucocorticoids contribute to the maintenance of plasma glucose levels. Glucocorticoids promote hepatic glucose production by increasing the rate of gluconeogenesis by stimulation of key enzymes, by releasing amino acids, the gluconeogenetic substrates, from peripheral tissues

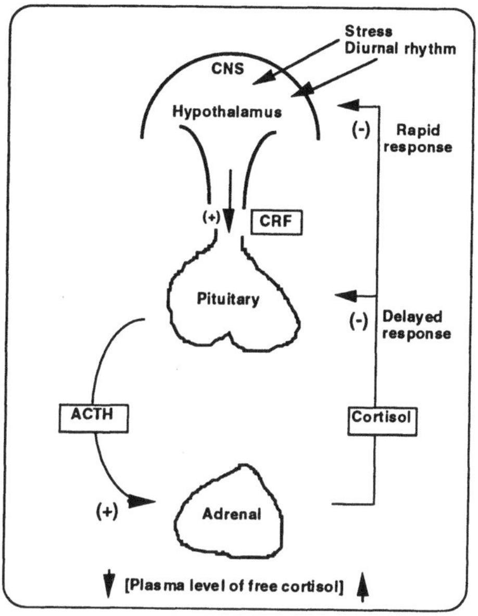

Fig. 5. Feedback regulation of cortisol biosynthesis. Low levels of free cortisol = stimulatory pathways. High levels of free cortisol = inhibitory pathways.

such as muscle through catabolic actions, and by "permitting" other hormones to stimulate key metabolic processes. Glucocorticoids also stimulate glycogen synthesis in the liver, and the effect in adipose tissue is an increased lipolysis with the release of glycerol and free fatty acids as a result. In spite of the lipolytic effect of glucocorticoids, an increased fat deposition is a classical manifestation of glucocorticoid excess. Fat is deposited centrally in the face, cervical area, trunk, and abdomen while the extremities are usually not affected (i.e. Cushing's syndrome). The reason for the abnormal fat deposition in connection with glucocorticoid excess is unknown.

Effects of glucocorticoids on host defense mechanisms

High concentrations of glucocorticoids suppress the host immune responses and inflammatory reactions. Most of these effects require glucocorticoids at supraphysiological levels, *e.g.* in doses used to treat autoimmune diseases or suppress transplant rejection. The role of physiological levels of these hormones in modulating immune function is not clear. At supraphysiological levels, these steroids cause involution of lymphoid tissue and induce apoptosis of lymphocytes in rodents, while in humans an inhibition of lymphocyte growth and function can be seen (Lipman, 1979). With regard to thymic involution, these effects are more dramatic in mice than in humans but in both species precursor cells seem not to be affected. Glucocorticoids also affect the proliferative response of lymphocytes to antigens and may also affect several other steps in the immune response.

Glucocorticoids are directly involved in the response to acute stress induced by surgery, trauma, or infection. The multiple actions of glucocorticoids on the cardiovascular system probably contribute to the steroidal stress-combating effect. These actions would improve vascular reactivity and cardiac performance. The suppressive actions of glucocorticoids on immunological and inflammatory responses together with the later effects described above have led to their wide-spread therapeutic use, for instance treatment of autoimmune diseases, asthma, and for prevention of transplant rejection. Alterations in the structure of glucocorticoids have led to the development of synthetic compounds with greater glucocorticoid activity *e.g.* dexamethasone (Dex), triamcinolone acetonide (TA) (see Fig. 6 for structural formulas). These synthetic steroids are more potent than the physiological glucocorticoids due to increased affinity for the receptor and a decreased plasma clearance. The delayed plasma clearance is due to decreased susceptibility to metabolism and an absence of binding of the analogs to CBG.

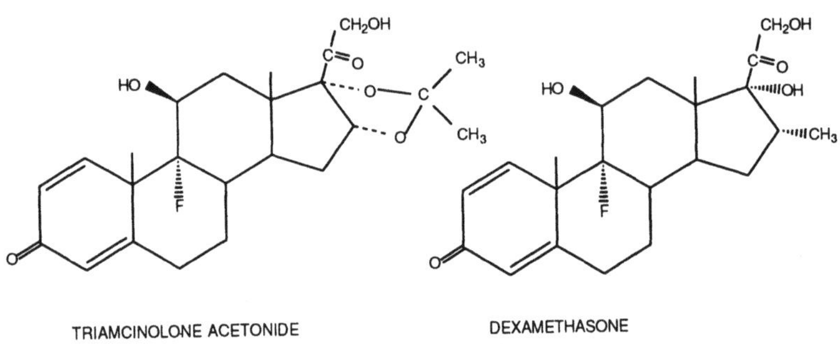

Fig. 6. Two widely used synthetic glucocorticoid analogs: triamicinolone acetonide (TA) and dexamethasone (Dex). Both analogs display high affinity binding to GR.

Other biological effects of glucocorticoids

Glucocorticoids affect many other tissues and functions in addition to the ones already described. Excess of glucocorticoids leads to inhibition of fibroblast growth, which leads to loss of collagen and connective tissue, and thus results in thinning of the skin and poor wound healing, common problems related to long-term therapeutic use of glucocorticoids. In addition, bone tissue is affected by an excess of glucocorticoids. Long term treatment with high doses of glucocorticoids can cause an increase in osteolysis (Bar-Shavit *et al.*, 1984; Hahn, 1989; Wong *et al.*, 1990), also termed steroid-induced osteoporosis. Glucocorticoids accelerate the development of a number of systems and organs in fetal and differentiating tissues (Baxter and Tyrrell, 1987).

Molecular mechanisms of action

A hypothetical model for glucocorticoid hormone action is depicted in Fig. 7. Glucocorticoid action is initiated by the entry of the steroid into the cell presumably by passive diffusion, although facilitated uptake has also been reported (Ballard, 1979; Rao, 1981; Furu *et al.*, 1987). The steroid binds to the cytosolic GR which prior to hormone binding is maintained in an inactive state as an oligomeric complex (Joab *et al.*, 1984; Rexin *et al.*, 1991). This oligomeric receptor complex contains a single molecule of GR (Okret *et al.*, 1985; Gehring and Arndt, 1985) in association with several heat shock proteins, including hsp90, hsp56 (also termed p56, and, in some reports, p59 or hsp59), and possibly hsp70 (Sanchez, 1990; Sanchez *et al.*, 1990a; Pratt *et al.*, 1992; Smith and Toft, 1993). None of the heat shock proteins thus far described are unique to a single receptor system or cell type. They are all common cellular proteins that are more abundant than steroid receptors (for instance, hsp90 generally constitutes ~1% of total cellular protein), and have been implicated to act as molecular chaperones that mediate efficient folding and assembly of proteins into native structures (Hartl *et al.*, 1992; Gething and Sambrook, 1992; Georgopoulos, 1992; Wiech *et al.*, 1992). Moreover, classical steroid receptors including PR, AR, ER and MR have all been shown to associate with at least hsp90 in a manner analogous to GR (Joab *et al.*, 1984; Renoir *et al.*, 1986; Redeuilh *et al.*, 1987; Rafestin-Oblin *et al.*, 1989; Pratt, 1990; Smith and Toft, 1993). In contrast, thyroid hormone and retinoic acid receptors do not interact with hsp90 (Dalman *et al.*, 1990, 1991).

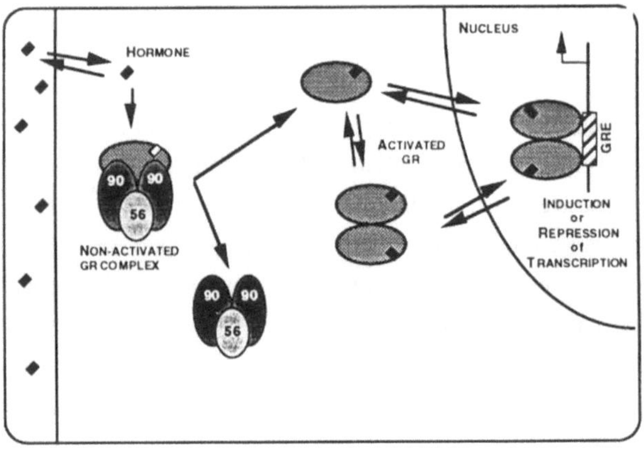

Fig. 7. Signal transduction pathway of glucocorticoid hormones. Only GR, hsp90, and hsp56 are outlined in the inactive oligomeric form of GR (see text for details).

As indicated above, the oligomeric receptor complex contains heat shock proteins other than hsp90. At least two distinct, additional heat shock proteins seem to be associated with steroid receptors: hsp70 and hsp56 (Tai *et al.*, 1992; for a review see Smith and Toft, 1993). Hsp56 has recently been cloned (Lebeau *et al.*, 1992) and shown to be a member of the immunophilin class of proteins (Yem *et al.*, 1992; Tai *et al.*, 1992). The immunophilins are proteins that bind in a high affinity manner immunosuppressive agents including cyclosporin A and rapamycin. This protein family has rotamase (peptidylprolyl cis-trans-isomerase) activity *in vitro*, and it is believed that the immunophilins, in a fashion similar to hsp70 and hsp90, play major roles in protein folding and protein trafficking in the cell. Hsp56 has sequence homologies to the near C-terminus of the FK506-binding proteins FKBP-12 and FKBP-13, and, more importantly, Tai *et al.*, (1992) have shown that FK506 binds directly to hsp56. In a recent report evidence was presented that upon coadministration of Dex and FK506, a potentiation of glucocorticoid induced gene expression is seen relative to Dex alone (Ning and Sanchez, 1993). The potentiation was suggested to be a consequence of an increased nuclear translocation rate of GR. However, Hutchison *et al.*, (1993a) have not observed any effects of FK506 on receptor folding or function. A heterocomplex containing all three heat shock proteins exists in reticulocyte lysate and in cytosolic extracts independent of the presence of steroid receptors (Sanchez et al. 1990b; Perdew and Whitelaw, 1991). The unfoldase activity of hsp70 is required for receptor-heat shock complex formation *in vitro* using the rabbit reticulocyte lysate systems (Hutchinson *et al.*, 1993b).

Upon hormone binding, the associated heat shock proteins dissociate from GR (Raaka *et al.*, 1985; Denis *et al.*, 1987; Mendel and Ortí, 1988), and GR acquires an increased affinity for DNA (Leach *et al.*, 1979; Sanchez *et al.*, 1987; Mendel *et al.*, 1986; Denis *et al.*, 1988a; Howard and Distelhorst, 1988). This process has been referred to as receptor activation or transformation (Grody *et al.*, 1982). The activated receptor dimerizes either prior to or in conjunction with DNA binding (Tsai *et al.*, 1988; Wrange *et al.*, 1989; Cairns *et al.*, 1992). In the nucleus, the activated GR complex can interact with nuclear chromatin acceptor sites (Becker *et al.*, 1986). Binding of the activated receptor to its specific DNA sequence, known as hormone responsive elements (HRE) (or more specifically: glucocorticoid responsive elements; GRE: see Introduction), leads to an increase or decrease in the rate of transcription of specific target genes (Yamamoto, 1985; Wahli and Martinez, 1991; Akerblom and Mellon, 1991). These regulatory sequences are located at a distance ranging from several 1000 base pairs upstream or downstream to close proximity of the transcription initiation site (Moore *et al.*, 1985; Slater *et al.*, 1985; Jantzen *et al.*, 1987; Hecht *et al.*, 1988; Chan *et al.*, 1991: see also Yamamoto, 1985; Gustafsson *et al.*, 1987).

The subcellular localization of GR is still a controversial issue. In the absence of hormone, GR has been reported to be predominantly cytoplasmic (Picard and Yamamoto, 1987; Wikström *et al.*, 1987; Cadepond *et al.*, 1992). Upon hormone binding, GR becomes tightly bound to the nuclear fraction, and this binding is resistant during cell fractionation *in vitro*. However, there are contradictory reports showing nuclear localization of GR even in the absence of hormone (Welshons *et al.*, 1985; Gasc *et al.*, 1989; Sanchez *et al.*, 1990b; Brink *et al.*, 1992). These contradictory results may depend on cell types and species used in the different studies and/or the techniques applied. In contrast, unliganded androgen, estrogen, and progesterone receptors appear to have a constitutive nuclear localization, where they remain in an inactive conformation until the binding of hormone triggers activation of the receptor (King and Greene, 1984; Perrot-Applanat *et al.*, 1985; Gasc *et al.*, 1989; Husmann *et al.*, 1990). Whatever the reason for the presence of GR in the cytoplasm, a most important issue would be to determine whether this putative compartmentalization of GR may be of any physiological significance.

Although GR appears to be identical in all tissues examined, the proteins affected by glucocorticoids vary widely as a result of varying patterns of expression or inhibition of specific batteries of genes in different cell types (also termed "gene networks", see Yamamoto, 1985 for review). There is also some evidence for alternative mechanisms of action of glucocorticoids. An example is the glucocorticoid-induced fast feedback inhibition of ACTH secretion (Fig. 5). The effects occur within minutes of glucocorticoid administration, and its rapidity suggests that it is not due to RNA and protein synthesis but rather to glucocorticoid-induced changes in secretory function or cell membrane permeability (Canny *et al.*, 1989; Childs and Unabia, 1990).

Steroid receptors function *in vivo* only following exposure to extracellular signals. As outlined above, in most cases this signal is represented by direct binding of the receptor to its cognate ligand. As will be discussed in the following sections, the hormonal ligand has profound influences on receptor structure and function. Furthermore, differences in mechanisms of action between agonists and antagonists are not completely understood. The aim of the present study was to learn more about the functional domain structure of the GR protein. In addition, we were interested in a detailed analysis of the ligand binding domain of GR in order to learn more about structural requirements for specific and high affinity binding of glucocorticoids. To this end, a more long term goal was to overexpress either the full length receptor protein or a derivative spanning the ligand binding domain. We wanted therefore to characterize several expression systems for production of large amounts of receptor protein faithfully mimicking *bona fide* ligand binding activity and, possibly, allowing advanced structural studies such as X-ray crystallographic analysis of liganded and unliganded receptors.

RESULTS AND DISCUSSION:

DOMAIN STRUCTURE OF THE GLUCOCORTICOID RECEPTOR

The glucocorticoid receptor monomer is a polypeptide which varies in length between species (Muller and Renkawitz, 1991), and heterogeneity in the length of GR has also recently been observed among different rat strains (Gearing *et al.*, 1993). The human GR consists of 777 amino acids while mouse and rat (rat hepatoma cell line J.2.17.2) GR consists of 783 and 795 amino acid residues, respectively. A comparison of the primary structures reveals a polyglutamine region in the N-terminal half of the receptor which varies in length between species and rat strains. The function of this polyglutamine tract is unknown. A similar polyglutamine heterogeneity has been observed in the androgen receptor between different species (Tilley *et al.*, 1989). Polyglutamine domains have been implicated as transactivating regions of other transcription factors such as SP1, Antennapedia, and transcription factor IID (Mitchell and Tijan, 1989; Kao *et al.*,1990). Furthermore, X-linked spinal and bulbar muscular atrophy has been speculated to be caused by an enlargement of the polyglutamine tract present in the hAR (LaSpada *et al.*,1991), indicating that the polyglutamine region may be important for receptor function. The overall amino acid identity for GR between species is more than 90 %.

It has been shown that defined proteolytic fragments of GR possess some, but not all of the characteristics of the intact molecule (Carlstedt-Duke *et al.*,1977; Wrange and Gustafsson, 1978; Carlstedt-Duke *et al.*,1982; Vedeckis, 1983; Wrange *et al.*, 1984) suggesting the presence of several autonomous functional domains. The availability of cDNAs coding for several steroid hormone receptor proteins allowed the detailed analysis of functional regions in these complex proteins. By expressing deletion mutants of the receptor as well as chimeric receptor proteins (domain swapping), discrete functions have been attributed to precise regions of the protein (Gigueré *et al.*, 1986; Hollenberg *et al.*, 1987; Green and Chambon, 1987; Godowski *et al.*, 1987, 1988) (Fig. 9).

The three major functional domains have been defined at the protein level by applying limited proteolysis of purified rGR followed by N-terminal sequence analysis (Carlstedt-Duke *et al.*, 1987). The ligand binding domain (LBD) was defined by cleavage with trypsin giving rise to a fragment of M_r 27-30 kDa. The N-terminus was at position 518. Digestion with α-chymotrypsin resulted in a fragment with M_r 39-41 kDa with maintained ligand and DNA binding activities (Carlstedt-Duke *et al.*, 1987 and references therein). Sequence analysis showed an NH_2-terminus corresponding to several cleavage points. However, two segments were identified starting at residues 410 and 414, respectively. Thus, chymotryptic fragment(s) also contained the entire LBD. Assuming that the border between regions A/B and C (Fig. 8) lies between residues 410 and 414 and the border between regions C and D-F lies between residues 517 and 518, this would mean that the size of DBD is ~12 kDa.

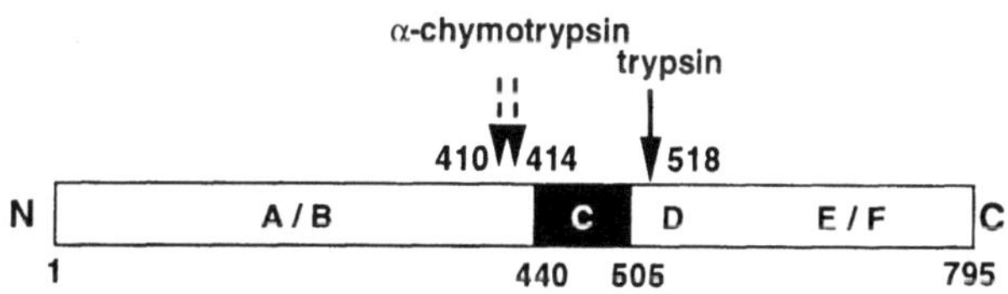

Fig. 8. Functional domains of GR as defined by limited proteolysis of purified rGR. Cleavage with α-chymotrypsin is indicated by dashed arrows for multiple cleavage sites. The solid arrow indicates the cleavage site for trypsin. Numbers refer to the amino acid residues as defined by the nucleotide sequence for the coding region of the cDNA. The filled box represents the DNA binding region as defined by analysis of the cDNA.

These results confirmed functional data that have emerged from mutagenesis of GR cDNA (Giguére *et al.*,1986; Danielsen *et al.*,1986). The chymotryptic cleavage site N-terminal of the DBD was later utilized in purification of recombinant DBD expressed in *E. coli* as a protein A fusion protein (Dahlman *et al.*,1988).

Posttranslational modifications

There is no evidence for extensive posttranslational proteolytic processing of GR since data from the cDNA and the CNBr-peptide analyses of purified GR are similar in all these respects (Carlstedt-Duke *et al.*, 1987) and furthermore, purified activated rGR has an estimated relative molecular weight of 90 kDa which is in agreement with that deduced from the cDNA open reading frame. The results obtained from total amino acid composition analysis of the purified receptor were in excellent agreement with those indirectly deduced from the cDNA clone (Carlstedt-Duke *et al.*, 1987). Thus, taken together, these results infer that the cDNA clone described (Miesfeld *et al.*, 1986) represents the intact receptor protein. However, the NH_2-terminus appears to be blocked (Carlstedt-Duke *et al.*, 1987), and when examining the N-terminal structure (Miesfeld *et al.*,1986) it resembles the known properties of N-acetyl-blocked proteins in general (Persson *et al.*,1985). The acetylated residue could be the initiator methionine itself.

There are, however, several reports where the M_r observed for GR on denaturating polyacrylamide gels is 7 - 10 kDa higher than the molecular weight deduced from the cDNA. Posttranslational glycosylation has been suggested to account for this difference in molecular weight. In support of this model the presence of sugar moieties within rGR has been reported (Blanchardie *et al.*,1986). In contrast, Yen and Simons (1991) have not observed any biochemical evidence for glycosylation of rGR after applying many different analytical methods. A conclusive answer to this question should be possible to obtain from mass-spectrometric analysis of the purified receptor protein.

In analogy to most, if not all other members of the nuclear receptor superfamily, the glucocorticoid receptor is a phosphoprotein (Ortí *et al.*,1992). Several of the classical steroid hormone receptors become hyperphosphorylated upon binding of their cognate ligands, suggesting important mechanistic connections between the levels of phosphorylation and the function of these receptors as transcriptional regulators. GR is constitutively phosphorylated, mainly on serine residues within the acidic region (t1) of the N-terminal domain (Fig. 9) (Dalman *et al.*, 1988; Smith *et al.*, 1989; Bodwell *et al.*, 1991). Early reports have indicated hormone-independent phosphorylation of GR (Tienrungroj *et al.*, 1987) but more recent work has shown hormone-dependent phosphorylation of GR (Ortí *et al.*, 1989; Hoeck *et al.*, 1989; Hoeck and Groner, 1990). Addition of antagonist does not induce receptor hyperphosphorylation (Ortí *et al.*, 1989; Hoeck *et al.*, 1989), it appears rather to slightly increase the overall rate of dephosphorylation of hyperphosphorylated (agonist-treated) receptors (Ortí *et al.*,1993). No decrease or increase in phosphorylation has been observed in relation to DNA binding by GR. Moreover, dephosphorylation *in vitro* of the receptor has not been found to affect the ability of GR to bind to DNA (Dalman *et al.*, 1988; Pongratz *et al.*, 1991). In analogy, PR and ER have multiple

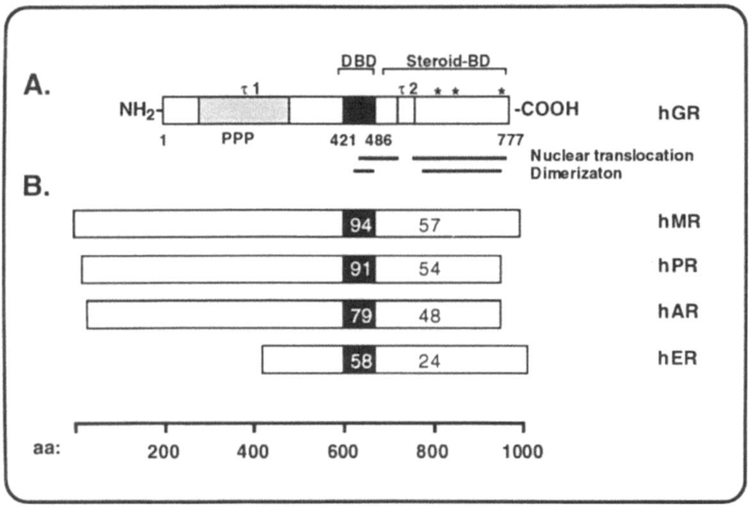

Fig. 9. The domain structure of members of the nuclear receptor superfamily. A, the major functional domains of the human glucocorticoid receptor (GR) protein are indicated. Major phosphorylation sites (P), and ligand cross-linked amino acids (*) are indicated. Two regions (τ1 and τ2) identified as being important for transactivation are indicated. DBD and LBD were defined by limited proteolysis of the protein. B, amino acid similarity within the classical steroid hormone receptors is indicated as a percentage of amino acid positional identity with the GR (see text for details).

phosphorylation sites, and hormone-dependent changes in the phosphorylation state have been reported (Rao *et al.*, 1987; Sullivan *et al.*, 1988; Migliaccio *et al.*, 1989; Sheridan *et al.*, 1989; Denner *et al.*, 1990a, 1990b).

The N-terminal Domain

The N-terminal domain is the most variable domain in size and amino acid composition among members of the nuclear receptor family (Evans, 1988). The N-terminal domain of GR is required for full transcriptional activity (Giguére *et al.*, 1986; Danielsen *et al.*, 1987; Hollenberg *et al.*, 1987; Miesfeld *et al.*, 1987) but is not required for hormone binding, receptor activation

(Gehring and Arndt, 1985), dimerization (Eriksson and Wrange, 1990) or DNA recognition (Freedman *et al.*, 1988), although there are reports indicating that the N-terminal domain affects the specificity in DNA binding (Payvar and Wrange, 1983; Danielsen *et al.*, 1987; Eriksson and Wrange, 1990). Removal of the N-terminus appears to result in a reduced ability by GR to distinguish between specific and nonspecific binding sites on DNA (Payvar and Wrange, 1983). The most antigenic regions of the receptor protein seem to be located within this domain, since most of the monoclonal and polyclonal antibodies raised against GR recognize epitopes within the N-terminus (Gustafsson *et al.*, 1987). The N-terminal domain is therefore often referred to as the immuno-dominant domain of GR (Carlstedt-Duke *et al.*, 1982).

At least one transcriptional activation region has been identified in the N-terminal domain of GR. This transactivating region has been shown to confer activity on a heterologous factor in chimeric protein constructs (Godowski *et al.*, 1988; Webster *et al.*,1988). For the human GR, this transactivation region ($\tau 1$) resides between amino acid residues 77-262 and is a highly acidic region (Fig. 9A) (Hollenberg and Evans, 1988). The transactivating domain is also the major site for phosphorylation although the importance of phosphorylation for transactivation activity is not fully understood. Furthemore, the $\tau 1$ region has been shown to be important for synergistic effects on transcriptional activation that are observed on tandem GRE sequences in front of a target promoter (Wright and Gustafsson, 1991). The $\tau 1$ region has also been proposed to interact with the basal transcription apparatus (Tasset *et al.*, 1990; Wright *et al.*, 1991; Ing *et al.*,1992; McEwen *et al.*,1993).

Transactivating domains have also been identified within the N-terminal domains of other steroid receptors. PR is the only steroid hormone receptor, to date, that exists in two distinct forms, PR_A and PR_B (~96 kDa and ~120 kDa, respectively; Lessey *et al.*,1983; Gronemeyer *et al.*,1985; Berkenstam *et al.*, 1988), presumably due to differential promoter usage which gives rise to two different mRNA species (Kastner *et al.*, 1990). The proteins produced differ in their N-termini: PR_A lacks the 164 most N-terminal residues present in the longer PR_B form. The chicken PR_A (cPR) seems to interact differently with the ovalbumin promoter than cPR_B (Tora *et al.*, 1988). Thus, the N-terminus seems to specify target gene activation. The importance of the N-terminal domain for cell/tissue specific transactivation responses has also been indicated for ER, AR, GR, and MR (Kumar *et al.*, 1987; Adler *et al.*, 1992; Pearce and Yamamoto, 1993). Since GR, MR, PR, and AR all recognize the same HRE, yet mediate very diverse effects *in vivo*, one possible mechanism for specificity could be differential protein-protein interaction patterns presumably dictated by N-terminal residues of the receptors. Thus, the activity of the transactivating domains in the receptor could be influenced in a cell type- and promoter-specific manner, resulting in pronounced differences in their ability to interact with distinct transcriptional co-factors (Tora *et al.*, 1988; Tasset *et al.*, 1990; Pearce and Yamamoto, 1993).

The DNA binding domain

The DNA binding domain (DBD) contains the strongest homology among the nuclear receptors (Figs. 9 and 10). This approximately 115 amino acid-long domain includes 9 cysteines which are all perfectly conserved among the members of the superfamily (Fig. 10). DBD is sufficient for selective interaction with specific response elements (Green and Chambon, 1987; Dahlman *et al.,* 1989). DBD contains two zinc atoms, each tetrahedrally coordinated by four cysteine residues forming two zinc fingers (Freedman *et al.,* 1988; Carlstedt-Duke *et al.,* 1988) analogous to the structures formed by transcription factor IIIA (Miller *et al.,* 1985).

```
                      First Finger                                    Second Finger
hGR 417 PPKLCLVCSDEASGCHYGVLTC GS CKVFFKRAV EGQHNYLC AG RND CIIDKIRRKNCPACRYRKCLQA
hMR 599 PSKICLVCGDEASGCHYGVVTC GS CKVFFKRAV EGQHNYLC AG RND CIIDKIRRKNCPACRLQKCLQA
hPR 563 PQKICLICGDEASGCHYGVLTC GS CKVFFKRAM EGQHNYLC AG RND CIVDKIRRKNCPACRLRKCCQA
hAR 553 PQKTCLICGDEASGCHYGALTC GS CKVFFKRAA EGKQKYLC AS RND CTIDKFRRKNCPSCRLRKCYEA
hER 181 ETRYCAVCNDYASGYHYGVWSC EG CKAFFKRSI QGHNDYMC PA TNQ CTIDKNRRKSCQACRLRKCYEV
```

Fig. 10. Primary structure of the core conserved region of DBD of several steroid hormone receptors. Amino acid residues are denoted by the one-letter code. For sequence references see: hGR (Hollenberg *et al.,*1985), hMR (Arriza *et al.,*1987), hPR (Misrahi *et al.,*1987), hAR (Lubahn *et al.,*1988; Tilley *et al.,*1989), and hER (Green *et al.,*1986; Greene *et al.,*1986). Conserved amino acids are in bold, and the conserved cysteines coordinating zinc are denoted by hollow letters. Amino acids determining the DNA-specificity (P box) are indicated by the shaded box. Amino acids framed by the shaded line represent the D box (see text for details).

The zinc fingers of the nuclear receptors act in a coordinated manner as a pair. Each finger domain contributes with different components to the specific DNA binding activity (Green *et al.,* 1988; Härd *et al.,* 1990; Schwabe *et al.,* 1990; Luisi *et al.,* 1991; Freedman, 1992). Each of these zinc fingers is encoded by separate exons (Encio and Detera-Wadleigh, 1991). The N-terminal finger is the principal determinant of specificity, but requires the contribution of the C-terminal finger for overall binding affinity (Fig. 10). Using site-directed mutagenesis to substitute different amino acids in DBD with glycines, specific amino acids required for DNA binding *in vitro* have been identified (Hollenberg and Evans, 1988; Severne *et al.,* 1988). Substitution of 3 amino acid residues within a segment termed the P box of GR (shaded box in Fig. 10) with the corresponding 3 residues in ER, converts the GR into an ERE (estrogen response element) binding protein. In contrast, any amino acid substitution within the loops of the zinc-fingers does not change the DNA binding specificity (Umesono and Evans, 1989). An identical amino acid substitution within the P box of ER with the corresponding residues of GR converts ER into a GRE binding protein (Mader *et al.,* 1989; Danielsen *et al.,* 1989). Moreover, a segment termed the D box located at the base of the second finger has been shown to be important for dimerization and cooperative binding of DBD to a GRE. Finally, the D box has also been implicated to be of importance in recognition of half-site spacing (GRE; see Fig. 3) (Umesono and Evans, 1989; Dahlman-Wright *et al.,* 1991; Dahlman-Wright *et al.,* 1993).

The three-dimensional structure of the DBD both in solution, using 2-D nuclear magnetic resonance spectroscopy (NMR) (Härd *et al.,* 1990), and for the DNA bound form using X-ray crystallography (Luisi *et al.,* 1991), has recently been solved. These analyses have confirmed and extended many of the findings from earlier studies of GR/DBD-DNA interactions using extensive mutagenesis of amino acid residues within DBD, complemented with mutagenesis and chemical modifications of bases and the phosphate backbone of the HRE (Scheidereit and Beato, 1984; Tsai *et al.,* 1988; Truss *et al.,* 1990, Nordeen *et al.,* 1990; Eriksson and Wrange, 1990; Cairns *et al.,* 1991). Both fingers provide contacts to the phosphate backbone of the DNA, and the C-terminal subdomain makes up the entire dimerization interface of the DBD.

Immediately C-terminally of the second finger there is a region enriched in basic amino acids. It has been suggested that this region is important for nuclear localization of GR. The amino acid sequence resembles the SV40 large T antigen nuclear translocation signal, and confers hormone independent nuclear localization when attached to a heterologous protein (Picard and Yamamoto, 1987). However, in the GR itself, it seems to be repressed by the presence of the hormone binding domain . Nuclear localization signals similar to that of GR have been identified in PR (Guiochon-Mantel *et al.,*1989) and ER (Picard *et al.,*1990a). Importantly, however, these nuclear localization signals in ER and PR have been termed "proto-signals" since none of them suffice on their own but cooperate to mediate nuclear accumulation of the receptors (Ylikomi *et al.,*1992)

The ligand binding domain

The ligand binding domain (LBD) of GR, comprising approximately 260 amino acids, is localized in the C-terminal portion of the protein (Figs. 8, 9 and 11) (Carlstedt-Duke *et al.,* 1987). *In vivo,* under physiological conditions, steroid receptors function only upon binding of their hormonal ligands. Deletion of amino acids from the C-terminus (Rusconi and Yamamoto, 1987), insertional mutations (Gigueré *et al.,* 1986) or point mutations (Danielsen *et al.,* 1987) within LBD result in the loss of hormone binding activity. C-terminal deletions of ER, PR and VDR also result in loss of hormone binding activity (Kumar *et al.*,1986; Carson *et al.*,1987; McDonnell *et al.*, 1989). These intial studies indicate the sensivity of the domain to single amino acid mutations anywhere in the region. Because of this sensitivity it has been very difficult to more closely define the domain by site-directed mutagenesis.

```
hGR   511 TSENPGNKTI VPATLP.... .......... .QLTPTLVSL LEVIEPEVLY
hMR   702 Q-PEE-TTY- A--KE-SVNT ALVPQLSTIS RA---SP-MV --N----I--
hPR   666 SQALSQRF-F S-GQ...... .........D I--I-P-IN- -MS---D-I-
hAR   650 -E-TTQKL-V SHIE...... .........G YECQ-IFLNV --A---G-VC
hER   284 GDMRAA-LWP S-LMIKRSKK NSLALSLTAD .....QM--A -LDA--PI--

hGR   546 AGYDSSVPDS TWRIMTTLNM LGGRQVIAAV KWAKAIPGFR NLHLDDQMTL
hMR   752 ------K--T AENLLS---R -A-K-M-QV- ----VL---K --P-E--I--
hPR   701 --H-NTK--T SSSLL-S--Q --E--LLSV- --S-SL---- ---I---I--
hAR   685 --H-NNQ--- FAALLSS--E --E--LVHV- -----L---- ---V----AV
hER   329 SE--PTR-F- EASM-GL-TN -AD-ELVHMI N---RV---V D-T-H--VH-

hGR   596 LQYSWMFLMA FALGWRSYRQ SSANLLCFAP DLIIN.EQRM TLPCMYDQCK
hMR   802 I-----C-SS ---S----KH TNSQF-Y--- --VF-.-EK- HQSA--EL-Q
hPR   751 I-----S-MV -G------KH V-GQM-Y--- ---L-.---- KESSF-SL-L
hAR   735 I-----G--V --M----FTN VNSRM-Y--- --VF-.-Y-- HKSR--S--V
hER   379 -ECA-LEILM IG-V---MEH PVK..-L--- N-LLDRN-GK CVEG-VEIFD

hGR   645 HMLYVSSELH RLQVSYEEYL CMKTLLLLS. ......SVPK DGLKSQELFD
hMR   851 G-HQI-LQFV ---LTF---T I--V-----. ......TI-- ------AA-E
hPR   800 T-WQIPQ-FV K----Q--F- ---V----N. ......TI-L E--R--TQ-E
hAR   784 R-RHL-Q-FG W--ITPQ-F- ---A---F-. ......II-V ----N-KF--
hER   427 ML-AT--RFR MMNLQG--FV -L-SII--NS GVYTFL-STL KS-EEKDHIH

hGR   688 EIRMTYIKEL GKAIVKREGN SSQNWQRFYQ LTKLLDSMHE VVENLLNYCF
hMR   894 -M-TN----- R-MVT-CPN- -G-S------ ---------D L-SD--EF--
hPR   843 -M-SS--R-- I---GL-QKG VVSSS----- ------NL-D L-KQ-HL--L
hAR   827 -L--N----- DRI-ACKRK- PTSCSR---- -------VQP IARE-HQFT-
hER   477 RVLDKITDT- IHLMA-AGLT LQ-QH--LA- -LLI-SHIRH MSNKGME...

hGR   738 QTFLDK.TMS IEFPEMLAEI ITNQIPKYSN GNIKKLLFHQ K.........
hMR   944 Y--RESHALK V---A--V-- -SD-L--VES --A-P-Y--R -.........
hPR   893 N--IQSRAL- V-----MS-V -AA-L--ILA -MV-P----K -.........
hAR   877 DLLIKSHMV- VD----M--- -SV-V--ILS -KV-PIY--T Q.........
hER   524 HLYSM-CKNV VPLYDL-L-M LDAHRLHAPT SRGGASVEET DQSHLATAGS

hGR       .......... .......... ..
hMR       .......... .......... ..
hPR       .......... .......... ..
hAR       .......... .......... ..
hER   574 TSSHSLQKYY ITGEAEGFPA TV
```

Fig. 11. Ligand binding domains of members of the steroid hormone receptor subfamily. Dashes represent identity with the hGR sequence; a dot is a gap used to align the sequences. Boxed amino acids represent affinity labeled amino acid-ligand adducts. The shaded regions indicate putative dimerization motifs. For sequence references see Fig. 10 (see text for details).

Various other methods have been applied in order to map the structure and function of this domain. For instance, using sterically small sulfhydryl-modifying (thiol-specific) reagents such as methyl methanethiosulfonate (MMTS), arsenite, cadmium(II) and selenite as probes of GR structure and function (Tienrungroj *et al.*, 1987; Miller and Simons, 1988; Tashima *et al.*, 1989; Simons *et al.*, 1990; Chakraborti *et al.*, 1992; Stancato *et al.*, 1993), vicinally spaced dithiol groups within the steroid binding domain have been suggested to be involved in ligand binding. Another approach has been to use protease digestion to create fragments of GR in order to map the minimal LBD required for ligand binding. Digestion with trypsin of the molybdate stabilized, oligomeric and liganded rat GR resulted in a fragment of M_r ~16,000 which still bound ligand albeit with a ~23-fold lower affinity (Simons *et al.*, 1989). This core hormone-binding unit encompasses amino acid residues 537 - 673 (rGR; corresponding to amino acids 518 - 655 of hGR) and is associated with hsp90 (Chakraborti and Simons, 1991). A model for the ligand binding cavity has been proposed based on the identification of the vicinal dithiols within LBD and the borders of the core hormone-binding unit; the binding cavity is thought to involve three thiols (Cys 640, 656, and 661; rGR) in a flexible cleft but thiol-steroid interactions are not essential for binding (Chakraborti *et al.*,1991a; see also Fig. 7. in Chakraborti *et al.*, 1992).

Affinity labeling using synthetic hormone analogs (tritiated steroid derivatives) has been widely used as a tool to functionally map LBD. Dexamethasone 21-mesylate (DM), the α-keto mesylate derivative of dexamethasone, is a synthetic steroid that binds to receptors in intact cells as well as in cell-free systems (Simons and Thompson, 1981) and remains attached to the receptor under denaturating conditions (Simons *et al.*,1983). DM has proven an invaluable tool in characterization of GR in various systems (Eisen *et al.*,1981; Simons *et al.*,1983; Smith and Harmon, 1985; Mendel *et al.*,1986; Simons *et al.*,1989; Chakraborti *et al.*,1992; Carlstedt-Duke *et al.*, 1988; Strömstedt *et al.*, 1990; Ohara-Nemoto *et al.*, 1990; Strömstedt *et al.*, 1993). DM has been shown to exclusively bind to cysteine residues (Simons, 1987). A single cysteine residue, located within the ligand binding domain of rGR: Cys656 (Simons *et al.*,1987; Carlstedt-Duke *et al.*, 1988), mGR: Cys 644 (Smith *et al.*,1988), and hGR: Cys638 (see Fig. 11) (Carlstedt-Duke *et al.*, 1988) has been identified as the residue covalently labeled by DM.

Using affinity labeling and radio-sequence analysis two additional amino acid residues were identified within LBD of rGR (Carlstedt-Duke *et al.*, 1988). A synthetic glucocorticoid, TA (see Fig. 6), which labels the receptor protein by photo-activation of the A-ring of the steroid was used. Two amino acids were identified as uniquely reactive in the photoaffinity labeling, Met-622 and Cys-754. All three amino acids identified are at the center of hydrophobic regions of LBD (Carlstedt-Duke *et al.*, 1988). Identification of both Met-622 and Cys-754 following photo-affinity labeling suggests that the protein folds with these two residues in close proximity to one another. Cys-754 was not labeled by DM (or any other cysteine residue beside Cys-656 within LBD) indicating that the steroid can bind in one orientation only, and, consequently, the distance between Met-622/Cys-754 and Cys-656 is approximately equivalent to the distance between the 3-keto group and the 21-mesylate group on the steroid molecule (Fig. 12). In sequence alignments, the sequence identity between GR, MR, and PR is more than 50 % within the LBD (Figs. 9 and 11, compare with ER) and the hydropathy profiles are almost identical (Carlstedt-Duke *et al.,* 1988). Furthermore, a considerable degree of overlap exists between GR, MR and PR with regard to steroid binding properties *in vitro* (Beaumont and Fanestil, 1983; Arriza *et al.,* 1987). Obviously, the three receptors are highly similar in the region of Met-604 and Cys-736 (hGR) as judged from the predicted secondary structure, since these residues interact with the A ring of the steroid, where their cognate ligands are identical.

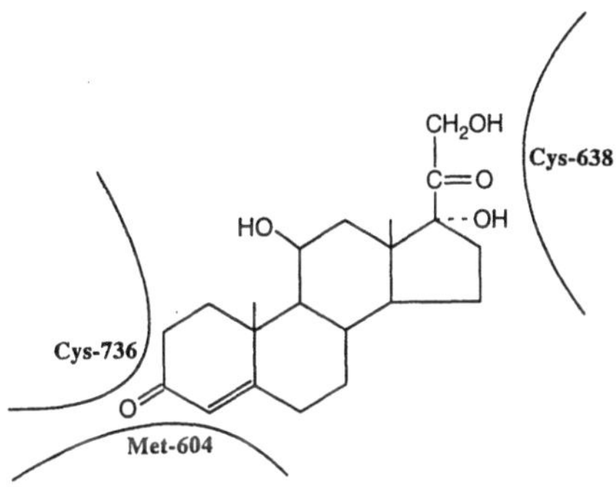

Fig. 12. Steroid binding site of hGR. Three segments that occur at the binding site have been identified by affinity labeling (Strömstedt *et al.,* 1990). Corresponding amino acid residues (Met-622, Cys-656, and Cys-754 was identified analysing rGR (Carlstedt-Duke *et al.,* 1988)

In a comparative study, affinity labeling using R5020 (promegestone) that labels both PR and GR by photo-activation of the A-ring of the steroid, resulted in a clear difference between the two receptors with regard to one of the two segments labeled (Strömstedt *et al.,* 1990). Analysis of rGR resulted in identification of the same two amino acid residues, Met-622 and Cys-754 as seen when labeling with TA (corresponding amino acid residues were labeled in hGR; Met-604 and Cys-736; Strömstedt *et al.,* 1990). In contrast, analysis of PR labeled with R5020 identified Met-759 and Met-909 (Strömstedt *et al.,* 1990). A comparison of the segments of the receptor proteins containing residues charged with R5020 showed that Met-759 in PR corresponds exactly to Met-604 in hGR (Fig. 11). The segment containing the second labeled residue in PR and GR shows a considerably lower degree of similarity (see Fig. 11). Moreover, Met-909 (PR) is located about 18 residues C-terminal of the residue corresponding to Cys-736 in GR (Fig. 11). Labeling with R5020 of this second fragment may represent a subtle difference between the two receptors concerning recognition of the ligand. A summary of the results from Carlstedt-Duke *et al.*, 1988;Strömstedt *et al.*, 1990 is given in Table I. Interestingly, Met-909 resides within a segment of PR LBD shown to be required for agonist binding but not for binding of RU486, an antagonist (Vegato *et al.*,1992). A similar analysis has been performed on ER. Aziridine affinity labeling of hER with ketononestrol aziridine and tamoxifen aziridine (agonist and antagonist, respectively) has identified one amino acid residue near the C-terminus of its LBD (Fig. 11) interacting with the ligand (Harlow *et al.*,1989). The same site, Cys-530 (hER), was labeled by either ligand (Harlow *et al.*,1989). Hence, the contact points of the ligands may be less important than the protein structural alterations induced by binding of the ligand or may not represent amino acid residues that play a direct role in steroid specificity.

Table I: Affinity labeled amino acid residues identified by radio-sequence analysis. For experimental details see Carlstedt-Duke *et al.*, 1988; Strömstedt *et al.*, 1990.

Receptor	Ligand	Amino acid residues labeled
rGR	TA	Met-622, Cys-754
	DM	Cys-656
	R5020	Met-622, Cys-754
hGR	TA	Met-604, Cys-736
	DM	Cys-638
hPR	R5020	Met-759, Met-909

Site directed mutagensis was used in order to analyze the importance of the amino acid adducts identified by affinity labeling and radio-sequence analysis of GR in Carlstedt-Duke *et al.*, 1988; Strömstedt *et al.*, 1990. hGR with six different point mutations was examined (Carlstedt-Duke, McGuire and Strömstedt, unpublished material): M601L (Met-601 mutated to Leu), M604L, C638S, C643S, C665S, and C736S. Cell-free studies were performed with extracts of COS-7 cells that had been transiently transfected with the corresponding cDNAs. The expression levels were determined by ligand binding assays and Western blot analysis. All mutant receptors were expressed at similar levels as wild type receptor. None of the mutations eliminated steroid binding. However C638S abolished binding of DM as expected from the affinity labeling data. Scatchard analysis of each of the receptors revealed that the mutations of Met-601 and Cys-736 resulted in a ~3-fold decrease in affinity for [^{3}H]Dex (Table II). M604L also showed a slight decrease in affinity for Dex. Chakraborti *et al.*,(1991b) has reported an increased affinity for Dex when substituting Cys-656 with Ser in rGR (corresponding to Cys-638 in hGR). In our system C638S displayed wild type affinity for Dex. However, it should be noted that the affinity of wild type receptor for Dex in that study was K_d ~5 nM which is 5-fold lower affinity than the wild-type levels obtained in our studies. All six receptor mutants were associated with hsp90 in the absence of ligand as assayed by a co-immunopreciptation assay using a monoclonal antibody against hsp90 to precipitate the oligomeric receptor complex (unpublished results). Further functional characterization will be carried out in order to determine the biological activity of the mutant receptors. Our results from affinity labeling experiments and site-directed mutagenesis of LBD are in agreement with previous data from mutational analysis of GR cDNA (Hollenberg *et al.*,1985; Giguere *et al.*,1986; Rusconi and Yamamoto, 1987) showing that the entire C-terminal domain is required for high affinity binding of ligand. Deletion of the last five C-terminal residues from rGR resulted in a receptor with a K_d ~30-fold higher than the wild type receptor, further deletions of an additional 20 residues results in a 1000-fold higher K_d, indicating that the C-terminal limit of LBD is near the C-terminal end. The difference in these results compared to the minimal LBD mapped by proteolysis by Simons et el., (1989) could be due to intrinsic differences in the design of these experiments. Thus, expression of mutant receptors require the ability of the mutant proteins to fold into a productive LBD, while the protease mapped minimal LBD was obtained by cleavage of a molecule that is already in the correct conformation for high affinity ligand binding.

Table II. Site-directed mutagensis of the ligand binding domain of hGR. Summary of ligand binding affinitites *in vitro*.

Mutation	**Steroid Affinity (K_d)** mean ± SEM	
wild type	0.72 ± 0.044 nM	(n=10)
M601L	1.93 ± 0.285 nM*	(n=5)
M604L	1.43 ± 0.166 nM	(n=2)
C638S	0.89 ± 0.130 nM ns	(n=5)
C643S	0.95 nM	(n=1)
C665S	0.84 ± 0.066 nM ns	(n=5)
C736S	2.16 ± 0.155 nM*	(n=6)

ns = not significant

* = significantly different from wild type (ANOVA; P>0.01)

The LBD contains the binding site for hsp90 and possibly other receptor associated proteins (Pratt *et al.*, 1988; Denis *et al.*, 1988b; Howard *et al.*, 1990; Dalman *et al.*, 1991). Receptors lacking the C-terminus activate transcription independently of hormone (Danielsen *et al.*, 1987; Miesfeld *et al.*, 1987; Hollenberg *et al.*, 1987; Godowski *et al.*, 1987) suggesting that, within the intact receptor, this domain confers an inactivation function in the absence of hormone (Cadepond *et al.*, 1991). This inhibition of DNA binding and transcription is possibly due to the interaction with hsp90, and the hormone overcomes this repression by binding to the receptor and thus inducing release of hsp90 from the receptor. It has been shown that the LBDs of GR or ER can act as transferable regulatory cassettes that can confer hormonal control onto heterologous proteins (Picard *et al.*, 1988; Superti-Furga *et al.*, 1991; Umek *et al.*, 1991; Jackson *et al.*, 1993). Furthermore, it has recently been shown that the hormone binding domains of steroid receptors may regulate the function of other proteins through hormone-regulated binding of chimeras to hsp90 (Scherrer *et al.*, 1993). The association of hsp90 to the hormone binding domain is also supported by the demonstration that hsp90 remains bound to the LBD when the receptor in the heterocomplex is cleaved with protease (Denis *et al.*, 1988b; Chakraborti and Simons, 1991).

Dimerization

The palindromic nature of many GREs and other steroid response elements (HREs) suggests that steroid hormone receptors can bind to DNA as symmetrical dimers. Using the isolated GR DBD, dimer formation is dependent on specific binding to DNA. One molecule of GR DBD facilitates the binding of a second molecule to the second half-site in a cooperative manner (Tsai *et al.*, 1988). The GR DBD does not form homodimers in solution even at millimolar concentrations. However, the full length receptor can form a homodimer in solution but has also been detected in solution as a monomer (Eriksson, and Wrange, 1990), although it appears to bind exclusively as a dimer to GREs (Perlman *et al.*, 1990). This suggests that an equilibrium exists between GR monomers and dimers in solution, and that either a preformed GR dimer binds to a GRE, or that it binds as a monomer to one half-site and strong positive cooperativity results in the binding of the second monomer, which means that the monomer-DNA complex is essentially undetectable (Perlman *et al.*, 1990). Several investigators have correlated homodimer formation with the acquirement of high affinity DNA binding. Stoichiometric analysis has shown that ER and PR also exist as dimers when bound to their respective responsive DNA element (Kumar and Chambon, 1988; Tsai *et al.*, 1988).

Kumar and Chambon (1988), have mapped the principal dimerization region of ER as part of LBD. It was further narrowed down by Parker and co-workers to be located between amino acid residues 500-540 (Fig. 11) (hER; Fawell *et al.*,1990a; Lees *et al.*,1990). This region shows a considerable degree of homology among other members of the nuclear receptors and might serve as a general dimerization interface. Homologous regions in thyroid hormone receptors (ThR) and retinoic acid receptors (RAR) seem to be important for the formation of heterodimers (Glass *et al.*, 1989; Forman *et al.*, 1989). Mutations within this segment disrupt dimerization of these receptors. In addition, there is evidence that other regions within the ER LBD might be involved in dimer formation (White *et al.*,1991)

In an alignment analysis of various members of the nuclear receptor superfamily, a second conserved motif has been identified, encompassing a stretch of approximately 60 amino acid residues in the N-terminal part of LBD (Fig. 11) (Maksymowych *et al.*, 1992). The conserved region contains heptad repeats of hydrophobic amino acids that have been suggested to form a putative dimerization interface with other conserved residues between various members of the superfamily (Maksymowych *et al.*, 1992). Thus, coiled-coil interactions could take place within this region that may mediate both homo- and heterodimer formation. This region has also been demonstrated to confer formation of heterodimers between RAR and ThR (O'Donnell and Koenig, 1990; O'Donnell *et al.*,1991; Darling *et al.*,1991).

The conservation of several regions within LBD amongst members of the nuclear receptor family may indicate that these regions have a similar function in other receptors and that either homo- or heterodimers can occur. In the case of steroid hormone receptors, dimerization does not appear to promote ligand binding *in vitro*, and site directed mutagenesis has demonstrated that these activities can be functionally separated although the hormone binding and dimerization surfaces are overlapping. This favors a model where the hormone allows dissociation of the heat shock protein complex and thereby unmasks the dimerization domain(s). Deletion of the entire LBD reduces the DNA binding activity of GR (Dahlman-Wright *et al.*,1992) and its transactivation capacity (Hollenberg, 1987). This has been suggested to be a consequence of an inability to form dimers. Although it is generally accepted that dimerization, at least among the steroid hormone receptors, is indispensable for high affinity binding of the receptor to an HRE, two recent reports present evidence for high affinity DNA binding activity by monomeric forms of ER (Furlow et al, 1993) and PR (Cohen-Solal et al, 1993).

Antagonists

Since steroid hormones have profound physiological effects, it has always been an important goal to develop pure and potent steroid antagonists which may block the given steroid effect both *in vitro* and *in vivo*. RU486 (17β-hydroxy-11α(4-dimethylaminophenyl)-17α-1-propanyl-estra-4,9-dien-3-one) (Fig, 13) was the first useful potent antagonist synthesized with both antiprogestin and antiglucocorticoid activities (Philibert *et al.*,1981; for reviews see Baulieu, 1991; Mao *et al.*,1992). RU486 is a pure antagonist *in vivo* devoid of any agonist activity (Jung-Testas and Baulieu, 1984; Moguilewsky and Philibert, 1984; Kalimi, 1989). The binding affinity of GR and PR for RU486 is higher than their binding affinities for cortisol and progesterone, respectively (Philibert *et al.*,1981; Agarwal *et al.*,1987). The antagonist competes for binding of an agonist and produces an antagonist-receptor complex which is transcriptionally inert, although it can bind to DNA. Mechanisms underlying these effects are not fully understood (Guichon-Mantel *et al.*,1988).

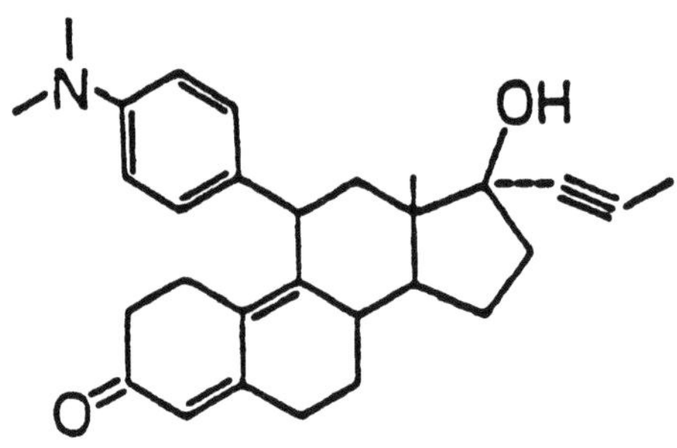

Fig. 13. The antihormone RU486 (mifepristone) is a 19-norsteroid which has a specific high affinity binding to PR and GR

It has been demonstrated that RU486-occupied PR exhibits DNA binding activity both *in vitro* (El-Ashry *et al.*,1989) and *in vivo* (Meyer *et al.*,1990). RU486 seems to block or slow down the dissociation of hsp90 from GR and hence, only a small fraction will bind DNA *in vitro* (Baulieu, 1987; Groyer *et al.*,1987; Agarwal and Lazer, 1987). The *in vivo* effect of RU486 on GR is unclear. *In vivo* footprinting experiments failed to show any binding of RU486 liganded receptor to the tyrosine aminotransferase gene (Becker *et al.*, 1986), and studies on the effects of GR in lymphoid cells indicated that RU486 prevents nuclear translocation of the receptor (Segnitz and Gehring, 1990). However, by using chimeric receptors (GR-Gal4) in *in vivo* competion experiments, RU486 inhibits GR transactivation by competition for DNA binding (Webster *et al.*, 1988).

Probing PR and GR with sulfhydryl modifying agents either in the presence of antagonist (RU486) or an agonist (R5020 and dexamethsone, respectively) has suggested that the difference in ligand structure between the antagonist and the agonists may result in changes of their interaction with the ligand binding domain of the corresponding receptor (Kalimi and Agarwal, 1988; Moudgil *et al.*, 1989). In gel shift assays, a difference in the relative migration between agonist bound and antagonist bound receptor-DNA complexes has been observed, indicating ligand-induced differences in receptor conformation (El-Ashry *et al.*, 1989; Meyer *et al.*, 1990). This conformational change induced by either class of ligand occurs in the absence of DNA and renders the entire ligand binding domain relatively resistant to digestion by proteases (Allan *et al.*, 1992). Both agonist and antagonist induce an equally dramatic, but distinct, structural alteration of the ligand binding domain as shown by the proteolytic cleavage patterns (Allan *et al.*, 1992). This structural alteration is also detected by an antipeptide monoclonal antibody directed against a 14-amino acid epitope in the carboxy-terminus of PR (Weigel *et al.*, 1992). Furthermore, mutagenesis of the PR LBD has generated a mutant receptor, lacking the 42 most C-terminal amino acids, that fails to bind agonists such as progesterone but still can bind RU486. Surprisingly, this mutant receptor activates transcription in the presence of antagonist both *in vitro* and *in vivo* (Vegato *et al.*, 1992), suggesting that the extreme C-terminal region of PR contains an inhibitory function that silences receptor transactivation in the absence of agonist and in the presence of antagonist. Thus, binding of an agonist is dependent on the extreme C-terminal region, while an antagonist (here: RU486) seems to bind to a segment more N-terminally within the LBD. Using affinity labeling, we showed labeling of a residue (Met-909) near the C-terminal of LBD (Strömstedt *et al.*, 1990) using R5020 (agonist).

Mechanisms of transcriptional regulation

Various GRE sequences have been identified *in vivo* by gene transfer techniques and *in vitro* by DNA binding studies with purified receptor preparations (Payvar *et al.*, 1983; Renkawitz *et al.*, 1984; Slater *et al.*, 1985; Becker *et al.*, 1986). Based on these functional studies, specific HREs have been formulated for the various nuclear receptors (i.e. a 15 bp consensus sequence has been formulated for GR-binding; see Fig. 3). Different mechanisms for steroid hormone receptor mediated transcriptional activation have been postulated. Based on studies of GR and PR mediated transcriptional activation in cell free model systems (reviewed by Allan *et al.*, 1991; Bagchi *et al.*, 1992), both PR and GR have been suggested to facilitate the binding of the basal transcription machinery, containing the RNA polymerase II preinitiation complex, by direct protein-protein interaction. Thus, basic or general transcription factors are recruited to assemble at the transcription start site of target promoters prior to initiation of transcription by RNA polymerase II (Tsai *et al.*, 1990; Klein-Hitpass *et al.*, 1990).

An alternative mechanism of transcriptional activation by GR has been postulated from studies of GR effects on chromatin structure. The binding of GR to target genes organized on nucleosomes is suggested to create a more open chromatin structure, enabling other factors to bind which in turn are required to initiate transcription (Zaret and Yamamoto, 1984; Cordingley *et al.*, 1987; Richard-Foy and Hager, 1987).

GR can repress transcription of several genes (Akerblom and Mellon, 1991), and again several mechanisms have been proposed. A simple model for negative regulation by steric hindrance has been demonstrated in several systems. For instance, GR binding sites overlapping the cAMP-responsive elements in the human glycoprotein α-subunit gene prevent binding of cAMP-responsive factors when GR is bound to DNA (Akerblom *et al.*, 1988). Similar mechanisms have been described concerning the negative regulation by GR of human osteocalcin gene expression (Morrison *et al.*, 1989; Strömstedt *et al.*, 1991). A more complex situation has been observed in the proliferin promoter which contains a GRE overlapping an AP-1 (i.e. Fos/Jun) binding site (Diamond *et al.*, 1990). A complex interplay between GR and c-fos/c-jun dictates the glucocorticoid response on this so called composite element (Lucas and Granner, 1992). A potentiation of activation by GR is seen in the presence of the c-jun homodimer while repression of GR-mediated activation occurs if c-jun/c-fos heterodimers are present (Miner and Yamamoto, 1991). The net GR response depends on the ratio of c-jun/c-fos in a given cell at a given time. GR has also been shown to modulate AP-1 activity on the collagenase I promoter through protein-protein interactions in solution without direct binding to DNA (Jonat *et al.*, 1990; Schule *et al.*, 1990; Lucibello *et al.*, 1990). Preliminary experiments have also indicated that GR can modulate the activity of oct transcription factors by a similar mechanism (Wieland *et al.*, 1991; Kutoh *et al.*, 1992). According to this model transcriptional repression is mediated through direct protein-protein interactions between GR and the target transcription factor resulting in an abortive complex unable to bind DNA.

Signal specificity and selectivity among steroid hormone receptors

How can transcriptional specificity be achieved for closely related receptors? Single or tandem GRE/PREs have been shown to function similarly with activated GR, PR, MR, and AR. Since GR appears to be expressed in all nucleated mammalian cells (Ballard *et al.*, 1974), there must be mechanisms other than DNA binding specificity determining the specificity of the response. Differential expression of receptors has been shown to be one way of controlling the specificity (Strähle *et al.*, 1989). In this case, the response is dependent on which receptor that is expressed in a particular cell type. This is true for MR, AR, ER and PR which are only expressed in a limited number of cells. A threshold level of GR seems to exist, that is, below this threshold level the cell does not respond to glucocorticoids (Dong *et al.*, 1990; Tanaka *et al.*, 1991). This together with the presence of inhibitory factors offers a degree of specificity in GR transactivation.

A more complex situation concerns the crossreactivity of mineralocorticoids and glucocorticoids. Cortisol (or corticosterone in rat and mouse) binds to MR with equal affinity as aldosterone, the physiological ligand of MR (Arriza et al, 1987; Funder, 1992). As mentioned earlier, GR and MR recognize the same DNA-response element, yet they can elicit opposing effects on ion transport within a single tissue, cell type, or within an individual cell (Turnamian and Binder, 1989; Bastl *et al.*, 1989; Joels and de Kloet, 1990; Laplace *et al.*, 1992). The presence of 11β-hydroxysteroid dehydrogenase (11-HSD) in a physiologically mineralocorticoid-responsive tissue, such as MR-containing cells in the kidney, inactivates glucocorticoids by metabolizing cortisol and corticosterone into their inactive 11-keto congeners. Aldosterone is protected from inactivation by the cyclation (aromatization) of the aldehyde at C-18 with the C-11 hydroxyl group. This will allow MR to bind aldosterone exclusively despite much higher plasma glucocorticoid levels (Funder *et al.*, 1988; Edwards *et al.*, 1988).

There are target tissues where this mechanism is less well established, *e.g.* the colon (Naray-Fejes-Toth and Fejes-Toth, 1990). Nevertheless, this tissue still shows differential effects of MR and GR on ion transport. Therefore a second mechanism to achieve specificity has been proposed. Studies of an HRE in the proliferin gene to which both GR/MR and AP1 can bind has revealed differences between MR and GR in their abilities to inhibit AP1 stimulated transcription (Pearce and Yamamoto, 1993). Under conditions where GR inhibits AP1 stimulated transcription, MR is inactive. By making GR-MR chimeric proteins a segment of the N-terminal domain of GR (amino acid residues 104 - 440; rGR) has been identified as necessary for repression of AP1 stimulated transcription. Hence, distinct physiological effects mediated by steroid receptors may be determined by differential interactions with non-receptor factors at composite HREs.

OVEREXPRESSION OF STEROID HORMONE RECEPTORS

Steroid hormone receptors are expressed in low amounts in target cells and are therefore difficult to purify in sufficient amounts to allow advanced structural studies such as X-ray crystallography. This has led various groups to search for expression systems that can provide the receptor quantities necessary for such studies. Both prokaryotic and eukaryotic systems have been utilized with various degrees of success (Srinivasan, 1992).

Expression in *Escherichia coli*

Both full length receptors as well as their autonomous functional domains have been overproduced in *E. coli*. Expression of DBDs from GR, ER, RAR and most recently RXR in these systems (Freedman et al 1988; Dahlman *et al.*,1989;) has faciliated three-dimensional structural determinations in solution by NMR (Härd *et al.*,1990; Schwabe *et al.*,1990; Knegtel *et al.*,1993; Lee *et al.*,1993), and the structural determination of protein-DNA co-crystals of GR DBD with the GRE target sequence has recently been achieved (Luisi *et al.*,1991). Expression of full length receptors in *E. coli* has proven to be more difficult since most of the receptors are fairly large proteins (50 -100 kDa) and as such difficult to express in *E. coli* (Marston, 1986). Expression of full length GR resulted in an insoluble product incapable of binding ligand (Bonifer *et al.*,1989). N-terminal truncated derivatives of hGR were expressed either alone or as C-terminal protein A (from *Staphylococcus aureus*) fusion proteins (Ohara-Nemoto *et al.*, 1990). One fragment comprising amino acids 477 - 777 expressed as a C-terminal fusion protein with a part of protein A resulted in a partially soluble protein. Less than 1% of the expressed material was recovered in the soluble fraction. The recombinant product bound steroids with normal specificity but with reduced affinity (K_d for TA ~70 nM) Further biochemical characterization of the recombinant product indicated an absence of heat shock proteins normally associated with the receptor in the absence of ligand (Ohara-Nemoto *et al.*, 1990). It has been shown that hsp90 is essential for the high affinity steroid binding ability of GR (Bresnick *et al.*,1989; Young *et al.*,1990; Nemoto *et al.*,1990; Picard *et al.*,1990b) and furthermore, Denis and Gustafsson, (1989) have shown that *in vitro* translation in rabbit reticulocyte lysate of GR mRNA yields an non-activated oligomeric receptor complex containing hsp90. PR, AR, and ER bind their ligand with high affinity independently of hsp90 binding (Bagchi *et al.*, 1990; Nemoto *et al.*, 1992; Xie *et al.*, 1992). The reason for this discrepancy between steroid hormone receptors is not known. The high affinity oligomeric GR complex can be reconstituted *in vitro* by incubating the hsp90-free low affinity GR with rabbit reticolucyte lysate at 37°C in the presence of an ATP generating system. Reconstitution of the inactive receptor occurs only in the absence of ligand (Scherrer *et al.*, 1990; Hutchison *et al.*, 1992; Pratt *et al.*, 1992). Reconstitution *in vitro* of ER and PR oligomeric complexes has also been carried out (Smith *et al.*, 1990; Inano *et al.*, 1990).

In contrast to the bacterially expressed material, *in vitro* translation of the same constructs used for the expression in *E. coli* produced oligomeric GR LBD complexes with high affinity ligand binding (K_d for TA ~2 nM) (Ohara-Nemoto *et al.*, 1990). There is an hsp90 equivalent present in *E. coli*, hsp83 which shows only 58% homology with the mammalian hsp90 (Bardwell and Craig, 1987). A highly charged sequence stretch is present in all eukaryotic hsp90 (Binart *et al.*, 1989) that has been suggested to be a putative region for interaction with GR and other steroid hormone receptors. There is no such homologous region in bacterial hsp83 which might be a reason why the assembly of the oligomeric receptor complexes in *E. coli* is hampered. Radio-sequence analysis of the expressed recombinant protein using [^{3}H]DM identified predominant labeling of Cys-638. However, all cysteines within LBD were affinity labeled to a certain degree indicating a more accessible structure compared to that of the native GR (Ohara-Nemoto *et al.*, 1990) again indicating the absence in *E. coli* of co-factors required for correct folding of GR LBD. However, LBDs from ER, PR, and AR produced in *E. coli* possess normal affinity for their cognate hormones (Eul *et al.*, 1989; Wittliff *et al.*, 1990; Power *et al.*, 1990; Young *et al.*, 1990; Ohara-Nemoto *et al.*, 1992). The only other steroid hormone receptor which seems to be dependent on hsp90 association for high affinity ligand binding activity is MR (Caamano *et al.*, 1993; Alnemri and Litwack, 1993). The *bona fide* DNA binding activity of DBDs produced in *E. coli* together with normal ligand binding activities by LBDs (except for GR LBD) has suggested that post-translational modifications such as phosphorylation are not necessary for these functional activities.

Expression in yeast

Expression in *Saccharomyces cerevisiae* has been widely used both for production of receptor proteins as well as for genetic dissection of the receptor signal transduction pathway. ER, PR, GR and VDR have been expressed in yeast systems (Metzger *et al.*, 1988; Mak *et al.*, 1989; Wright *et al.*, 1990; Sone *et al.*, 1990). An aberrant ligand specificity has been observed for GR expressed in yeast (Schena and Yamamoto, 1988; Wright *et al.*, 1990). This may, at least to a certain extent, be due to an artifact in the assay system and was the result of GR expression levels rather than an abnormal receptor conformation (Wright and Gustafsson, 1992). At low expression levels, a perfectly normal ligand specificity profile for GR has been observed (Wright and Gustafsson, 1991). However, very low ligand binding activity is obtained in extracts prepared from yeast cells expressing GR. Disruption of the oligomeric hsp90-GR receptor complex during the preparation of the extract could be one explanation for this low ligand binding activity.

Expression in mammalian cells

Stably transfected Chinese hamster ovary (CHO) cells have been used for high level expression of GR and ER. In both systems, the receptor has been stably transfected into the cells under control of a highly active SV40 early region promoter and coamplified with either selectable dihydrofolate reductase cDNA (Bellingham *et al.*, 1989; Hirst *et al.*, 1990) or coamplified with the human metallothionein IIA gene in the presence of increasing concentrations of metal (Cd) (Kushner *et al.*, 1990; Alksnis *et al.*, 1991). For GR expression levels of 0.5 - 1 x 10^6 receptors per cell have been obtained using these methods and in the case for ER, expression levels corresponding to approximately 3 - 6 x 10^6 receptors per cell have been produced. CHO cells stably transfected with ER display some curious phenotypes. After exposure to physiological levels of estrogen, the cells cease to grow and seem to flatten and eventually lyse. The reason for these toxic effects of estrogen is not known, but one possibility might be that the large amount of ER could illegitimately activate and/or inhibit endogenous CHO genes with a lethal effect. In spite of the reasonably high expression levels obtained in these systems these cells are usually laborious and costly to maintain.

Viral-based expression systems

The Baculovirus expression vector system (BEVS) has the capacity to express heterologous gene products in very large quantities and has the potential for proper posttranslational modifications (Fraser, 1989), although glycosylation is different in insect cells relative to mammalian cells. The expression system is based on the life cycle of the *Autographa californica* nuclear polyhedrosis virus in *Spodoptera frugiperda* (sf9) cells (Summers and Smith, 1985). Near the end of the viral life cycle, polyhedrin protein, the major structural component of the viral occlusions, is expressed providing the means for horizontal transmission of the virus. The polyhedrin protein accumulates and can make up as much as 50 to 75% of the total protein of the cell. By replacing the polyhedrin structural gene (nonessential for infection or replication of the virus) with a foreign gene of choice, high levels of the foreign gene product are then expressed under the control of the powerful polyhedrin promoter. The bacterial β-galactosidase gene has been expressed at levels of approximately 500 mg/liter culture (Pennock *et al.*,1984), while other proteins are expressed at much lower levels. The reason for the variable expression levels are not fully understood, but differences in translational efficiencies between individual genes might cause the discrepancy in expression levels (Summers and Smith, 1985).

High level expression of GR (Srinivasan and Thompson, 1990; Alnemri *et al.*,1991), ER (Brown and Sharp, 1990), thyroid hormone receptor β1 (ThRβ1) (Barkhem *et al.*,1991; zu Putlitz *et al.*,1991), VDR (Ross *et al.*,1991; MacDonald *et al.*,1991), retinoic acid receptor γ (RARγ) (Ross *et al.*,1992), PR (Christensen *et al.*,1991), AR (Xie *et al.*,1991; Chang *et al.*,1992; reviewed by Jänne *et al.*,1993), and MR (Alnemri *et al.*,1991) have been achieved in BEVS. Every steroid/ thyroid hormone receptor expressed and characterized in BEVS so far shows *bona fide* ligand affinity and specificity characteristics. Expression levels reported are in the range of 0.5 to 5 mg/ liter culture with the exception of AR which constitutes nearly 30% of total cellular protein when expressed in sf9 cells. In the case of AR, less than 10% of the expressed protein is soluble. The remaining 90% is localized as insoluble aggregates in the cytoplasm. Similar results, although at lower levels of expression, have been obtained when expressing ER (Brown and Sharp, 1990) and have also been observed expressing GR (Srinivasan, 1992; our unpublished observations). Alnemri and Litwack (1993) reported that less than 1% of GR and MR expressed in BEVS are assembled into cytoplasmic oligomeric receptor complexes, thus leaving a major fraction of the expressed receptor protein as self-assembled insoluble aggregates which co-fractionated with nuclear material. Furthermore, deletion of the entire LBD results in overexpressed receptors that do not form insoluble aggregates. Thus, LBD influences the intracellular solubility of the receptors. Incubation of the insoluble aggregates with rabbit reiculocyte lysate reconstitutes the oligomeric receptor complex to a limited degree (Alnemri and Litwack, 1993). The authors hypothesize that the interaction of hsp90 with GR or MR is complex and highly regulated and requires the participation of other cellular factors which are limiting in insect cells.

A novel viral expression system based on the Semliki forest virus replicon (SFV) has recently been developed by Liljeström and Garoff (1991). The life cycle of SFV has several features which make the virus an attractive protein expression system. First, the SFV RNA genome is infectious due to its positive polarity. Hence, it functions directly as an mRNA. Furthermore, the infecting RNA molecule encodes its own RNA replicase, which in turn drives an efficient RNA replication and transcription. This results in the effect that practically all ribosomes of the infected host cell will be occupied with the synthesis of virus encoded proteins. Moreover, SFV replication occurs in the cytoplasm, where the virus encoded replicase transcribes and caps the RNA for production of the viral proteins. SFV has a very broad host cell range and can infect almost any cell line in culture and hence, this will make it possible to work in homologous systems circumventing the problems associated with heterologous systems (*e.g.* sf9 cells; see Alnemri and Litwack, 1993), namely, proper posttranslational modifications and specific association with cellular co-factors. Construction of the expression vector is non-laborious, straight-forward subcloning, does not require any recombination and thus omits time-consuming screening experiments usually associated with viral systems. Several proteins have been expressed using the SFV system including the human transferrin receptor, mouse dihydrofolate reductase, chick lysozyme and bacterial β-galactosidase. The expression levels of the heterologous proteins are generally high, constituting nearly as much as 25 % of the total cellular protein (Liljeström and Garoff, 1991). High level expression of functional full length hGR as well as an N-terminally truncated hGR derivative was obtained using SFV expression vectors and Baby hamster kidney cells (BHK-21) (Strömstedt *et al.,* 1993). Expression levels equivalent to 4 - 6 x 10^6 binding sites/cell were obtained for full length hGR while the expression levels was one order of magnitude lower for hGR LBD. Importantly, full length hGR and the LBD were capable of binding glucocorticoids with correct specificity and high affinity. Moreover, partially purified hGR bound specifically to its cognate DNA response element. Thus, the SFV expression system produces receptors exhibiting *bona fide* ligand binding and DNA binding properties (Strömstedt *et al.,* 1993). An efficient shut down of host cell protein synthesis in favour of the production of the heterologous protein enables efficient metabolic labeling of the recombinant protein without interfering background activity from labeling of endogenous proteins. For instance, metabolic labeling in BHK cells after 18-24 h of infection with SFV-particles carrying full length GR showed no labeling of other proteins but GR (Strömstedt *et al.,* 1993).

GENERAL DISCUSSION AND CONCLUSIONS

Steroid hormone receptors are ligand-dependent transcription factors. Binding of the ligand to the receptor results in conformational changes of the protein, acquisition of the ability to bind to DNA and, thereby, the ability to modulate the transcriptional activity of target genes. An understanding of the mechanisms involved in ligand binding and specificity is therefore of paramount importance for the understanding of steroid signal transduction.

The ligand binding function of the glucocorticoid receptor is located in an autonomous functional domain of the protein that can be separated from the rest of the protein by limited proteolysis and purified in an isolated form. Sequence analysis of the isolated domain (Carlstedt-Duke *et al.*) demonstrated that the ligand binding domain consists of the C-terminal portion of the protein. The situation of LBD in the protein was confirmed by affinity labeling and genetic experiments (Wrange *et al.*,1984; Hollenberg etal., 1987; Godowski et el., 1987). However, LBD is a relatively large domain consisting of about 260 amino acid residues and insertions or deletions virtually throughout the entire domain have been reported to obliterate steroid-dependent transcription (Giguere *et al.*,1986). Thus, more subtle approaches were required to probe the structural elements of the domain involved in steroid binding.

Normally, the steroid binds to its cognate receptor in an equilibrium reaction. However, irreversible covalent binding of the steroid can be induced by either photo-activation of the steroid or the introduction of electrophilic groups into the steroid. This results in the formation of adducts between the steroid and amino acid residues in close proximity following the binding of the steroid at the steroid-binding surface of LBD. By taking advantage of this property, together with the application of radio-sequence analysis, we could identify three specific residues that form adducts with the steroid, Met-604 (hGR), Cys-638 and Cys-736 (Carlstedt-Duke *et .al.*; Strömstedt P.-E. *et al.*) (Fig. 14). All three residues are located within hydrophobic segments of LBD. The ability to form adducts with the steroid demonstrate that these three residues are located in segments constituting the steroid-binding surface but does not necessarily mean that they participate directly in steroid recognition. In order to test this possibility, site-directed mutagenesis of these residues, as well as adjacent Met and Cys residues, was carried out. In light of the previous results obtained by random mutation, insertions or deletions within this domain, usually resulting in a total disruption of the steroid-binding (Fig. 14) function, as minimal changes as possible were made. Mutation of two of the affinity labeled residues, Met-604 and Cys-736, as well as an adjacent residue, Met-601, resulted in a moderate decrease in affinity for glucocorticoids, thus indicating that these residues appear to play a role in steroid-binding. Mutation of the third residue, Cys-638, did not affect steroid-binding. Similar studies have been performed with ER (Harlow *et al.*,1989) resulting in affinity labeling of a single Cys residue. Mutation of this residue to Ser did not have any effect on steroid binding, similar to our results concerning Cys-638 (hGR).

An alternative approach to identify functionally relevant mutations of steroid receptors has been to study receptor proteins associated with familial steroid resistance. To date, two mutations of GR have been described in patients with familial cortisol resistance (glucocorticoid resistance syndrome) (Hurley *et al.*,1991; Malchoff *et al.*,1993). The two mutations reported, D641V and V729I respectively (Fig. 14), give rise to mutant GR with 2 - 3-fold lower affinity for Dex. Thus, the change in affinity described in these clinical reports is in the same order of magnitude as that obtained in our studies with M601L, M604L and C736S. In these patients, the serum cortisol levels are greatly increased without causing clinical or biochemical stigmata of glucocorticoid

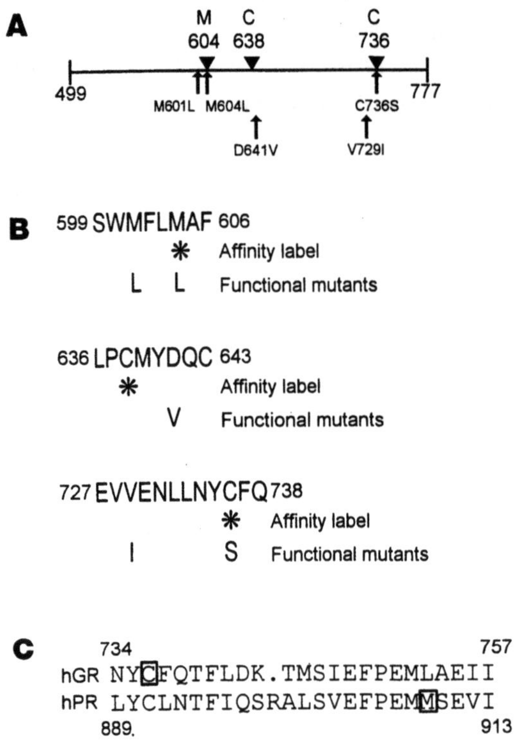

Fig. 14. Schematic presentation of point mutations that affect hormone binding and residues that form adducts with the steroid. (see text for details)

excess. Interestingly, the two mutations described in familial cortisol resistance each occur in close proximity of affinity labeled residues (Fig. 14). This lends further support to the notion that the affinity labeled residues identify segments of LBD that are located within the steroid-binding surface and that appear to play a role in steroid binding.

Familial androgen resistance is much more common than glucocorticoid resistance and there are numerous reports of mutations within AR LBD associated with this syndrome. The mutations are wide spread throughout the domain but the majority of the mutations are clustered within two regions. Both these regions correspond to regions in GR in proximity of the affinity labeled residues.

Affinity labeling of hPR and radio-sequence analysis showed some clear differences between GR and PR with regard to binding of the synthetic progestin R5020 (Strömstedt, P.-E. *et al.* 1990). Although a Cys residue is found in the identical position in the two receptors, Cys-736 (hGR) is affinity labelled whereas Cys-891 (hPR) is not. Instead, a residue located 18 positions more C-terminally, Met-909 (hPR), is affinity labeled (Fig. 14). This region of PR LBD has been shown to play a role in agonist/antagonist function with regard to transactivation (Vegeto *et al.*,1992). Clearly there are subtle differences in the structure of the two receptors within this region. The differences in affinity labeling could represent structural elements involved in steroid specificity. Another possibility is that the same ligand, R5020, induces different conformational changes in GR versus PR thus giving rise to different receptor-steroid adducts.

An understanding of the structural elements involved in steroid binding will be possible first when the tertiary structure of this domain has been solved. In order to achieve this, GR will have to be overexpressed since it normally occurs in very low concentrations in target cells. Attempts to overexpress GR or the isolated LBD in various systems have revealed an additional important issue influencing the steroid-binding process. As shown in Ohara-Nemoto *et al.* 1990, bacterially expressed GR binds ligand with normal specificity but with greatly reduced affinity. Affinity labeling of this product indicated a more open structure or variety in folding of the protein as all Cys residues in LBD were affinity labelled, although Cys-638 was the predominant adduct. Expression of the same construct by *in vitro* translation in rabbit reticulocyte lysate resulted in a protein with essentially normal affinity for Dex. The functional difference between the two preparations has been correlated with the association of GR with heat shock proteins (Ohara-Nemoto *et al.* 1990). Similar results have been obtained with MR but not with PR, ER or AR (Eul *et al.*,1989; Wittliff *et al.*,1990; Power *et al.*,1990; Young *et al.*,1990; Caamano *et al.*,1993). Thus, the association of GR with the hsp90 complex may facilitate steroid binding, and it will be necessary to determine the quaternary structure to fully understand this function in GR.

Attempts to co-express hsp90 in *E. coli*, together with various derivatives of GR, did not resolve this problem (Ohara-Nemoto, unpublished results). Although this problem is most obvious in bacterial expression systems due to the apparent lack of appropriate functional heat shock chaperone proteins, the role of co-factors in the steroid binding function of GR is probably the limiting factor in most systems for overexpression of this protein. Clearly, it is a limiting factor in BVES since a very small fraction of the expressed GR is soluble and functional (Brown and Sharp, 1990; Xie *et al.,* 1992; Alnemri and Litwack, 1993). The steroid binding function of the insoluble fraction can be partially reconstituted by incubation with rabbit reticulocyte lysate in the presence of an ATP-generating system (Caamano *et al.,* 1993; Alnemri and Litwack, 1993). Co-expression of either hsp90 or hsp70 together with MR or GR in BVES did not improve the solubility of the receptors indicating that additional factor(s) is needed (Alnemri and Litwack, 1993) A disadvantage of BVES is that one is limited to expression in insect Sf9 cells. Little is known about the function of insect heat shock proteins and other co-factors with regard to GR. The expression system used in this study, the Semliki forest virus system, has the advantage that it can be used with virtually any mammalian cell system. A characterisation of the role of the individual heat shock proteins and other co-factors in GR function will be a pre-requisite for opitimal overexpression of functional GR. Using the SFV system, it should be possible to select a cell system which is optimal with regard to these factors or in which the limiting co-factor can be induced or co-expressed.

Acknowledgements:

"These studies were supported by a grant from The Swedish Medical Research Council (No. 13X-2829)"

REFERENCES

Adler AJ, Danielsen M, Robins DM (1992) Androgen-specific gene activation via a consensus glucocorticoid response element is determined by interaction with nonreceptor factors. Proc Natl Acad Sci USA 89:11660-11663.

Agarwal M, Lazer G(1987) Analysis of the structure and function of steroid receptors with the aid of the antihormone RU486. In: Receptor mediated antisteroid action. (ed. Agarwal M) 43-75.

Akerblom IE, Slater EP, Beato M, Baxter JD, Mellon PL (1988) Negative regulation by glucocorticoids through interference with cAMP responsive enhancers. Science 241:350-353.

Akerblom IE, Mellon PL (1991) Repression of gene expression by steroid and thyroid hormones. In: Nuclear hormone receptors: Molecular mechanisms, cellular functions, clinical abnormalities. (ed. Parker MG) Academic Press, London. 175-196.

Alksnis M, Barkhem T, Strömstedt P-E, Ahola H, Kutoh E, Gustafsson J-Å, Poellinger L, Nilsson S (1991) High level expression of functional full length and truncated glucocorticoid receptor in Chinese hamster ovary cells. Demonstration of ligand-induced down-regulation of expressed receptor mRNA and protein. J Biol Chem 266:10078-10085.

Allan GF, Tsai SY, O'Malley BW, Tsai M-J (1991) Steroid hormone receptors and *in vitro* transcription. BioEssays 13:73-78.

Allan GF, Leng X, Tsai SY, Weigel NL, Edwards DP, Tsai M- J, O'Malley BW (1992) Hormone and antihormone induce distinct conformational changes which are central to steroid receptor activation. J Biol Chem 267:19513- 19520.

Alnemri ES, Maksymowych AB, Robertson NM, LitwackG (1991a) Characterization and purification of a functional rat glucocorticoid receptor overexpressed in a Baculovirus system. J Biol Chem 266:3925-3936.

Alnemri ES, Maksymowych AB, Robertson NM, Litwack G (1991b) Overexpression and characterization of locorticoid receptor. J Biol Chem 266:18072- 18081.

Alnemri ES, Litwack G (1993) The steroid binding domain influences intracellular solubility of the Baculovirus overexpressed glucocorticoid and mineralocorticoid receptors. Biochemistry 32:5387-5393.

Amero SA, Kretsinger RH, Moncrief ND, Yamamoto KR, Pearson WR (1992) The origin of nuclear receptor proteins: A single precursor distinct from other transcription factors. Mol Endo 6:3-7.

Arriza JL, Weinberger C, Cerelli G, Glaser TM, Handelin BL, Housman DE, Evans RM (1987) Cloning of human mineralocorticoid receptor complementary DNA: Structural and functional kinship with the glucocorticoid receptor. Science 237:268-275.

Bagchi MK, Tsai M-J, O'Malley BW, Tsai SY (1992) Analysis of the mechanism of steroid hormone receptor-dependent gene activation in cell-free systems. Endocrine Rev 13:525-535.

Bagchi MK, Tsai SY, Tsai M-J, O'Malley BW (1992) Ligand and DNA- dependent phosphorylation of human progesterone receptor *in vitro*. Proc Natl Acad Sci USA 89:2664-2668.

Ballard PL (1979) Delivery and transport of glucocorticoids to target cells. In: Glucocorticoid Hormone Action, Monographs on Endocrinology (eds: Baxter JD and Rousseau GG). Springer Verlag, Berlin. 12:25-48.

Bardwell JCA, Craig EA (1987) Eukaryotic Mr 83,000 heat shock protein has a homologue in Escherichia coli. Proc Natl Acad Sci USA 84:5177-5181.

Barkhem T, Carlsson B, Simons J, Möller B, Berkenstam A, Gustafsson J-Å, Nilsson S (1991) High level expression of functional full length human thyroid hormone receptor b1 in insect cells using a recombinant Baculovirus. J. Steroid Biochem. Mol Biol 38:667-675.

Bar-Shavit Z, Kahn AJ, Pegg LE, Stone KR, Teitelbaum SL (1984) Glucocorticoids modulate macrophage surface oligosaccharides and their bone binding activity. J Clin Invest 73:1277-1283.

Bastl CP, Schulman G, Cragoe EJ (1989) Low-dose glucocorticoids stimulate electroneutral NaCl absorption in rat colon. Am J Physiol 257:F1027-1038.

Baulieu E (1987) Antihormone-steroid hormonal activity, heat-shock protein hsp90 and receptors. Horm Res 28:181-195.

Baulieu EE (1991) The antisteroid RU486: Its cellular and molecular mode of action. Trends Endocrin Metab 2:233-239.

Baxter JD, Tyrrell JB (1987) The adrenal cortex. In: Endocrinology and metabolism, 2nd ed. (eds. Felig P *et al.*) 511-650.

Beato M (1989) Gene regulation by steroid hormones. Cell 56:335-344.

Beaumont K, Fanestil DD (1983) Characterization of rat brain aldosterone receptors reveals high affinity for corticosterone. Endocrinology 113:2043-2047.

Becker PB, Gloss B, Schmid W, Strähle U, Schütz G (1986) *In vivo* protein-DNA interactions in a glucocorticoid response element require the presence of the hormone. Nature 324:686-688.

Bellingham DL, Cidlowaski JA (1989) Stable overproduction of intact glucocorticoid receptors in mammalian cells using a selectable glucocorticoid responsive dihydrofolate reductase gene. Mol Endocrinol 3:1733-1747.

Berkenstam A, Glaumann H, Gustafson J-Å (1988) Unspecific and sequence- specific deoxyribonucleic acid binding of the partially purified human progesterone receptor. Mol Endocrinol 2:571-578.

Binart N, Chambraud B, Dumas B, Rowlands D, Bigogne C, Levin JM, Garnie J, Baulieu EE, Catelli,MG (1989) The cDNA-derived amino acid sequence of chick heat shock protein Mr 90,000 (HSP90) reveals a "DNA like" structure: potential binding site of interaction with steroid receptors. Biochem Biophys Res Commun 159:140-147.

Blanchardie P, Lustenberger P, Denis M, Orsonneau JL, Bernard S (1986) Interaction of rat liver glucocorticoid receptor with lectins: Is the glucocorticoid receptor a glycoprotein? J Steroid Biochem 24:263—267.

Bodwell JE, Ort E., Coull JM, Pappin DJC, Smith LI, Swift F (1991) Identification of phosphorylated sites in the mouse glucocorticoid receptor. J Biol Chem 266:7549-7555.

Bonifer C, Dahlman K, Strömstedt P-E, Flock J-I, Gustafsson J-Å (1989) DNA binding of glucocorticoid receptor protien A fusion proteins expressed in E. coli. J Steroid Biochem 32:5-11.

Boyet JD, Hofert JF (1972) Studies concerning the inhibition of glucose metabolism in thymus lymphocytes by cortisol and epinephrine. Endocrinology 91:233-239.

Bresnick EH, Dalman FC, Sanchez ER, Pratt WB (1989) Evidence that the 90-kDa heat shock protein is necessary for the steroid binding conformation of the L cell glucocorticoid receptor. J Biol Chem 264:4992-4997.

Brink M, Humbel BM, DeKloet ER, van Driel R (1992) The unliganded glucocoreticoid receptor is localized in the nucleus, not in the cytoplasm. Endocrinology 130:3575-3581.

Brown M, Sharp PA (1990) Human estrogen receptor forms multiple protein-DNA complexes. J Biol Chem 265:11238-11243.

Burnside J, Darling DS, Chin WW (1990) A nuclear factor that enhances binding of thyroid hormone receptors to thyroid hormone response elements. J Biol Chem 265:2500-2504.

Caamano CA, Morano MI, Watson SJ, Akil H (1993) The mammalian 90- kDa heat shock protein forms functional complexes in vitro with corticosteroid receptors expressed in bacteria, The Endocrine Society Abstract 1958B.

Cadepond F, Schweizer-Groyer G, Segard-Maurel I, Jibard N, Hollenberg SM, Giguére V, Evans RM, Bauilieu E-E (1991) Heat shock protein 90 as a critical factor in maintaining glucocorticoid receptor in a nonfunctional state. J Biol Chem 266, 5834-5841.

Cadepond F, Gasc J-M, Delahaye F, Jibard N, Schweizer-Groyer G, Segard-Maurel I, Evans E, Baulieu E-E (1992) Hormonal regulation of the nuclear localization signals of the human glucocorticoid receptor. Exp Cell Res 201:99-108.

Cairns C, Gustafsson J-Å, Carlstedt-Duke J (1991) Identification of protein contact sites within the glucocorticoid/progestin response element. Mol Endocrinol 5:598-604.

Cairns W, Cairns C, Pongratz I, Poellinger L, Okret S (1991) Assembly of a glucocorticoid receptor complex prior to DNA binding enhances its specific interaction with glucocorticoid response element. J Biol Chem 266:11221-11226.

Canny BJ, Funder JW, Clarke IJ (1989) Glucocorticoids regulate ovine hypophysial portal levelss of corticotropin-releasing factor and arginine vasopresin in a stress-specific manner. Endocrinology 125:2532-2539.

Carlstedt-Duke J, Gustafsson J-Å, Wrange Ö (1977) Formation and characteristics of hepatic dexamethasone-receptor complexes of different molecular weight. Biochim Biophys Acta 497:507-524.

Carlstedt-Duke J, Okret S, Wrange Ö, Gustafsson J-Å (1982) Immunochemical analysis of the glucocorticoid receptor: Identification of a third domain separate from the steroid-binding and DNA-binding domains. Proc Natl Acad Sci USA 84, 4437-4440.

Carlstedt-Duke J, Strömstedt P-E, Wrange Ö, Bergman T, Gustafsson J-Å, and Jörnvall H (1987) Domain structure of the glucocorticoid receptor protein. Proc Natl Acad Sci USA 84:4437-4440.

Carlstedt-Duke J, Strömstedt P-E, Persson B, Cederlund E, Gustafsson J-Å, and Jörnvall H (1988) Identification of hormone-interacting amino acid residues within the steroid-binding domain of the glucocorticoid receptor in relation to other steroid hormone receptors. J Biol Chem 263:6842-6846.

Carlstedt-Duke J, Strömstedt P-E, Dahlman K, Rae C, Berkenstam A, Hapgood J, Jörnvall H, Gustafsson J-Å (1989) Structural analysis of the glucocorticoid receptor protein. In: The steroid/thyroid hormone receptor family and gene regulation. (eds. Carlstedt-Duke J, Eriksson H, Gustafsson J-Å) 93-108.

Carson MA, Tsai M-J, Conneely OM, Maxwell BL, Clark JH, Dobson ADW, Elbrecht A, Toft DO, Schrader WT, O'Malley BW (1987) Structure-function properties of the chicken progesterone receptor A syntesized from complementary deoxyribonucleic acid. Mol Endcrinol 1:791-798.

Carson-Jurica MA, Schrader WT, O'Malley BW (1990) Steroid receptor family: structure and function. Endocrine Rev 11:201-220.

Chakraborti PK, Simons Jr, SS (1991) Association of heat shock protein 90 with the 16 kDa steroid binding core fragment of rat glucocorticoid receptor. Biochem Biophys Res Commun 176:1338-1344.

Chakraborti PK, Garabedian MJ, Yamamoto KR, Simons Jr SS (1991a) Role of cysteines 640, 656, and 661 in steroid binding to rat glucocorticoid receptors. J Biol Chem 267:11366-11373.

Chakraborti PK, Garabedian MJ, Yamamoto KR, Simons Jr, SS (1991b) Creation of "super" glucocorticoid receptors by point mutations in the steroid binding domain. J Biol Chem 266:22075-22078.

Chakraborti PK, Garabedian MJ, Yamamoto KR, Simons Jr, SS (1992) Role of cysteines 640, 656, and 661 in steroid binding to rat glucocorticoid receptor. J Biol Chem 267:11366-11373.

Chan GC, Hess P, Meenakshi T, Carlstedt-Duke J, Gustafsson J.-Å, Payvar F (1991) Delayed secondary glucocorticoid response elements. Unusual nucleotide motifs specify glucocorticoid receptor binding to transcribed regions of alpha 2u-globulin DNA. J Biol Chem 266:22634-22644.

Chandler VL, Maler BA, Yamamoto KR (1983) DNA sequences bound specifically by glucocorticoid receptor *in vitro* render a heterologous promoter hormone responsive *in vivo*. Cell 33:489-499.

Chang C, Wang C, DeLuca HF, Ross TK, Shih CC-Y (1992) Characterization of human androgen receptor overexpressed in the baculovirus system. Proc Natl Acad Sci USA 89:5946-5950.

Childs GV, Unabia G (1990) Rapid corticosterone inhibition of corticotropin-releasing hormone binding and adrenocorticotropin release by enriched populations of corticotropes: counteractions by arginine vasopressin and its second messangers. Endocrinology 126:1967-1975.

Choen-Solal K, Bailly A, Rauch C, Quesne M, Milgrom E (1993) Specific binding of progesterone receptor to progesterone-responsive elements does not require prior dimerization. Eur J Biochem 214:189-195.

Christensen K, Estes PA, Onate SA, Beck CA, DeMarzo A, Altman M, Lieberman BA, John J, Nordeen S, Edwards DP (1991) Characterization and functional properties of the A and B forms of human progesterone receptors synthesized in a Baculovirus system. Mol Endocrinol 5:1755-1770.

Cordingley MG, Reigel AT, Hager GL (1987) Steroid-dependent interaction of transcription factors with the inducible promoter of mouse mammary tumor virus *in vivo*. Cell 48:261-270.

Dahlman K, Strömstedt P-E, Rae C, Jörnvall H, Flock J-I, Carlstedt-Duke J, Gustafsson J-Å (1989) high level expression in *Escherichia coli* of the DNA-binding domain of the glucocorticoid receptor in a functional form utilizing domain specific cleavage of a fusion protein. J Biol Chem 264:804-809.

Dahlman-Wright K, Wright APH, Gustafsson J-Å, Carlstedt-Duke J (1991) Interaction of the glucocorticoid receptor DNA-binding domain with DNA as a dimer is mediated via a short segment of five amino acids. J Biol Chem 266:3107- 3112.

Dahlman-Wright K, Wright APH, Gustafsson J-Å (1992) Determinants of high affinity DNA binding by the glucocorticoid receptor: evaluation of receptor domains outside the DNA-binding domain. Biochemistry 31:9040-9044.

Dahlman-Wright K,Grandien K, Nilsson S, Gustafsson J-Å, Carlstedt- Duke J (1993) Protein-protein interactions between the DNA-binding domains of nuclear receptors: influence on DNA-binding. J Steroid Biochem Molec Biol 45:239-250.

Dalman FC, Sanchez ER, Lin AL-Y, Perini F, Pratt WB (1988) Localization of phosphorylation sites with respect ot the functional domains of the mouse L cell glucocorticoid receptor. J Biol Chem 263:12259-12267.

Dalman FC, Koenig RJ, Perdew GH, Massa E, Pratt WB (1990) In contrast to the glucocorticoid receptor, the thyroid hormone receptor is translated in the DNA-binding state and is not associated with hsp90. J Biol Chem 265:3615-3618.

Dalman FC, Sturzenbecker LJ, Levin AA, Lucas DA, Perdew GH, Petkovitch M, Chambon P, Grippo JF, Pratt WB (1991) Retinoic acid receptor belongs to a subclass of nuclear receptors that do not form "docking" complexes with hsp90. Biochemistry 30:5605-5608.

Dallman M, Akana S, Cascio C (1987) Regulation of ACTH secretion: variations on the theme of B. Recent Progress in Hormone Research 43:113-172.

Danielsen M, Northrop JP, Ringold GM (1986) The mouse glucocorticoid receptor: mapping of functional domains by cloning, sequencing and expression of wild-type and mutant receptor proteins. EMBO J 5:2513-2522.

Danielsen M, Northrop JP, Jonklaas J, Ringold GM (1987) Domains of the glucocorticoid receptor involved in specific and nonspecific deoxyribonucleic acid binding, hormone activation, and transcriptional enhancement. Mol Endocrinol 1:816-822.

Danielsen M, Hinck L, Ringold GM (1989) Two amino acids within the knuckle of the first zinc finger specify DNA response element activation by the glucocorticoid receptor. Cell 57:1131-1138.

Darling DS, Beebe JS, Burnside J, Winslow ER, Chin WW 1991 Mol Endocrinol 5:73-84.

Dellweg H-G, Hotz A, Mugele K, Gehring U (1982) Active domains in wild- type and mutant glucocorticoid receptors. EMBO J 1:285-289.

DeLuca HF (1991) New concepts of vitamin D functions. Ann NY Acad Sci 669:59- 68.

Denis M, Wikström A-C, Gustafsson J-Å (1987) The molybdate-stabilized nonactivated glucocorticoid receptor contains a dimer of Mr 90,000 non-hormone- binding protein. J Biol Chem 262:11803-11806.

Denis M, Poellinger L, Wikström A-C, Gustafsson J-Å (1988a) Requirement of hormone for thermal conversion of the glucocorticoid receptor to a DNA- binding state. Nature 333:686-688.

Denis M, Gustafsson J-Å, Wikström, A-C (1988b) Interaction of the Mr=90,000 heat shock protein with the steroid-binding domain of the glucocorticoid receptor. J Biol Chem 263:18520-18523.

Denis M, Gustafsson J-Å (1989) Translation of glucocorticoid receptor mRNA in vitro yields a non-activated protein. J Biol Chem 264:6005-6008.

Denner LA, Schrader WT, O'Malley BW, Weigel NL (1990a) Hormonal regulation and identification of chicken progesterone receptor phosphoryation sites. J Biol Chem 265:16548-16555.

Denner LA, Weigel NL, Maxwell BL, Schrader WT, O'Malley BW (1990b) Regulation of progesterone receptor-mediated transcription by phosphorylation. Science 250:1740-1743.

Diamond MI, Miner JN, Yoshinaga SK, Yamamoto KR (1990) Transcription factor interactions: Selectors of positive or negative regulation from a single DNA element. Science 249:1266-1272

Dong Y, Cairns W, Okret S, Gustafsson J-Å (1990) A glucocorticoid-resistant rat hepatoma cell variant contains a functional glucocorticoid receptor. J Biol Chem 265:7526-7531.

Edwards CRW, Stewart PM, Burt D, Brett L, McIntyre MA, Sutanto WS, DeKloet ER, Monder C (1988) Localisation of 11β- hydroxysteroid dehydrogenase: tissue specific protector of the mineralocorticoid receptor. Lancet 2:986-989.

Eisen HJ, Schleenbaker RE, Simons Jr SS (1981) Affinity labeling of the rat liver glucocorticoid receptor with dexamethasone 21-mesylate. Identification of covalently labeled receptor by immunochemical methods. J Biol Chem 256:12920- 12925.

El-Ashry D, Onate SA, Nordeen SK, Edwards DP (1989) Human progesterone receptor complexed with the antagonist RU486 binds to hormone response elements in a structurally altered form. Mol Endocrinol 3:1545-1558.

Encio IJ, Detera-Wadleigh SD (1991) The genomic structure of the human glucocorticoid receptor. J Biol Chem 266:7182-7188.

Eriksson P, Wrange Ö (1990) Protein-protein contacts in the glucocorticoid receptor homodimer influence its DNA binding properties. J Biol Chem 265:3535-3542.

Eul J, Meyer ME, Tora L, Bocquel MT, Quirin-Stricker C, Chambon P, Gronemeyer H (1989) Expression of active hormone and DNA-binding domains of the chicken progesterone receptor in E. coli.. EMBO J 8:83-90.

Evans RM (1988) The steroid and thyroid hormone receptor superfamily. Science 240:889- 895.
Fawell SE, Lees JA, White R, Parker MG (1990a) Characterization and colocalization of steroid binding and dimerization activities in the mouse estrogen receptor. Cell 60:953-962.
Forman BM, Yang C-R, Au M, Casanova J, Ghysdael J, Samuels HH (1989) A domain containing leucine-zipper-like motifs mediate novel *in vivo* interactions between the thyroid hormone and retinoic acid receptors. Mol Endocrinol 3:1610-1626.
Fraser MJ (1989) Expression of eukaryotic genes in insect cell cultures. In Vitro Cell Dev Biol 25:225-235.
Freedman LP, Luisi BF, Korszun KR, Basavappa R, Sigler PJ, Yamamoto KR (1988) The function and structure of the metal coordination sites within the glucocorticoid receptor DNA binding domain. Nature 334:543-546.
Freedman LP (1992) Anatomy of the steroid receptor zinc finger region. Endocrine Rev 13:129-145.
Funder JW, Pearce PT, Smith R, Smith IA (1988) Mineralocorticoid action: Target tissue specificity is enzyme, not receptor, mediated. Science 242:583-585.
Funder JW (1992) Glucocorticoid receptors. J Steroid Biochem Molec Biol 43:389-394.
Furlow JD, Murdoch FE, Gorski J (1993) High affinity binding of the estrogen receptor to a DNA response element does not require homodimer formatation or estrogen. J Biol Chem 268, 12519-12525.
Furu K, Kilvik K, Gautvik KM, Haug E (1987) The mechanism of 3H dexamethasone uptake into prolactin producing rat pituitary cells (GH3) in culture. J Steroid Biochem 28:587-591.
Gasc J-M, DeLahaye F, Baulieu E-E (1989) Compared intracellular localization of the glucocorticosteroid and progesterone receptors: an immunocytochemical study. Exp Cell Res 181:492-504.
Gearing KL, Göttlicher M, Teboul M, Widmark E, Gustafsson J-Å (1993) Interaction of the peroxisome-proliferator-activated receptor and retinoid X receptor. Proc Natl Acad Sci USA 90:1440-1444.
Gearing KL, Gustafsson J-Å, Okret S (1993) Heterogeneity in the polyglutamine tract of the glucocorticoid receptor from different rat strains. Nucleic Acids Res 21:2014.
Gehring U, Arndt H (1985) Heteromeric nature of glucocorticoid receptor. FEBS Lett 179:138-142.
Georgopoulos C (1992) The emergence of the chaperone machines. Trends Biochem Sci 17:295-299.
Gething M-J, Sambrook J (1992) Protein folding in the cell. Nature 355:33-45.
Giguère V, Hollenberg SM, Rosenfeld MG, Evans RM (1986) Functional domains of the human glucocorticoid receptor. Cell 46:645-652.
Glass CK, Lipkin SM, Devary OV, Rosenfeld MG (1989) Positive and negative regulation of gene transcription by a retinoic acid-thyroid hormone receptor heterodimer. Cell 59:697-708.
Godowski PJ, Rusconi S, Miesfeld R, Yamamoto KR (1987) Glucocorticoid receptor mutants that are constitutive activators of transcriptional enhancement. Nature 325:365-368.
Godowski PJ, Picard D, Yamamoto KR (1988) Signal transduction and transcriptional regulation by glucocorticoid receptor-LexA fusion proteins. Science 241:812-816.
Göttlicher M, Widmark E, Li Q, Gustafsson J-Å (1992) Fatty acids activate chimera of the clofibric acid-activated receptor and the glucocorticoid receptor. Proc Natl Acad Sci USA 89:4653-4657.

Gower DB (1988) The biosynthesis of steroid hormones: an up-date. In: Hormones and their Actions, Part I (eds: Cooke GA, King RJB, and van der Molen HJ) 3-28.

Granner DK (1979) The role of glucocorticoid hormones as biological amplifiers. In: Glucocorticoid hormone action (eds: Baxter JD, Rousseau G) 12:593-609.

Green S, Walter P, Krust V, Bornert JM, Argos P, Chambon P (1986) Human estrogen receptor cDNA: Sequence, expression, and homology to v-erb A. Nature 320:134-139.

Green S, Chambon P (1987) Oestradiol induction of a glucocorticoid-responsive gene by a chimeric receptor. Nature 325:75-78.

Green S, Kumar V, Theula I, Wahli W, Chambon P (1988) The N-terminal DNA-binding 'zinc finger' of the oestrogen and glucocorticoid receptors determines target gene specificity. EMBO J. 7:3037-3044.

Green S (1993) Peroxisome proliferators: a model for receptor mediated carcinogenesis. Cancer Surveys 14:221-232.

Green GL, Gilna P, Waterfield M, Baker A, Hort Y, Shine J (1986) Sequence and expression of a functional estrogen cDNA. Sceience 231:1150-1154.

Grody WW, Schrader WT, O'Malley BW (1982) Activation, transformation, and subunit structure of steroid hormone receptors. Endocrine Rev 3:141-163.

Gronemeyer H, Govindan MV, Chambon P (1985) Immunological similarity between chick oviduct progesterone receptor forms A and B. J Biol Chem 260:6916-6925.

Gronemeyer H (1992) Control of transcription activation by steroid hormone receptors. FASEB J 6:2524-2529.

Groyer A, Schweizer-Groyer G, Cadepond F, Mariller M, Baulieu E-E (1987) Antiglucocorticosteroid effects suggest why steroid hormone is required for receptors to bind DNA *in vivo* but not *in vitro*. Nature 328:624-626.

Guiochon-Mantel A, Loosfelt H, Ragot T, Bailly A, Atger M, Misrahi M, Perricaudet M, Milgrom E (1988) Receptors bound to antiprogestin form abortive complexes with hormone responsive elements. Nature 336:695-698.

Guiochon-Mantel A, Loosfelt H, Lescop P, Sar S, Atger M, Perrot-Applanat M, Milgrom E (1989) Mechanism of nuclear localization of the progesterone receptor: evidence for interaction between monomers. Cell 57:1147-1154.

Gustafsson J-Å, Carlstedt-Duke J, Poellinger L, Okret S, Wikström A-C, Brönnegård M, Gillner M, Dong Y, Fuxe K, Cintra A, Härfstrand A, Agnati L (1987) Biochemistry, molecular biology, and physiology of the glucocorticoid receptor. Endocrine Rev 8:185-234.

Hahn TJ (1989) Steroid hormones and the skeleton. In: Metabolic bone disease: Approach based on cellular and tissue mechanisms. (eds. Tam CS, Heersche JNH, Murray TM) 223-231.

Hammond GL (1990) Molecular properties of corticosteroid-binding globulin and the sex-steroid binding proteins. Endocrine Rev. 11:65-79.

Härd T, Kellenbach E, Boelens R, Maler BA, Dahlman K, Freedman LP, Carlstedt-Duke J, Yamamoto KR, Gustafsson J-Å, Kaptein R (1990) Solution structure of the glucocorticoid receptor DNA-binding domain. Science 249:157-160.

Harlow KW, Smith DN, Katzenellenbogen JA, Greene GL, Katzenellenbogen BS (1989) Identification of cysteine 530 as the covalent attachement site of an affinity-labeling estrogen (Ketononestrol Aziridine) and antiestrogen (Tamoxifen Aziridine) in the human estrogen receptor. J Biol Chem 264:17476-17485.

Hartl FU, Martin J, Neupert W (1992) Protein folding in the cell: the role of molecular chaperones hsp70 and hsp60. Annu. Rev. Biophys. Biomol. Struct. 21:293-322.

Hecht A, Berkenstam A, Strömstedt P-E, Gustafsson J-Å, Sippel AE (1988) A progesterone responsive element maps to the far upstream steroid dependent DNase hypersensitive site of chicken lysozyme chromatin. EMBO J 7:2063-2073.

Hirst MA, Northrop JP, Ringold GM (1990) High level expression of wild type and variant mouse glucocorticoid receptors in Chinese hamster ovary cells. Mol. Endocrinol 4:162-170.

Hoeck W, Rusconi S, Groner B (1989) Down-regulation and phosphorylation of glucocortiocid recptors in cultured cells. J Biol Chem 264:14396-14402.

Hoeck W, Groner B (1990) Hormone-dependent phosphorylation of the glucocorticiod receptor occurs mainly in the amino-terminal transactivation domain. J Biol Chem 265:5403-5408.

Hollenberg SM, Weinberger C, Ong ES, Cerelli G, Oro A, Lebo R, Thompson EB, Rosenfeld MG, Evans RM (1985) Primary structure and expression of a functional glucocorticoid receptor cDNA. Nature 318:635-641.

Hollenberg SM Giguère V, Segui P, Evans RM (1987) Colcalization of DNA-binding and transcriptional activation functions in the human glucocorticoid receptor. Cell 49:39-46.

Hollenberg SM, Evans RM (1988) Multiple and cooperative trans-activation domains of the human glucocorticoid receptor. Cell 55:899-906.

Howard KJ, Distelhorst CW (1988) Effect of the 90 kDa heat shock protein, hsp90, on glucocorticoid receptor binding to DNA-cellulose. Biochem Biophys Res Commun 151:1226-1232.

Howard KJ, Holley SJ, Yamamoto KR, Distelhorst CW (1990) Mapping the hsp90 binding region of the glucocorticoid receptor. J Biol Chem 265:11928-11935.

Hurley DM, Accili D, Stratakis CA, Karl M, Vamvakopoulos N, Rorer E, Constantine K, Taylor SI, Chrousos GP (1991) Point mutation causing a single amino acid substitution in the hormone binding domain of the glucocorticoid receptor in familial glucocorticoid resistance. J Clin Invest 87:680-686.

Husmann DA, Wilson CM, McPhaul MJ, Tilley WD, Wilson JD (1990) Antipeptide antibodies to two distinct regions of the androgen receptor localize the receptor protein to the nuclei of target cells in the rat and human prostate. Endocrinology 126:2359-2368.

Hutchison KA, Czar MJ, Scherrer LC, Pratt WB (1992) Monovalent cation selectivity for ATP-dependent association of the glucocorticoid receptor with hsp70 and hsp90. J Biol Chem 267:14047-14053.

Hutchison KA, Dittmar KD, Pratt WB (1993b) Proof that hsp70 is required for formation of the glucocorticoid receptor-hsp90 heterocomplex. Abstract: The Endocrine Society 74th meeting Abstract 586B.

Inano K, Haino M, Iwasaki M, Ono N, Horigome T, Sugano H (1990) Reconstitution of the 9S estrogen receptor with heat shock protein 90. FEBS Lett 267:157-159.

Ing NH, Beekman JM, Tsai SY, Tsai M-J, O'Malley BW (1992) Members of the steroid hormone receptor superfamily interact with TFIIB (S300-II). J Biol Chem 267:17617-17623.

Jackson P, Baltimore D, Picard D (1993) Hormone-conditional transformation by fusion proteins of c-Abl and its transforminf variants. EMBO J 12:2809-2819.

Jänne OA, Palvimo JJ, Kallio P, Mehto M, Xie Y-B, Sui Y-P (1993) Production of recombinant androgen receptor in a heterologous expression system. Clin Chem 39, 346-352.

Jantzen H-M, Strähle U, Gloss B, Stewart F, Schmid W, Boshart M, Miksicek R, Schütz G (1987) Cooperativity of glucocorticoid response elements located far upstream of the tyrosine aminotransferase gene. Cell 49:29-38.

Joab I, Radanyi C, Renoir M, Buchou T, Catelli M-G, Binart N, Mester J, Baulieu E-E (1984) Common non-hormone binding component in non-transformed chick oviduct receptors of four steroid hormones. Nature 308:850-853

Joels M, deKloet ER (1990) Mineralocorticoid receptor-mediated changes in membrane properties of rat CA1 pyramidal neurons in vitro. Proc Natl Acad Sci USA 87:4495.

Jonat C, Rahmsdorf HJ, Park K-K, Cato ACB, Gebel S, Ponta H, Herrlich P (1990) Antitumor promotion and antiinflammmation: down-modulation of AP-1 (Fos/Jun) activity by glucocorticoid hormone. Cell 62:1189-1204.

Jones MT, Gillham B (1988) Factors involved in the regulation of adrenocorticotropic hormone/ β-lipotropic hormone. Physiol Rev 68:743-818.

Jung-Testas I, Baulieu E-E (1984) Antisteroid action in cultured L-29 mouse fibroblasts. J Steroid Biochem 20:301-306.

Kalimi M (1989) Role of antiglucocorticoid RU486 on dexamethasone-induced hypertension in rats. Am J Physiol 256:682-685.

Kalimi M, Agarwal M (1988) Interaction of antiglucocorticoid RU486 with rat kidney glucocorticoid receptor. Biochem Biophys Res Commun 153:365-371.

Kao CC, Lieberman PM, Schmidt MC, Zhou Q, Pei R, Berk AJ (1990) Cloning of a transcriptionslly active human TATA binding factor. Science 248:1646-1649.

Kastner P, Krust A, Turcotte B, Stropp U, Tora L, Gronemeyer H, Chambon P (1990) Two distinct estrogen-regulated promoters generate transcripts encoding the two functionally different human progesterone receptor forms A and B. EMBO J 9:1603-1614.

Katzenellenbogen BS (1980) Dynamics of steroid hormone receptor action. Ann Rev Physiol 42:17-35.

King WJ, Greene GL (1984) Monoclonal antibodies localize oestrogen receptor in the nuclei of target cells. Nature 307:745-747.

Klein-Hitpass L, Tsai SY, Weigel NL, Allan GF, Riley D, Rodriguez R, Schrader WT, Tsai M-J, O'Malley BW (1990) The progesterone receptor stimulates cell-free transcription by enhancing the formation of a stable preinitiation complex. Cell 60:247-257.

Kliewer SA, Umesono K, Mangelsdorf DJ, Evans RM (1992a) Retinoid X receptor interacts with nuclear receptors in retinoic acid, thyroid hormone and vitamin D3 signalling. Nature 355:446-449.

Kliewer SA, Umesono K, Noonan DJ, Heyman RA, Evans RM (1992b) Convergence of 9-cis retinoic acid and peroxisome proliferator signalling pathways through heterodimer formation of their receptors. Nature 358:771-774.

Klock G, Strähle U, Schütz G (1987) Oestrogen and glucocorticoid responsive elements are closely related but distinct. Nature 329:734-736.

Knegtel RMA, Katahira M, Schiluis JG, Bonvin AMJJ, Boelens R, Eib D, van der Saag PT, Kaptein R (1993) The solution structure of the human retinoic acid receptor-b DNA-binding domain. J Biomol NMR 3:1-17.

Krust A, Green S, Argos P, Kumar V, Walter P, Bornert J-M, Chambon P (1986) The chicken oestrogen receptor sequence: homology with v-erb A and the human glucocorticoid receptors. EMBO J 5:891-897.

Kumar V, Green S, Staub A, Chambon P(1986) Localization of the oestradiol-binding and putative DNA-binding domains of the human estrogen receptor. EMBO J 5:2231-223.

Kumar V, Green S, Stack G, Berry M, Jin J-R, Chambon P (1987) Functional domains of the human estrogen receptor. Cell 51:941-951.

Kumar V, Chambon P (1988) The estrogen receptor binds tightly to its responsive element as a ligand-induced homodimer. Cell 55:145-156.

Kushner PJ, Hort E, Shine J, Baxter JD, Greene GL 1990 Construction of cell lines that express high levels of human estrogen receptor and are killed by estrogens. Mol Endocrinol 4:1465-1473.

Kutoh E, Strömstedt P-E, Poellinger L (1992) Functional interference between the ubiquitous and constitutive octamer transcription factor 1 (OTF-1) and the glucocorticoid receptor by direct protein-protein interaction involving the homeo subdomain of OTF-1. Mol Cell Biol 12:4960-4969.

Laplace JR, Husted RF, Stokes JB (1992) Cellular responses to steroids in the enhancement of Na+ transport by rat collecting duct cells in culture. Differences between glucocorticoid and mineralocorticoid hormones.J Clin Invest 90:1370-1378.

LaSpada AR, Wilson EM, Lubahn DB, Harding AE, Fischbeck KH (1991) Androgen receptor mutation in X-linked spinal and bulbar muscular atrophy. Nature 352:77-79.

Laudet V, Hänni C, Coll J, Catzeflis F, Stéhelin D (1992) Evolution of the nuclear receptor gene superfamily. EMBO J 11:1003-1013.

Lazar MA, Berrodin TK (1990) Thyroid hormone receptors form distinct nuclear proteins-dependent and independent complexes with thyroid hormone response element. Mol Endocrinol 4:1627-1635.

Leach KL, Dahmer MK, Hammond ND, Sando JJ, Pratt WB (1979) Molybdate inhibition of glucocorticoid receptor inactivation and transformation. J Biol Chem 254:11884-11890.

Lebeau MC, Masso IN, Herrick J, Faber L, Renoir J-M, Radanyi C, Baulieu EE J (1992) P59, an hsp 90-binding protien. Cloning and sequencing of its cDNA and preparation of a peptide-directed polyclonal antibody. J Biol Chem 267:4281-4284.

Lee MS, Kliewer SA, Provencal J, Wright PE, Evans RM (1993) Structure ofthe retionoid X receptor a DNA binding domain: a helix required for homodimeric DNA binding.Science 260:1117-1121.

Lees JA, Fawell SE, White R, Parker MG (1990) A 22-amino-acid peptide restores DNA binding activity to dimerization-defective mutants of the estrogen receptor. Mol Cell Biol 10:5529-5531.

Leid MP, Kastner P, Lyons, R, Nakshatri H, Saunders M, Zacharewski T, Chen J-Y, Staub A, Garnier J-M, Mader S, Chambon P (1992) Purification, cloning and RXR identity of the HeLa cell factor with which RAR or TR heterodimerizes to bind target sequences specifically. Cell 66:377-395.

Lessey BA, Alexander PS, Horwitz KB (1983) The subunit structure of human breast cancer progesterone receptors: characterization by chromatography and photoaffinity labeling. Endocrinology 112:1267-1274.

Liljeström P, Garoff H (1991) A new generation of animal cell expression vectors based on the semliki forest virus replicon. BioTechnology 9:1356-1361.

Liljeström P, Lusa S, Huylebroeck S, Garoff H (1991) In vitro mutagenesis of a full-length cDNA clone of Semliki Forest virus: the 6,000-molecular-weight membrane protein modulates virus release. J Virol 65:4107-4113.

Lipman ME (1979) Glucocorticoid receptors and effects in human lymphoid and leukemic cells. In: Glucocorticoid hormone action. (eds. Baxter JD, Rousseau GG) 377-401.

Lubahn DB, Joseph DR, Sullivan PM, Willard HF, French FS, Wilson EM (1988) Cloning of human androgen receptor complementary DNA and localization to the X chromosome. Science 240:327-330.

Lucas PC, Granner DK (1992) Hormone response domains in gene transcription. Ann Rev Biochem 61:1131-1173.

Lucibello FC, Slater EP, Jooss KU, Beato M, Müller R (1990) Mutual transrepression of Fos and the glucocorticoid receptor: involvement of a functional domain in Fos which is absent in FosB. EMBO J 9:2827-2834.

Luisi BF, Xu WX, Otwinowski Z, Freedman LP, Yamamoto KR, Sigler PB (1991) Crystallographic analysis of the interaction of the glucocorticoid receptor with DNA. Nature 352:497-505.

Lydon JP, Power RF, Conneely OM (1992) Dfferential modes of activation define orphan subclasses within the steroid/thyroid receptor superfamily. Gene Expr 2:273-283.

MacDonald PN, Haussler CA, Terpening CM, Galligan MA, Reeder MC, Whitfield GK, Haussler MR (1991) Baculovirus-mediated expression of the human vitamin D receptor. Functional characterization, vitamin D response element interactions, and evidence for a receptor auxiliary factor. J Biol Chem 266:18808- 18813.

Mader S, Kumar V, de Verneuil H, Chambon P (1989) Three amino acids of the oestrogen receptor are essential to its ability to distinguish an oestrogen from a glucocorticoid responsive element. Nature 338:271-274.

Mak P, McDonnell DP, Weigel NL, Schrader WT, O'Malley BW (1989) Expression of functional chicken oviduct progesterone receptors in yeast (Saccharomyces cerevisiae) J Biol Chem 264:21613-21618.

Maksymowych AB, Hsu T-C, Litwack (1992) A novel, highly conserved structural motif is present in all members of the steroid receptor superfamily. Receptor 2:225-240.

Malchoff DM, Brufsky A, Reardon G, McDermott P., Javier EC, Bergh C-H, Rowe D, Malchoff CD (1993) A mutation of the glucocorticoid receptor in primary cortisol resistance. J Clin Invest 9:1918-1925.

Mao J, Regelson W, Kalimi M (1992) Molecular mechanism of RU486 action: a review. Mol Cell Biochem 109:1-8.

Marks MS, Hallenbeck PL, Nagata T, Segars JH, Appella E, Nikodem VM, Ozato K (1992) H-2RIIBP (RXRb) heterodimerization provides a mechanism for combinatorial diversity in the regulation of retinoic acid and thyroid hormone responsive genes. EMBO J 11:1419-1435.

Marston FAO (1986) The purification of eukaryotic polypeptides synthesized in Escherichia coli. Biochem J 240:1-12.

Martinez E, Givel F, Wahli W (1991) A common ancestor DNA motif for invertebrate and vertebrate hormone response elements. EMBO J 10, 263-268.

McDonnell DP, Scott RA, Kerner SA, O'Malley BW, Pike JW (1989) Functional domains of the human vitamin D3 receptor regulate osteocalcin gene expression. Mol Endocrinol 3:635-

McEwan IJ, Wright APH, Dahlman-Wright K, Carlstedt-Duke J, Gustafsson J-Å (1993) Direct interaction of the tau_1 transactivation domain of the human glucocorticoid receptor with the basal transcriptional machinery. Mol Cell Biol 13:399-407.

Mendel DB, Bodwell JE, Gametchu B, Harrison RW, Munck A (1986) Molybdate-stabilized nonactivated glucocorticoid-receptor complexes contain a 90-kDa non-steroid-binding phosphoprotein that is lost on activation. J Biol Chem 261:3758-3763.

Mendel DB, Orti E (1988) Isoform composition and stoichiometry of the ~90-kDa heat shock protein associated with glucocorticoid receptors. J Biol Chem 263:6695-6702.

Metzger D, White JH, Chambon P (1988) The human oestrogen receptor function in yeast. Nature 334:31-36.

Meyer M-E, Pornon A, Ji J, Bocquel M-T, Chambon P, Gronemeyer H (1990) Agonistic and antagonistic activities of RU486 on the functions of the human progesterone receptor. EMBO J 9:3923-3932.

Miesfeld R, Okret S, Wikström A-C, Wrange Ö, Gustafsson J-Å, Yamamoto KR (1984) Characterization of a steroid hormone receptor gene and mRNA in wild-type and mutant cells.Nature 312:779-781.

Miesfeld R, Rusconi S, Godowski PJ, Maler BM, Okret S, Wikström A-C, Gustafsson J-Å, Yamamoto KR (1986) Genetic complementation of a glucocorticoid receptor deficiency by expression of cloned receptor cDNA. Cell 46:389-399.

Miesfeld R, Godowski PJ, Maler BA, Yamamoto KR (1987) Glucocorticoid receptor mutants that define a small region sufficient for enhancer activation. Science 236:423-427.

Miller J, McClachlin AD, Klug A (1985) Repetitive zinc-binding domains in the protein transcription factor IIIA from *Xenopus* oocytes. EMBO J 4:1609-1614.

Miller NR, Simons Jr SS (1988) Steroid binding to hepatoma tissue culture cell glucocorticoid receptors involves at least two sulfhydryl groups. J Biol Chem 263:15217-15225.

Miller WL (1988) Molecular biology of steroid hormone synthesis. Endocrine Rev 9:295-318.

Miner JN, Yamamot KR (1991) Regulatory crosstalk at composite response elements. Trends Biochem Sci 16:423-426.

Migliaccio A, Di Domenico M, Green S, de Falco A, Kajtaniak EL, Blasi F, Chambon P, Auricchio F (1989) Phosphorylation on tyrosine of the *in vitro* synthesised human estrogen receptor activates its hormone binding. Mol Endocrinol 3:1061-1069.

Misrahi M, Atger M, d'Auriol L, Loosfelt H, Meriel C, Fridlansky F, Guiochon-Mantel A, Galibert F, Milgrom E (1987) Complete amino acid sequence of the human progesterone receptor deduced from cloned cDNA. Biochem Biophys Res Commun 143:740-748.

Mitchell PJ, Tijan R (1989) Transcriptional regulation in mammalian cells by sequence-specific DNA binding proteins. Science 245:371-378.

Moguilewsky M, Philibert D (1984) RU486: potent antiglucocorticoid activity correlated with strong binding to the cytosolic glucocorticoid receptors followed by impaired activation. J SteoridBiochem 20:271.276.

Moore DD (1990) Diversity and unity in the nuclear hormone receptors: a terpenoid receptor superfamily. The New Biologist 2:100-105.

Moore DD, Marks AR, Buchley DI, Kapler G, Payvar F, Goodman HM (1985) The first intron of the human growth hormone gene contains a binding site for glucocorticoid receptor. Proc Natl Acad Sci USA 82:699-702.

Morrison NA, Shine J, Fragonas J-C, Verkest V, McMenemy ML, Eisman JA (1989) 1,25-dihydroxyvitamin D-responsive element and glucocorticoid repression in the osteocalcin gene. Science 246:1158-1161.

Moudgil V, JoAnter M, Hurd C (1989) Mammalian progesterone receptor shows differential sensivity to sulfhydryl group modifying agents when bound to agonist and antagonist ligands. J Biol Chem 264:2203-2211.

Muller M, Renkawitz R (1991) The glucocorticoid receptor. Biochim Biophys Acta 1088:171-182.

Murray MB, Towle HC (1989) Identification of nuclear factors that enhance binding of the thyroid hormone receptor to a thyroid hormone response element. Mol Endocrinol 3:1434-1442.

Naray-Fejes-Toth A, Fejes-Toth G (1990) Glucocorticoid receptors mediate mineralocorticoid-like effects in culturued collecting duct cells. Am J Physiol 259:F672

Nemoto T, Ohara-Nemoto Y, Denis M, Gustafsson J-Å (1990) The transformed glucocorticoid receptor has a lower steroid-binding affinity than the nontransformed receptor. Biochemistry 29, 1880-1886.

Nemoto T, Ohara-Nemoto Y, Ota M (1992) Association of the 90-kDa heat shock protein does not affect the ligand-binding ability of the androgen receptor. J Steroid Biochem.Mol Biol 42, 803-812.

Ning Y-M, Sanchez ER (1993) Potentiation of glucocorticoid receptor-mediated gene expression by immunophilin ligands FK506 and rapamycin. J Biol Chem 268:6073-6076.

Nordeen SK, Suh BJ, Kühnel B, Hutchison III CA (1990) Structural determinants of a glucocorticoid receptor recognition element. Mol Endocrinol 4:1866-1873.

O'Donnell AL, Koenig RJ (1990) Mutational analysis identifies a new functional domain of the thyroid hormone receptor. Mol Endocrinol 4:715-720.

O'Donnell AL, Rosen ED, Darling DS, Koenig RJ (1991) Thyroid hormone receptor mutations that interfere with transcriptional activation also interfere with receptor interaction with nuclear protein. Mol Endocrinol 5:94-99.

Ohara-Nemoto Y, Strömstedt P-E, Dahlman-Wright K, Nemoto T, Gustafsson J-Å, and Carlstedt-Duke J (1990) The steroid-binding properties of recombinant glucocorticoid receptor: A putative role for heat shock protein hsp90. J Steroid Biochem Molec Biol 37:481-490.

Ohara-Nemoto Y, Nemoto T, Ota M- 1991 The Mr 90,000 heat shock protein-free androgen receptor has high affinity for steroid, in contrast to the glucocorticoid receptor. J Biochem 109:113-119.

Okret S, Wikström A-C, Gustafsson J-Å (1985) Molybdate-stabilized glucocortcoid receptor: Evidence for a receptor heteromer. Biochemistry 24:6581-6586.

Ortí E, Mendel DB, Smith LI, Munck A (1989) Agonist-dependent phosphorylation and nuclear dephosphorylation of glucocorticoid receptors in intact cells. J Biol Chem 264:9728-9731.

Ortí E, Bodwell JE, Munck A (1992) Phosphorylation of steroid hormone receptors. Endocrine Rev 13:105-128.

Ortí E, Hu L-M, Munck A (1993) Kinetics of glucocorticoid receptor phosphorylation in intact cells. Evidence for hormone-induced hyperphosphorylation after activation and recycling of hyperphophorylated receptors. J Biol Chem 268:7779-7784.

Payvar F, Wrange Ö (1983) Relative selectivities and efficiencies of DNA binding by purified intact and protease-cleaved glucocorticoid receptor. In: Steroid hormone receptors: structure and function. (eds. Eriksson H, Gustafsson J-Å) 267-282.

Payvar F, DeFranco D, Firestone GL, Edgar B, Wrange Ö, Okret S, Gustafsson J-Å, Yamamoto KR (1983) Sequence-specific binding of glucocorticoid receptor to MTV DNA at sites within and upstream of the transcribed region. Cell 35:381-392.

Pearce D, Yamamoto KR (1993) Mineralocorticoid and glucocorticoid receptor activities distinguished by nonreceptor factors at a composite response element. Science 259:1161-1165.

Pennock GD, Shoemaker C, Miller LK (1984) Strong and regulated expresson of E. coli. beta-galactosidase in insect cells using a baculovirus vector for insect cell culture procedure. Texas Ag. Exp. Station, College Station, Bulletin 1555.

Perdew GH, Whitelaw ML (1991) Evidence that the 90-kDa heat shock protein (hsp90) exists in cytool in heteromeric complexes containing hsp70 and three other proteins with Mr 63,000, 56,000 and 50,000. J Biol Chem 266:6708-6713.

Perlmann T, Eriksson P, Wrange Ö (1990) Quantitative analysis of the glucocorticoid receptor DNA interaction at the mouse mammary tumor virus glucocorticoid response element. J Biol Chem 265:17222-17229.

Perrot-Applanat M, Logeat F, Groyer-Picard MT, Milgrom E (1985) Immunocytochemical study of mammalian progesterone receptor using monoclonal antibodies. Endocrinology 116:1473-1484.

Persson B, Flinta C, von Heijne G, Jörnvall H (1985) Eur J Biochem 152:523-527.

Philibert D, Deraedt R, Teutsch G (1981) RU38486: a potent antiglucocorticoid in vitro. (Abstr.) 8th International congress of pharmacology 14631.

Picard D, Yamamoto KR (1987) Two signals mediate hormone-dependent nuclear localization of the glucocorticoid receptor. EMBO J 6:3333-3340.

Picard D, Salser SJ, Yamamoto KR (1988) A movable and regulable inactivation function within the steroid binding domain of the glucocorticoid receptor. Cell 54:1073-1080.

Picard D, Kumar V, Chambon P, Yamamoto KR (1990a) Signal transduction by steroid hormones: nuclear localization is differentially regulated in estrogen and glucocorticoid receptors. Cell Regul 1:291-299.

Picard D, Khursheed B, Garabedian MJ, Fortin MG, Lindquist S, Yamamoto KR (1990b) Reduced levels of hsp90 compromise steroid receptor action *in vivo*. Nature 348:166-168.

Poellinger L, Göttlicher M, Gustafsson J-Å (1992) The dioxin and peroxisome proliferator-activated receptors: nuclear receptors in search of endogenous ligands. Trends Pharmacol Sci 13:241-245.

Pongratz I, Strömstedt P-E, Mason GGF, Poellinger L (1991) Inhibition of the specific DNA binding activity of the dioxin receptor by phosphatase treatment. J Biol Chem 266:16813-16817.

Ponta H, Kennedy N, Skroch P, Hynes NE, Groner B (1985) Hormonal response region in the mouse mammary tumor virus long terminal repeat can be dissociated from the proviral promoter and has enhancer properties. Proc Natl Acad Sci USA 82:1020-1024.

Power RF, Conneely OM, McDonnell DP, Clark JH, Butt TR, Schrader WT, O'Malley BW (1990) High level expression of a truncated chicken progesterone receptor in Escherichia coli. J Biol Chem 265:1419-1424.

Power RF, Lydon JP, Conneely OM, O'Malley BW (1991a) Dopamine activation of an orphan of the steroid receptor superfamily. Science 252:1546-1548.

Power RF, Mani SK, Codina J, Conneely OM, O'Malley BW (1991b) Dopaminergic and ligand-independent activation of steroid hormone receptors. Science 254:1636-1639.

Power RF, Conneely OM, O'Malley BW (1992) New insights into activation of the steroid hormone receptor superfamily. Trends Pharmacol Sci 13:318-323.

Pratt WB, Jolly DJ, Pratt DV, Hollenberg SM, Giguère V, Cadepond FM, Schweizer-Groyer G, Catelli M-G, Evans R, Baulieu E-E (1988) A region in the steroid binding domain determines formation of the non-DNA binding, 9S glucocorticoid receptor complex. J Biol Chem 263:267-273.

Pratt WB (1990) Interaction of hsp 90 with steroid receptors: organizing some diverse observations and presenting the newest concepts. Mol Cell Endocrinol 74:C69-C76.

Pratt WB, Hutchison KA, Scherrer LC (1992a) Steroid receptor folding by heat-shock proteins and composition of the receptor heterocomplex. Trends Endocrinol Metab 3:326-333.

Pratt WB, Scherrer LC, Hutchison KA, Dalman FC (1992b) A model of glucocorticoid receptor unfolding and stabilization by a heat shock complex. J Steroid Biochem Molec Biol 41:223-229.

Putlitz J, Datta S, Madison LD, Macchia E, Jameson JL (1991) Human thyroid hormone b1 receptor produced by recombinant baculovirus-infected insect cells. Biochem Biophys Res Commun 175:285-290.

Raaka BM, Finnerty M, Sun E, Samuels HH (1985) Effects of molybdate on steroid receptors in intact GH1 cells. Evidence for dissociation of an intracellular 10 S receptor oligomer prior to nuclear accumulation. J Biol Chem 260:14009-14015.

Rafestin-Oblin M-E, Couette B, Radanyi C, Lombes M, Baulieu E-E (1989) Mineralocorticosteroid receptor of the chick intestine: oligomeric structure and transformation. J Biol Chem 264:9304-9309.

Rao GS (1981) Mode of entry of steroid and thyroid hormones into cells. Mol Cell Endocrinol 21:97-108.

Rao GS, Lemoch H, Cronrath C (1987) Interaction of calmodulin with iodothyronines: effect of iodothyronines on the calmodulin activation of cyclic AMP phosphodiestrase. Mol Cell Endocrinol 53:45-52.

Redeuilh G, Moncharmont B, Secco C, Baulieu E-E (1987) Subunit composition of the molybdate-stabilized "8-9S" nontransformed estradiol receptor purified from calf uterus. J Biol Chem 262:6969-6975.

Renkawitz R, Schütz G, von der Ahe D, Beato M (1984) Sequences in the promoter region of the chicken lysozyme gene required for steroid regulation and receptor binding. Cell 37:503-510.

Renoir J-M, Buchou T, Baulieu E-E (1986) Involvement of a non-hormone-binding 90-kilodalton protein in the nontransformed 8S form of the rabbit uterus progesterone receptor. Biochemistry 25:6405-6413.

Reshef L, Shapiro B (1961) Effect of epinephrine, cortisone and growth hormone on release of unestrified fatty acids by adipose tissue *in vitro*. Metabolism 9:551-555.

Rexin M, Busch W, Gehring U (1991) Protein components of the nonactivated glucocorticoid receptor. J Biol Chem 266:24601-24605.

Richard-Foy H, Hager GL (1987) Sequence-specific positioning of nucleosomes over the steroid-inducible MMTV promoter. EMBO J 6:2321-2328.

Rosner W (1990) The functions of corticosteroid-binding globulin and sex hormone-binding globulin: Recent advances. Endocrine Rev 11:80-91.

Ross TK, Prahl JM, DeLuca HF (1991) Overproduction of rat 1,25-dihydroxyvitamin D3 receptor in insect cells using the baculovirus system. Proc Natl Acad Sci USA 88:6555-6559.

Rusconi S, Yamamoto KR (1987) Functional dissection of the hormone and DNA binding activities of the glucocorticoid receptor. EMBO J 6:1309-1315.

Sanchez ER, Meshinchi S, Tienrungroj W, Schlesinger MJ, Toft DO, Pratt WB (1987) Relationship of the 90-kDa heat shock protein to the untransformed and transformed states of the L cell glucocorticoid receptor. J Biol Chem 262:6986-6991.

Sanchez ER (1990) Hsp56: A novel heat shock protein associated with untransformed steroid receptor complexes. J Biol Chem 265:22067-22070.

Sanchez ER, Hirst M, Scherrer LC, Tang H-Y, Welsh MJ, Harmon JM, Simons SS, Ringold GM, Pratt WB (1990a) Hormone-free mouse glucocorticoid receptors overexpressed in chinese hamster ovary cells are localized to the nucleus and are associated with both hsp 70 and hsp 90. J Biol Chem 265:20123-20130.

Sanchez ER, Faber LE, Henzel WJ, Pratt WB (1990b) The 56-59 kilodalton protein identified in untransformed steroid receptor complexes is a unique protein that exists in cytosol in a complex with both the 70- and 90-kilodalton heat shock proteins. Biochemistry 29:5145-5152.

Scheidereit C, Beato M (1984) Contacts between hormone receptor and DNA double helix within a glucocorticoid regulatory element of mouse mammary tumour virus. Proc Natl Acad Sci USA 81:3029-3033.

Schena M, Yamamoto KR (1988) Mammalian glucocorticoid receptor derivatives enhance transcription in yeast. Science 241:965-967.

Scherrer LC, Dalman FC, Massa E, Meshinchi S, Pratt WB (1990) J Biol Chem 265:20123-20130.

Scherrer LC, Picard D, Massa E, Harmon JM, Simons Jr SS, Yamamoto KR, Pratt WB (1993) Evidence that the hormone binding domain of steroid receptors confers hormonal control on chimeric proteins by determining their hormone-regulated binding to heat-shock protein 90. Biochemistry 32:5381-5386.

Schüle R, Rangarajan P, Kliewer S, Ransone LJ, Bolado J, Yang N, Verma IM, Evans RM (1990) Functional antagonism between oncoprotein c-Jun and the glucocorticoid receptor. Cell 62:1217-1226.

Schwabe JWR, Neuhaus D, Rhodes D (1990) Solution structure of the DNA-binding domain of the oestrogen receptor. Nature 348: 458-461.

Segnitz B, Gehring U (1990) Mechanism of action of a steroidal antiglucocorticoid in lymphoid cells. J Biol Chem 265:2789-2796.

Segraves WA (1991) Something old, some things new: The steroid receptor superfamily in *Drosophila*. Cell 67:225-228.

Severne Y, Wieland S, Schaffner W, Rusconi S (1988) Metal binding "finger" structure in the glucocorticoid receptor defined by site-directed mutagenesis. EMBO J 7:2503-2508.

Sheridan P, Evans R M, Horwitz KB (1989) Phosphotryptic peptide analysis of human progesterone receptors- new phosphorylation sites formed in nuclei after hormone treatment. J Biol Chem 264:6520-6528.

Simons Jr SS, Thompson EB (1981) Dexamethasone 21-mesylate: an affinity label of glucocorticoid receptors from rat hepatoma tissue culture cells. Proc Natl Acad Sci USA 78:3541-3545.

Simons Jr SS, Schleenbaker RE, Eisen HJ (1983) Activation of covalent affinity labeled glucocorticoid receptor-steroid complexes. J Biol Chem 258:2229-2238.

Simons Jr SS (1987) Selective covalent labeling of cysteines in bovine serum albumin and hepatoma tissue culture cell glucocorticoid receptors by dexamethsone 21-mesylate. J Biol Chem 262:9669-9675.

Simons Jr SS, Pumphrey JG, Rudikoff S, Eisen HJ (1987) Identification of cysteine 656 as the amino acid of hepatoma tissue culture cell glucocorticoid receptors that is covalently labeled by dexamethasone 21-mesylate. J Biol Chem 262:9676-9680.

Simons Jr SS, Sistare FD, Chakraborti PK (1989) Steroid binding activity is retained in a 16-kDa fragment of the steroid binding domain of rat glucocorticoid receptors. J Biol Chem 264:14493-14497.

Simons Jr SS, Chakraborti PK, Cavanaugh AH (1990) Arsenite and cadmium(II) as probes of glucocorticoid receptor structure and function. J Biol Chem 265:1938-1945.

Simpson ER, Waterman MR (1988) Regulation of the synthesis of steroidogenic enzymes in adrenal cortical cells by ACTH. Ann Rev Physiol 50:427-440.

Slater EP, Rabenau O, Karin M, Baxter JD, Beato M (1985) Glucocorticoid receptor binding and activation of a heterologous promoter in response to dexamethasone by the first intron of the human growth hormone gene. Mol Cell Biol 5:2984-2992.
Smith AC, Harmon JM (1985) Multiple forms of the glucocorticoid receptor steroid binding protein identified by affinity labeling and high-reolution two-domensional electrophoresis. Biochemistry 24:4946-4951.
Smith DF, Schowalter DB, Kost SL, Toft DO (1990) Reconstitution of progesterone receptor with heat shock proteins. Mol Endocrinol 4:1704-1711.
Smith DF, Toft DO (1993) Steroid receptors and their associated proteins. Molec Endocrinol 7:4-11.
Smith, L.I., Bodwell, J.E., Mendel, D.B., Ciardelli, T., North, W.G., Munck, A. 1988 Identification of cysteine-644 as the covalent site of attachment of dexamethasone 21-mesylate to murine glucocorticoid receptor in WEHI-7 cells. Biochemistry 27:3747-3753.
Smith LI, Mendel D.B, Bodwell JE, Munck A (1989) Phosphorylated sites within the functional domains of the ~100-kDa steroid-binding subunit of glucocorticoid receptors. Biochemistry 28:4490-4498.
Smith CL, Conneely OM, O'Malley BW (1993) Modulation of the ligand-independent activation of the human estrogen receptor bu hormone and antihormone. Proc Natl Acad Sci USA 90:6120-6124.
Sone T, McDonnell DP, O'Malley BW, Pike JW (1990) Expression of human vitamin D receptor in Saccharomyces cerevisiae. purification, properties and generation of polyclonal antibodies. J Biol Chem 265:21997-22003.
Srinivasan G, Thompson EB (1990) Overexpression of full-length human glucocorticoid receptor in *Spodoptera frugiperda* cells using Baculovirus expression vector system. Mol Endocrinol 4:209-216.
Srinivasan G (1992) Overexpression of receptors of the steroid/thyroid family. Mol Endocrinol 6:857-860.
Stancato LF, Hutchison KA, Chakraborti PK, Simons Jr SS, Pratt WB (1993) Differential effects of the reversible thiol-reactive agents aresenite and methyl methanethiosulfonate on steroid binding by the glucocorticoid receptor. Biochemistry 32:3729-3736.
Strähle U, Boshart M, Klock G, Stewart F, Schütz G (1989) Glucocorticoid- and progesterone-specific effects are determined by differential expression of the respective hormone receptors. Nature 339:629-632.
Strömstedt P-E, Berkenstam, A, Jörnvall H, Gustafsson J-Å, Carlstedt-Duke J (1990) Radiosequence analysis of the human progestin receptor charged with [^{3}H]promegestone: A comparison with the glucocorticoid receptor. J Biol Chem 265:12973-12977.
Strömstedt P-E, Poellinger L, Gustafsson J-Å, Carlstedt-Duke J (1991) The glucocorticoid receptor binds to a sequence overlapping the TATA box of the human osteocalcin promoter: a potential mechanism for negative regulation. Mol Cell Biol 11:3379-3383.
Strömstedt P-E, Barkhem T, Carlstedt-Duke J, Nilsson S, and Gustafsson J-Å (1993) Semliki forest virus replicon-mediated expression of the human glucocorticoid receptor. functional characterization of full length and truncated receptor derivatives. (submitted)
Sullivan WP, Madden BJ, McCormick DJ, Toft DO (1988) Hormone-dependent phosphorylation of the avian progesterone receptor. J Biol Chem 263:14717-14723.
Summers MD, Smith GE (1985) Genetic engineering of the genome of the Autographa Californica nuclear Polyhydrosis virus. In: Genetically altered viruses and the environment. Cold spring harbor laboratory report 22 (eds. Fields B, Martin M.A, Kamely D) 319-351.

Superti-Furga G, Bergers G, Picard D, Busslinger, M. 1991 Hormone-dependent transcriptional regulation and cellular transformation by Fos-steroid receptor fusion proteins. Proc Natl Acad Sci USA 88:5114-5118.

Tai PK, Alberts MW, Chang H, Faber LE, Schreiber SL (1992) Association of a 59-kilodalton immunophilin with glucocorticooid receptor complex. Science 256:1315-1318.

Tanaka H, Dong Y, Li Q, Okret S, Gustafsson J-Å (1991) Identification and characterization of a cis-acting element that interferes with glucocorticoid-inducible activation of the mouse mammary tumor virus promoter. Proc Natl Acad Sci USA 88:5393-5397.

Tashima Y, Terui M, Itoh H, Mizunuma H, Kobayashi,R, Marumo F (1989) Effect of selenite on glucocorticoid receptor. J Biochem 105:358-361.

Tasset D, Tora L, Fromental C, Scheer E, Chambon P (1990) Distinct classes of transcriptional activating domains function by different mechanisms. Cell 62:1177-1187.

Tienrungroy W, Sanchez ER, Housley PR, Harrison RW, Pratt WB (1987) Glucocorticoid receptor phosphorylation, transformation, and DNA binding. J Biol Chem 262:17342-17349.

Tilley WD, Marcelli M, Wilson JD, McPhaul MJ (1989) Characterization and expression of a cDNA encoding th human androgen receptor. Proc Natl Acad Sci USA 86:327-331.

Tora L, Gronemeyer H, Turcotte B, Gaub M-P, Chambon P (1988) The N-terminal region of the chicken progesterone receptor specifies target gene activation. Nature 333:185-188.

Truss M, Chalepakis G, Beato M (1990) Contacts between steroid hormone receptors and thymines in DNA: An interference method. Proc Natl Acad Sci USA 87:7180-7184.

Tsai SY, Carlstedt-Duke J, Weigel NL, Dahlman K, Gustafsson J-Å, Tsai M-J, O'Malley BW (1988) Molecular interactions of steroid hormone receptors with its enhancer element: Evidence for receptor dimer formation. Cell 55:361-369.

Tsai SY, Tsai M-J, O'Malley BW (1989b) Cooperative binding of steroid hormone receptors contributes to transcriptional synergism at target enhancer elements. Cell 57:443-448.

Turnamian SG, Binder HJ (1989) Regulation of active sodium and potassium transport in the distal colon of the rat. Role of the aldosterone and glucocorticoid receptors. J Clin Invest 84:1924-29.

Umek R.M, Friedman AD, McKnight SL (1991) CCAAT-enhancer binding protein:a component of a differentiation switch. Science 251:288-292.

Umesono K, Murakami KK, Thompson CC, Evans RM (1991) Direct repeats as selective response elements for the thyroid hormone, retinoic acid, and vitamin D3 receptors.Cell 65:1255-1266.

Umesono K, Evans RM (1989) Determinants of target gene specificity for steroid/thyroid hormone receptors. Cell 57:1139-1146.

Vedeckis WV (1983) Limites proteolysis of the mouse liver glucocorticoid receptor. Biochemistry 22:1975-1983.

Vegeto E, Allan GF, Schrader WT, Tsai M-J, McDonnell DP, O'Malley BW (1992) The mechanism of RU486 antagonism is dependent on the conformation of the carboxy-terminal tail of the human progesterone receptor. Cell 69:703-713.

von der Ahe D, Janich S, Scheidereit C, Renkawitz R, Schütz G, Beato M (1985) Glucocorticoid and progesterone receptors bind to the same sites in two hormonally regulated promoters. Nature 313:706-709.

Wahli W, Martinez E (1991) Superfamily of steroid nuclear receptors: positive and negative regulators of gene expression. FASEB J 5:2243-2249.

Webster NJG, Green S, Rui Jin J, Chambon P (1988) The hormone-binding domains of the estrogen and glucocorticoid receptors contain an inducible transcription activation function. Cell 54:199-207.

Weigel NL, Beck CA, Estes PA, Prendergast P, Altmann M, Christensen K, Edwards DP (1992a) Ligands induce conformational changes in the carboxyl-terminus of progesterone receptors which are detected by a site-directed antipeptide monoclonal antibody. Mol Endocrinol 6:1585-1597.

Welshons WV, Krummel BM, Gorski J (1985) Nuclear localization of unoccupied receptors for glucocorticoids, estrogens, and progesterone in GH_3 cells. Endocrinology 117:2140-2147.

White R, Fawell SE, Parker MG (1991) Analysis of oestrogen receptor dimerisation using chimeric receptors. J Steroid Biochem Mol Biol 40:333-341.

Wiech H, Buchner J, Zimmermann R, Jakob U (1992) Hsp90 chaperones protein folding *in vitro*. Nature 358:169-170.

Wieland S, Döbbeling U, Rusconi S (1991) Interference and synergism of glucocorticoid receptor and octamer factors. EMBO J 10:2513-2521.

Wikström A-C, Bakke O, Okret S, Brönnegård M, Gustafsson J-Å (1987) Intracellular localization of the glucocorticoid receptor: evidence for cytoplasmic and nuclear localization. Endocrinology 120:1232-1242.

Wilson TE, Mouw AR, Weaver CA, Milbrandt J, Parker KL (1993) The orphan receptor NGFI-B regulates expression of the gene encoding steroid 21-hydroxylase. Mol Cell Biol 13:861-868.

Wittliff JL, Wenz LL, Dong J, Nawaz Z, Butt TR (1990) Expression and characterization of an active human estrogen receptor as a ubiquitin fusion protein from *Escherichia coli*. J Biol Chem 265:22016-22022.

Wong M-M, Rao LG, Ly H, Hamilton L, Tong J, Sturtridge W, McBroom R, Aubin JE, Murray TM (1990) Long-term effects of physiologic concentrations of dexamethasone on human bone-derived cells. J Bone Min Res 5:803-813.

Wrange Ö, Gustafsson J-Å (1978) Separation of the hormone- and DNA-binding sites of the hepatic glucocorticoid receptor by means of proteolysis. J Biol Chem 253:856-865.

Wrange Ö, Okret S, Radojcic M, Carlstedt-Duke J, Gustafsson J-Å (1984) Characterization of the purified activated glucocorticoid receptor from rat liver cytosol. J Biol Chem 259:4534-4541.

Wrange Ö, Eriksson P, Perlmann T (1989) The purified activated glucocorticoid receptor is a homodimer. J Biol Chem 264:5253-5259.

Wright APH, Carlstedt-Duke J, Gustafsson J-Å (1990) Ligand-specific transactivation of gene expression by a derivative of the human glucocorticoid receptor expressed in yeast. J Biol Chem 265:14763-14769.

Wright APH, Gustafsson J-Å (1991) Mechanism of synergistic transcriptional transactivation by the human glucocorticoid receptor. Proc Natl Acad Sci USA 88:8283-8287.

Wright APH, McEwan IJ, Dahlman-Wright K, Gustafsson J-Å (1991) High level expression of the major transactivation domain of the human glucocorticoid receptor in yeast cells inhibits endogenous gene expression and cell growth. Mol Endocrinol 5:1366-1372.

Wright APH, Gustafsson J-Å (1992) Glucocorticoid-specific gene activation by the intact human glucocorticoid receptor expressed in yeast. glucocorticoid specificity depends on low level receptor expression. J Biol Chem 267:11191-11195.

Xie Y-B, Sui Y-P, Shan L-X, Palvimo JJ, Phillips DM, Jänne OA (1992) Expression of androgen receptor in insect cells: purification of the receptor and renaturation of its steroid- and DNA-binding functions. J Biol Chem 267, 4939-4948.

Yamamoto KR (1985) Steroid receptor regulated transcription of specific genes and gene networks. Ann Rev Genet 19:209-252.

Yamamoto KR, Alberts BM (1976) Steroid receptors: elements for modulation of eukaryotic transcription. Ann Rev Biochem 45:721-746.

Yem AW, Tomasselli AG, Heinrikson L, Zurcher-Neely H, Ruff VA, Johnson RA, Deibel MR J (1992) The Hsp56 component of steroid receptor complexes binds to immobilized FK506 and shows homology to FKBP-12 and FKBP-13. J Biol Chem 267:2868-2871.

Yen PM, Simons Jr SS (1991)Evidence against posttranslational glycosylation of rat glucocorticoid receptors. Receptor 1:191-205.

Ylikomi T, Bocquel MT, Berry M, Gronemeyer H, Chambon P (1992) Cooperation of proto-signals for nuclear accumulation of estrogen and progesterone receptors. EMBO J 11:3681-3694.

Young CY-F, Qiu S, Prescott JL, Tindall DJ (1990) Overexpression of partial human androgen receptor in E. coli: characterization of steroid binding, and immunological properties. Mol Endocrinol 4:1841-1849.

Yu VC, Delsert C, Andersen B, Holloway JM, Devary OX, Näär AM, Kim SY, Boutin JM, Glass EK, Rosenfeld MG (1991) RXRb: acoregulator that enhances binding of retinoic acid, thyroid hormone, and vitamin D receptors to their cognate response elements. Cell 67:1251-1266.

Zaret KS, Yamamoto KR (1984) Reversible and persistent changes in chromatin structure accompany activation of a glucocorticoid-dependent enhancer element. Cell 38:29-38.

Zhang X-K, Lehmann J, Hoffman B, Dawson MI, Cameron J, Graupner G, Hermann T, Tran P, Pfahl M (1992) Retinoid X receptor is an auxiliary protein for thyroid hormone and retinoic acid receptors. Nature 355:441-446.

ADIPOCYTE DIFFERENTIATION IS DEPENDENT ON THE INDUCTION OF THE ACYL-COA BINDING PROTEIN

Trausti Baldursson*, Connie Gram°, Jens Knudsen°, Karsten Kristiansen* and Susanne Mandrup*
Department of Molecular Biology* and Institute of Biochemistry°
University of Odense
Campusvej 55
DK-5230 Odense M
Denmark

INTRODUCTION

The 3T3-L1 cell line was selected from an established Swiss mouse 3T3 cell lines by virtue of its ability to differentiate into cells possessing the morphological and biochemical characteristics of adipose cells (Green and Kehinde, 1974). Preconfluent growing 3T3-L1 cells maintain a fibroblastic phenotype, but upon confluence treatment with an appropriate combination of adipogenic factors induces a highly synchronous adipose conversion (reviewed in Cornelius *et al.*, 1994). Several lines of evidence have indicated that the transcription factor CCAAT/enhancer binding protein α (C/EBPα) plays a decisive role in the differentiation process (Samuelsson *et al.*, 1991; Lin and Lane, 1992; Lin and Lane, 1994; Freytag *et al.*, 1994).
Four different interdependent signal transduction pathways have been implicated in adipose conversion of 3T3-L1 adipoblasts. One involves the IGF-1 receptor or, at unphysiologically high concentrations of insulin, the insulin receptor. In either case, tyrosine phosphorylation of IRS-1 or other tyrosine phosphorylated docking proteins leads via SH-2/SH-3 adaptor proteins to activation of downstream signalling pathways. The second pathway involves a glucocorticoid/prostaglandin dependent pathway. This pathway is via PGI_2 and cAMP linked to the third PKA dependent pathway. Finally, recent experiments have indicated a fourth important signalling pathway depending on fatty acids or their metabolites (Amri *et al.*, 1994; Safonova *et al.*, 1994; Chawla and Lazar, 1994; Tontonoz *et al.*, 1994; Ibrahim *et al.*, 1995). This signalling pathway ultimately induces genes responsive to members of the peroxisome proliferator activated receptor (PPAR) subfamily of the nuclear hormone receptor superfamily (Safonova *et al.*, 1994;

NATO ASI Series, Vol. H 92
Signalling Mechanisms – from Transcription Factors to Oxidative Stress
Edited by L. Packer, K. Wirtz

Chawla and Lazar, 1994; Tontonoz *et al.*, 1994; Ibrahim *et al.*, 1995). PPARs function as ligand activated transcription factors that require heterodimerization with members of the retinoid X receptor subfamily in order to bind to peroxisome proliferator responsive elements (PPREs) and activate transcription of responsive genes (Gearing *et al.*, 1993; Keller *et al.*, 1993). *In vivo* PPARs are activated by a surprisingly large variety of substances including certain hypolipidaemic drugs, plasticizers, herbicides, and naturally occurring fatty acids (Issemann and Green, 1991; Göttlicher *et al.*, 1992; Dreyer *et al.*, 1992). It has been speculated that the effects of these substances converge on a perturbation of the intracellular fatty acid metabolism leading to PPAR activation via interaction with free fatty acid themselves (Hertz *et al.*, 1994), their dicarboxylic derivatives (Kaikaus *et al.*, 1993) or their CoA esters (Tomaszewski and Melnick, 1994). Indeed, we have recently demonstrated that rat PPARα can be specifically labelled *in vitro* by a photoreactive fatty acid (Albrektsen *et al.*, 1994), and that free fatty acid and their CoA esters influence the binding of rat PPARα/ultraspiracle like heterodimers to PPREs (Elholm *et al.*, in preparation). Intracellular transport of fatty acids and acyl-CoA esters may depend on lipid carrier proteins like fatty acid binding proteins and the acyl-CoA binding protein (ACBP) (Bass, 1993; Knudsen *et al.*, 1993). Synthesis of ACBP is induced during adipose conversion of 3T3-L1 cells (Hansen *et al.*, 1991). Consequently, it was of interest to examine whether induction of ACBP synthesis was necessary for parts of the differentiation program, and to what extent ectopic expression of ACBP would affect the phenotype of 3T3-L1 cells.

RESULTS

Construction of expression vectors for high level expression of ACBP cDNA in the sense and the antisense orientation.

Three different cloning vectors were used. (1) The pMAMneo vector (Clontech) in which the transcription of the foreign gene is driven by the RSV enhancer and the glucocorticoid inducible MMTV LTR/promoter. (2) The pUBI vector which was constructed by deleting the *Eco*RI site at position 5780 in pMAMneo and replacing the *Eco*RI-*Nhe*I fragment containing the RSV enhancer and the MMTV LTR/promoter by the *Eco*RI-*Nhe*I PCR fragment containing the human ubiquitin C promoter (position -1464 to -15) (Wiborg *et al.*, 1985). (3) The pMT vector which was constructed by deleting the *Eco*RI site at position 5780 in pMAMneo and replacing the *Eco*RI-*Nhe*I fragment containing the RSV enhancer and the MMTV LTR/promoter by an *Eco*RI-*Nhe*I PCR fragment containing part of the human metallothionein promoter (position -209 to +28, Karin *et al.*, 1987). This fragment contains the elements responsible for basic transcriptional regulation and regulation by heavy metal ions but not the glucocorticoid responsive element of the promoter.

The *Xba*I-*Kpn*I fragment from pRLcACBP (Hansen *et al.*, 1991) containing rat ACBP cDNA was inserted in M13mp19 to generate M13mp19-E2b. The ACBP cDNA fragment of pRLcACBP lacks the first two first nucleotides of the ATG start codon. Site directed mutagenesis was employed to generate an ATG start codon by deletion of the TCCG sequence preceding the cDNA sequence in M13mp19-E2b. The sequence around the start codon in the resulting plasmid M13mp19-E2bdel was 5'-GCAGCAGGA**ATG**TCT-3' which is compatible with efficient translation initiation (Kozak, 1989). The 312 bp *Xba*I-*Bal*I fragment (including the entire coding region of the ACBP gene, 17 bp 3' trailer with no polyadenylation signal and 31 bp polylinker from pBluescript) of M13mp19-E2bdel was cloned between the *Nhe*I site and the *Xho*I site of the polylinker in each of the cloning vectors described above. This generated pMMTV-rACBP, pUBI-rACBP and pMT-rACBP, respectively.

The 504 bp *Eco*RI fragment of pGEM-3/cDBI (containing the entire murine ACBP coding region, 39 bp 5' leader, 180 bp 3' trailer and 21 bp poly(A) tail (DBI-7, Owens et al., 1989)) was cloned in pBluescript SK+ (Stratagene) to make pBls.SK+/cDBI. The *Xba*I-*Xho*I fragment from this clone was inserted between the *Nhe*I site and the *Xho*I site of the cloning vectors described above to generate pMMTV-ASmACBP, pUBI-ASmACBP and pMT-ASmACBP, respectively.

Isolation of stably transfected clones.

The cell culture conditions and the differentiation protocol were as described previously (Hansen et al., 1991). Briefly, cells were cultured in Dulbecco's modified Eagle's medium (DMEM)/F12 (1:1, v/v) containing 10% (v/v) calf serum, 2.5 μg/ml amphotericin and 100 μg/ml gentamycin in a humidified atmosphere of 6.4% CO_2 at 37°C. For differentiation, the cells were grown to confluence, which was designated day 0, the medium was replaced with one containing 10% (v/v) fetal calf serum and the cells were treated for 48 h with 1 μM dexamethasone. After removal of the dexamethasone, 10 μg/ml insulin was added to half of the cells for the rest of the culture period.

For stable transfections, exponentially growing cells were trypsinized, seeded at 7.5×10^4

cells per 3.5 cm well and incubated for 24 h. The cells were transfected with 3.7 µg vector/well according to the modified calcium phosphate-mediated transfection procedure originally described by Chen and Okayama (Sambrook *et al.*, 1989). Stably transfected clones were selected for 2-3 weeks in growth medium containing 400 µg/ml G418.

PCR analysis and Southern blot analysis of DNA isolated from clones resistant to G418 identified 3 clones which had integrated vectors for expression of rat ACBP cDNA (two stably transfected with pMT-rACBP and one stably transfected with pUBI-rACBP) and 9 clones which had integrated vectors for expression of murine ACBP cDNA in the antisense orientation (one stably transfected with pMMTV-ASmACBP and 8 stably transfected with pUBI-ASmACBP). These clones were propagated for further analysis.

Characterization of stably transfected clones.

In all clones which had been stably transfected with vectors for expression of the rat ACBP cDNA this sequence was actively transcribed as determined by Northern blotting. Compared with untransfected 3T3-L1 cells there was a 2 to 3 fold increase in the level of ACBP at the preadipocyte stage. This increase was most significant in the 3T3-L1/UBI-rACBP clone. In comparison with untransfected 3T3-L1 cells all clones expressing rACBP had an increased tendency to accumulate triglyceride droplets at the preconfluent stage, and the droplets that accumulated during the postconfluent differentiation tended to be larger and more unevenly distributed in the different cells (Fig. 1).

The 3T3-L1/UBI-rACBP clone was selected for further analysis. The ACBP level at the preconfluent stage was increased 3 fold compared with the level in untransfected 3T3-L1 cells. After induction of the endogenous ACBP gene at day 5, the level of ACBP in the untransfected cells approached that of the transfected cells. Thus, in the period from day 5 to 9 the level of ACBP in the insulin treated 3T3-L1/UBI-rACBP clone was only 10-35% above that of the control 3T3-L1 cells. Densitometric scanning of the Northern blots revealed that the ratio of endogenous ACBP mRNA to actin mRNA was increased in the 3T3-L1/UBI-rACBP clone relative to untransfected 3T3-L1 cells prior to the addition of insulin (i.e. day 0 to 2). This indicates that the increased level of ACBP at the preadipocyte stage in the 3T3-L1/UBI-rACBP clone was due to an increased expression from the endogenous ACBP gene as well as expression from the exogenous ACBP gene. Following addition of insulin at day 2 there was no significant difference in the ratio of endogenous ACBP mRNA to actin mRNA in 3T3-L1/UBI-rACBP cells as compared to 3T3-L1 cells. However, in cells maintained in medium without insulin the ratio of endogenous

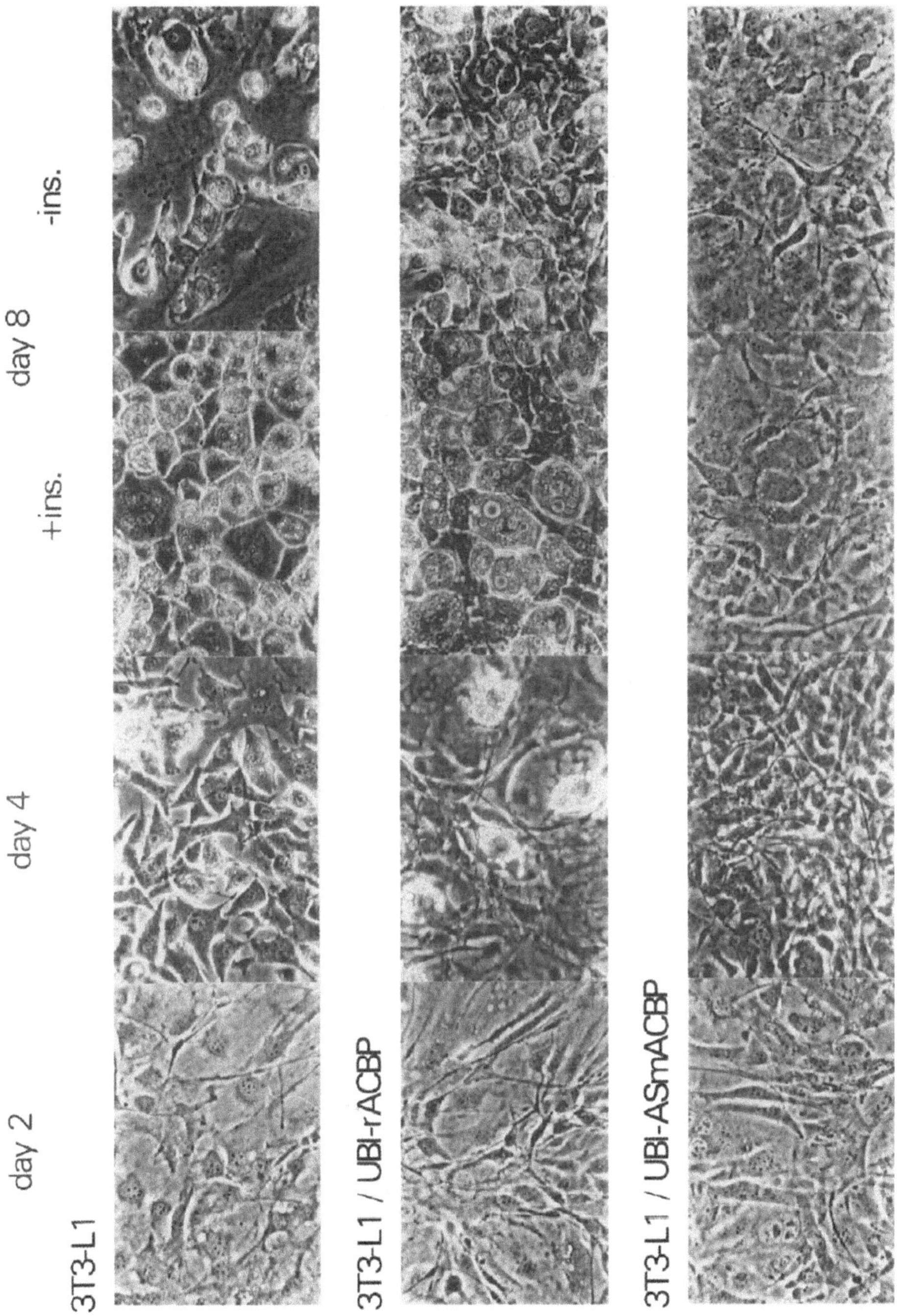
day 2
day 4
day 8
+ins.
-ins.
3T3-L1
3T3-L1 / UBI-rACBP
3T3-L1 / UBI-ASmACBP

Figure 1 (previous page): Effect on 3T3-L1 cell morphology of an increased and a reduced intracellular ACBP level.
The ability to develop a differentiated morphology and accumulate triglycerides were compared for untransfected 3T3-L1 cells and cells stably transfected with pUBI-rACBP and pUBI-ASmACBP, respectively. Days refer to days after confluence at which time the treatment with differentiation inducers was initiated.

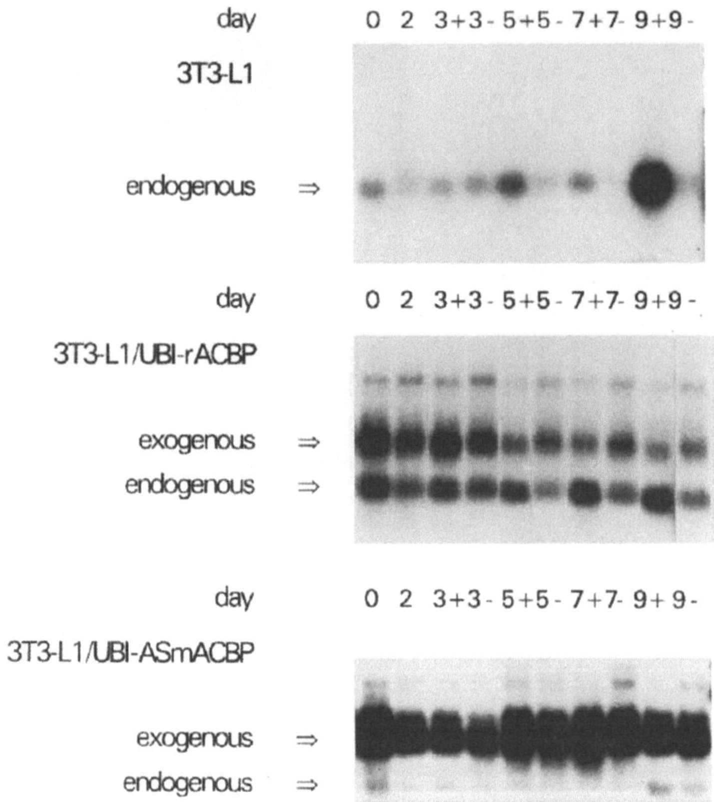

Figure 2: Expression of ACBP transcripts in untransfected 3T3-L1 cells and cells stably transfected with pUBI-rACBP and pUBI-ASmACBP.
mRNA from the 3 cell types was purified at the days indicated using the Pharmacia QuickPrep Micro mRNA Purification Kit, and 0.5 µg RNA from each sample was used for analysis by Northern blotting (Sambrook *et al.*, 1989). The probe used for hybridization was the 458 bp *Eco*RI fragment of the rat ACBP cDNA clone pRLcACBP-E2b (Hansen *et al.*,1991). + = + insulin

ACBP mRNA to actin mRNA was significantly higher in 3T3-L1/UBI-rACBP cells than in 3T3-L1 cells. This indicates that ectopic expression of ACBP stimulates the expression from the endogenous gene in the absence of insulin.

In the clones stably transfected with vectors for expression of murine ACBP cDNA in the antisense orientation there was a reduced level of ACBP at confluence compared to the level found in untransfected 3T3-L1 cells. The ACBP level was reduced ~40% in the 3T3-L1/UBI-ASmACBP clones and ~20% in the 3T3-L1/MMTV-ASmACBP clone. In 6 of the 8 3T3-L1/UBI-ASmACBP clones and in the 3T3-L1/MMTV-ASmACBP clone there was no induction of ACBP upon addition of differentiation inducers, and none of these 7 clones was able to accumulate triglycerides to any significant degree (Fig. 1). Under our culture conditions the fatty acids used for triglyceride synthesis are mainly *de novo* synthesized indicating that ACBP is necessary for triglyceride synthesis based on *de novo* fatty acid synthesis.

One 3T3-L1/UBI-ASmACBP clone with a low level of ACBP was selected for further analysis. In comparison with control 3T3-L1 cells, the level of ACBP was markedly decreased and was not significantly affected by the addition of insulin (data to be published elsewhere). The endogenous ACBP transcript is barely detectable on the Northern blot, whereas the antisense transcript remained at a high level throughout the period.

The enzyme glycerol-3-phosphate dehydrogenase (GPDH) is induced late in adipocyte differentiation, and this induction is independent of a simultaneous accumulation of triglycerides (Kuri-Harcuch *et al.*, 1978) making GPDH an appropriate marker for terminal differentiation *per se*. Northern blot analysis and enzymatic assays showed that there was a complete absence of the GPDH transcript and enzymatic activity in the clone expressing high levels of antisense transcript indicating that an increased level of ACBP is necessary not only for lipid accumulation based on *de novo* fatty acid synthesis but also for terminal differentiation. The ratios of GPDH mRNA to actin mRNA in 3T3-L1 and 3T3-L1/UBI-rACBP cells was compared using densitometric scanning of Northern blots. In keeping with preliminary enzymatic assays showing a decreased GPDH activity in the 3T3-L1/UBI-rACBP clone, the GPDH/ACBP mRNA ratio was 2-4 fold lower in the 3T3-L1/UBI-rACBP clone than in untransfected 3T3-L1 cells.

DISCUSSION

The results presented in this report indicate that ACBP plays an important role in adipocyte differentiation. The two main results indicating that induction of ACBP is necessary for triglyceride synthesis based on *de novo* fatty acid synthesis are: (1) In 7 out of 9 clones which had been stably transfected with vectors for high level expression of murine ACBP cDNA in the antisense orientation there was no induction of ACBP and no accumulation of triglyceride droplets following addition of differentiation inducers. The two antisense clones which showed an induction of ACBP after treatment with the differentiation inducers did also accumulate triglyceride droplets. (2) All clones which had been stably transfected with vectors for high level expression of rat ACBP and consequently had an increased level of ACBP at the preadipocyte stage showed an increased tendency to accumulate triglyceride droplets before confluence.

These observations do not prove a causal relation between ACBP and triglyceride accumulation based on *de novo* fatty acid synthesis. However, the fact that none of the antisense expressing clones displayed a phenotype similar to the clones with overexpression of ACBP and *vice versa* strongly indicates that the observed changes in phenotype relative to untransfected 3T3-L1 are the result of an increased and decreased level of ACBP, respectively.

To investigate whether changes in the expression of ACBP affected differentiation *per se* and not only triglyceride accumulation, we selected a few clones for further analyses. Our results indicate that the increased level of ACBP stimulated the expression from the endogenous ACBP gene, suggesting that ectopic expression of ACBP may stimulate the expression of some adipocyte specific proteins in parallel with the stimulation of triglyceride accumulation based on *de novo* fatty acid synthesis. However, the increased level of ACBP did not appear to accelerate the expression of all adipocyte specific genes as the extent of induction of the GPDH gene was reduced in the 3T3-L1/UBI-rACBP clone. In the 3T3-L1/UBI-ASmACBP clone there was no expression of GPDH.

Thus, the present results indicate that abrogation of ACBP expression inhibits terminal differentiation including the ability to increase triglyceride synthesis based on *de novo* fatty acid synthesis. In addition, the results indicate that an increased level of ACBP at the preadipocyte stage accelerates *de novo* fatty acid synthesis based accumulation of triglycerides and possibly also the induction of a subset of adipocyte specific proteins. This suggests that ACBP participates in pathways needed for terminal differentiation of adipocytes. It is conceivable that one such pathway involves activation of PPAR responsive genes.

REFERENCES

Albrektsen, T., Schröder, L., Göttlicher, M., Elholm, M., Gustafsson, J.-Å., Kristiansen, K. and Knudsen, J. (1995) Ligand binding of the peroxisome proliferator activated receptor (PPAR) studied by photoaffinity labelling. Submitted.

Amri, E.-Z., Ailhaud, G. and Grimaldi, P.-A. (1994) Fatty acids as signal transducing molecules of preadipose to adipose cells. J. Lipid Res. 35: 930-937.

Bass, N.M. (1993) Cellular binding proteins for fatty acids and retinoids: Similar or specialized functions? Mol. Cell. Biochem. 123: 191-202.

Chawla, A. and Lazar, M.A. (1994) Peroxisome proliferator and retinoid signalling pathways co-regulate preadipocyte phenotype and survival. Proc. Natl. Acad. Sci. USA 91: 1786-1790.

Cornelius, P., MacDougald, O.A., Lane M.D. (1994) Regulation of adipocyte development. Ann. Rev. Nutr. 14: 99-129.

Dreyer, C., Krey, G., Keller, H., Givel, F., Helftenbein, G. and Wahli, W. (1992) Control of the peroxisomal β-oxidation pathway by a novel family of nuclear hormone receptors. Cell 68: 879-887.

Freytag, S.O., Paielli, D.L. and Gilbert, J.D. (1994) Ectopic expression of the CCAAT/enhancer binding protein α promotes the adipogenic program in a variety of mouse fibroblastic cells. Genes Dev. 8: 1654-1663.

Gearing, K.L., Göttlicher, M., Widmark, E. and Gustafsson, J.-Å. (1993) Interaction of the peroxisome-proliferator-activated receptor and retinoid X receptor. Proc. Natl. Acad. Sci. USA 90: 1440-1444.

Göttlicher, M., Widmark, E., Li, Q. and Gustafsson, J.-Å. (1992) Fatty acids activate a chimera of the clofibric acid-activated receptor and the glucocorticoid receptor. Proc. Natl. Acad. Sci. USA 89: 4653-4657.

Green, H. and Kehinde, O. (1974) Sublines of mouse 3T3-L1 cells that accumulate lipid. Cell 1: 113-116.

Hansen, H.O., Andreasen, P.H., Mandrup, S., Kristiansen, K. and Knudsen, J. (1991) Induction of acyl-CoA binding protein and its mRNA in 3T3-L1 cells by insulin during preadipocyte-to-adipocyte differentiation. Biochem. J. 277: 341-344.

Hertz, R., Berman, I. and Bar-Tana, J. (1994) Transcriptional activation by amphipathic carboxylic peroxisomal proliferators is induced by the free fatty acid rather than the acyl-CoA derivative. Eur. J. Biochem. 221: 611-615.

Ibrahimi, A., Teboul, L., Gaillard, D., Amri, E.-Z., Ailhaud, G., Young, P., Cawthorne, M.A. and Grimaldi, P.A. (1995) Evidence for a common mechanism of action for fatty acids and thiazolidinedione antidiabetic agents on gene expression in preadipose cells. Mol. Pharnacol., in the press.

Issemann, I. and Green, S. (1990) Activation of a member of the steroid hormon receptor superfamily by peroxisome proliferators. Nature 347: 645-650.

Kaikaus, R., Chan, W.K., Lysenko, N., Ray, R., Ortiz de Montellano, P.R. and Bass, N.M. (1993) Induction of peroxisomal fatty acid β-oxidation and liver fatty acid-binding

protein by peroxisome proliferators. Mediation via the cytochrome P-450IVa1 ω-hydroxylase pathway. J. Biol. Chem. 13: 9593-9603.
Karin, M., Haslinger, A., Heguy, A., Dietlin, T. and Cooke, T. (1987) Metal-responsive elements act as positive modulators of human metallothionein-II_A enhancer activity. Mol. Cell. Biol. 7: 606-613.
Keller, H., Dreyer, C., Medin, J., Mahfoudi, A., Ozato, K. and Wahli, W. (1993) Fatty acids and retinoids control lipid metabolism through activation of peroxisome proliferator-activated receptor-retinoid X receptor heterodimers. Proc. Natl. Acad. Sci. USA 90: 2160-2164.
Knudsen, J., Mandrup, S., Rasmussen, J.T., Andreasen P.H., Poulsen, P. and Kristiansen, K. (1993) The function of acyl-CoA binding protein/Diazepam binding inhibitor (DBI). Mol. Cell. Biochem. 123: 129-138.
Kozak, M. (1989) The scanning model for translation: An update. J. Cell Biol. 108: 229-241.
Kuri-Harcuch, W., Wise, L.S. and Green, H. (1978) Interruption of the adipose conversion of 3T3 cells by biotin deficiency: Differentiation without triglyceride accumulation. Cell 14: 53-59.
Lin, F.-T. and Lane, M.D. (1992) Antisense CCAAT/enhancer-binding protein RNA suppresses coordinate gene expression and triglyceride accumulation during differentiation. Genes Dev. 6: 533-544.
Lin, F.-T. and Lane , M.D. (1994) CCAAT/enhancer binding protein α is sufficient to initiate the 3T3-L1 adipocyte differentiation program. Proc. Natl. Acad. Sci. USA 91: 8757-8761.
Owens, G.P., Sinha, A.K., Sikela, J.M. and Hahn, W.E. (1989) Sequence and expression of the murine diazepam binding inhibitor. Mol. Brain Res. 6: 101-108.
Safonova, I., Reichert, U., Shroot, B., Ailhaud, G. and Grimaldi, P. (1994) Fatty acids and retinoids act synergistically on adipose cell differentiation. Biochem. Biophys. Comm. 204: 498-504.
Sambrook, J., Fritsch, E.F. and Maniatis, T. (1989) Molecular cloning. A laboratory manual., 2nd edit. Cold Spring Harbour Laboratory Press, Cold Spring Harbour, N.Y.
Samuelsson, L., Strömberg, K., Vikman, K., Bjursell, G. and Enerbäck, S. (1991) The CCAAT/enhancer binding protein and its role in adipocyte differentiation: Evidence for direct involvement in terminal adipocyte development. EMBO J. 10: 3787-3793.
Tomaszewski, K.E. and Melnick, R.L. (1994) *In vitro* evidence for involvement of acyl-CoA thioesters in peroxisome proliferation and hypolipidaemia. Biochim. Biophys. Acta 1120: 118-124.
Wiborg, O., Pedersen, M.S., Wind, A., Berglund, L.E., Marcker, K.A. and Vuust, J. (1985) The human ubiquitin coding sequences. EMBO J. 4: 775-759.

TNF-induced Mechanisms for IL6 Gene Induction

Guy Haegeman & Walter Fiers
Laboratorium voor Moleculaire Biologie
K.L. Ledeganckstraat 35
9000 Gent
Belgium

1. Introduction

Tumor Necrosis Factor (TNF) is a pleiotropic cytokine, originally detected and at present still mostly important because of its cytostatic/cytotoxic potential towards several transformed cell lines (Fiers, 1993). One of the prototypic cell lines for studying the cell killing effects of TNF is the mouse fibrosarcoma line L929. TNF treatment of L929 cells not only leads to cellular toxicity, but also results in the induction of several genes, among which the gene coding for Interleukin-6 (IL6). However, the mechanism of both TNF effects, i.e. gene induction and cell killing, is not yet fully understood, nor are the corresponding signaling intermediates at present known in great detail. In an effort to elucidate the cellular processes, which take place following TNF induction, numerous studies are being pursued regarding different levels of the signaling pathway(s), i.e. from intracellular interactions with the membrane bound receptor, down to the actual site of action. In this paper, we will briefly review and mainly focus on the molecular mechanisms for activation of the IL6 gene promoter in response to TNF.

2. The major transcription factor: NFκB

One of the most studied and widely known transcription factors is the nuclear factor κB (NFκB). Although originally described as a specific factor for activation of κ

NATO ASI Series, Vol. H 92
Signalling Mechanisms – from Transcription Factors
to Oxidative Stress
Edited by L. Packer, K. Wirtz

light chain immunoglobulin (Ig) genes in B cells (Sen & Baltimore, 1986), it is now known as a major transcription factor involved in expression of a wide variety of genes, including the IL6 gene (Libermann & Baltimore, 1990; Shimizu et al., 1990; Zhang et al., 1990). NFκB is activated by various chemical and biological compounds and functions as the main mediator of immunological and inflammatory responses (Baeuerle & Henkel, 1994). Also in the cell line L929, a factor complex specifically binding to the original Ig-κB sequence, is activated in response to TNF (Patestos et al., 1993), which mainly consists of the classical p50 (NFKB1) and p65 (relA) subunits (S. Plaisance, personal communication). In unstimulated cells, NFκB resides in the cytoplasm bound to the inhibitory molecule IκB, from which it is released to migrate into the nucleus for promoter activation (Baeuerle & Baltimore, 1988; Baeuerle & Baltimore, 1989). Decomposition of the cytoplasmic NFκB-IκB complex is marked by specific phosphorylation (Beg et al., 1993; Cordle et al., 1993; Koong et al., 1994; Mellits et al., 1993; Naumann & Scheidereit, 1994; Sun et al., 1994) and probably ubiquitination (Palombella et al., 1994) of the IκB molecule, followed by a rapid and extensive degradation process (Beg et al., 1993; Brown et al., 1993; Henkel et al., 1993; Kanno et al., 1994; Mellits et al., 1993; Rice & Ernst, 1993; Sun et al., 1993; Sun et al., 1994), which thus sets the activating compound free to fulfil its function.

The precise signaling pathway leading to the specific phosphorylation step of IκB is not known. At least, in several cells of mouse and man analysed, no "classical" (i.e. PMA-dependent) PKC activity seems to be involved in activation of NFκB in response to TNF (Meichle et al., 1990; Boone et al., 1995) or to LPS (Cordle et al., 1993), nor could NFκB activation in L929 cells be blocked by inhibitors of protein tyrosine kinase activity (Boone et al., 1995). As a matter of fact, Diaz-Meco et al. (1994) recently reported on the existence of a novel ζPKC-linked kinase that would be responsible for the specific phosphorylation of IκB; the actual nature of this kinase remains, however, to be established.

Furthermore, production of reactive oxygen intermediates (ROI) is directly involved in NFκB activation (Schreck et al., 1991) and various authors have reported on the inhibition of NFκB by thiol compounds, antioxidants and other radical scavengers (Galter et al., 1994; Israël et al., 1992; Renard et al., 1994; Schenk et al., 1994; Schreck et al., 1992; Staal et al., 1990; Suzuki & Packer, 1993). In L929 cells we have shown that blocking the mitochondrial electron transport system directly correlates with inhibition of NFκB activation and L929 variants, in which the mitochondrial functions are knocked out, are no longer able to respond to TNF in terms of IL6 gene induction (Schulze-Osthoff et al., 1993).

Other mechanisms of TNF signaling and NFκB triggering involve the activation of different sphingomyelinases, that release ceramide as a presumptive key molecule and signaling compound (Schütze et al., 1992; Wiegmann et al., 1994). Whether ceramide released by acid sphingomyelinase thus serves as a mediator of NFκB activation, is not fully proven for intact cells. At least in our hands, the cell permeable C6-ceramide (kindly provided by Dr. Y. Hannun) added to L929 cells did not stimulate NFκB, nor did it contribute in any way to the amounts of NFκB activated by TNF (S. Plaisance, unpublished results).

In conclusion, various signaling mechanisms have been reported to be responsible for, or at least be connected with, activation of NFκB in response to cytokine induction. How these different systems are linked to each other and whether they act in a consecutive rather than in a parallel manner is at present not known and need to be further explored, taking into account the different nature of the target cell lines analysed.

3. *Cooperative factors*

Among the various effects exerted by TNF, we already have mentioned that in L929 cells TNF-mediated gene induction and cell killing are most prominent. Both effects are not independent from each other, but were shown to be closely linked, pointing to an (at least) partial common signaling pathway (Vandevoorde et al., 1991; Vandevoorde et al., 1992). The abundance of the major activation and transcription factor NFκB does, however, not correlate with the variations of IL6 gene expression; therefore, other cooperative factors are believed to be necessary for modulating the levels of IL6 mRNA production in concert with the observed levels of cellular toxicity, exerted by TNF (Patestos et al., 1993).

The promoter of the IL6 gene displays a variety of responsive sequence elements (Tanabe et al., 1988); various transcription factors other than NFκB and coactivated by the same or different TNF-directed signaling pathways, can bind onto the IL6 promoter and synergize with NFκB for transcriptional activation. One of the most prominent cofactors for IL6 gene induction is NF-IL6, a member of the C/EBP family (Akira et al., 1990); this factor was shown to be absolutely required for IL6 promoter activation in response to IL1 (Isshiki et al., 1990). In the promoter of IL8, in which the binding sites for NF-IL6 and NFκB are juxtapositioned, it was shown that the intactness of both binding sites acting together is a prerequisite for IL8 promoter activity in response to various stimuli, including TNF and IL1 (Mukaida et al., 1990); similar conclusions were also drawn for activation of the IL6 promoter and synergistic cooperation between NF-IL6 and NFκB was postulated, although here the respective binding sites are more than 70 bp apart (Matsusaka et al., 1992). Further support for this hypothesis was given by the fact that, at least in vitro, actual complex formation was observed between either NFκB p50 and NF-IL6 (LeClair et al., 1992; Stein et al., 1993) or NFκB p65 and NF-IL6 (Matsusaka et al., 1992; Stein et al., 1993), suggesting direct interactions between NF-IL6 and NFκB subunits as a possible mechanism for concerted stimulatory activity.

The proposed interactions between NF-IL6 and NFκB are, however, not an exclusive prerequisite for IL6 promoter activity in different cells and in different kinds of stimulatory conditions, and protein-protein interactions of the NFκB p65 subunit with other transcription factors have also been described (Ray & Prefontaine, 1994; Stein et al., 1993). Furthermore, Dendorfer et al. (1994) demonstrated that, upon mutation of one or more particular sequence elements within the IL6 promoter, also including the NF-IL6 binding site, promoter activity was only partially reduced. Apparently, the factors of which the respective binding sites had remained intact, still contributed and synergized with each other for promoter stimulation. Zhang et al. (1994) reached about the same conclusions and found that alteration of the NF-IL6 site in a relatively short deletion variant of the IL6 promoter, did not result in decrease of promoter activity in response to LPS.

In L929 cells, we have introduced various IL6 promoter CAT reporter gene constructs and found the recombinant promoter equally responsive to various stimuli including TNF as the endogeneous one (Vanden Berghe et al., 1993). Upon truncation of the recombinant construction from the 5'-end, gradually decreasing promoter activity was seen with decreasing promoter length, mainly due to a reduced background and stimulatory potential as more and more responsive elements were sequentially removed. In principle, this observation fits the presence of multiple, partially redundant regulatory elements, as described by Dendorfer et al. (1994).

Anyhow, the smallest deletion variant, which distinctly lacks the NF-IL6 binding site, was still inducible by TNF and the corresponding promoter activity could be upregulated or repressed in correlation with an increased or reduced cell killing effect by TNF. Whether these regulatory effects can be ascribed to the synergistic activity of NFκB with a sofar undefined cooperative factor binding to the small IL6 promoter section left, or else, whether these effects are the direct result of a different phosphorylation status of the various NFκB subunits binding to this minimal promoter area, is presently under investigation (W. Vanden Berghe, personal communication).

4. Final considerations

Although apparently not necessary for induction of all TNF-responsive genes (Klampfer et al., 1994), the transcription factor NFκB is certainly a key element for transcriptional activation of the IL6 gene in response to TNF. Various signaling pathways and mechanisms for NFκB activation have already been described, but their actual link and their respective importance for a given target cell line still awaits elucidation.

Cytokine-induced IL6 gene induction is, however, the result of a concerted action of more than just one (i.e. NFκB) transcription factor. The redundancy of sequence elements, as present in the IL6 promoter, opens many variations for gene activation in response to external stimuli. It now remains to be analysed how in a given cell type different factors, triggered by the same or by different signaling pathways, synergize to affect IL6 promoter activation in response to a defined stimulus, such as e.g. TNF.

G.H. is a Research Director with the Belgian "Nationaal Fonds voor Wetenschappelijk Onderzoek". Research in the authors' laboratory was supported by grants from the FGWO, the IUAP and an EC Biotech Programme.

References

Akira S, Isshiki H, Sugita T, Tanabe O, Kinoshita S, Nishio Y, Nakajima T, Hirano T, Kishimoto T (1990) A nuclear factor for IL-6 expression (NF-IL6) is a member of a C/EBP family. EMBO J 9:1897-1906

Baeuerle PA, Baltimore D (1989) A 65-kD subunit of active NF-κB is required for inhibition of NF-κB by IκB. Genes Development 3:1689-1698

Baeuerle PA, Baltimore D (1988) IκB: A specific inhibitor of the NF-κB transcription factor. Science 242:540-546

Baeuerle PA, Henkel T (1994) Function and activation of NF-κB in the immune system. Annu Rev Immunol 12:141-179

Beg AA, Finco TS, Nantermet PV, Baldwin Jr AS (1993) Tumor necrosis factor and interleukin-1 lead to phosphorylation and loss of IκBα: A mechanism for NF-κB activation. Mol Cell Biol 13:3301-3310

Boone E, Fiers W, Haegeman G (1995) TNF-mediated gene induction involves protein kinase C activity and tyrosine kinase phosphorylation. In preparation.

Brown K, Park S, Kanno T, Franzoso G, Siebenlist U (1993) Mutual regulation of the transcriptional activator NF-κB and its inhibitor, IκB-α. Proc Natl Acad Sci USA 90:2532-2536

Cordle SR, Donald R, Read MA, Hawiger J (1993) Lipopolysaccharide induces phosphorylation of MAD3 and activation of c-Rel and related NF-κB proteins in human monocytic THP-1 cells. J Biol Chem 268:11803-11810

Dendorfer U, Oettgen P, Libermann TA (1994) Multiple regulatory elements in the interleukin-6 gene mediate induction by prostaglandins, cyclic AMP, and lipopolysaccharide. Mol Cell Biol 14:4443-4454

Diaz-Meco M, Dominguez I, Sanz L, Dent P, Lozano J, Municio MM, Berra E, Hay RT, Sturgill TW, Moscat J (1994) ζPKC induces phosphorylation and inactivation of IκB-α in vitro. EMBO J 13:2842-2848

Fiers W (1993) Tumour necrosis factor. In The Natural Immune System: Humoral Factors (Sim E, ed), 65-119. IRL Press Oxford

Galter D, Mihm S, Dröge W (1994) Distinct effects of glutathione disulphide on the nuclear transcription factors κB and the activator protein-1. Eur J Biochem 221:639-648

Henkel T, Machleidt T, Alkalay I, Krönke M, Ben-Neriah Y, Baeuerle PA (1993) Rapid proteolysis of IκB-α is necessary for activation of transcription factor NF-κB. Nature 365:182-185

Israël N, Gougerot-Pocidalo M-A, Aillet F, Virelizier J-L (1992) Redox status of cells influences constitutive or induced NF-κB translocation and HIV long terminal repeat activity in human T and monocytic cell lines. J Immunol 149:3386-3393

Isshiki H, Akira S, Tanabe O, Nakajima T, Shimamoto T, Hirano T, Kishimoto T (1990) Constitutive and interleukin-1 (IL-1)-inducible factors interact with the IL-1-responsive element in the IL-6 gene. Mol Cell Biol 10:2757-2764

Kanno T, Brown K, Franzoso G, Siebenlist U (1994) Kinetic analysis of human T-cell leukemia virus type I Tax-mediated activation of NF-κB. Mol Cell Biol 14:6443-6451

Klampfer L, Lee TH, Hsu W, Vilček J, Chen-Kiang S (1994) NF-IL6 and AP-1 cooperatively modulate the activation of the TSG-6 gene by tumor necrosis factor alpha and interleukin-1. Mol Cell Biol 14:6561-6569

Koong AC, Chen EY, Giaccia AJ (1994) Hypoxia causes the activation of nuclear factor κB through the phosphorylation of IκBα on tyrosine residues. Cancer Res 54:1425-1430

LeClair KP, Blanar MA, Sharp PA (1992) The p50 subunit of NF-κB associates with the NF-IL6 transcription factor. Proc Natl Acad Sci USA 89:8145-8149

Libermann TA, Baltimore D (1990) Activation of interleukin-6 gene expression through the NF-κB transcription factor. Mol Cell Biol 10:2327-2334

Matsusaka T, Fujikawa K, Nishio Y, Mukaida N, Matsushima K, Kishimoto T, Akira S (1993) Transcription factors NF-IL6 and NF-κB synergistically activate transcription of the inflammatory cytokines, interleukin 6 and interleukin 8. Proc Natl Acad Sci USA 90:10193-10197

Meichle A, Schütze S, Hensel G, Brunsing D, Krönke M (1990) Protein kinase C-independent activation of nuclear factor κB by tumor necrosis factor. J Biol Chem 265:8339-8343

Mellits KH, Hay RT, Goodbourn S (1993) Proteolytic degradation of MAD3 (IκBα) and enhanced processing of the NF-κB precursor p105 are obligatory steps in the activation of NF-κB. Nucleic Acids Res 21:5059-5066

Mukaida N, Mahe Y, Matsushima K (1990) Cooperative interaction of nuclear factor-κB- and cis-regulatory enhancer binding protein-like factor binding elements in activating the interleukin-8 gene by pro-inflammatory cytokines. J Biol Chem 265:21128-21133

Naumann M, Scheidereit C (1994) Activation of NF-κB in vivo is regulated by multiple phosphorylations. EMBO J 13:4597-4607

Palombella VJ, Rando OJ, Goldberg AL, Maniatis T (1994) The ubiquitin-proteasome pathway is required for processing the NF-κB1 precursor protein and the activation of NF-κB. Cell 78:773-785

Patestos NP, Haegeman G, Vandevoorde V, Fiers W (1993) Activation of the nuclear factor κB is not sufficient for regulation of tumor necrosis factor-induced interleukin-6 gene expression. Biochimie 75:1007-1018

Ray A, Prefontaine KE (1994) Physical association and functional antagonism between the p65 subunit of transcription factor NF-κB and the glucocorticoid receptor. Proc Natl Acad Sci USA 91:752-756

Renard P, Zachary M-D, Bougelet C, Remacle J, Raes M (1994) Modulation of human cultured fibroblasts cell redox potential: Effects on the transcriptional factor NFκB activated by IL-1. International Summer School on Molecular Mechanisms of Transcellular Signaling: From the Membrane to the Gene. Spetsai, Poster Abstracts, 65

Rice NR, Ernst MK (1993) In vivo control of NF-κB activation by IκBα. EMBO J 12:4685-4695

Schenk H, Klein M, Erdbrügger W, Dröge W, Schulze-Osthoff K (1994) Distinct effects of thioredoxin and antioxidants on the activation of transcription factors NF-κB and AP-1. Proc Natl Acad Sci USA 91:1672-1676

Schreck R, Meier B, Männel DN, Dröge W, Baeuerle PA (1992) Dithiocarbamates as potent inhibitors of nuclear factor κB activation in intact cells. J Exp Med 175:1181-1194

Schreck R, Rieber P, Baeuerle PA (1991) Reactive oxygen intermediates as apparently widely used messengers in the activation of the NF-κB transcription factor and HIV-1. EMBO J 10:2247-2258

Schulze-Osthoff K, Beyaert R, Vandevoorde V, Haegeman G, Fiers W (1993) Depletion of the mitochondrial electron transport abrogates the cytotoxic and gene-inductive effects of TNF. EMBO J 12:3095-3104

Schütze S, Potthoff K, Machleidt T, Berkovic D, Wiegmann K, Krönke M (1992) TNF activates NF-κB by phosphatidylcholine-specific phospholipase C-induced "acidic" sphingomyelin breakdown. Cell 71:765-776

Sen R, Baltimore D (1986) Multiple nuclear factors interact with the immunoglobulin enhancer sequences. Cell 46:705-716

Shimizu H, Mitomo K, Watanabe T, Okamoto S, Yamamoto K (1990) Involvement of a NF-κB-like transcription factor in the activation of the interleukin-6 gene by inflammatory lymphokines. Mol Cell Biol 10:561-568

Staal FJT, Roederer M, Herzenberg LA, Herzenberg LA (1990) Intracellular thiols regulate activation of nuclear factor κB and transcription of human immunodeficiency virus. Proc Natl Acad Sci USA 87:9943-9947

Stein B, Cogswell PC, Baldwin Jr AS (1993) Functional and physical associations between NF-κB and C/EBP family members: A Rel domain-bZIP interaction. Mol Cell Biol 13:3964-3974

Sun S-C, Elwood J, Béraud C, Greene WC (1994) Human T-cell leukemia virus type I Tax activation of NF-κB/Rel involves phosphorylation and degradation of IκBα and RelA (p65)-mediated induction of the c-rel gene. Mol Cell Biol 14:7377-7384

Sun S-C, Ganchi PA, Ballard DW, Greene WC (1993) NF-κB controls expression of inhibitor IκBα: Evidence for an inducible autoregulatory pathway. Science 259:1912-1915

Suzuki YJ, Packer L (1993) Inhibition of NF-κB activation by vitamin E derivatives. Biochem Biophys Res Commun 193:277-283

Tanabe O, Akira S, Kamiya T, Wong GG, Hirano T, Kishimoto T (1988) Genomic structure of the murine IL-6 gene. High degree conservation of potential regulatory sequences between mouse and human. J Immunol 141:3875-3881

Vanden Berghe W, Haegeman G, Fiers W (1993) Studies on the inducibility of the IL6 promoter using a reporter gene construction. Arch Intern Physiol Biochim Biophys 101:B35

Vandevoorde V, Haegeman G, Fiers W (1992) TNF-mediated IL6 gene expression and cytotoxicity are co-inducible in TNF-resistant L929 cells. FEBS Lett 302:235-238

Vandevoorde V, Haegeman G, Fiers W (1991) Tumor necrosis factor-induced interleukin-6 expression and cytotoxicity follow a common signal transduction pathway in L929 cells. Biochem Biophys Res Commun 178:993-1001

Wiegmann K, Schütze S, Machleidt T, Witte D, Krönke M (1994) Functional dichotomy of neutral and acidic sphingomyelinases in tumor necrosis factor signaling. Cell 78:1005-1015

Zhang Y, Broser M, Rom WN (1994) Activation of the interleukin 6 gene by Mycobacterium tuberculosis or lipopolysaccharide is mediated by nuclear factors NF-IL6 and NF-κB. Proc Natl Acad Sci USA 91:2225-2229

Zhang Y, Lin J-X, Vilček J (1990) Interleukin-6 induction by tumor necrosis factor and interleukin-1 in human fibroblasts involves activation of a nuclear factor binding to a κB-like sequence. Mol Cell Biol 10:3818-3823

Regulation of 14 kDa Group II PLA_2 in Rat Mesangial Cells

Margriet J.B.M. Vervoordeldonk[1], Casper G. Schalkwijk[1], Rosa M. Sanchez[1], Josef Pfeilschifter[2] and Henk van den Bosch[1]

[1] Centre for Biomembranes and Lipid Enzymology
Utrecht University
Padualaan 8
3584 CH Utrecht
The Netherlands

INTRODUCTION

Phospholipase A_2 (PLA_2) is believed to play an essential role in inflammation through the release of arachidonic acid from membrane phospholipids for the production of important lipid mediators such as eicosanoids and platelet activating factor (Van den Bosch, 1980; Glaser *et al.*, 1993). The last years it has become clear that PLA_2s are a heterogeneous family of enzymes that can be classified in two classes based on their molecular weight. There is a class of low molecular weight PLA_2s (14 kDa) and one of the more recently discovered high molecular mass enzymes (85 kDa). The high molecular weight PLA_2, also referred to as $cPLA_2$, is mainly located in the cytosolic fraction of cells and tissues including human platelets (Takayamo *et al.*, 1991), rat renal mesangial cells (Gronich *et al.*, 1988; Bonventre *et al.*, 1990), and the human monoblast U937 cell line (Clark *et al.*, 1990; Kramer *et al.*, 1991). Although the enzyme has been shown to be 85 kDa by sequence and cloning (Clark *et al.*, 1991; Sharp *et al.*, 1991), it shows a molecular weight on SDS-PAGE of about M_r 110,000 (Leslie *et al.*,1988; Clark *et al.*, 1990; Kramer *et al.*, 1991). This enzyme preferentially hydrolyzes arachidonic acid from the sn-2-position of phospholipids (Clark *et al.*, 1990), is insensitive for dithiotreitol and has optimal activity at micromolar Ca^{2+}-concentrations (Gronich *et al.*, 1990). $cPLA_2$ also comprises multiple phosphorylation sites, among which a MAP-kinase phosphorylation site (serine-505) that appears to play an important role in enzyme activation (Liscovitch and Cantley, 1994). The 14 kDa PLA_2s can be further divided in two groups, based on their amino acid sequence (Heinrikson *et al.*, 1977). Mammalian group I PLA_2 comprises the pancreatic type of PLA_2 and is characterized by the presence of cys 11. Homologous non-pancreatic group II phospholipase A_2 is lacking cys 11. Type II PLA_2 is often found as a membrane-

[2] Department of Pharmacology, Biocenter, University of Basel, Klingelbergstrasse 70, CH-4056 Basel, Switzerland

NATO ASI Series, Vol. H 92
Signalling Mechanisms – from Transcription Factors to Oxidative Stress
Edited by L. Packer, K. Wirtz

bound enzyme (Aarsman *et al.*, 1989; Ono *et al.*, 1988) which has an optimal activity at millimolar Ca^{2+}-concentrations (Mizushima *et al.*, 1989) and has no selectivity for arachidonic acid (Schalkwijk *et al.*, 1990). This type has been purified and characterized from various cellular sources such as rat (Hayakawa *et al.*,1988), rabbit (Mizushima *et al.*, 1989) and human platelets (Kramer *et al.*,1989), rat liver mitochondria (Aarsman *et al.*, 1989) and rat spleen (Ono *et al.*, 1988). Mammalian group II PLA_2 is also found in soluble form at inflammatory sites (Forst *et al.*, 1986; Hara *et al.*, 1989; Seilhamer *et al.*, 1988) and circulating in blood of patients with various diseases (Baek *et al.*, 1991) suggesting its role in inflammatory processes. In line with this hypothesis, the PLA_2 purified from synovial fluid is pro-inflammatory (Vadas *et al.*, 1985).

INDUCTION OF $sPLA_2$ IN RAT MESANGIAL CELLS

Interleukin-1ß (IL-1ß) and tumor necrosis factor (TNF) are potent pro-inflammatory cytokines which play an important role in the mediation of inflammatory processes. Rat glomerular mesangial cells are one of the target cells for these cytokines. We and others have recently shown that IL-1ß, TNF and also the cAMP elevating agent forskolin enhanced the synthesis and secretion of phospholipase A_2 in parallel to the formation of prostaglandin E_2 (PGE_2) (Pfeilschifter *et al.*, 1989a; Schalkwijk *et al.*,1991a). Using monoclonal antibodies this enzyme was identified as 14 kDa group II PLA_2 and this was confirmed by sequence analysis (Schalkwijk *et al.*, 1992a). Immunoblot experiments of control and stimulated cells indicated that group II PLA_2 was hardly present in unstimulated cells and became detectable after a lag period of approximately 8 hours of stimulation in both cells and culture medium (Schalkwijk *et al.*, 1991a). These elevated protein levels were accompanied by a time-dependent increase in PLA_2-activity. Over 85% of this *de novo* synthesized group II PLA_2 appears to be rapidly secreted from cytokine- and forskolin-stimulated cells. The effects of the cytokines on $sPLA_2$ are blocked by antinomycin D, indicating that *de novo* mRNA synthesis is involved.

LOCALIZATION OF $sPLA_2$ IN RAT MESANGIAL CELLS

We investigated the intracellular localization of group II PLA_2 in control and IL-1ß stimulated cells with Western blotting and immunofluorescence microscopy. Western blot experiments showed that $sPLA_2$ is hardly present in control cells (Schalkwijk, 1992b). In agreement with previous observations, the enzyme can not be detected in the homogenate of control cells, only in the 200,000·g pellet fraction a small amount of protein is visible (Fig.1). However, after

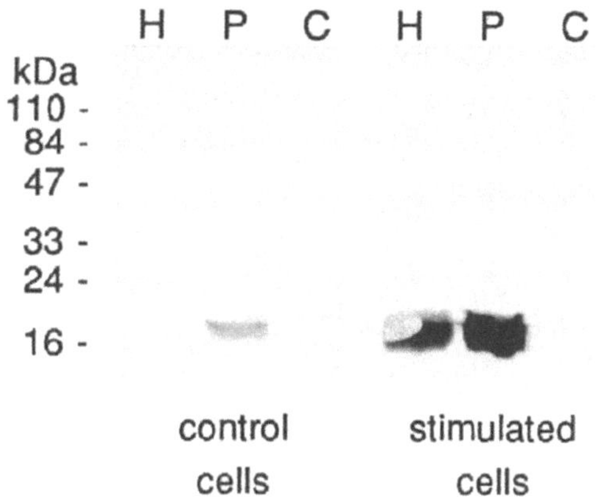

Figure 1. Immunoblot of control- and Il-1ß-stimulated mesangial cells using homogenate (H), particulate fraction (P) and cytosolic fraction (C).

stimulation of the cells with IL-1ß group II PLA_2 became detectable in the homogenate. The enzyme appeared to be localized in the particulate fraction whereas no $sPLA_2$ could be detected in the cytosol. The localization of $sPLA_2$ in stimulated mesangial cells was investigated in more detail by immunofluorescence microscopy (Vervoordeldonk *et al.*, 1994). In unstimulated cells negligible labelling was obtained (results not shown). The immuno-fluorescence staining

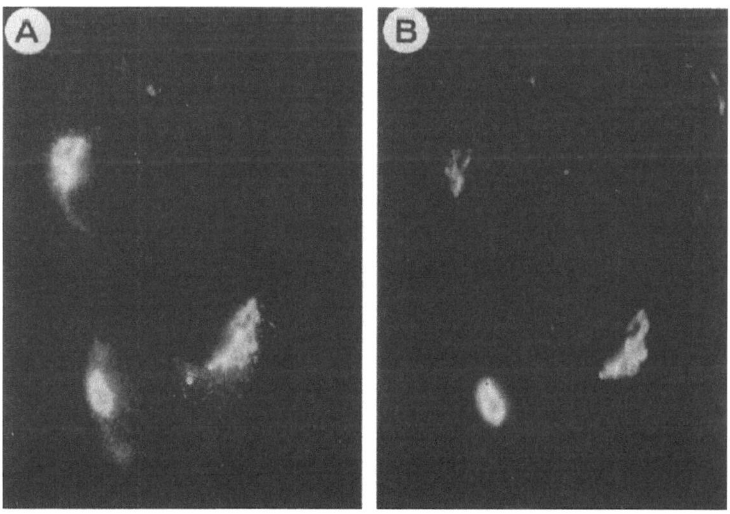

Figure 2. Colocalization of group II PLA_2 (A) with a Golgi marker (B) in rat mesangial cells.

pattern of $sPLA_2$ in IL-1ß-treated cells included punctate structures, presumable secretory vesicles, and staining of structures around the nucleus (Fig. 2). Double labelling experiments using both TRITC-labelled ricin as Golgi marker and GAM-FITC to localize $sPLA_2$ confirmed that $sPLA_2$ is predominantly localized in the Golgi-complex. This fits well with the common pathway known for secretory proteins.

EFFECT OF BREFELDIN A ON IL-1ß-INDUCED $sPLA_2$

In order to study the contribution of $sPLA_2$ to the arachidonic acid release and PGE_2 synthesis, we tried to accumulate the enzyme in IL-1ß-stimulated cells by blocking the secretion of $sPLA_2$ (Sanchez *et al.*, 1993). For this purpose we used a known inhibitor of protein secretion, brefeldin A (BFA). BFA can block protein secretion by causing dismantling of the Golgi cisternae in many cultured cells (Lippincott-Schwarz *et al.*, 1991). As expected BFA blocked the secretion of $sPLA_2$ from mesangial cells completely at a concentration of 0.5 µg/ml BFA, but Western blot analysis showed that this was not accompanied by an increase in cellular $sPLA_2$ protein level (Fig. 3). In contrast, the drug prevented *de novo* synthesis of $sPLA_2$ protein, as induced by IL-1ß, in a dose-dependent manner. These results were confirmed by immunofluorescence experiments. We then studied the effect of BFA on [3H]-leucine incorporation into cellular proteins and confirmed a recent report (Fishman and Curran, 1992)

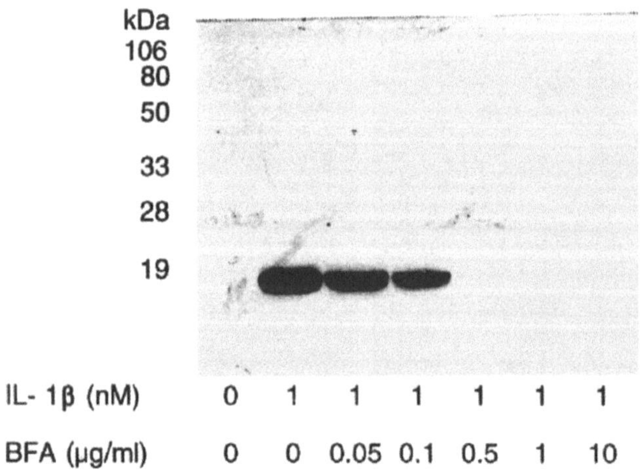

Figure 3. Dose-dependent inhibition of IL-1ß-induced $sPLA_2$ synthesis by Brefeldin A.

that this drug inhibits protein synthesis in general. However, the IL-1ß-induced $sPLA_2$ synthesis was more sensitive to BFA than that of protein synthesis in general.

INHIBITION OF INDUCED $sPLA_2$ SYNTHESIS AND SECRETION BY DEXAMETHASONE

Glucocorticosteriods, which are potent anti-inflammatory drugs, inhibit prostaglandin formation in many cells and animal models of inflammation. The mechanism by which glucocorticosteriods inhibit the synthesis and secretion of pro-inflammatory arachidonate metabolites is still controversial. The cytokine-stimulated formation of PGE_2 in rat mesangial cells could be prevented by dexamethasone, a synthetic glucocorticosteriod (Pfeilschifter *et al.*, 1989b). In view of this fact we investigated if dexamethasone could also prevent the induction of $sPLA_2$ in rat mesangial cells (Schalkwijk *et al.*, 1991b). We showed that the cytokine-induced secretion of $sPLA_2$ activity and protein into the culture medium is dose-dependently blocked by pretreatment of the cells with dexamethasone (Fig. 4). A direct correlation between $sPLA_2$ activity and protein levels is shown. For the corresponding cells a completely similar pattern compared to that of the culture medium was found. Thus, not only the secretion but also the induced synthesis was inhibited and this could explain the decrease in $sPLA_2$ activity. However, initially it has been postulated that glucocorticosteriods can induce the formation of

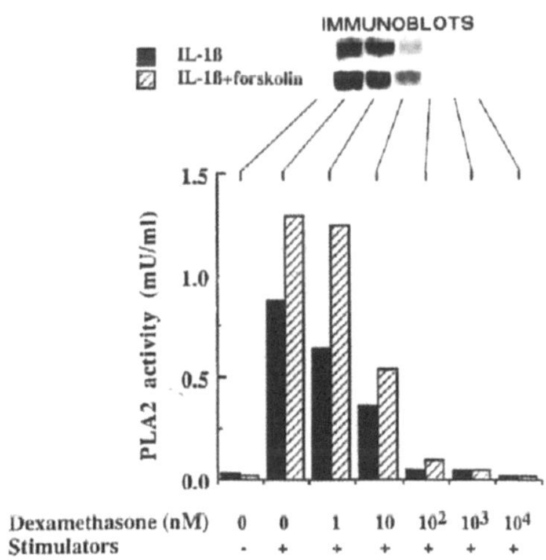

Figure 4. Dose-dependent inhibition by dexamethasone of the release of PLA_2 from stimulated mesangial cells.

PLA_2 inhibitory proteins, termed annexins (Flower *et al.*, 1988; Davidson, 1989). Later on it was found that the inhibition *in vitro* could be overcome by increasing the substrate concentration and the inhibitory action of annexins towards PLA_2 was ascribed to substrate sequestration (Davidson *et al.*, 1987; Aarsman *et al.*, 1987) rather than to direct interaction with the enzyme. Although in rat mesangial cells the decrease in PLA_2 protein levels in itself could explain the dexamethasone-induced decrease in PLA_2 activity, a contribution of the glucocorticosteriod-induced anti-phospholipase protein annexin cannot be ruled out *a priori*. A prerequisite for this is that dexamethasone affects the levels of annexin in mesangial cells or changes the cellular localization of annexin in order to inhibit PLA_2 through substrate sequestration or direct protein-protein interaction. We investigated the levels of annexin I in rat mesangial cells because this member of the annexin family is the most likely candidate for inhibiting the action of PLA_2. We demonstrated with a quantitative immunoblot procedure that dexamethasone had no effect on the level of annexin I (Fig. 5). Western blot experiments from control, IL-1β- and dexamethasone-treated cells showed now detectable change in annexin I levels at a dexamethasone concentration that completely prevented induction of $sPLA_2$ by IL-1β (Vervoordeldonk *et al.*, 1994). From this we concluded that the synthesis of annexin I is not induced in rat mesangial cells. We also established the absence of extracellular annexin I indicating that it is also not involved in the regulation of the secreted $sPLA_2$. We showed with Western blot experiments and immunofluorescence microscopy that annexin I was located mainly in the cytoplasm with a small amount being present in the nucleus. Dexamethasone

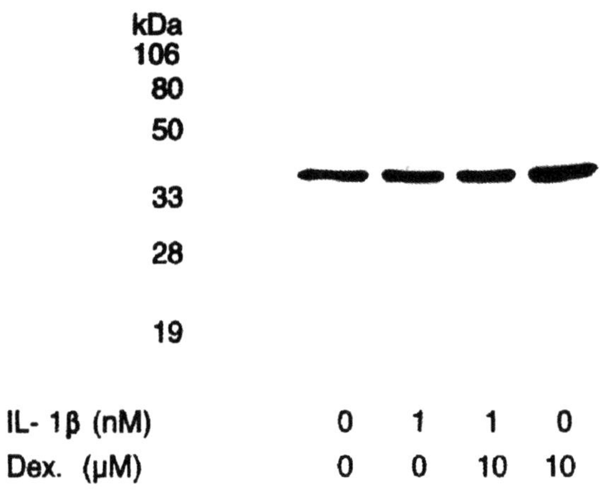

Figure 5. Levels of annexin I in control, IL-1ß-, IL-1ß- plus dexamethasone- and dexamethasone-treated cells.

caused no change in the distribution of annexin I over the cytosolic- and particulate- fraction. We conclude from these experiments that the dexamethasone induced inhibition of PLA_2-activity in rat mesangial cells is neither mediated by induced synthesis, nor by induced translocation of annexin I.

THE INDUCED $sPLA_2$ SYNTHESIS AND SECRETION IS INHIBITED BY TRANSFORMING GROWTH FACTOR-ß2.

Transforming growth factor-ß2 (TGF-ß2) exerts diverse activities on a variety of cells. We investigated the effect of TGF-ß2 on $sPLA_2$ activity and protein expression as induced by IL-1ß and forskolin because the presence of large quantities of TGF-ß2-receptors of high affinity have been reported on mesangial cells (Mackay *et al.*,1989). Furthermore, the presence of TGF-ß2 in chronic inflammatory conditions has been demonstrated (Wahl *et al.*, 1989) and TGF-ß2 antagonizes in many systems the effect caused by mitogens such as IL-1ß. Pretreatment of the cells with TGF-ß2 prior to stimulation dramatically affected the $sPLA_2$ secretion (Schalkwijk *et al.*, 1992). The IL-1ß-induced secretion of $sPLA_2$ is completely prevented by TGF-ß2 (Fig. 6). However, induction of $sPLA_2$ by IL-1ß plus forskolin is not completely inhibited. In the cell homogenates the same pattern is found. Experiments with mesangial cells stimulated with forskolin alone, however, showed that TGF-ß2 can also prevent the forskolin-induced PLA_2 protein synthesis completely. Under conditions where TGF-ß2 fully suppresses $sPLA_2$ it still

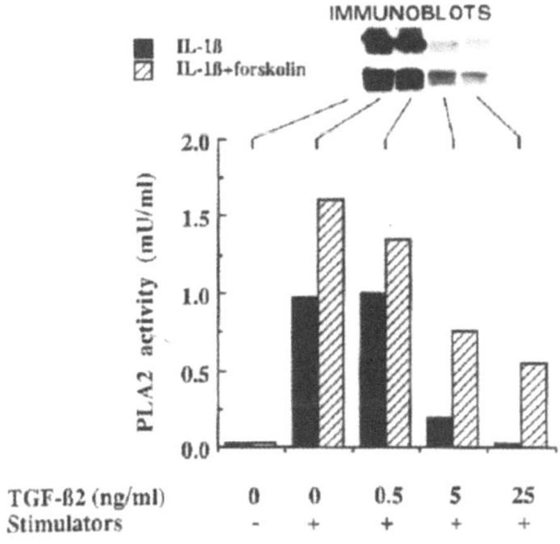

Figure 6. Dose-dependence of TGF-ß2 pretreatment of mesangial cells on the induced secretion of $sPLA_2$.

enhances PGE_2 formation compared to control cells. This can be explained because of the fact that TGF-ß2 causes an increase in $cPLA_2$ activity (Schalkwijk *et al.*, 1992).

THE EFFECT OF DEXAMETHASONE AND TRANSFORMING GROWTH FACTOR-ß2 ON THE mRNA LEVELS OF $sPLA_2$

To elucidate whether dexamethasone and TGF-ß2 exert their effect at transcriptional or translational level we analyzed the mRNA levels of $sPLA_2$ in IL-1ß- and forskolin-stimulated mesangial cells. Previous experiments (Muhl *et al.*,1992) had shown that combined stimulation of mesangial cells with IL-1ß plus forskolin resulted in increased $sPLA_2$ mRNA levels which became partially suppressed by dexamethasone under conditions were the $sPLA_2$ synthesis was totally inhibited. Northern blot analysis using a ^{32}P-labelled cDNA-probe for group II PLA_2 (van Schaik *et al.*, 1993) revealed that dexamethasone exerts its effect on transcriptional or translational level dependent on the stimulus used (Vervoordeldonk, unpublished results). IL-1ß alone induced the synthesis of $sPLA_2$ mRNA but this level was hardly affected by dexamethasone, implicating post-transcriptional inhibition of $sPLA_2$ synthesis (Fig. 7A). In contrast, the forskolin-induced mRNA elevation was completely blocked by dexamethasone (Fig. 7B).These results are in agreement with the results obtained by Nakano in rat smooth muscle cells (Nakano *et al.*, 1990). They reported the inhibition of tumor necrosis factor-induced PLA_2 release in rat smooth muscle cells by glucocorticosteriods without inhibition of the $sPLA_2$ mRNA synthesis whereas in forskolin-stimulated cells the increase in mRNA was completely blocked. It appears that similar events as observed for tumor necrosis factor are also involved in the suppression of IL-1ß-induced $sPLA_2$ synthesis by dexamethasone. In case of TGF-ß2, both IL-1ß- and forskolin -induced group II PLA_2 mRNA elevation can be completely

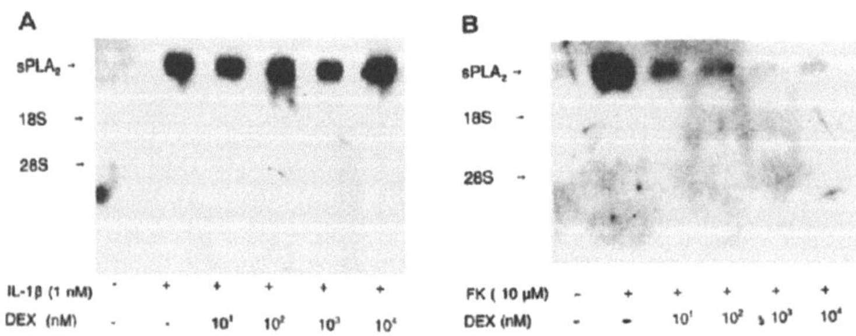

Figure 7. Effect of dexamethasone on $sPLA_2$ mRNA levels in (A) IL-1ß- and (B) forskolin-stimulated cells.

blocked in a concentration dependent manner (data not shown). From these results we conclude that TGF-ß2 inhibits both the IL-1ß- and forskolin-induced $sPLA_2$ synthesis primarily at the transcriptional level.

CONCLUDING REMARKS

Phospholipase A_2 activity is generally implicated as the rate-limiting step in the biochemical transformations leading to the production of bioactive lipids such as eicosanoids and platelet-activating factor. The present manuscript documents in a summarizing manner our studies on the regulation of phospholipase A_2 activity in relation to prostaglandin E_2 formation in rat glomerular mesangial cells. Initial experiments indicated that pro-inflammatory cytokines, such as interleukin-1β and tumor necrosis factor, induced the secretion of phospholipase A_2 activity from these cells in parallel to enhanced prostaglandin E_2 formation. Using monoclonal antibodies the secreted enzyme was identified as 14 kDa group II phospholipase A_2 and this was confirmed by sequence analysis. The enzyme is not secreted from a pre-existing cellular pool (as is present in *e.g.* platelets) but appears to be *de novo* synthesized upon cytokine treatment prior to secretion. Immunofluorescence localization studies are in line with this interpretation. Brefeldin A was found to block not only the secretion but also the synthesis of this PLA_2.

The cytokine-induced synthesis of prostaglandins can be prevented by dexamethasone. In parallel, PLA_2 activity becomes diminished. Western blot analysis indicated that this is not due to inhibition of the enzyme, but is caused by suppression of the cytokine-induced synthesis of group II PLA_2. These results provide a novel mechanism for the anti-inflammatory action of glucocorticosteroids not involving annexin induction, *i.e.* direct suppression of group II PLA_2 gene expression. Independent experiments confirm that annexins are not additionally involved in the observed decrease in PLA_2 activity upon dexamethasone treatment of mesangial cells. Using a cDNA for group II PLA_2, generated from rat liver by PCR technology, indicated that the suppression of interleukin-induced PLA_2 synthesis by dexamethasone is mainly at the post-transcriptional level. By contrast, the suppression of forskolin-induced PLA_2 synthesis by dexamethasone appears to be caused mainly by inhibition of mRNA synthesis.

The interleukin- and forskolin-induced synthesis of group II PLA_2 can also be completely suppressed by transforming growth factor-β2. In both cases the suppression is due to decreased mRNA levels. This suppression leads also to an inhibition of interleukin-induced prostaglandin E_2 formation, but in contrast to dexamethasone, the addition of TGF-β2 does not completely block prostaglandin formation. This can be explained by the observation that TGF-β2 at one hand suppresses group II PLA_2 mRNA levels and enzyme synthesis but at the same time increases the activity of a high molecular weight cytosolic phospholipase A_2. The contribution of the secreted group II PLA_2 to arachidonate release for prostaglandin formation has been assessed in experiments using either stimulation of cells in the presence of a neutralizing antibody or by addition of immunopurified enzyme to control cells (data not shown; Pfeilschifter *et al.*, 1993).

REFERENCES

Aarsman A, de Jong JGM, Arnoldussen E, Neys FW, van Wassenaar PD, van den Bosch H (1989) Immunoaffinity purification, partial sequence and subcellular localization of rat liver phospholipase A_2. J Biol Chem 264:10008–10014

Baek SH, Takayoma K, Kudo I, Inoue K, Lee HW, Do JY, Chang MW (1991) Detection and characterization of extracellular phospholipase A_2 in pleural effusion of patients with tuberculosis. Life Sci 49:1095–1102

Bonventre JV, Gronich JH, Nemenoff RA (1990) Epidermal growth factor enhances glomerular mesangial cell soluble phospholipase A_2 activity. J Biol Chem 265:4934–4938

Clark JD, Lin L-L, Uriz RW, Ramesha CS, Sultzman LA, Lin AY, Milona N, Knopf JL (1991) A novel arachidonic acid-selective cytosolic PLA_2 contains a Ca^{2+}-dependent translocation domain with homology to PKC and GAP. Cell 65:1043–1051

Clark JD, Milona N, Knopf JL (1990) Purification of a 100-kilodalton cytosolic phospholipase A_2 from the human monocytic cell line U937. Proc Natl Acad Sci USA 87:7708–7712

Fishman PH, Curran PK (1992) Brefeldin A inhibits protein synthesis in cultured cells. FEBS Lett 314:371–374

Forst S, Weiss J, Elsbach P, Managanore JM, Reardon I, Heinrikson RL (1986) Structure and functional properties of a phospholipase A_2 purified from an inflammatory escudate. Biochemistry 25:8381–8385

Glaser KD, Mobilio D, Chang JY, Senko N (1993) Phospholipase A_2 enzymes, regulation and inhibition. Trends Pharmacol Sci 14:92–98

Gronich JH, Bonventre JV, Nemenoff RA (1988) Identification and characterization of a harmonally regulated form of phospholipase A_2 in rat renal mesangial cells. J Biol Chem 263:16645–16651

Gronich JH, Bonventre JV, Nemenoff RA (1990) Purification of a high molecular mass form of phospholipase A_2 from rat kidney activated at physiological calcium concentrations. Biochem J 271:37–43

Hara S, Kudo I, Chang H, Matsuta K, Miyamoto Y, Inoue K (1989) Purification and characterization of extracellular phospholipase A_2 from synovial fluid in rheumatoid arthritis. J Biochem 105:395–399

Hayakawa M, Kudo I, Tomita M, Nojima S, Inoue K (1988) The primary structure of rat platelet phospholipase A_2. J Biochem 104:767–772

Heinrikson RL, Krueger ET, Keim DS (1977) Amino acid sequence of phospholipase A_2-α from the venom of *cratalus adamenteus.* J Biol Chem 252:4913–4921

Kramer RH, Hession C, Johansen B, Hayes G, McGray P, Pinchang Chow E, Tizard R and Pepinsky RB (1989) Structure and properties of a human non-pancreatic phospholipase A_2. J. Biol Chem 264:5768–5775

Kramer RM, Roberts EF, Manetta J, Putman JE (1991) The Ca^{2+}-sensitive cytosolic phospholipase A_2 is a 100-kDa protein in human monoblast U937 cells. J Biol Chem 266:5268–5272

Leslie CC, Voelker DR, Channon JY, Wall MW, Zalarney PT (1988) Properties and purification of an arachidonoyl hydrolyzing phospholipase A_2 from a macrophage cell line, Raw 264.7. Biochim Biophys Acta 963:476–492

Lippincott-Schwarz J, Yuan L, Tipper C, Amhardt M, Orci L, Klausner RD (1991) Brefeldin A's effects on endosomes lysosomes, and TGN suggests a general mechanism for regulating organelle structure and membrane traffic. Cell 67:601–616

Liscovitch M, Cantley LC (1994) Lipid second messengers. Cell 77, 329–334

Mizushima H, Kudo I, Horigomo K, Murakami M, Hayakawa M, Kim DK, Kondo E, Tomita M and Inoue K (1989) Purification of rabbit platelet secretory phospholipase A_2 and its characteristics. J Biochem 105:520–525

Ono I, Tojo H, Kuramitsu S, Kagamiyama H, Okamoto M (1988) Purification and characterization of membrane-associated phospholipase A_2 from rat spleen. J Biol Chem 263:5732–5738

Pfeilschifter J, Pignat W, Vosbeck K, Märki F (1989a) Interleukin 1 and tumor necrosis factor synergistically stimulate prostaglandin synthesis and phospholipase A_2 release from rat renal mesangial cells. Biochem Biophys Res Commun 159:385–394

Pfeilschifter J, Pignat W, Vosbeck U, Märki F, Wiesenberg I (1989b) Susceptibility of interleukin 1- and tumor necrosis factor-induced prostaglandin E_2 and phospholipase A_2 release from rat renal mesangial cells to different drugs. Biochem Soc Trans 17:916–917

Pfeilschifter J, Schalkwijk C, Briner VA, van den Bosch H (1993) Cytokine-stimulated secretion of group II phospholipase A_2 by rat mesangial cells. Its contribution to arachidonic acid release and prostaglandin synthesis by cultured rat glomerular cells. J Clin Invest 92:2516–2523

Sánchez, R, Vervoordeldonk MJBM, Schalkwijk CG, van den Bosch H (1993) Prevention of the induced synthesis and secretion of group II phospholipase A_2 by brefeldin A. FEBS Lett 332:99–104

Schalkwijk CG, Märki F, van den Bosch H (1990) Studieson acyl-chain selectivity of cellular phospholipases A_2. Biochem Biophys Acta 1044:139–146

Schalkwijk C, Pfeilschifter J, Märki F, van den Bosch H (1991a) Interleukin-1β, tumor necrosis factor and forskolin stimulate the synthesis and secretion of group II phospholipase A_2 in rat mesangial cells. Biochem Biophys Res Commun 174:268–275

Schalkwijk CG, Vervoordeldonk MJBM, Pfeilschifter J, Märki F, van den Bosch H (1991b) Cytokine- and forskolin-induced synthesis of group II phospholipase A_2 and prostaglandin E_2 in rat mesangial cells is prevented by dexamethasone. Biochem Biophys Res Commun 180:46–52

Schalkwijk C, Pfeilschifter J, Märki F, van den Bosch H (1992a) Interleukin-1β- and forskolin-induced synthesis and secretion of group II phospholipase A_2 and prostaglandin E_2 in rat mesangial cells is prevented by transforming growth factor-β2. J Biol Chem 267:8846–8851

Schalkwijk CG (1992b) Characterization of cellular phospholipase A_2 and their role in prostaglandin E_2 formation. Ph.D. thesis, Utrecht University

Seilhamer JJ, Pruzanski W, Vadas P, Plant S, Miller JA, Kloss J, Johnson LK (1988) Cloning and recombinant expression of phospholipase A_2 present in rheumatoid arthritic synovial fluid. J Biol Chem 264:5335–5338

Sharp JD, White DL, Chiou XG, Goodson T, Gamboa GC, McClure D, Burgett S, Hoskins J, Skatrud PL, Sportsman JR, Becker GW, Kang LH, Roberts EF, Kramer RM (1991)Molecular cloning and expression of human Ca^{2+}-sensitive cytosolic phospholipase A_2. J Biol Chem 266:14850–14853

Takayama K, Kudo I, Kim DK, Nagata K, Nozawa Y and Inoue K (1991) Purification and characterization of human platelet phospholipase A_2 which preferentially hydrolyzes an arachidonoyl residue. FEBS Lett 282:326–330

Vadas P, Stefanski E, Pruzanski W (1985) Characterization of extracellular phospholipase A_2 in rheumatoid synovial fluid. Life Sci 36:579–587

Van den Bosch H (1980) Intracellular phospholipases A. Biochem Biophys Acta 604:191–246

Vervoordeldonk MJBM, Schalkwijk C, Viswanatah BS, Aarsman AJ, van den Bosch H (1994) Levels and localization of group II phospholipase A_2 and annexin I in interleukin- and dexamethasone-treated rat mesangial cells: Evidence against annexin mediation of the dexamethasone-induced inhibition of group II phospholipase A_2. Biochim Biophys Acta, in press

Multifunctional growth factors in morphogenesis and tumor progression

Jean Paul THIERY
Laboratoire de Physiopathologie du Développement
CNRS URA 1337 and Ecole Normale Supérieure
46, rue d'Ulm
75230 Paris cedex 05, France.

Inducers and morphogens

Over the last few years, major progress has been made in the understanding of cellular and molecular processes of morphogenesis. Studies of vertebrate embryos have particularly benefited from research in amphibians. The pioneering work of Spemann and Mangold (1924) has allowed an understanding of the basic principle of neural induction. More recently, the sequence of events leading to the formation of the body plan has been established. In the Xenopus early blastula stage, the vegetal cells interact with the animal pole cells to induce the mesoderm in the marginal zone which soon becomes regionalized through the graded action of several growth factors, including members of the TGFβ superfamily and FGF (Slack, 1994; Smith, 1993). The Spemann's organizer, corresponding to a territory located in the dorsal marginal zone, is established progressively at the end of the blastula stage. This territory acquires the autonomous capacity to involute through the complex movements of convergence-extension during gastrulation. Goosecoid, a homeotic gene, has been found to be of crucial importance in the acquisition of the morphogenetic properties of the Spemann's organizer (Niehrs et al., 1993). In an elegant series of experiments, De Robertis and colleagues have shown that goosecoid can confer the properties of the Spemann's organizer to ventral blastomeres. Embryos in which C4 ventral blastomeres are injected with goosecoid mRNA develop with a duplicated axis. Furthermore, the migratory behavior of cells deriving from the involuting marginal zone is particularly enhanced when over-expressing goosecoid. The blastula stage embryo is subsequently reorganized into a three-layered embryo in which the body plan is progressively established through a series of inductive events. Different assays have been designed for the study of neural induction and the subsequent regionalization of the neural primordium. Folistatin was discovered

NATO ASI Series, Vol. H 92
Signalling Mechanisms – from Transcription Factors to Oxidative Stress
Edited by L. Packer, K. Wirtz

recently to engage the ectoderm into neurogenesis through the local inhibition of activin, a potent mesodermal inducer; the neurogenic differentiation program must now be considered as a default pathway (Hemmati-Brivanlou et al., 1994).

These recent studies have allowed the identification of growth factors and transcription factors as two major classes of inducers. A similar situation is found at later stages of development in many tissues including the nervous system. Perhaps one of the most striking examples is the establishment of dorso-ventral polarity in the neural tube. Very recent studies have shown that the notochord induces the formation of the floor plate in the ventral neural tube, in turn the floor plate establishes a ventro-dorsal polarity later reflected by the positions of motoneurons, intermediate and commissural neurons; A novel growth factor: sonic hedgehog was identified as one of the key inducer in the notochord and the floor plate (Ingham, 1994; Roelink et al., 1994). Dorsalin, a novel member of the TGFβ superfamily, was found to be expressed transiently in the dorsal neural tube (Basler et al., 1993); Dorsalin marks the presumptive neural crest territory and perhaps creates a dorso-ventral gradient which may antagonize the ventralizing activity of sonic hedgehog. Most interestingly, the neural crest, which is composed of neural epithelial cells, converts into mesenchymal migratory cells (Erickson and Perris, 1993). These cells will populate many different territories outside the nervous system to give rise to the peripheral nervous system, many cranio-facial structures and melanocytes. Very recently, slug, a zinc finger transcription factor, was found to be specifically expressed in the dorsal neural tube in the presumptive territory of the neural crest (Nieto et al., 1994). Slug is also found to be specifically associated with cells undergoing an epithelial-mesenchymal transition (EMT) in the primitive streak during gastrulation. Slug remains expressed, at least transitorily, in the migratory cells both at the primitive streak level during gastrulation and in all the neural crest cells. Slug is closely related to the transcription factor snail originally identified in Drosophila as a gene specifying the territory of the ventral furrow. The homologue of snail has been identified in a number of vertebrates where it is expressed in several mesenchymes.

Scatter/growth factors in epithelial morphogenesis

These recent findings point to the crucial need to understand the morphoregulatory activity of multifunctional growth factors. In our laboratory, we have focussed our work on the analysis of the action of growth factors that can induce the in vitro conversion of epithelial cells into mesenchymal cells. The process of epithelial-mesenchymal transition (EMT) operates repeatedly during embryogenesis, including in gastrulation of many vertebrates, the formation of neural crest cells and the differentiation of somites.

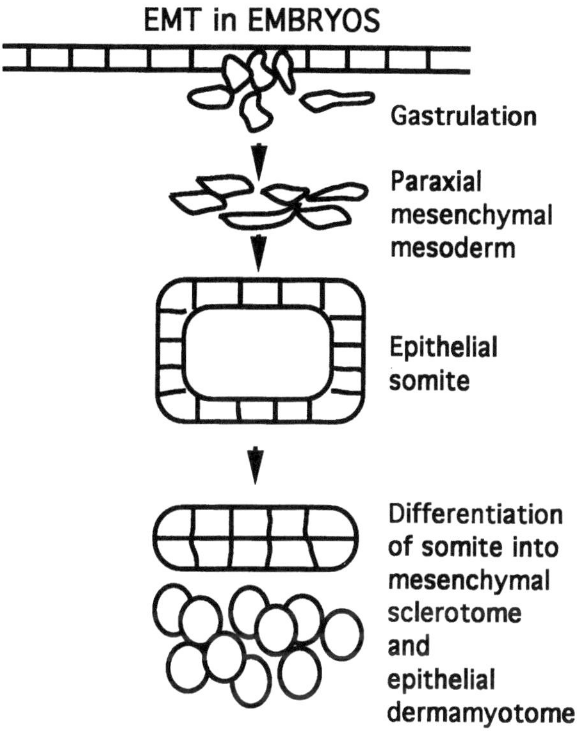

We have studied the rat bladder carcinoma cell line: NBT-II which can convert into a fibroblastic-like phenotype upon exposure to growth factors such as Fibroblast Growth Factor 1 (FGF-1), EGF, TGFα, Scatter Factor/Hepatocyte Growth Factor (SF/HGF) (Bellusci et al., 1994a; Gavrilovic et al., 1990; Vallés et al., 1990a). EMT is also obtained when NBT-II cells are grown on substrates coated with native collagens (Tucker et al., 1990). FGF-1 induces the dissociation of epithelial clusters within 3 to 5 hours and specialized junctions such as desmosomes are internalized. During the first hours of FGF-1 induction, the soluble pool of desmoplakins (DP) transiently increases, while the insoluble pool of DP molecules, which corresponds to cytoskeletal-linked proteins decreases. These results suggest that solubilization of DP proteins may correspond to an early step in desmosome destabilization leading to cell dissociation. Cytoskeletal proteins are also targets of EMT-inducing molecules. The reorganization of the cytokeratin and the actin filament systems and the appearance of vimentin, an intermediate filament protein characteristic of mesenchymal cells, is another feature characterizing the EMT of NBT-II cells (Boyer et al., 1989; Vallés et al., 1990a).
A similar kinetic of dissociation is observed with the other growth factors. It is noteworthy that all these growth factors recognize tyrosine kinase surface receptors.

In order to confirm whether the scatter activity of these growth factors requires the activation of their cognate receptors, we have treated cells with tyrphostins and genistein, two known inhibitors of tyrosine phosphorylation. In both cases, the EMT was measured quantitatively by determining the percentage of cortical DP negative cells. In contrast, treatment with sodium orthovanadate which inhibits tyrosine phosphatases, induces an EMT and cell locomotion in the NBT-II carcinoma cell line, strongly suggesting that the first step in signal transduction is likely to be the autophosphorylation of the tyrosine kinase receptors possibly followed by the activation of other kinases (Boyer and Thiery, 1993). NBT-II cells respond to FGF-1 but not to FGF-2 (basic FGF). Recently, several types of FGF receptors have been characterized. The prototype receptors flg (FGFR-1) or bek (FGFR-2) bind with similar affinities to FGF-1 and FGF-2. However, a splice variant of bek named K-SAM was found to recognize FGF-1 and FGF-7, formely called keratinocyte growth factor, rather than FGF-2. This receptor now designated FGFR2b was found to be expressed by subconfluent NBT-II cells. Indeed, these cells respond also to FGF-7, albeit transiently. We have discovered that the inability to maintain a dissociated state was a consequence of a down-regulation of the FGFR2b; this receptor was progressively replaced by its spliced variant FGF2c. After one week of treatment with FGF-7, the culture appeared heterogeneous since epithelial clusters reacquiring the FGFR2b now respond to FGF-7, and dissociate while the already dissociated cells return to the epithelial state since they lose FGFR2b (Savagner et al., 1994).

A careful analysis of the ability of NBT-II cells to respond to FGF-1, EGF, TGFα and SF/HGF revealed that cells in subconfluent cultures were able to scatter but not to proliferate. In contrast, the same growth factors induced DNA synthesis in cells maintained at high density. In large colonies, peripheral cells disperse rapidly while cells in the center enter mitosis (Vallés et al, 1990b). This finding may be relevant to morphogenetic processes in which most cells remaining in the tissue rudiment proliferate, however cells at its edge may be released. It is also an important concept to be considered for the mechanism of carcinoma invasion and metastasis; the primary tumor continues to grow in its center while peripheral cells can be released in the stroma and invade the surrounding tissues or metastasize into distant organs. The opposing activities of these growth factors depend on the cell status and is not accounted for by differences in the prevalence or affinity of the receptors on the surface of confluent versus subconfluent cells. Furthermore, the stimulation of each of the tyrosine kinase receptors, expressed by NBT-II cells, leads to these two different responses and is strictly dependent on the cell status. It is therefore intriguing to consider that there are two distinct transduction pathways in low and high density culture cells. The first evidence that this is the case comes from an experiment using TGFβ, a potent suppressor of epithelial cell proliferation. TGFβ abolished the mitogenic response of NBT-II cells treated with FGF-1. However, it did not perturb the EMT when cells were maintained at low density. To further discriminate the two transduction pathways, we have studied the biological responses of NBT-II cells stimulated by FGF-1 after treatment with agents known to synergize or antagonize second messengers. The mitogenic response was considerably increased when cells were treated with cAMP elevating agents including 8-bromo cAMP, dibutyryl cAMP, forskolin and cholera toxin. These agents potentiated the action of FGF-1, even at sub-threshold concentrations of the growth factor. In contrast, the EMT was dramatically inhibited by these drugs. Migratory mesenchymal NBT-II cells maintained in the presence of the dissociating growth factor for weeks returned to the epithelial state. The vimentin intermediate filament network collapsed around the nucleus while cells stop locomoting and small desmosomes form de novo at the cell surface of cells in contact (Boyer and Thiery, 1993). We are currently investigating possible mechanism of action regulating these two different cellular responses to FGF-1. We are attempting to identify which of the signalosome components may be specifically associated with an early event in one of these two pathways.

Role of scatter/growth factors in vivo

Perhaps one of the most impressive properties of the SF/HGF, which is a prototypic growth/scatter factor, is its ability to induce branching morphogenesis in vitro 3D cultures of MDCK cells (Matsumoto and Nakamura, 1993; Montesano et al., 1991). It is perhaps not surprising to observe a differential effect in vitro on a 2D tissue culture substrate and in 3D culture. We can easily reconcile these two apparently different responses since the 2D substrate may strongly favor separation of cells from small epithelial islands and promote locomotion, while the 3D culture of organoids may permit growth at the same time as transient dissociation and reepithelialization in buds forming along original tubules, allowing branching morphogenesis to occur. A fine regulation of EMT and its reversal to the epithelial state (MET) may be ensured either by the control of the prevalence of the surface receptors or by the biodisponibility of the growth factors. Recently we have found that a normal mammary epithelial cell line was able to convert transiently to the mesenchymal state in the presence of EGF, but reverted spontaneously to the epithelial state; the MET correlated with the down-regulation of the EGF receptor after several days of exposure to the growth factor. (Matthay et al., 1993). In vivo branching morphogenesis occurs in many epithelia which are in close contact with mesenchymes expressing high amount of SH/HGF as well as other growth factors. There are good reasons to believe that SF/HGF or members of this family are one of the key components in epithelial-mesenchymal interactions, although these factors may not provide the specificity of induction which is a well known property of many types of mesenchymes.

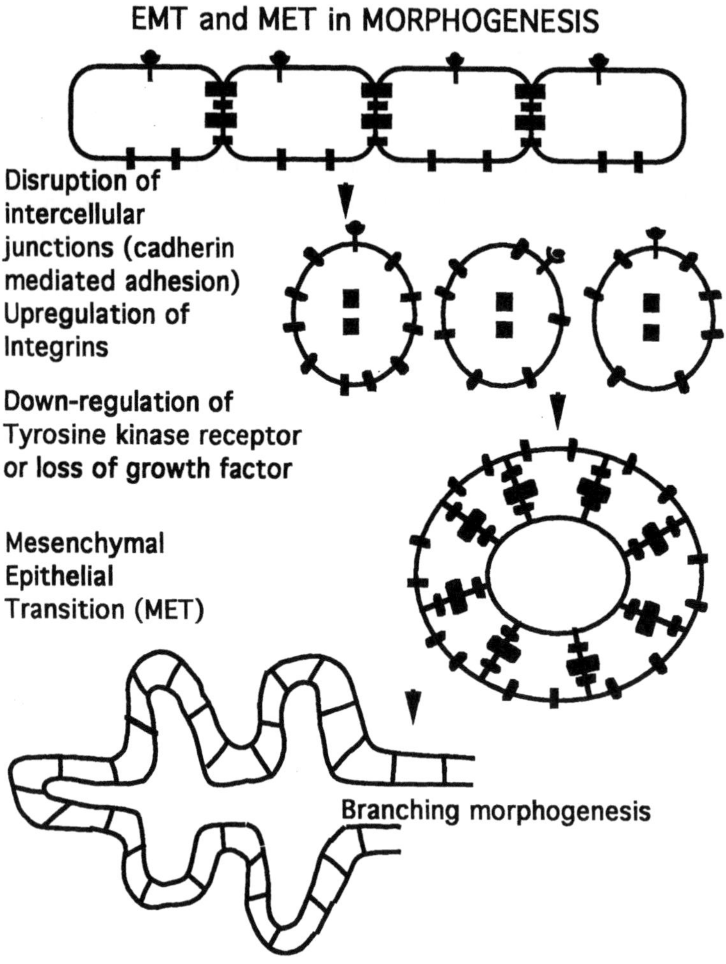

Synergy between collagens and growth factors in EMT

Early on, our attention was focussed on NBT-II cells since they were found to locomote when grown on collagen type 1. We have extended these studies showing that NBT-II can interact with many extracelullar matrix components including laminin and fibronectin. However, an EMT was obtained only in the presence of native collagens (type I, III, IV, V). Growth/scatter factors (FGF-1, TGFα) considerably accelerated the dispersion and increased the speed of locomotion of NBT-II cells grown on collagen (Tucker et al., 1991a). The findings may be particularly relevant to the in vivo situation where cells detaching from an epithelium will be exposed to the extracellular matrix (ECM). The combined action of the scatter/growth factor and of the ECM will greatly favor the dispersal of newly formed mesenchymal cells. Work in progress in our laboratory shows that FGF-1 induces an over-expression of one integrin which specifically recognizes collagen type 1 (Vallés et al., unpublished). This mechanism may well explain the synergy of action of the two EMT inducers.

Role of scatter/growth factors in invasion of carcinoma cells

To test whether the bladder carcinoma cells will have an increased invasiveness in the presence of growth factors, we first cultured these cells on the surface of the three dimensional collagen gel. A dramatic increase in the invasive properties of NBT-II cells was observed in the presence of FGF-1. Interestingly, gelatinase activities are induced early after FGF-1 or TGFα treatment of NBT-II cells facilitating the invasion into the three D-collagen gel (Gavrilovic et al., 1990)

We then confronted NBT-II cells-spheroids with rat bladder explants maintained in organotypic cultures. This orthotopic confrontation method allows the reproduction of some of the critical steps in tumor invasion in vivo. NBT-II clusters progressively replaced the urothelium at the site of contact with the explant. In the absence of growth factors, the carcinoma cells grew in a pattern suggestive of in situ carcinoma. In contrast, several members of the FGF growth factor family, including FGF-1, FGF-4 and TGFα, promoted the invasive ability of NBT-II cells. The most dramatic figures of invasion were obtained with FGF-4 autocrinally expressed by NBT-II cells following transfection with the appropriate cDNA. These data strongly support the idea that scatter/growth factors can potentiate invasiveness in the complex environment of the bladder tissues (Tucker et al., 1991b).

Role of growth/scatter factor in tumor progression

We have compared the tumorigenic and metastastatic potential of untransfected NBT-II cells and NBT-II cells expressing one growth factor: FGF-1, FGF-1 linked to a signal peptide, FGF-4, TGFα, SF/HGF. NBT-II cells formed detectable tumors seven weeks after subcutaneous injections of 3.5. 10^6 cells in nude mice. Micrometastases were also detected at that time. In sharp contrast, transfected NBT-II cells, producing either the normal form of FGF-1 or a secreted form, developed tumors within one week and in most cases, mice bearing very large tumors died within 2 weeks (Bellusci et al., 1994a; Jouanneau et al., 1991).

The tumors deriving from NBT-II producing one of two forms of FGF-1 were more vascularized. However, the tumor derived from NBT-II secreting FGF-1 was more abnormally vascularized than those obtained with untransfected NBT-II or NBT-II cells producing other growth factors. A careful histological analysis revealed that the tumors were massively infiltrated with blood cells in large well delimited cisternal spaces, often partially or totally devoid of an endothelial lining. This abnormal vascularization was not found in tumors expressing FGF-4 which is normally secreted by cells since its gene, in contrast to FGF-1 and FGF-2, encodes a signal peptide (Jouanneau et al., submitted). Similarly, tumors deriving from TGFα or SF/HGF expressing NBT-II were characterized by a normal vascularization. The tumors expressing FGF-4, SF/HGF or TGFα started to develop with a lag-time which was however reduced compared to untransfected NBT-II cell inoculate. These studies led to the conclusion that FGF-1, produced autocrinally by NBT-II cells considerably increased their tumorigenicity, and furthermore, micrometastases were detected at a very early stage of the primary tumor formation. One of the critical issue is to determine the relative importance of angiogenesis in the early development of tumors. Experiments are in progress in our laboratory to answer this question.

Tumors usually develop as a heterogeneous collection of cells. Accumulation of genetic alterations and epigenetic mechanisms contribute to the maintainance of this heterogeneity. However, in experimental tumors, cells of a given genotype may compete efficiently with the other cells and may become dominant within the primary tumor. Clonal dominance may be ensured by the ability of a minor population of cells to produce its own growth/scatter factors. In order to challenge this hypothesis, we have inoculated NBT-II cells containing a defined ratio of FGF-1 producing cells (14% to 0.02%). Surprisingly, tumors originating from inoculate containing 14% of cells producing FGF, develop almost as rapidly as those deriving from a pure population of FGF-1 producing cells. However, the primary tumor and the metastases were found to be composed of similar proportions of the two cell populations, indicating that the FGF producing cells had no growth or tumorigenic advantage in the presence of untransfected cells. In fact, these cells confer an increased tumorigenic potential to the untransfected NBT-II cells. Even a very small percentage of FGF-1 producing cells (less than 0.01%) could confer this advantage. We made the hypothesis that the development of primary tumors, with a heterogenous population, may occur through a community effect (Jouanneau et al., 1994). This mechanism has been discovered and analyzed extensively in the formation of muscle in Xenopus embryos (Gurdon et al., 1993). As mentioned at the beginning of this article, the induction and regionalization of the mesoderm is a result of a combinatorial effect of growth factors expressed in time and space in a graded manner. However, a large territory in the newly induced heterogenous mesoderm has to differentiate into the striated muscle cell lineage. The molecular basis of the community effect is not known, but it is assumed to involve the local production of a growth factor which will contribute to the establishment of a homogenous territory.

Identification of a novel growth/scatter factor implicated in bladder carcinoma tumor progression by a community effect.

Epithelial NBT-II cells inoculated in nude mice give rise to primary tumors and micrometastases in lymph nodes after a long latency period. Carcinoma cells were rescued from the lymph nodes and expanded in culture prior to their inoculation. These cultures were often heterogeneous containing epithelial colonies and mesenchymal cells. Tumors deriving from these metastatic cells develop faster than E-NBT-II-cells. Cells from lymph node micrometastases had essentially a mesenchymal phenotype in culture. When injected in nude mice, they produced tumors with the same kinetics as the FGF-1 producing cells. We hypothesized that we had selected a clone of M-NBT-II (M stands for mesenchymal or metastatic) which produces, autocrinally, a growth/scatter factor having very similar properties to FGF-1. The culture supernatant of M-NBT-II was able to dissociate subconfluent culture of E-NBT-II. It also scattered MDCK cell colonies. Partial purification of the factor indicated that it shared some common properties with SF/HGF, including heparin binding and chromatographic behavior on anion and cation exchange resins. We suspect that this factor also has a similar molecular weight. Since it was also able to scatter MDCK cells, we have named it scatter factor-like (SFL). However, we have demonstrated that SFL is not immunologically related to SF/HGF. In addition, we could not detect by Northern blot or PCR detectable amount of SF/HGF transcript. In addition, SFL could not induce tubulogenesis in MDCK cell grown in 3D collagen gels (Bellusci et al., 1994b). E-NBT-II cells, inoculated with SFL producing cells (M-NBT-II) in a 7 to 1 ratio, were able to develop tumors with a greatly reduced lag-time, similarly to the effects obtained with FGF-1 producing cells. SFL can therefore be a naturally produced factor mediating a community effect in bladder tumors (Bellusci et al., 1994c).

Concluding remarks.

This short review provides additional evidence that growth factors are multifunctionals. During early embryogenesis members of the FGF and TGFβ families can induce pluripotent cells to become assigned to a mesodermal fate. These molecules later have diverse functions such as growth factors, angiogens, differentiation factors and survival factors. Several growth factors can also induce scattering of epithelial colonies and cell motility. In vivo, they may be involved in EMT and branching morphogenesis. During tumor progression these factors may contribute to the growth of the primary tumors, to its vascularization and to the dispersal of cells during the transition from in situ to invasive carcinoma. One critical issue is to try and elucidate the transduction pathways involved in each of these functions; each of these pathways may already be established in a given cell type according to its status in cell collectives.

Acknowledgements

I thank Dr. Nish Patel for critical reading of this manuscript. Works from the author's laboratory is supported by the CNRS, ARC 6455, the Ligue contre le Cancer (National Committee and Committee of Paris) and the N.I.H. (Grant R01 CA 49417-05).

References

Basler, K., T. Edlund, T. M. Jessell and T. Yamada. (1993). Control of cell pattern in - neural tube: regulation of cell differentiation by dorsalin-1, a novel TGFβ family member. Cell. 73: 687-702.

Bellusci, S., G. Moens, A. Delouvée, J. P. Thiery and J. Jouanneau. (1994c). SFL production by carcinoma cells induces the aggressive properties of non-producing cells in vivo by way of a community effect. Invasion Metastasis. in press:

Bellusci, S., G. Moens, G. Gaudino, P. Comoglio, T. Nakamura, J. P. Thiery and J. Jouanneau. (1994a). Creation of an hepatocyte growth factor/scatter factor autocrine loop in carcinoma cells induces invasive properties associated with increased oncogenic potential. Oncogene. 9: 1091-1099.

Bellusci, S., G. Moens, J. P. Thiery and J. Jouanneau. (1994b). A scatter factor-like factor is produced by a metastatic variant of a bladder carcinoma line. J. Cell Science. 107: 1277-1287.

Boyer, B. and J. P. Thiery. (1993). Cyclic AMP Distinguishes between two functions of acidic FGF in a rat bladder carcinoma cell line. J. Cell Biol. 120: 767-776.

Boyer, B., G. C. Tucker, A. M. Vallés, W. W. Franke and J. P. Thiery. (1989). Rearrangements of desmosomal and cytoskeletal proteins during the transition from epithelial to fibroblastoid organization in cultured rat bladder carcinoma cells. J. Cell Biol. 109: 1495-1509.

Erickson, C. A. and R. Perris. (1993). The role of cell-cell and cell-matrix interactions in the morphogenesis of the neural crest. Dev. Biol. 159: 60-74.

Gavrilovic, J., G. Moens, J. P. Thiery and J. Jouanneau. (1990). Expression of TGFα induces a motile fibroblast-like phenotype with extracellular matrix-degrading potential in a rat bladder carcinoma cell line. Cell Regulation. 1: 1003-1014.

Gurdon, J. B., E. Tiller, J. Roberts and K. Kato. (1993). A community effect in muscle development. 3: 1-11.

Hemmati-Brivanlou, A., O. G. Kelly and D. A. Melton. (1994). Follistatin, an antagonist of activin, is expressed in the Spemann organizer and displays direct neuralizing activity. Cell. 77: 283-295.

Ingham, P. W. (1994). Hedgehog points the way. Curr. Biol. 4: 347-350.

Jouanneau, J., J. Gavrilovic, D. Caruelle, M. Jaye, G. Moens, J. P. Caruelle and J. P. Thiery. (1991). Secreted or non-secreted forms of acidic fibroblast growth factor produced by transfected epithelial cells influence cell morphology, motility and invasive potential. Proc. Natl. Acad. Sci. USA. 88: 2893-2897.

Jouanneau, J., G. Moens, Y. Bourgeois, M. F. Poupon and J. P. Thiery. (1994). A minority of carcinoma cells producing acidic fibroblast growth factor induces a community effect for tumor progression. Proc. Natl. Acad. Sci. USA. USA 91: 286-290.

Matsumoto, K. and T. Nakamura. (1993). Roles of HGF as a pleiotropic factor in organ regeneration. 225-249.

Matthay, M. A., J. P. Thiery, F. Lafont, M. F. Stampfer and B. Boyer. (1993). Transient effect of epidermal growth factor on the motility of an immortalized mamary epithelial cell line. J. Cell Science. 106: 869-878.

Montesano, R., K. Matsumoto, T. Nakamura and L. Orci. (1991). Identification of a fibroblast-derived epithelial morphogen as hepatocyte growth factor. Cell. 67: 901-908.

Niehrs, C., R. Keller, K. W. Y. Cho and E. M. De Robertis. (1993). The homeobox gene goosecoid controls cell migration in Xenopus embryos. Cell. 72: 491-503.

Nieto, M. A., M. G. Sargent, D. G. Wilkinson and J. Cooke. (1994). Control of cell behavior during vertebrate development by slug, a zinc finger gene. Science. 264: 835-839.

Roelink, H., A. Augsbrurger, J. Heemskerk, V. Korzh, S. Norlin, A. Ruiz i Altaba, Y. Tanabe, M. Placzek, T. Edlund, T. M. Jessell and J. Dodd. (1994). Floor plate and motor neuron induction by vhh-1, a vertebrate homolog of hedgehog expressed by the notochord. Cell. 76: 761-775.

Savagner, P., A. M. Vallés, J. Jouanneau, K. Yamada and J. P. Thiery. (1994). Alternative splicing in Fibroblast Growth Factor receptor 2 is associated with induced epithelial-mesenchymal transition in rat bladder carcinoma cells. Mol. Biol. Cell. 5: in press.

Slack, J. M. W. (1994). Inducing factors in Xenopus early embryos. Curr. Biol. 4: 116-126.

Smith, J. C. (1993). Mesoderm-inducing factors in early vertebrate development. EMBO J.. 12: 4463-4470.

Spemann, H. and H. Mangold. (1924). Ueber induktion von embryonalanlagen durch implantation artfremder organisatoren. Arch. mikrosk. Anat. Entwmech. 100: 599-638.

Tucker, G. C., B. Boyer, J. Gavrilovic, H. Emonard and J. P. Thiery. (1990). Collagen-mediated dispersion of NBT-II rat bladder carcinoma cells. Cancer Res. 50: 129-137.

Tucker, G. C., B. Boyer, A. M. Vallés and J. P. Thiery. (1991a). Combined effects of extracellular matrix and growth factors on NBT-II rat bladder carcinoma cell dispersion. J. Cell Science. 100: 371-380.

Tucker, G. C., A. Delouvée, J. Jouanneau, J. Gavrilovic, G. Moens, A. M. Vallés and J. P. Thiery. (1991b). Amplification of invasiveness in organotypic cultures after NBT-II rat bladder carcinoma stimulation with in vitro scattering factors. Invasion Metastasis. 11: 297-309.

Vallés, A. M., B. Boyer, G. C. Tucker, J. Badet, D. Barritault and J. P. Thiery. (1990a). Acidic fibroblast growth factor is a modulator of epithelial plasticity in a rat bladder carcinoma cell line. Proc. Natl. Acad. Sci. USA. 87: 1124-1128.

Vallés, A.M., Tucker, G., Thiery, J.P. and Boyer, B. (1990) Alternative patterns of mitogenesis and cell scattering induced by acedic FGF as a function of cell density in a rat bladder carcinoma cell line. Cell Regulation 1, 975-988 .

GENE TARGETING OF THE RECEPTOR-LIKE PROTEIN TYROSINE PHOSPHATASE LAR BY HOMOLOGOUS RECOMBINATION IN MOUSE EMBRYONIC STEM CELLS

Roel Schaapveld, Jan Schepens, Frank Oerlemans, Michel Streuli*, Bé Wieringa, and Wiljan Hendriks

Department of Cell Biology & Histology, University of Nijmegen, Adelbertusplein 1, 6525 EK Nijmegen, The Netherlands

*Division of Tumor Immunology, Dana-Farber Cancer Institute, 44 Binney Street, Boston, MA 02115, USA

SUMMARY

Receptor-like protein tyrosine phosphatases (RPTPases) comprise an extracellular ligand-binding region, a transmembrane domain, and as a rule two cytoplasmic tyrosine phosphatase domains. *In vitro* studies using the cytoplasmic parts of RPTPases and artificial substrates have suggested that the first, membrane proximal phosphatase domain exhibits catalytic activity, whereas the second, C-terminal phosphatase domain may regulate the phosphatase activity of the first domain. Further studies, however, are hampered by the fact that RPTPase-specific ligands and substrates still remain to be identified. Also, the complexity of transmembrane signalling is difficult to mimick *in vitro*. To circumvent these problems, the individual functions of the two phosphatase domains in RPTPases can be studied *in vivo* by means of homologous recombination in mouse embryonic stem (ES) cells. Here, we decribe the use of 'double replacement' gene targeting in mouse embryonic stem cells to generate cell and animal models for studying the individual role of both phosphatase domains of the RPTPase Leukocyte common Antigen-Related molecule LAR. In addition, exploiting the process of gene conversion, LAR-negative ES cells were generated to enable structure-function analysis of LAR mutants on a null background.

NATO ASI Series, Vol. H 92
Signalling Mechanisms – from Transcription Factors to Oxidative Stress
Edited by L. Packer, K. Wirtz

RECEPTOR-LIKE PROTEIN TYROSINE PHOSPHATASES

Protein-tyrosine phosphatases (PTPases) are the natural antagonists of the well-known protein tyrosine kinases (PTKs) and have an important role in the control of many cellular signalling pathways, cell growth, and differentiation (Fischer et al., 1991; Tonks et al., 1992; Mourey and Dixon, 1994). The identification of many distinct PTPase genes and the discovery of multiple isoforms resulting from alternative splicing and post-translational modifications have revealed a suprising complexity for this protein family. On the basis of their overall structures, two classes of PTPases can be distinguished, namely (i) those found in the cytoplasm and the nucleus that have only one tyrosine phosphatase domain and (ii) membrane-bound, receptor-like PTPases (RPTPases) that have an extracellular ligand-binding region, a transmembrane domain, and, with a few exceptions, two repeated cytoplasmic tyrosine phosphatase domains (Fig. 1).

One of the best studied RPTPases is the Leukocyte common Antigen-Related molecule LAR (Streuli et al., 1988). LAR is composed of two cytoplasmic phosphatase domains, a transmembrane segment, and an extracellular part which shares homology to

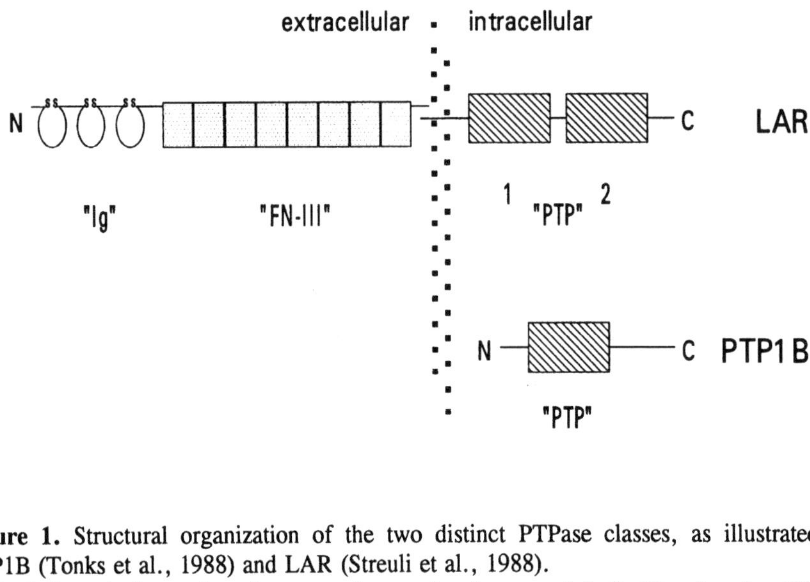

Figure 1. Structural organization of the two distinct PTPase classes, as illustrated by PTP1B (Tonks et al., 1988) and LAR (Streuli et al., 1988).
Dashed lines indicate the plasmamembrane. Ig, Immunoglobulin-like domain; FN-III, fibronectin type III repeat; PTP, protein tyrosine phosphatase domain; N, amino terminus; C, carboxy-terminal end.

immunoglobulin (Ig)-like and fibronectin type III (FN-III) domains (Fig. 1). Such an arrangement of these extracellular motifs is commonly found in cell adhesion molecules like N-CAM (Edelman and Crossin, 1991). It is tempting to speculate that LAR may play a role in cell adhesion, converting extracellular signals into intracellular responses by ligand-mediated dephosphorylation of intracellular substrates.

LAR is expressed on the cell surface as a complex of two non-covalently associated subunits derived from a pro-protein (Streuli et al., 1992; Yu et al., 1992). A ~200 kDa precursor is intracellularly processed at a paired basic amino acid site by a subtilisin-like endoprotease generating 150 kDa and 85 kDa fragments (Fig. 1). The N-linked glycosylated 150 kDa fragment represents the amino terminus of the protein, and is shed during growth. Whether this shedding has an effect on the intracellular phosphatase activity remains to be investigated. The 85 kDa fragment contains the transmembrane segment and the two phosphatase domains of approximately 260 amino acid residues each. *In vitro* site-directed mutagenesis studies have suggested that the first phosphatase domain in RPTPases exhibits catalytic activity, for which a cysteine residue in the 'signature sequence' (I/V)HCXAGXXR(S/T)G is essential (reviewed in Mourey and Dixon, 1994). It is still a point of controversy whether the second, C-terminal phosphatase domain has only a regulatory function in modulating the substrate specificity of the first domain (Streuli et al., 1990; Pot et al., 1991; Krueger and Saito, 1992), or has catalytic activity by itself (Wang and Pallen, 1991; Tan et al., 1993). This problem is hard to address because of the complexity of transmembrane signalling. Furthermore, extracellular ligands and downstream substrates, that are specific for a single RPTPase, are currently not known. Therefore, we set out to study the loss of function for the two individual phosphatase domains of LAR *in vivo* by means of gene targeting using homologous recombination in mouse embryonic stem (ES) cells.

GENE TARGETING BY HOMOLOGOUS RECOMBINATION

Gene targeting, the homologous recombination of DNA sequences residing in the chromosome with newly introduced DNA sequences (Thomas and Capecchi, 1987), is now widely used to study gene function *in vivo* (for review see Koller and Smithies, 1992). Pluripotent ES cells, containing the desired alteration of the genome, are microinjected into blastocysts and subsequently transferred into pseudo-pregnant foster mothers. These ES cells can contribute to the germline of the resulting chimaeric mice, which then can transmit the mutated allele to their offspring. In this way, mutations can be transferred from the culture dish to the whole animal. Subsequent inbreeding can give rise to mice carrying the mutation in both alleles, allowing the analysis of the mutation in the heterozygous and homozygous state.

Thusfar, gene inactivation is the major application of gene targeting and more than a hundred different so-called knock-out mice have been created and studied. Since phenotypic alterations in null mutants are often very complex or embryonic lethal, gene function may be better understood by introducing subtle mutations in functional domains of a protein while leaving the remainder of the protein intact. To date, three procedures have been described to introduce specific mutations in the genome, namely (i) the 'hit and run' or 'in-out' procedure (Hasty et al., 1991; Valancius and Smithies, 1991), (ii) a one-step recombination strategy using the Cre-*lox*P recombination system (Gu et al., 1993), and (iii) the 'double replacement', 'two-step', or 'tag-and-exchange' strategy (Askew et al., 1993; Gondo et al., 1994; Stacey et al., 1994; Wu et al., 1994). This latter procedure requires two rounds of homologous recombination using replacement-type targeting vectors. In the first step, a cassette containing both a positive and a negative selectable marker is introduced into the gene locus to mark the region of interest by selection for the positive selectable marker. In the second step, using the negative selectable marker to enrich for recombinants, the cassette is replaced again by the original endogenous sequences carrying any desired subtle mutation (see also Fig. 2). This procedure can be used to generate cell lines and subsequently mice in which either the LAR phosphatase domains 1 or 2, or both, are inactivated.

TARGETING OF THE RPTPase LAR GENE

Using a human LAR cDNA clone as a probe, the relevant segments of the structural and genomic murine LAR DNAs have been isolated (Schaapveld et al., 1994). A replacement-type targeting vector was constructed in which a dominant positive/ negative selection marker, a hygromycin phosphotransferase-thymidine kinase (HyTK) fusion gene (Lupton et al., 1991), is flanked by 3.2 and 2.3 kb of endogenous LAR

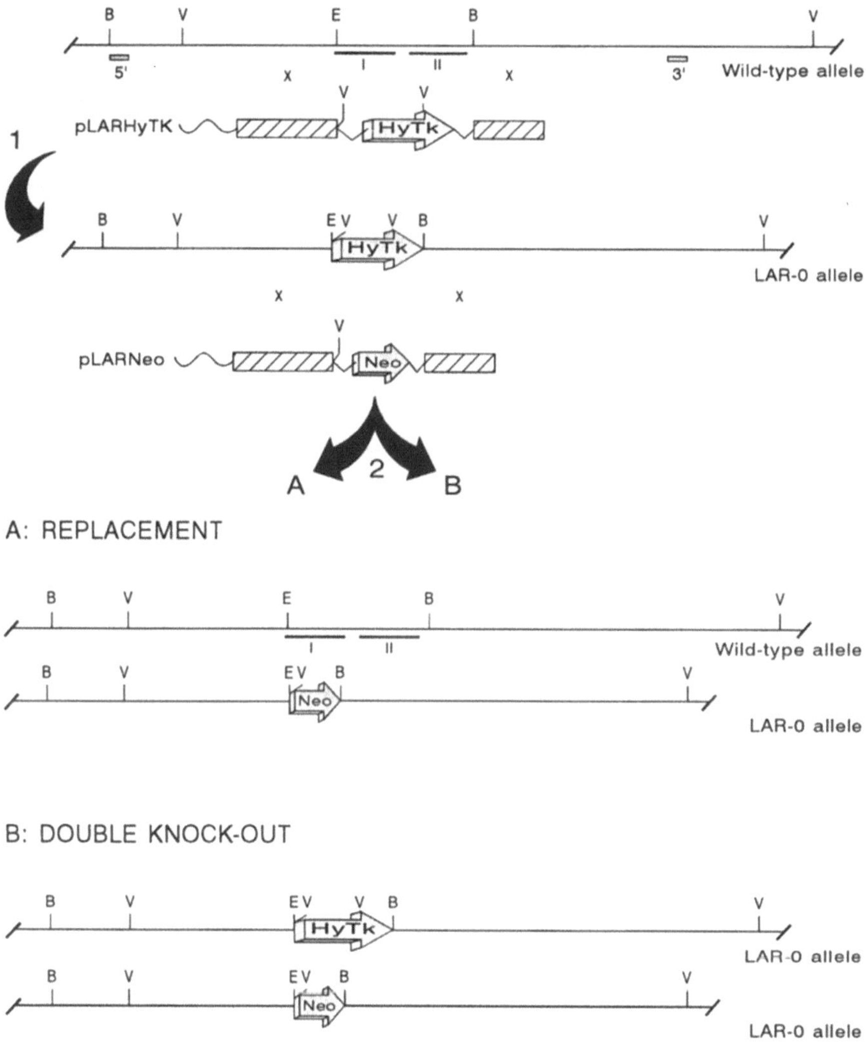

Figure 2. Disruption of the LAR gene by double replacement targeting.
A recombination event replaces the genomic sequences comprising both phosphatase domains by the HyTK fusion cassette (step 1). The resulting HygroB^r/FIAU^s homologous recombinant cell line is then used in a second round of targeting (step 2) in which either the HyTK fusion cassette (A) or the LAR phosphatase domains of the remaining wild-type allele (B) are replaced by the Neo gene.
Hatched bars represent the endogenous LAR genomic segments and the grey arrows symbolize the selection cassettes. Small grey bars indicate the 5'- and 3'-diagnostic probes, and the solid bars I and II mark the genomic segments encoding the phosphatase domains 1 and 2, respectively. B, *Bam*HI; V, *Eco*RV; E, *Eco*RI.

genomic segments which provide the necessary homology for targeted integration. Upon homologous recombination, a 4.5 kb genomic segment, containing all exons encoding the cytoplasmic part of LAR, will be replaced by the HyTK cassette (Fig. 2, step 1). The targeting construct was introduced into E14 ES cells by electroporation and 272 Hygromycin B resistant (HygroB^{r}) clones were screened for homologous recombination by Southern blot analysis using 5'- and 3'-diagnostic probes derived from genomic parts flanking the targeting vector region. Using the 3'-probe, ten clones were found to display the diagnostic *Eco*RV fragment, indicative of a targeting event. Eight out of these ten clones also revealed the proper *Bam*HI digestion pattern with a 5'-probe, demonstrating correct homologous recombination at both ends (Fig. 3). Unfortunately, these cell lines were unsuitable for germline transmission, probably due to high expression of the thymidine kinase fusion protein in germ cells (Braun et al., 1990; Ramirez-Solis et al., 1993). However, they can serve as starting cell lines for the introduction of inactivating point mutations in the phosphatase domains by a second recombination step.

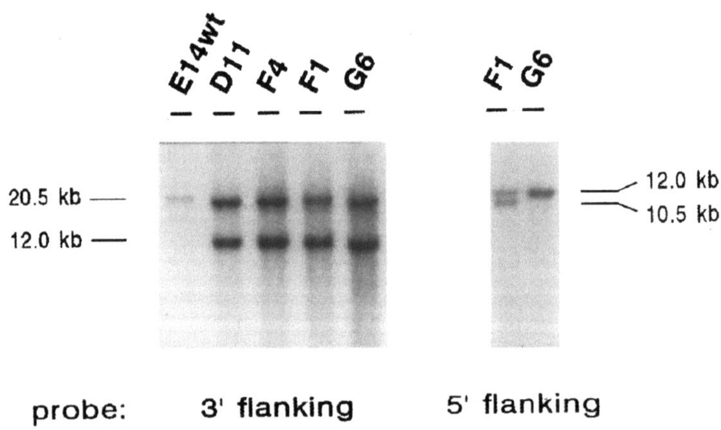

Figure 3. Screening for homologous recombinants.
Southern blot autoradiograms of ± 10 μg genomic DNA from individual HygroB^{r} ES clones. The 3'-flanking probe reveals the diagnostic 12.0 kb fragment in *Eco*RV digested DNA (left panel), indicating targeting at the LAR locus. Use of a 5'-flanking probe on *Bam*HI digested DNA (right panel) is necessary to confirm proper homologous recombination (clone G6 is a result of an integration event at the LAR locus).

To test the feasability of this 'double replacement' strategy, and to obtain germline transmission of a LAR null allele and consequently produce LAR knock-out mice, one of the targeted ES cell lines (E14.F4) was again transfected, but now with a gene targeting vector in which all exons encoding the cytoplasmic part of LAR were replaced by the neomycin phosphotransferase (Neo) gene (Fig. 2, step 2). Applying different combinations of the selective agents HygroB, G418 and 1-[2 deoxy, 2-fluoro-ß-D-arabinofuranosyl]-5-iodouracil (FIAU) resulted in 1) G418^r/FIAU^r ES cells, in which the HyTK cassette is exchanged for the Neo cassette, and 2) G418^r/HygroB^r ES cells, in which both LAR alleles are targeted (Table 1, Fig. 4). Highly chimaeric mice were generated upon injection into blastocysts of either 'replacement' clones or 'double knock-out' clones. Chimaeras resulting from the replacement clones were tested in a breeding program and germline transmission of the desired genotype has been obtained (Schaapveld et al., manuscript in preparation), demonstrating that a second round of homologous recombination in the E14.F4 cell line is possible without affecting the germline competence. This opens the way to introduce inactivating point mutations in each of the two LAR phosphatase domains using a replacement-type targeting vector carrying the mutated genomic segment.

Table 1. Targeting of the LAR gene in E14.F4 cells (step 2).

Selection	# clones screened	Targeting events	Homologous recombinants
G418 & HygroB	185	3	1 (DK)
G418	189	2	2 (DK & RC)
G418 & FIAU	17	2	2 (RC)

DK=Double Knock-out; RC=Replacement Clone.

As mentioned above, as a result of the second replacement step also cell lines were obtained in which both LAR alleles were inactivated. Such a double knock-out cell line would already be a very useful tool in assessing the function of a gene, especially when the mutation is lethal in early development. Others have already used double knock-out cell lines, generated by two subsequent rounds of targeting with two different vectors, to study genes involved in cell growth and differentiation (te Riele et al., 1990;

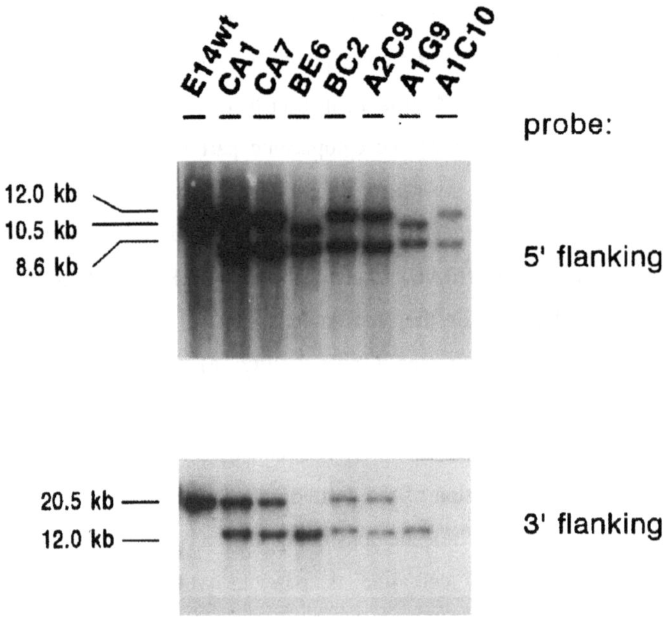

Figure 4. Detection of 'replacement' and 'double knock-out' clones.
Southern blot autoradiograms of ±10 μg genomic DNA from individual clones resulting from the second round of targeting in the E14.F4 cell line (see Fig. 3). In DNA digested with *Bam*HI (upper panel) the 5'-flanking probe reveals the diagnostic 8.6 kb and/or 10.5 kb fragments indicating targeting at the LAR locus. Replacement (BC2, CA1, and CA7) and double knock-out (A1G9, and BE6) cell lines are shown. Use of a 3'-flanking probe on *Eco*RV digested DNA (lower panel) reveals the 12 kb diagnostic fragment for both events. Screening with HyTK and Neo probes was necessary to discriminate between integration (A1C10, and A2C9) and recombination events (data not shown).

Mortensen et al., 1991). Recently, a method has been described that requires only a single targeting construct to generate double knock-out cell lines (Mortensen et al., 1992). it is based on the phenomenon that cells heterozygous for a given locus can be turned into homozygous cells by mitotic recombination or gene conversion. Considering the fact that cells bearing two copies of a Neo gene are more resistant to G418 selective effects, homozygous cells can be generated from a heterozygous cell line (containing only one copy of the Neo gene) by culturing in sublethal concentrations of G418. We tested this approach by taking advantage of the presence of the HyTK fusion cassette and the Neo gene in the LAR double knock-out HygroB^r/G418^r/FIAU^s cell line, E14.BE6. This cell line was cultured not only in the presence of a high G418 concentration, but also in the presence of FIAU. As a result, only cells that mutate or delete the HyTK cassette and at the same time increase their neomycin phosphotransferase levels will survive. Gene conversion resulting in the copying of the Neo containing allele onto the HyTK containing allele will have both these effects. In this way, we indeed were able to generate double knock-out cell lines containing two copies of the Neo gene (Table 2, Fig. 5).

Table 2. Use of high G418 concentrations to generate homozygous mutant ES cells.

Selection G418 (mg/ml)	Selection FIAU (μM)	# Clones picked	% Homozygous (# analyzed)
1.0	0.2	19	14 (7)
1.5	0.2	15	16 (12)
2.0	0.2	2	- (0)

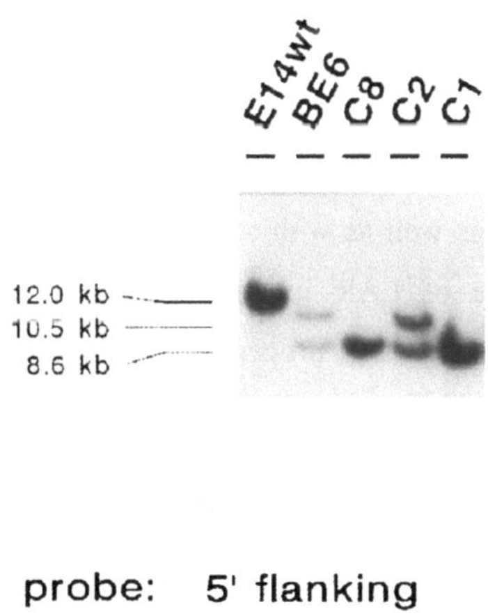

Figure 5. Detection of gene conversion events in LAR-negative ES cells.
Southern blot autoradiograms of ± 10 μg genomic DNA from individual clones surviving high concentration of G418. In DNA digested with *Bam*HI the 5'-flanking probe reveals the two diagnostic 8.6 kb fragments indicating loss of the HyTK cassette by gene conversion using the Neo allele as template (cell lines C1, and C8).

IN CONCLUSION

Using replacement-type targeting vectors we have shown various ways of generating cell and animal models to study the role of the RPTPase LAR in signal transduction. First, using a double replacement strategy involving two subsequent homologous recombination events, ES cell lines lacking the gene sequences essential for LAR phosphatase activity were generated. Second, using one such line as a model, the feasability to select for gene conversion events by raising the concentration of the selective agent was confirmed. Third, the E14.F4 line that resulted from the first targeting step involving the HyTK positive/negative selection cassette was succesfully used for a second round of homologous recombination without loss of germline competence.

We have observed homologous recombination events involving the region that spans both phosphatase domains of LAR at an average frequency of about 1 out of 50 clones for both step 1 and step 2. An insertion rather than a replacement mutation was seen in 20% of our clones. In agreement with others, this underlines the necessity to use probes not only from both sides of the targeted locus, but also probes spanning the selectable markers and the deleted region, to screen for genuine homologous recombination events. However, this frequency of illegtimate targeting events may not be general and might depend on intrinsic properties of the locus under study.

Future experiments will be aimed at the reconstitution of the LAR-negative ES cell line with wild-type human LAR as well as with several mutants. Generated cell lines can then be used to study the role of LAR in cell proliferation and differentiation by analyzing, for example, their cell growth properties, phopshotyrosine protein patterns, and the LAR phosphatase activity itself during ES cell differentiation. The strategies and techniques used here to create $LAR^{-/-}$ models are generally applicable and provide a unique means to study many 'orphan' signal transduction molecules.

ACKNOWLEDGMENTS

We would like to thank Dr. S. Lupton for providing the HyTK fusion gene. This work was supported by the Dutch Organization for Scientific Research (NWO).

REFERENCES

Askew GR, Doetschman T, and Lingrel JB (1993) Site-directed point mutations in embryonic stem cells: a gene targeting tag-and-exchange strategy. Mol. Cell. Biol. 13:4115-4124

Brady-Kalnay SM, and Tonks NK (1994) Protein tyrosine phosphatases: From structure to function. Trends Cell Biol. 4:73-76

Braun RE, Lo D, Pinkert CA, Widera G, Flavell RA, Palmiter RD, and Brinster RL (1990) Infertility in male transgenic mice: disruption of sperm development by HSV-tk expression in postmeiotic germ cells. Biol. Reprod. 43:684-693

Edelman, GM, and Crossin KL (1991) Cell adhesion molecules: Implications for a molecular histology. Annu. Rev. Biochem. 60:155-190

Fischer EH, Charbonneau H, Tonks NK (1991). Protein tyrosine phosphatases: A diverse family of intracellular and transmembrane enzymes. Science 253:401-406

Gondo Y, Nakamura K, Nakao K, Sasaoka T, Ito K-I, Kimura M, and Katsuki M (1994) Gene replacement of the p53 gene with the *lac*Z gene in mouse embryonic stem cells and mice by using two steps of homologous recombination. Biochem. Biophys. Res. Comm. 202:830-837

Gu H, Zou Y-R, Rajewski K (1993) Independent control of immunoglobulin switch recombination at individual switch regions evidenced through Cre-*lox*P-mediated gene targeting. Cell 73:1155-1164

Hasty P, Ramirez-Solis R, Krumlauf R, and Bradley A (1991) Introduction of a subtle mutation into the *Hox*-2.6 locus in embryonic stem cells. Nature 350:243-246

Koller BH, and Smithies O (1992) Altering genes in animals by gene targeting. Annu. Rev. Immunol. 10:705-730

Krueger NX, and Saito H (1992) A human transmembrane protein-tyrosine phosphatase, PTPζ, is expressed in brain and has an N-terminal receptor domain homologous to carbonic anhydrases. Proc. Natl. Acad. Sci. USA 89:7417-7421

Lupton SD, Brunton LL, Kalberg VA, and Overell RW (1991) Dominant positive and negative selection using a hygromycin phosphotransferase-thymidine kinase fusion gene. Mol. Cell. Biol. 11:3374-3378

Mortensen RM, Conner DA, Chao S, Geisterfer-Lowrance AAT, and Seidman JG (1992) Production of homozygous mutant ES cells with a single targeting construct. Mol. Cell. Biol. 12:2391-2395

Mortensen RM, Zubiaur M, Neer EJ, and Seidman JG (1991) Embryonic stem cells lacking a functional inhibitory G-protein subunit (α_{i2}) produced by gene targeting of both alleles. Proc. Natl. Acad. Sci. USA 88:7036-7040

Mourey RJ, and Dixon JE (1994) Protein tyrosine phosphatases: Characterization of cellular and intracellular domains. Curr. Op. Gen. Dev. 4:31-39

Pot DA, Woodford TA, Remboutsika E, Haun RS, and Dixon JE (1991) Cloning, bacterial expression, purification, and characterization of the cytoplasmic domain of rat LAR, a receptor-like protein tyrosine phosphatase. J. Biol. Chem. 266: 19688-19696

Ramirez-Solis R, Zheng H, Whiting J, Krumlauf R, and Bradley A (1993) *Hoxb*-4 (*Hox*-2.6) mutant mice show homeotic transformation of a cervical vertebra and defects in the closure of the sternal rudiments. Cell 73:279-294

te Riele H, Robanus Maandag E, Clarke A, Hooper M, and Berns A (1990) Consecutive inactivation of both alleles of the *pim*-1 proto-oncogene by homologous recombination in embryonic stem cells. Nature 348:649-651

Schaapveld RQJ, Maagdenberg van den AMJM, Schepens JTG, Olde Weghuis D, Geurts van Kessel A, Wieringa B, and Hendriks WJAJ (1994) The murine gene for leukocyte common antigen-related molecule LAR: Cloning, characterization, and chromosomal localization. Submitted for publication

Stacey A, Schnieke A, McWhir J, Cooper J, Colman A, and Melton DW (1994) Use of double-replacement gene targeting to replace the murine α-lactalbumin gene with its human counterpart in embryonic stem cells and mice. Mol. Cell. Biol. 14:1009-1016

Streuli M, Krueger NX, Ariniello PD, Tang M, Munro JM, Blattler WA, Adler DA, Disteche CM, and Saito H (1992) Expression of the receptor-linked protein tyrosine phosphatase LAR: proteolytic cleavage and shedding of the CAM-like extracellular structure. EMBO J. 11:897-907

Streuli M, Krueger NX, Hall LR, Schlossman SF, and Saito H (1988) A new member of the immunoglobulin superfamily that has a cytoplasmic region homologous to the leukocyte common antigen. J. Exp. Med. 168:1553-1562

Streuli M, Krueger NX, Thai T, Tang M, and Saito H (1990) Distinct functional roles of the two intracellular phosphatase like domains of the receptor-linked protein tyrosine phosphatases LCA and LAR. EMBO J. 9:2399-2407

Tan X, Stover DR, and Walsh KA (1993) Demonstration of protein tyrosine phosphatase activity in the second of two homologous domains of CD45. J. Biol. Chem. 268:6835-6838

Thomas KR, and Capecchi MR (1987) Site-directed mutagenesis by gene targeting in mouse embryo-derived stem cells. Cell 51:503-512

Tonks NK, Diltz CD, and Fischer EH (1988) Purification of the major protein-tyrosine-phosphatases of human placenta. J. Biol. Chem. 263:6722-6730

Tonks NK, Yang Q, Flint AJ, Gebbink MFBG, Franza Jr. BR, Hill DE, Sun H, and Brady-Kalnay S (1992) Protein tyrosine phosphatases: The problems of a growing family. Cold Spring Harb. Symp. Quant. Biol. 57:87-94

Wang Y, and Pallen CJ (1991) The receptor-linked protein tyrosine phosphatase HPTPα has two catalytic domains with distinct substrate specificities. EMBO J. 10:3231-3237

Valancius V, and Smithies O (1991) Testing an "in-out" targeting procedure for making subtle genomic modifications in mouse embryonic stem cells. Mol. Cell. Biol. 11:1402-1408

Wu H, Liu X, and Jaenisch R (1994) Double replacement: Strategy for efficient introduction of subtle mutations into murine *Col1a-1* gene by homologous recombination in embryonic stem cells. Proc. Natl. Acad. Sci. USA 91:2819-2823

LIPID MODIFICATIONS OF EUKARYOTIC PROTEINS: BIOCHEMISTRY AND FUNCTION

A.I. Magee
Laboratory of Eukaryotic Molecular Genetics 'C'
National Institute for Medical Research
The Ridgeway, Mill Hill
LONDON NW7 1AA, United Kingdom.

The most familiar mode of membrane anchoring is via transmembrane peptide sequences. However, mainly over the last fifteen years several types of lipid-based membrane anchors have been identified. The best known of these are glycosylphosphatidylinositol (GPI) anchoring, fatty acylation and prenylation. These can provide the means to localise proteins to different sites such as the outer cell surface or the inner surface of cellular membranes, as well as to sub-domains within those membranes. In addition, the reversibility of some of the modifications affords the potential for dynamic regulation of protein localisation and function. In this chapter I will review the biochemistry of each of these modifications and discuss their function in a variety of systems.

GPI Anchors

GPI anchors are found attached almost exclusively to a subset of cell surface proteins and are ubiquitous in eukaryotes down to yeasts. They are attached to the C-terminus of the protein via an ethanolamine phosphate that is linked to a glycan core consisting of three mannose residues and a non-N-acetylated glucosamine residue. This is linked alpha 1,6 to the inositol ring of phosphatidylinositol (PI). The basic structure can undergo considerable modification by the addition of various substituents to the glycan, including mannose, galactose, N-acetyl glucosamine, N-acetyl galactosamine, sialic acid and ethanolamine phosphate. In addition, the inositol ring can be palmitoylated. The lipid moiety of the PI can be diacyl, monoacyl or alkyl/acyl. GPI anchors can therefore be metabolically labelled with several precursors including [^{3}H]inositol, [^{3}H]ethanolamine and [^{3}H]fatty acids. Many structures have now been determined using a combination of compositional analysis, enzymatic digestion, NMR and

NATO ASI Series, Vol. H 92
Signalling Mechanisms – from Transcription Factors to Oxidative Stress
Edited by L. Packer, K. Wirtz

mass spectroscopy (Ferguson, 1992; McConville and Ferguson, 1993; see also Hooper and Turner, 1992).

GPI-anchored proteins are initially synthesised with an N-terminal signal sequence and translocated into the ER. They also contain a C-terminal signal for GPI addition. This consists of a 15-20 residue hydrophobic sequence and a processing site of 2-3 small side-chain amino acids positioned approximately 10-12 residues upstream. This provides a short-lived peptide anchor that is rapidly post-translationally cleaved and replaced with a preformed GPI anchor. During transport to the cell surface the lipid moiety can undergo fatty acid remodelling. No enzymes catalysing steps in this pathway have been purified to date but thymoma mutants defective in several steps have been isolated and these should help in the cloning of the relevant genes. Several inhibitors of the pathway are available including mannosamine and 2-fluoro-2-deoxy-D-glucose that block glycan synthesis and phenylmethanesulphonyl fluoride that blocks ethanolamine addition to GPI intermediates and inositol acylation.

Several functions have been proposed for GPI anchors. Clearly they bind proteins to the plasma membrane but a transmembrane peptide would serve this purpose. One advantage of the GPI anchor could be the ability selectively to release the protein by activation of a phospholipase. Although such enzymes do exist in eukaryotic cells and serum there is little evidence that this process occurs physiologically. The suggestion has also been made that GPI anchors may allow increased lateral mobility of proteins but actual measurements have given conflicting results. The most convincing role for GPI anchoring seems to be in the formation of membrane microdomains. GPI-anchored proteins form lateral interactions with themselves and with glycolipids resulting in "rafts" that partially exclude other molecules. These rafts form in the Golgi complex of cells and are somehow selectively packaged and targeted in polarised cells, e.g. to the apical domain of epithelial cells or the axons of neurones. GPI-anchored proteins also selectively enter caveolae, probably by a similar mechanism. Caveolae are plasma membrane invaginations of 50-100nm diameter that are involved in the uptake of small molecules and may also play a role in signal transduction since they also contain many G proteins and protein kinases (Lisanti *et al.*, 1994). Interestingly these signalling molecules also localise in caveolae due to lipid modifications (see below).

Fatty Acylation

Two types of fatty acylation have been well described. Myristoylation is the attachment of the rare fourteen carbon saturated fatty acid myristate to the N-terminus of proteins by amide linkage. This occurs only on proteins lacking transmembrane peptides and is co-translational. A highly specific N-myristoyl transferase (NMT) attaches myristate to a N-terminal glycine residue using myristoyl coenzyme A (CoA) as the acyl donor after removal of the initiator methionine. The enzyme is soluble, has been purified and cloned from a number of sources, and is highly specific for chain length but not for side-chain substituents or unsaturation; thus 2-hydroxy myristate becomes a useful inhibitor after conversion to the CoA ester *in vitro* or *in vivo*. The N-terminal sequence of the protein substrate is also important, glycine being essential at position 1 and a small amino acid (optimally Ser) at position 5. Substitutions in the other residues up to position 8 can influence substrate activity also (Gordon *et al.*, 1991).

The other common fatty acylation involves the attachment of longer chain acids, primarily the common sixteen carbon saturated fatty acid palmitate, hence the term "palmitoylation". However, the enzyme is not stringent in its requirements for chain length or unsaturation. Linkage is through a thioester bond to the side chain of cysteine which can occur at any position in the primary sequence of the protein, there being no recognisable sequence motif that specifies palmitoylation. No enzyme has been purified to date, but activity is membrane-bound and requires detergent for solubilisation. The co-substrate appears to be acyl CoA. Palmitoylation can occur on transmembrane or cytoplasmic proteins, the only apparent requirement being that they contain cysteine residues that are presented to the enzyme active site near the cytoplasmic surface of the membrane. In the case of cytoplasmic proteins this can be achieved by a prior lipid modification (see below) or by protein-protein interactions with a membrane-bound protein. Palmitoylation is a post-translational modification that occurs during passage through the ER and Golgi in the case of transmembrane proteins. The intracellular site of palmitoylation of cytoplasmic proteins is not known but it may be that these substrates are acylated by the same enzyme, whose active site is on the cytoplasmic side of the membrane. Palmitoylation can be dynamic *in vivo* (see below) and a specific thioesterase has been described that may co-operate with an acyl transferase to catalyse this cycle (Camp and Hofmann, 1993). No reliable inhibitors of palmitoylation were available until recently but some isomers of the glycosylation inhibitor tunicamycin have now been shown quite specifically to inhibit this process (Patterson and Skene, 1994). These should prove invaluable in studies of the function of palmitoylation.

These two fatty acylations can be distinguished by metabolic labelling with high specific activity [9,10-^{3}H]palmitate or myristate, that selectively label the appropriate class of protein. In addition the linkage to protein can easily be determined by neutral hydroxylamine treatment which cleaves thioesters but not amides. Acid hydrolysis will remove the remaining amide-linked fatty acid. If amounts allow the chain length of the released fatty acid should be confirmed by TLC or HPLC analysis (Magee *et al.*, 1994; Masterson and Magee, 1992).

Prenylation

Two types of prenyl groups, derived from the ubiquitous isoprenoid biosynthetic pathway, have been found attached to proteins, farnesyl (fifteen carbon) and geranylgeranyl (GG, twenty carbon) (Giannakouros and Magee, 1993). They are bound in very stable thioether linkage and appear to be irreversible. This type of modification can be metabolically labelled using [^{3}H]mevalonic acid, the precursor to all isoprenoids. Pretreatment of cells with inhibitors of HMG-CoA reductase such as mevinolin depletes the endogenous mevalonate pool and greatly enhances the incorporation of label (Magee, 1994). After isolation of the protein the nature of the prenyl group can be determined by cleavage using methyl iodide or Raney nickel followed by HPLC or GC/MS analysis (Gelb *et al.*, 1992). The co-substrates are the prenyl diphosphates and prenylation is catalysed by a family of heterodimeric soluble enzymes that recognise specific C-terminal motifs. Cysteines located in a CysX_1 X_2 X_3 motif where X_3 is a small amino acid e.g. Ser, Ala or Met become farnesylated. If X_3 is a large hydrophobic amino acid such as Leu or Phe then GG is attached. These reactions are performed by farnesyl transferase (FT) and GG transferase I (GGTI) respectively. The C-terminal motifs CysCys and CysXCys are geranylgeranylated by GG transferase II. The alpha and beta subunits of all these enzymes are related to each other; in fact FT and GGTI share the same alpha subunit. There is great interest in inhibitors of these enzymes, especially FT which is responsible for prenylation of the Ras proteins and whose inhibition could have antitumour effects. The most useful *in vivo* inhibitors appear to be peptidomimetic analogues of CysXXX, in which the peptide bonds have been stabilised by reduction or methylation, and some natural products. Despite the potentially pleiotropic effects of such compounds promising results have been obtained both in tissue culture and in animal tumour models and it appears that they might be developed into clinically efficacious cytostatic agents (Tamanoi, 1993; Gibbs *et al.*, 1994).

Subsequent to prenylation CysXXX motifs are further modified by proteolytic cleavage of the XXX residues and carboxyl-methylation of the C-terminus. CysXCys motifs also undergo carboxyl-methylation but CysCys motifs do not appear to be further modified. Less is known about the enzymes catalysing these reactions except that they are membrane-bound and may form a processing complex. The methyl transferase in *S. cerevisiae* is the product of the *sterile14* gene and is predicted to be a multi-spanning membrane protein. The methyl transferase is inhibited by N-acetyl farnesyl cysteine and the protease is inhibited by CysXXX analogues with a non-cleavable Cys-X_1 bond or N-Boc-S-farnesyl-L-cysteine aldehyde.

Functions of Lipid Modifications: Co-operating Signals

As might be expected from the lipid nature of these modifications they play a role in interactions with lipid bilayers. Whether this is due to direct insertion into the bilayer or to binding with protein "receptors" is not clear. Nevertheless, the relative and absolute hydrophobicity of these lipid substituents (myristate $<$ palmitate $\sim$ farnesyl $<$ geranylgeranyl) indicates that single moieties would not provide sufficient affinity directly to bind a protein to a cellular membrane, with the possible exception of GG (Black and Mould, 1991; Black, 1992; Peitzsch and McLaughlin, 1993). Indeed many myristoylated proteins are soluble and in some cases it appears that myristoylation plays a role in mediating protein-protein rather than protein-bilayer interactions (Zheng *et al.*, 1993; Moscufo and Chow, 1992. Most palmitoylated and prenylated proteins, however, are membrane-associated, at least for part of their lifetime.

It has emerged that membrane binding is often achieved by modification of proteins with more than one lipid group. Thus proteins with a CysXCys motif (e.g. Rabs) are doubly geranylgeranylated on the cysteines whilst some proteins farnesylated or geranylgeranylated at CysXXX motifs (e.g. H- and N-Ras) are also palmitoylated on one or more upstream cysteines. Recently an N-terminal motif that specifies both myristoylation and palmitoylation has been identified in G-protein alpha subunits and Src family members (Parenti *et al.*, 1993; Resh, 1994). This usually consists of a palmitoylated cysteine residue immediately following the myristoylated glycine, although palmitoylation sites further downstream can be used. In yet other examples (e.g. Ras and Src) a single lipid modification co-operates with a nearby polybasic amino acid sequence (Newman and Magee, 1993; Resh, 1994). Mutational analysis has shown that any of these signals operating in isolation is insufficient to achieve membrane binding.

Not only do co-operating signals translocate the proteins to membranes but they also somehow target to sub-domains. Thus Src family members (e.g. Fyn and Lck) carrying myristate and palmitate double modification are directed to the same glycolipid-enriched domains (rafts) as GPI-anchored proteins whereas Src itself which is only myristoylated is not (Rodgers *et al.*, 1994; Shenoy-Scaria *et al.*, 1994). The attachment of these signals to heterologous proteins using recombinant DNA techniques now allows their use to alter the subcellular distribution of any protein of interest.

Reversible membrane binding of lipid-modified proteins can be achieved in several ways (Newman and Magee, 1992). Some myristoylated proteins undergo conformational changes dependent on their ligand binding or post-translational modification that result in reversible exposure of the myristate (Zozulya and Stryer, 1992; Kobayashi *et al.*, 1993). Proteins that are stably modified with the more hydrophobic single or double geranylgeranyl moiety can be removed from membranes by lipid-binding carrier proteins to form soluble 1:1 complexes that can cycle between compartments (Takai *et al.*, 1992). Finally, the reversible palmitoylation of proteins can mediate their dynamic association with membranes (James and Olson, 1989; Skene and Virag, 1989).

In conclusion, the modification of proteins with a wide range of lipids gives the cell great flexibility in the ways in which it localises them and regulates their activities. The number and importance of such modifications is likely to increase in the coming years.

References

Black SD and Mould DR (1991) Development of hydrophobicity parameters to analyze proteins which bear post- or cotranslational modifications. Analyt. Biochem. 193:72-82.

Black SD (1992) Development of hydrophobicity parameters for prenylated proteins. Biochem. Biophys. Res. Comm. 186:1437-1442.

Camp LA and Hofmann SL (1993) Purification and properties of a palmitoyl-protein thioesterase that cleaves palmitate from H-ras. J. Biol. Chem. 268:22566-22574.

Ferguson MAJ (1992) In Lipid Modification of Proteins: A Practical Approach (eds) NM Hooper and AJ Turner, pp. 191-230. IRL Press, Oxford.

Gelb MH, Farnsworth CC and Glomset JA (1992) Structural analysis of prenylated proteins. In Lipid Modification of Proteins: A Practical Approach (eds) NM Hooper and AJ Turner, pp. 231-259. IRL Press, Oxford

Giannakouros T and Magee AI (1993) Protein prenylation and associated modifications. In Lipid Modifications of Proteins. (ed) MJ Schlesinger. CRC Press Inc,. 135-162.

Gibbs JB, Oliff A and Kohl NE (1994) Farnesyltransferase inhibitors: Ras research yields a potential cancer therapeutic. Cell 77:175-178.

Gordon JI, Duronio RJ, Rudnick DA, Adams SP and Goke, GW (1991) Protein *N*-Myristoylation. J. Biol. Chem. 266:8647-8650.

Hooper NM and Turner AJ (eds) (1992) Lipid Modification of Proteins: A Practical Approach, In The Practical Approach Series. IRL Press, Oxford.

James G and Olson EN (1989) Identification of a novel fatty acylated protein that partitions between the plasma membrane and cytosol and is deacylated in response to serum and growth factor stimulation. J. Biol. Chem. 264:20998-21006.

Kobayashi M, Takamatsu K, Saitoh S and Noguchi T (1993) Myristoylation of hippocalcin is linked to its calcium-dependent membrane association properties. J. Biol. Chem. 268:18898-18904.

Lisanti MP, Scherer PE, Tang Z and Sargiacomo M (1994) Caveolae, caveolin and caveolin-rich membrane domains: a signalling hypothesis. Trends Cell Biol. 4:231-235.

Magee AI and Newman CMH (1992) The role of lipid anchors for small G-binding proteins in membrane trafficking. Trends in Cell Biology 11:318-323.

Magee AI, Wootton J and de Bony J (1994) Optimised methods for detecting radiolabelled lipid-modified proteins in polyacrylamide gels. In Methods in Enzymology, in press.

Magee AI (1995) Prenylation and carboxylmethylation of proteins. In *Current Protocols in Protein Science*, in press.

Moscufo N and Chow M (1992) Myristate-protein interactions in poliovirus: interactions of VP4 threonine 28 contribute to the structural conformation of assembly intermediates and the stability of assembled virions. J. Virol. 66:6849-6857.

Newman CMH and Magee AI (1992) Post-translational processing of the ras superfamily of small GTP-binding proteins. BBA Reviews on Cancer 1155:79-96.

Masterson WJ and Magee AI (1992) Lipid modifications involved in protein targeting. In Protein Targeting: A Practical Approach. pp. 233-259 (eds. AI Magee and T Wileman) IRL Press, Oxford.

McConville MJ and Ferguson MAJ (1993) The structure, biosynthesis and function of glycosylated phosphatidylinositols in the parasitic protozoa and higher eukaryotes. Biochem. J. 294:305-324.

Newman CMH and Magee AI (1993) Post-translational processing of the ras superfamily of small GTP-binding proteins. BBA Reviews on Cancer 1155:79-96.

Parenti M, Viganó A, Newman CMH, Milligan G and Magee AI (1993) A novel N-terminal motif for palmitoylation of G-protein α-subunits. Biochem. J. 291:349-353.

Patterson SE and Skene JHP (1994) Novel inhibitory action of tunicamycin homologues suggests a role for dynamic protein fatty acylation in growth cone-mediated neurite extension. J. Cell Biol. 124:521-536.

Peitzsch RM and McLaughlin S (1993) Binding of acylated peptides and fatty acids to phospholipid vesicles: pertinence to myristoylated proteins. Biochemistry 32:10436-10443.

Resh MD (1994) Myristylation and palmitylation of Src family members: the fats of the matter. Cell 76:411-413.

Rodgers W, Crise B and Rose JK (1994) Signals determining protein tyrosine kinase and glycosyl-phosphatidylinositol-anchored protein targeting to a glycolipid-enriched membrane fraction. Mol. Cell. Biol. 14:5384-5391.

Shenoy-Scaria AM, Dietzen DJ, Kwong J, Link DC and Lublin DM (1994) Cysteine^3 of Src family protein tyrosine kinases determines palmitoylation and localization in caveolae. J. Cell Biol. 126:353-363.

Skene, JHP and Virág I. (1989) Posttranslational membrane attachment and dynamic fatty acylation of a neuronal growth cone protein, GAP-43. J. Cell Biol. 108:613-624.

Takai Y, Kaibuchi K, Kikuchi A and Kawata M (1992) Small GTP-binding proteins. Int. Rev. Cytol. 133:187-230.

Tamanoi F. (1993) Inhibitors of Ras farnesyltransferases. TIBS 18:349-353.

Zheng J, Knighton DR, Xuong N-H, Taylor SS, Sowadski JM and Eyck LFT (1993) Crystal structures of the myristylated catalytic subunit of cAMP-dependent protein kinase reveal open and closed conformations. Protein Science 2:1559-1573.

Zozulya S and Stryer L (1992) Calcium-myristoyl protein switch. Proc. Natl. Acad. Sci. USA 89:11569-11573.

Intracellular Synthesis, Transport and Sorting of Glycosphingolipids.

Gerrit van Meer and Petra van der Bijl[1]
Department of Cell Biology
Medical School AZU H02.314
University of Utrecht
3584 CX Utrecht
The Netherlands[2]

Each cellular membrane is a unique mixture of some 50 out of about 500 different membrane lipids, but the function of these complex lipid compositions is still far from clear. While the major lipid classes are the phospho(glycero)lipids and cholesterol, another important class of membrane lipids in mammalian cells are the glycosphingolipids. Whereas usually they make up some 10% of the lipids on the cell surface, their headgroups appear to completely cover the apical surface of epithelial cells (Simons and van Meer, 1988).

The function of glycosphingolipids on the epithelial cell surface may be to protect the cells from enzymes in the external environment, like e.g. from pancreatic phospholipase A_2 in the intestinal lumen. Perhaps more interestingly, specific glycolipids mediate recognition events on the cell surface, that we have started to understand only recently (Hakomori, 1990; Schnaar, 1991; Zeller and Marchase, 1992). The use of the neutral glycolipid galactosylceramide (GalCer) as a receptor by HIV (Harouse et al., 1991; Yahi et al., 1992), Gb3 (Figure 1) by Shiga toxin and the Shiga-like verotoxin (Maloney and Lingwood, 1994; Sandvig et al., 1994), and the ganglioside GM1 by cholera toxin (Fishman, 1982) are particularly detrimental examples. Moreover, while gangliosides may interact with proteins to modulate signal transduction (Zeller and Marchase, 1992), ceramide, the backbone of all glycosphingolipids, and derivatives have been reported to act as lipid second messengers in signal transduction at the plasma membrane (Hannun, 1994; Kolesnick and Golde, 1994)[3]. In yeast, glycophosphoinositol-sphingolipids constitute the covalent 'GPI' membrane anchors for proteins (Conzelmann et al., 1992). Finally, as we will describe below, glycosphingolipids have been assigned an important role in sorting membrane lipids and GPI-anchored proteins to the two surface domains of epithelial cells (Simons and van Meer, 1988; Lisanti and Rodriguez-Boulan, 1990).

---

[1] also Department of Veterinary Biochemistry, Veterinary School, Utrecht University.
[2] tel. 31-30506480; fax 31-30541797
[3] Most of this ceramide in the plasma membrane is apparently produced from the phospho-sphingolipid sphingomyelin (SM) by a receptor-coupled sphingomyelinase. Although it has been argued that ceramide may act completely independently from the other major lipid second messenger in the plasma membrane diacylglycerol, our own evidence that ceramide and diacylglycerol exchange on the plasma membrane via the action of sphingomyelin synthase seems to refute this (van Helvoort et al., 1994).

While the major phospho(glycero)lipids, phosphatidylcholine (PC), -ethanolamine (PE), -serine (PS) and -inositol (PI), are synthesized at the cytosolic surface of the ER, complex sphingolipids in the Golgi and cholesterol in ER/peroxisomes, their intracellular distribution is very different. Most cholesterol and sphingolipids reside in the plasma membrane, which is also enriched in PS. Lipids diffuse in the plane of the membrane, across membrane bilayers and to some extent through the aqueous phase. In addition, vesicular pathways carry lipids between organelles of the vacuolar system. All these processes tend to randomize the lipid compositions of the cellular membranes. A major challenge in cell biology is to find out how the cell succeeds in the complex task of maintaining the concentration gradients of lipids. The present paper focuses on the phospho(sphingo)lipid sphingomyelin (SM) and the glyco(sphingo)lipids, because most is known on their synthesis in the ER/Golgi and their transport by membrane vesicles.

Excellent reviews on lipid traffic have appeared (Koval and Pagano, 1991; Hoekstra and Kok, 1992; van Meer and Burger, 1992; van Echten and Sandhoff, 1993; van Meer, 1993; Voelker, 1993). A further basis for the present discussion are topical reviews on enzymes of sphingolipid metabolism (Merrill and Jones, 1990; Dennis and Vance, 1992; van Echten and Sandhoff, 1993; Vos et al., 1994), on alternative roles for transfer proteins specific for glycosphingolipids (van Echten and Sandhoff, 1993), on cell regulation by membrane lipids (see (Nishizuka, 1992; Hannun, 1994; Kolesnick and Golde, 1994), and on transmembrane lipid asymmetry (Devaux, 1993).

Sites of sphingolipid synthesis.

Ceramide is synthesized from sphinganine on the cytosolic face of the endoplasmic reticulum (ER; Mandon et al., 1992; Hirschberg et al., 1993), and is transported to the Golgi for conversion to SM and glycosphingolipids[4]. SM synthase transfers a phosphocholine headgroup from PC onto ceramide. Studies that carefully discriminated Golgi from plasma membrane localized the major part of the SM synthase activity to the cis-Golgi (Futerman et al., 1990; Jeckel et al., 1990). Still, some activity was found on the plasma membrane (Futerman et al., 1990), where it is limited to the basolateral surface of epithelial MDCK cells (van Helvoort et al., 1994). While the enzyme in the cis-Golgi is involved in the net synthesis of SM from newly synthesized ceramide, the plasma membrane activity probably only metabolizes ceramide produced by the plasma membrane sphingomyelinase during signal transduction[5].

---

[4] Due to the lack of a polar headgroup the hydrophobic ceramide is supposedly deeply embedded in the membrane and easily flips across membranes. It most likely follows the vesicular route, but an exchange mechanism has also been proposed (Collins and Warren, 1992; Moreau et al., 1993).

[5] SM synthesis is stimulated some 5-fold in cells treated with brefeldin A (Brüning et al., 1992; Hatch and

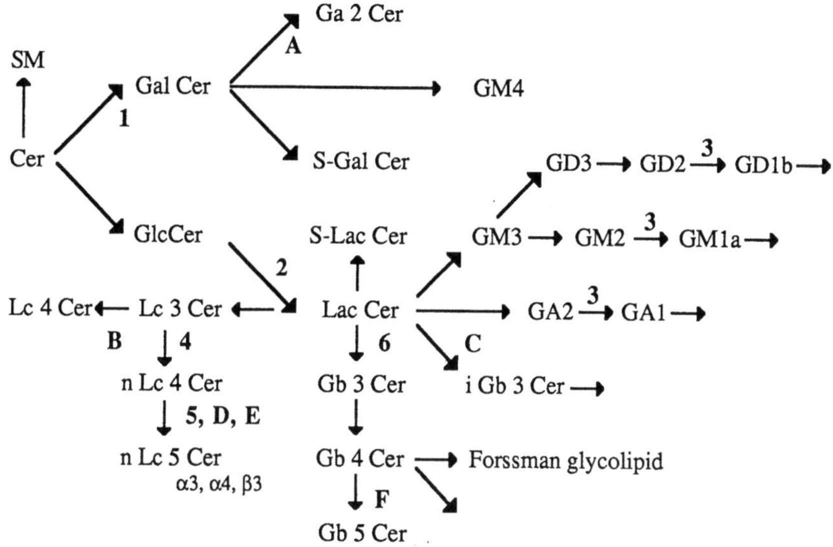

Figure 1. Biosynthetic pathways of a number of sphingolipids from ceramide, with special emphasis on the galactosyltransferases. Numbers refer to Table I. Nomenclature of neutral glycosphingolipids (Nomenclature, 1977), gangliosides (Svennerholm, 1963), enzymes (Basu et al., 1987).

Also some of the first reactions in glycosphingolipid biosynthesis cofractionated with cis-Golgi markers on sucrose gradients: the transfer of glucose to ceramide by glucosyltransferase (GlcT) to yield glucosylceramide (GlcCer; Futerman and Pagano, 1991; Jeckel et al., 1992)[6], and the sialyl transferases that convert LacCer (Gal-GlcCer) to GM3 and GD3 (Iber et al., 1992)[7]. In the presence of BFA these lipids and also Lc3, a three sugar sphingolipid, were still produced suggesting redistribution of the respective transferases to the ER[8]. Redistribution has been shown in a direct experiment for GlcT (Strous et al., 1993).

Vance, 1992; Kallen et al., 1993a). The activation strictly correlates with relocation of the SM synthase to the ER by brefeldin A, bringing up the question whether regulation of SM synthase by recycling from the cis-Golgi to the ER is a physiological event (Giudici et al., submitted).

[6] GlcT activity was also found in a so far unknown compartment (Futerman and Pagano, 1991), while its activity was in two distinct peaks after subfractionation of the Golgi on sucrose gradients (Jeckel et al., 1992).

[7] Whereas the activity of sialyltransferase I (SAT-1 yields GM3) decreases along the cis-trans-Golgi axis, that of the late transferases SAT-4 and SAT-5 increases, while SAT-2 (giving GD3) was evenly distributed. Similar data have been reported for liver Golgi (see (Trinchera et al., 1990). In addition, GalNAc-T (giving GA2, GM2 and GD2) colocalized with a medial Golgi and GalT-3 (giving GA1, GM1 and GD1b) with a trans-Golgi marker.

[8] BFA did not inhibit addition of sialic acid or N-acetylglucosamine to LacCer (SAT-1 giving GM3, GlcNAc-T giving Lc3). BFA blocked galactose addition to LacCer (GalT-6 giving Gb3) and to Lc3 (GalT-4 giving nLc4)(Sherwood and Holmes, 1992). Indeed, sucrose gradient fractionation located 50% of GlcNAc-T to a more proximal compartment than the bulk of GalT-4 (Holmes, 1989). Earlier BFA studies localized GlcT, GalT-2, SAT-1 and SAT-2 into the BFA-sensitive compartment and GalNAc-T past the BFA block (Holmes, 1989; van Echten et al., 1990).

Whether the glycosyl transferases that no longer act on newly synthesized substrates, like GalT-4, GalT-6 and GalNAcT (Table I), are situated in the trans Golgi network that resists the action of BFA, remains to be clarified. Recently, the α-hydroxy (HFA) galactosyl transferase GalT-1 that generates HFA-GalCer has been localized to the ER. The C-terminal ER retention motif -K(X)KXX (Nilsson and Warren, 1994) is present in GalT-1 as -KKXK (Schulte and Stoffel, 1993; Stahl et al., 1994; Schaeren-Wiemers et al., 1994, submitted). The location of the non-hydroxy (NFA) GalT-1 may be different (Sato et al., 1988). On a sucrose gradient it colocalized with Golgi in epithelial cells (van der Bijl et al., unpublished).

Table I. Mammalian sphingolipid galactosyltransferases.

Enzyme		Bond	Acceptor	Product
GalT-1		Gal ß 1-1	Cer	GalCer
HFA- specific (Schulte and Stoffel, 1993; Stahl et al., 1994)				
NFA- specific (Sato et al., 1988).				
GalT-2		Gal ß 1-4	Glc	LacCer
GalT-3		Gal ß 1-3	GalNAc	GA1, GM1a, GD1b
	F	Gal ß 1-3	GalNAc	Gb5 Cer
GalT-4		Gal ß 1-4	GlcNAc	nLc4 Cer
	B	Gal ß 1-3	GlcNAc	Lc4 Cer
GalT-5		Gal α 1-3	Gal ß1	α3 nLc5 Cer
	C	Gal α 1-3	Gal ß1	iGb3 Cer
GalT-6		Gal α 1-4	Gal ß1	Gb3 Cer
	A	Gal α 1-4	Gal ß1	Ga2 Cer
	D	Gal α 1-4	Gal ß1	α4 nLc5 Cer
	E	Gal ß 1-3	Gal ß1	ß3 nLc5 Cer

For nomenclature of enzymes and lipids see Figure 1.

Sidedness of sphingolipid synthesis in the Golgi.

In intact Golgi membranes SM synthase is protected against protease digestion and newly synthesized SM is situated in the lumenal leaflet of the Golgi membrane (Futerman et al., 1990; Helms et al., 1990; Karrenbauer et al., 1990; Jeckel et al., 1992). Accordingly, on the plasma membrane the enzyme is oriented towards the surface (van Helvoort et al., 1994). In contrast, in intact Golgi membranes GlcT is sensitive to proteases and non-membrane permeant protein modifying reagents (Coste et al., 1986; Futerman and Pagano, 1991; Trinchera et al., 1991b; Jeckel et al., 1992). After synthesis two short chain GlcCer analogs are accessible on the cytosolic surface of isolated Golgi membranes and in permeabilized cells (Jeckel et al., 1992). Although data from Trinchera *et al.* (Trinchera et al., 1991b) suggested that LacCer is synthesized in the cytosolic leaflet of the Golgi as well, mutant CHO cells lacking the UDP-galactose carrier but having normal levels of GalT-2 displayed strongly reduced LacCer synthesis implying a lumenal GalT-2 disposition (see (Deutscher and Hirschberg, 1986)). Indeed, newly synthesized short chain LacCer was protected against BSA extraction (Lannert et al., 1994); Burger et al., submitted). Surprisingly, just like GlcCer (Jeckel et al., 1992) newly synthesized short chain NFA-GalCer was extractable by BSA. The cytosolic orientation of NFA-GalT-1 was confirmed by the observation that the mutant cells without UDP-Gal carrier still synthesized GalCer. Because GlcCer is processed to LacCer (see above) and GalCer to sulfatide (Tennekoon et al., 1983) in the lumenal leaflet of the Golgi, both GlcCer and GalCer must translocate across the Golgi membrane. The orientation of HFA-GalT-1 in the ER membrane is not yet clear (Schulte and Stoffel, 1993; Stahl et al., 1994; Schaeren-Wiemers et al., 1994, submitted). Newly synthesized HFA-GalCer was found to be accessible at the cytosolic surface. Translocator proteins for the monohexosyl-glycosphingolipids can thus be expected to be present in ER and Golgi, and recently we have established that short chain analogs of GlcCer and GalCer when added to a post-nuclear supernatant can be converted to LacCer and Ga2Cer, that were protected against BSA extraction in the lumen of the Golgi (Burger et al., submitted). Whether or not complex glycosphingolipids can translocate back to the cytosolic surface, where they could bind, and thereby regulate, specific cytosolic proteins (Higashi et al., 1992), remains an important issue to be solved[9].

Mechanisms of sphingolipid transport.

The lumenal synthesis of both SM and the complex glycosphingolipids would predict that subsequent transport to the plasma membrane occurs by vesicles. Indeed, transport of LacCer between Golgi cisternae (Wattenberg, 1990), of gangliosides to the cell surface (Miller-Podraza and Fishman, 1982; Young et al., 1992), and of short chain SM's to the cell surface (Helms et al., 1990; van 't Hof et al., 1992; van Meer and van 't Hof, 1993) have been found to display the typical characteristics of vesicular traffic. In line with this, a block in transport of native SM to the cell surface has been observed in the presence of BFA (Warnock et al., 1994) and in

[9] In digitonin-permeabilized cells glycosphingolipids (e.g. GalCer and GM1) displayed dissimilar distributions by immuno-fluorescence (Gillard et al., 1993). The problem with this approach is the potential lipid redistribution during permeabilization with detergent. Although 12% of the gangliosides in rat liver were found in mitochondrial cell fractions (Matyas and Morré, 1987), no Forssman glycolipid was found in mitochondria by a non-invasive technique (van Genderen et al., 1991).

mitotic cells (van Helvoort et al., man. in prep.). During endocytosis, short chain SM remained sequestered on the inside of endocytotic organelles (Koval and Pagano, 1991).

The mechanism of transport of GlcCer and GalCer is less clear. Their cytosolic orientation after synthesis opens the possibility for (protein-mediated) exchange through the aqueous phase of the cytosol (Sasaki, 1990). So far, most evidence favors a vesicular mechanism of transport for the last transport step to the surface of the plasma membrane. Transport of short chain GlcCer to the cell surface was reduced to the same extent as that of short chain SM by monensin (Lipsky and Pagano, 1985; van Meer and van 't Hof, 1993), and by microtubule-depolymerization and low extracellular pH in epithelial cells ((van Meer and van 't Hof, 1993); van Meer, unpublished data). Also, short chain GlcCer was found to be protected in isolated Golgi (Karrenbauer et al., 1990; Kobayashi et al., 1992a) and transport vesicles (Babia et al., 1994). However, the experiments to show sequestration were not devised to provide quantitative data. They do not show that sequestration in vesicles is an obligatory step in all GlcCer transport to the cell surface. As an independent argument for vesicular traffic of GlcCer to the cell surface, it has been reported that newly synthesized C_6-NBD-GlcCer did not arrive at the surface of mitotic CHO cells (Kobayashi and Pagano, 1989) where the normal Golgi route of lipid exocytosis appears to be inhibited (Collins and Warren, 1992). However, we have been unable to confirm this (van Helvoort, man. in prep.). In fact, transport of native GlcCer to the plasma membrane was found to continue under conditions (BFA) where SM transport was inhibited (Warnock et al., 1994). Although this argues against an obligatory vesicular transport step in GlcCer transport to the plasma membrane, the cell fractionation technique used leaves open the possibility that these experiments monitored transport to the cytosolic leaflet of the plasma membrane and not to the cell surface.

While Kok and colleagues (Kok et al., 1992) observed that GlcCer remained sequestered in the exoplasmic leaflet during endocytotic recycling, a recent paper concluded that translocation of GlcCer towards the cytosolic leaflet of the plasma membrane may to some extent occur (Martin and Pagano, 1994). A full answer to this question requires a quantitative assessment of these events.

It appears that sphingolipids follow all vesicular routes that exist in the cell. From the Golgi they reach the cell surface (Quinn and Allan, 1992; van 't Hof et al., 1992; Young et al., 1992; Kallen et al., 1993a; Kallen et al., 1993b; van Meer and van 't Hof, 1993), from where they recycle through early and late endosomes (Koval and Pagano, 1991; Kok et al., 1992) and to some extent end up in the lysosomes (van Echten and Sandhoff, 1993). It is unclear at present whether they follow the direct pathway from the Golgi to the endosomes. After endocytosis, GlcCer has been observed to reach the Golgi (Kok et al., 1991), but it has been argued that this occurred via a non-vesicular route (Martin and Pagano, 1994). Retrograde transport of sphingolipids through the Golgi stack and to the ER has been inferred from (i) (re)utilization of endocytosed GlcCer for LacCer synthesis (Trinchera et al., 1991a), (ii) the presence of endocytosed Gb3 throughout Golgi and ER, as monitored by its ligand Shiga toxin (Sandvig et al., 1994), and (iii) the presence of gangliosides and Forssman glycolipid in the ER and nuclear membrane (Matyas and Morré, 1987; van Genderen et al., 1991).

Sphingolipid transport and sorting by microdomains.

Individual sphingolipids do not partition randomly into the various vesicular routes. Arguments for three sphingolipid sorting steps have been provided:

(1) From the facts that SM is synthesized in the cis-Golgi, that there is extensive vesicle recycling between the cis-Golgi and the ER and that the concentration of SM in the ER and intermediate compartment is very low (Schweizer et al., 1994), SM apparently follows preferentially the anterograde route from cis-Golgi instead of the retrograde pathway to the ER[10]. In most cell types, newly synthesized GlcCer reaches the cell surface earlier than SM (Karrenbauer et al., 1990; van Meer and van 't Hof, 1993). This may be due to a more efficient incorporation of GlcCer than of SM in anterograde vesicles, or alternatively it may imply that GlcCer enters the vesicular pathway in a more distal compartment (Karrenbauer et al., 1990).

(2) In epithelial cells newly synthesized short chain GlcCer was enriched on the apical surface 2-9 fold over SM (van Meer et al., 1987; van 't Hof et al., 1992; van Meer and van 't Hof, 1993), GalCer and sulfatide (van der Bijl et al., submitted). These, in turn, were preferentially delivered to the basolateral surface. Assuming that the last transport step of these lipids to the cell surface is vesicular, a model was formulated in which lipid segregation occurs by the formation of microdomains in the lumenal leaflet of the membrane of the trans Golgi network (van Meer et al., 1987; Simons and van Meer, 1988).

(3) Finally, endocytosed C_6-NBD-GlcCer reached the Golgi, whereas C_6-NBD-SM, -GalCer and -LacCer preferentially recycled to the cell surface in (undifferentiated) epithelial HT29 cells (Kok et al., 1991) but this may have occurred via a non-vesicular route (Martin and Pagano, 1994). In a recent study we have observed no sorting between C_6-NBD-lipids during epithelial transcytosis (van Genderen and van Meer, submitted).

The only way by which a lumenal lipid can be enriched into one vesicular pathway relative to another is by an increase and a decrease of its surface density at the respective sites of vesicle budding. This implies the segregation of lipids into at least two membrane domains that differ in lipid composition (van Meer and Burger, 1992; van Genderen and van Meer, 1993). It

[10] We have suggested that by the high affinity of SM for cholesterol, this SM sorting might deplete cholesterol from the ER (van Meer, 1989).

has been demonstrated in innumerable model membrane studies that this process could be driven by simple lipid immiscibility and phase separation. However, a domain structure of membranes would only seem of biological relevance if the various domains would be recognized by specific proteins (reviewed in (Jacobson and Vaz, 1992)). Specific lipid and protein compositions have been demonstrated for the (macro)domains of the plasma membrane of epithelial and neuronal cells that are separated by a diffusion barrier at the tight junction (see (Kobayashi et al., 1992b))[11].

A first example of specific lipids and proteins assembling into the same microdomain may be the combination of certain (glyco)sphingolipids and a class of proteins attached to the exoplasmic side of the membrane by a glyco-phospho-inositol anchor (GPI-proteins). In most epithelial cells, both glycosphingolipids and GPI-proteins are enriched on the apical cell surface relative to the basolateral surface. It has been proposed that glycosphingolipids are sorted in the trans Golgi network by self-aggregation into apical precursor domains (van Meer et al., 1987) and that GPI-proteins preferentially associate with these putative glycosphingolipid microdomains (Lisanti and Rodriguez-Boulan, 1990). Indeed, when thyroid FRT cells were found to exhibit a reversed polarity of GPI-proteins, epithelial lipid sorting also appeared to be reversed (Zurzolo et al., 1994). In addition, both GPI-proteins and sphingolipids could be recovered from epithelial cells in detergent-insoluble membrane fragments, and newly synthesized GPI-protein only became insoluble after entering the Golgi (Brown and Rose, 1992)[12]. Sphingolipid-GPI-protein clustering was also inferred from a block in transport of GPI-proteins into the Golgi after inhibition of ceramide biosynthesis in yeast by myriocin (Horvath et al., 1994). Independently, the involvement of membrane microdomains in GPI-protein sorting has recently been proposed based on the observation that newly arrived GPI-proteins at the apical surface were clustered and immobile (Hannan et al., 1993). Some GPI-proteins have been observed to self-associate in a low pH- and ion-dependent manner (Fukuoka et al., 1992).

An interesting development is the finding in the detergent-resistant fraction of a membrane-spanning protein with homology to plant lectins (Fiedler et al., 1994), which may imply a role for lectin-like molecules in sorting by recognition of the carbohydrate portion of

---

[11] Interestingly, the tight junction was also observed to act as a barrier to lipid diffusion between neighboring cells. Forssman glycolipid did not pass from one epithelial cell to the next at the tight junction between the cells (Nichols et al., 1986; van Meer et al., 1986; van Genderen et al., 1991). A different conclusion was reached in a recent paper (Grebenkämper and Galla, 1994), but the C_6-NBD-PC that was used readily exchanges through the aqueous phase and is therefore not a reliable reporter molecule.

[12] After cold triton-X-100 extraction GPI-proteins and sphingolipids were in the same triton-resistant vesicles. Since it is unclear how triton-X-100 extracts lipids from an asymmetric membrane and since virtually all SM (a large part of which is in the basolateral plasma membrane) ended up in the vesicles, it cannot be concluded that GPI-proteins and sphingolipids were in close contact before detergent addition.

glycosphingolipids and/or GPI-proteins. Indeed, a lactose binding lectin has been found to be secreted apically by MDCK cells (Lindstedt et al., 1993), and a family of such lectins with different specificities has been found (Barondes et al., 1994). However, the mechanism of secretion does not seem to involve the vesicular pathway from the ER to the TGN (Lindstedt et al., 1993), so that it remains to be determined whether they pass through the TGN, which would be required for a role of these molecules in sorting.

On the cell surface some GPI-protein rich microdomains may persist, as GPI-proteins seem concentrated in caveolae that are also enriched in gangliosides (Anderson, 1993; Parton, 1994)[13]. Caveolar integrity is disrupted by cholesterol depletion (Anderson, 1993). However, although the possibility of domain formation of cholesterol in plasma membranes has been suggested (El Yandouzi and Le Grimellec, 1992; Rothblatt et al., 1992), there is no direct evidence that cholesterol is enriched in caveolae. Finally, a tyrosine kinase on the cytosolic surface has been found associated with GPI-proteins in the detergent-resistant complex (Arreaza et al., 1994). All this has led to much speculation concerning a possible role of caveolae in signaling (Lisanti et al., 1994). Interestingly, a membrane protein present in caveolae, caveolin, has been localized to transport vesicles arising from the trans Golgi network (Fiedler et al., 1993). Caveolin was found absent from the FRT cells that displayed reversed sorting of glycosphingolipids and GPI-proteins (Sargiacomo et al., 1993; Zurzolo et al., 1994), suggesting a possible role for caveolin in epithelial sorting of these components. Undoubtedly, studies on the protein components of the microdomains especially in epithelial cells will provide important information on GPI-protein/sphingolipid domain formation and the interaction of the microdomain with the cytosolic sorting machinery which in the end confers the specificity on the sorting process.

References.

Anderson RGW (1993) Plasmalemmal caveolae and GPI-anchored membrane proteins. Curr Opin Cell Biol 5: 647-652

Arreaza G, Melkonian KA, LaFevre-Bernt M, Brown DA (1994) Triton X-100-resistant membrane complexes from cultured kidney epithelial cells contain the Src family protein tyrosine kinase $p62^{yes}$. J Biol Chem 269: 19123-19127

Babia T, Kok JW, van der Haar M, Kalicharan R, Hoekstra D (1994) Transport of biosynthetic sphingolipids from Golgi to plasma membrane in HT29 cells: Involvement of different carrier vesicle populations Eur J Cell Biol 63: 172-181

Barondes SH, Cooper DNW, Gitt MA, Leffler H (1994) Galectins - Structure and function of a large family of animal lectins. J Biol Chem 269: 20807-20810

Basu M, De T, Das KK, Kyle JW, Chon HC, Schaeper RJ, Basu S (1987) Glycolipids.

[13] However, possibly the aggregation of GPI-proteins was a fixation artefact (Mayor et al., 1994). In addition, the published isolation procedures for caveolae (Chang et al., 1994; Lisanti et al., 1994) include a triton-X-100 extraction in the cold, which generates the problems with interpretation discussed under note 12.

Methods Enzymol 138: 575-607
Brown DA, Rose JK (1992) Sorting of GPI-anchored proteins to glycolipid-enriched membrane subdomains during transport to the apical cell surface. Cell 68: 533-544
Brüning A, Karrenbauer A, Schnabel E, Wieland FT (1992) Brefeldin A-induced increase of sphingomyelin synthesis. J Biol Chem 267: 5052-5055
Chang WJ, Ying YS, Rothberg KG, Hooper NM, Turner AJ, Gambliel HA, Degunzburg J, Mumby SM, Gilman AG, Anderson RGW (1994) Purification and characterization of smooth muscle cell caveolae. J Cell Biol 126: 127-138
Collins RN, Warren G (1992) Sphingolipid transport in mitotic HeLa cells. J Biol Chem 267: 24906-24911
Conzelmann A, Puoti A, Lester RL, Desponds C (1992) Two different types of lipid moieties are present in glycophosphoinositol-anchored membrane proteins of *Saccharomyces cerevisiae*. EMBO J 11: 457-466
Coste H, Martel MB, Got R (1986) Topology of glucosylceramide synthesis in Golgi membranes from porcine submaxillary glands. Biochim Biophys Acta 858: 6-12
Dennis EA, Vance DE Eds. (1992) Phospholipid biosynthesis. Methods Enzymol 209: 1-544
Deutscher SL, Hirschberg CB (1986) Mechanism of galactosylation in the Golgi apparatus. A Chinese hamster ovary cell mutant deficient in translocation of UDP-galactose across Golgi vesicle membranes. J Biol Chem 261: 96-100
Devaux PF (1993) Lipid transmembrane asymmetry and flip-flop in biological membranes and in lipid bilayers. Curr Opin Struct Biol 3: 489-494
El Yandouzi EH, Le Grimellec C (1992) Cholesterol heterogeneity in the plasma membrane of epithelial cells. Biochemistry 31: 547-551
Fiedler K, Kobayashi T, Kurzchalia TV, Simons K (1993) Glycosphingolipid-enriched, detergent-insoluble complexes in protein sorting in epithelial cells. Biochemistry 32: 6365-6373
Fiedler K, Parton RG, Kellner R, Etzold T, Simons K (1994) Vip36, a novel component of glycolipid rafts and exocytic carrier vesicles in epithelial cells. EMBO J 13: 1729-1740
Fishman PH (1982) Role of membrane gangliosides in the binding and activation of bacterial toxins. J Membr Biol 69: 85-97
Fukuoka S-I, Freedman SD, Yu H, Sukhatme VP, Scheele GA (1992) GP-2/THP gene family encodes self-binding glycosylphosphatidylinositol-anchored proteins in apical secretory compartments of pancreas and kidney. Proc Natl Acad Sci USA 89: 1189-1193
Futerman AH, Pagano RE (1991) Determination of the intracellular sites and topology of glucosylceramide synthesis in rat liver. Biochem J 280: 295-302
Futerman AH, Stieger B, Hubbard AL, Pagano RE (1990) Sphingomyelin synthesis in rat liver occurs predominantly at the cis and medial cisternae of the Golgi apparatus. J Biol Chem 265: 8650-8657
Gillard BK, Thurmon LT, Marcus DM (1993) Variable subcellular localization of glycosphingolipids. Glycobiology 3: 57-67
Grebenkämper K, Galla H-J (1994) Translational diffusion measurements of a fluorescent phospholipid between MDCK-I cells support the lipid model of the tight junctions. Chem Phys Lipids 71: 133-143
Hakomori S-i (1990) Bifunctional role of glycosphingolipids. J Biol Chem 265: 18713-18716
Hannan LA, Lisanti MP, Rodriguez-Boulan E, Edidin M (1993) Correctly sorted molecules of a GPI-anchored protein are clustered and immobile when they arrive at the apical surface of MDCK cells. J Cell Biol 120: 353-358
Hannun YA (1994) The sphingomyelin cycle and the second messenger function of ceramide. J Biol Chem 269: 3125-3128
Harouse JM, Bhat S, Spitalnik SL, Laughlin M, Stefano K, Silberberg DH, Gonzalez-Scarano F (1991) Inhibition of entry of HIV-1 in neural cell lines by antibodies against galactosyl ceramide. Science USA 253: 320-323
Hatch GM, Vance DE (1992) Stimulation of sphingomyelin biosynthesis by brefeldin A and sphingomyelin breakdown by okadaic acid treatment of rat hepatocytes. J Biol Chem 267: 12443-12451
Helms JB, Karrenbauer A, Wirtz KWA, Rothman JE, Wieland FT (1990) Reconstitution of

steps in the constitutive secretory pathway in permeabilized cells. Secretion of glycosylated tripeptide and truncated sphingomyelin. J Biol Chem 265: 20027-20032
Higashi H, Omori A, Yamagata T (1992) Calmodulin, a ganglioside-binding protein. Binding of gangliosides to calmodulin in the presence of calcium. J Biol Chem 267: 9831-9838
Hirschberg K, Rodger J, Futerman AH (1993) The long-chain sphingoid base of sphingolipids is acylated at the cytosolic surface of the endoplasmic reticulum in rat liver. Biochem J 290: 751-757
Hoekstra D, Kok JW (1992) Trafficking of glycosphingolipids in eukaryotic cells; sorting and recycling of lipids. Biochim Biophys Acta 1113: 277-294
Holmes EH (1989) Characterization and membrane organization of ß1→3- and ß1→4-galactosyltransferases from human colonic adenocarcinoma cell lines Colo 205 and SW403: Basis for preferential synthesis of type 1 chain lacto-series carbohydrate structures. Arch Biochem Biophys 270: 630-646
Horvath A, Sütterlin C, Manning-Krieg U, Movva NR, Riezman H (1994) Ceramide synthesis enhances transport of GPI-anchored proteins to the Golgi apparatus in yeast. EMBO J 13: 3687-3695
Iber H, van Echten G, Sandhoff K (1992) Fractionation of primary cultured cerebellar neurons: Distribution of sialyltransferases involved in ganglioside biosynthesis. J Neurochem 58: 1533-1537
IUPAC-IUB Commision on Biochemical Nomenclature (1977) The nomenclature of lipids. Recommendations 1976. Eur J Biochem 79: 11-21
Jacobson K, Vaz WLC Eds. (1992) Domains in Biological Membranes. Comm Mol Cell Biophys 8: 1-114
Jeckel D, Karrenbauer A, Birk R, Schmidt RR, Wieland F (1990) Sphingomyelin is synthesized in the cis Golgi. FEBS Lett 261: 155-157
Jeckel D, Karrenbauer A, Burger KNJ, van Meer G, Wieland F (1992) Glucosylceramide is synthesized at the cytosolic surface of various Golgi subfractions. J Cell Biol 117: 259-267
Kallen K-J, Quinn P, Allan D (1993a) Effects of brefeldin A on sphingomyelin transport and lipid synthesis in BHK21 cells. Biochem J 289: 307-312
Kallen K-J, Quinn P, Allan D (1993b) Monensin inhibits synthesis of plasma membrane sphingomyelin by blocking transport of ceramide through the Golgi: evidence for two sites of sphingomyelin synthesis in BHK cells. Biochim Biophys Acta 1166: 305-308
Karrenbauer A, Jeckel D, Just W, Birk R, Schmidt RR, Rothman JE, Wieland FT (1990) The rate of bulk flow from the Golgi to the plasma membrane. Cell 63: 259-267
Kobayashi T, Pagano RE (1989) Lipid transport during mitosis. Alternative pathways for delivery of newly synthesized lipids to the cell surface. J Biol Chem 264: 5966-5973
Kobayashi T, Pimplikar SW, Parton RG, Bhakdi S, Simons K (1992a) Sphingolipid transport from the trans-Golgi network to the apical surface in permeabilized MDCK cells. FEBS Lett 300: 227-231
Kobayashi T, Storrie B, Simons K, Dotti CG (1992b) A functional barrier to movement of lipids in polarized neurons. Nature 359: 647-650
Kok JW, Babia T, Hoekstra D (1991) Sorting of sphingolipids in the endocytic pathway of HT29 cells. J Cell Biol 114: 231-239
Kok JW, Hoekstra K, Eskelinen S, Hoekstra D (1992) Recycling pathways of glucosylceramide in BHK cells: distinct involvement of early and late endosomes. J Cell Sci 103: 1139-1152
Kolesnick R, Golde DW (1994) The sphingomyelin pathway in tumor necrosis factor and interleukin-1 signaling. Cell 77: 325-328
Koval M, Pagano RE (1991) Intracellular transport and metabolism of sphingomyelin. Biochim Biophys Acta 1082: 113-125
Lannert H, Bünning C, Jeckel D, Wieland FT (1994) Lactosylceramide is synthesized in the lumen of the Golgi apparatus. FEBS Lett 342: 91-96
Lindstedt R, Apodaca G, Barondes SH, Mostov KE, Leffler H (1993) Apical secretion of a cytosolic protein by Madin-Darby canine kidney cells Evidence for polarized release of an endogenous lectin by a non-classical secretory pathway. J Biol Chem 268: 11750-11757
Lipsky NG, Pagano RE (1985) Intracellular translocation of fluorescent sphingolipids in

cultured fibroblasts: Endogenously synthesized sphingomyelin and glucocerebroside analogues pass through the Golgi apparatus en route to the plasma membrane. J Cell Biol 100: 27-34

Lisanti MP, Rodriguez-Boulan E (1990) Glycophospholipid membrane anchoring provides clues to the mechanism of protein sorting in polarized epithelial cells. TIBS 15: 113-118

Lisanti MP, Scherer PE, Vidugiriene J, Tang ZL, Hermanowskivosatka A, Tu YH, Cook RF, Sargiacomo M (1994) Characterization of caveolin-rich membrane domains isolated from an endothelial-rich source: Implications for human disease. J Cell Biol 126: 111-126

Maloney MD, Lingwood CA (1994) CD19 has a potential CD77 (globotriaosyl ceramide)-binding site with sequence similarity to verotoxin B-subunits: Implications of molecular mimicry for B cell adhesion and enterohemorrhagic Escherichia coli pathogenesis. J Exp Med 180: 191-201

Mandon EC, Ehses I, Rother J, van Echten G, Sandhoff K (1992) Subcellular localization and membrane topology of serine palmitoyltransferase, 3-dehydrosphinganine reductase, and sphinganine N-acyltransferase in mouse liver. J Biol Chem 267: 11144-11148

Martin OC, Pagano RE (1994) Internalization and sorting of a fluorescent analogue of glucosylceramide to the Golgi apparatus of human skin fibroblasts: Utilization of endocytic and nonendocytic transport mechanisms. J Cell Biol 125: 769-781

Matyas GR, Morré DJ (1987) Subcellular distribution and biosynthesis of rat liver gangliosides. Biochim Biophys Acta 921: 599-614

Mayor S, Rothberg KG, Maxfield FR (1994) Sequestration of GPI-anchored proteins in caveolae triggered by cross-linking. Science 264: 1948-1951

Merrill AH Jr., Jones DD (1990) An update of the enzymology and regulation of sphingomyelin metabolism. Biochim Biophys Acta 1044: 1-12

Miller-Podraza H, Fishman PH (1982) Translocation of newly synthesized gangliosides to the cell surface. Biochemistry 21: 3265-3270

Moreau P, Cassagne C, Keenan TW, Morré DJ (1993) Ceramide excluded from cell-free vesicular lipid transfer from endoplasmic reticulum to Golgi apparatus-Evidence for lipid sorting. Biochim Biophys Acta 1146: 9-16

Nichols GE, Borgman CA, Young Jr. WW (1986) On tight junction structure: Forssman glycolipid does not flow between MDCK cells in an intact epithelial monolayer. Biochem Biophys Res Commun 138: 1163-1169

Nilsson T, Warren GB (1994) Retention and retrieval in the endoplasmic reticulum and the Golgi apparatus. Curr Opin Cell Biol 6: 517-521

Nishizuka Y (1992) Intracellular signaling by hydrolysis of phspholipids and activation of protein kinase C. Science 258: 607-614

Parton RG (1994) Ultrastructural localization of gangliosides; GM_1 is concentrated in caveolae. J Histochem Cytochem 42: 155-166

Quinn P, Allan D (1992) Two separate pools of sphingomyelin in BHK cells. Biochim Biophys Acta 1124: 95-100

Rothblatt GH, Mahlberg FH, Johnson WJ, Phillips MC (1992) Apolipoproteins, membrane cholesterol domains, and the regulation of cholesterol efflux. J Lipid Res 33: 1091-1097

Sandvig K, Ryd M, Garred Ø, Schweda E, Holm PK, Van Deurs B (1994) Retrograde transport from the Golgi complex to the ER of both Shiga toxin and the nontoxic Shiga B-fragment is regulated by butyric acid and cAMP. J Cell Biol 126: 53-64

Sargiacomo M, Sudol M, Tang ZL, Lisanti MP (1993) Signal transducing molecules and glycosyl-phosphatidylinositol-linked proteins form a caveolin-rich insoluble complex in MDCK cells. J Cell Biol 122: 789-807

Sasaki T (1990) Glycolipid transfer protein and intracellular traffic of glucosylceramide. Experientia 46: 611-616

Sato C, Black JA, Yu RK (1988) Subcellular distribution of UDP-galactose:ceramide galactosyltransferase in rat brain oligodendroglia. J Neurochem 50: 1887-1893

Schnaar RL (1991) Glycosphingolipids in cell surface recognition. Glycobiology 1: 477-485

Schulte S, Stoffel W (1993) Ceramide UDPgalactosyltransferase from myelinating rat brain: Purification, cloning, and expression. Proc Natl Acad Sci USA 90: 10265-10269

Schweizer A, Clausen H, van Meer G, Hauri H-P (1994) Localization of O-glycan initiation, sphingomyelin synthesis, and glucosylceramide synthesis in Vero cells with respect to the endoplasmic reticulum-Golgi intermediate compartment. J Biol Chem 269: 4035-4041

Sherwood AL, Holmes EH (1992) Brefeldin A induced inhibition of *de Novo* globo- and neolacto-series glycolipid core chain biosynthesis in human cells. J Biol Chem 267: 25328-25336

Simons K, van Meer G (1988) Lipid sorting in epithelial cells. Biochemistry 27: 6197-6202

Stahl N, Jurevics H, Morell P, Suzuki K, Popko B (1994) Isolation, characterization, and expression of cDNA clones that encode rat UDP-galactose: Ceramide galactosyltransferase. J Neurosci Res 38: 234-242

Strous GJ, van Kerkhof P, van Meer G, Rijnboutt S, Stoorvogel W (1993) Differential effects of brefeldin A on transport of secretory and lysosomal proteins. J Biol Chem 268: 2341-2347

Svennerholm L (1963) Chromatographic separation of human brain gangliosides. J Neurochem 10: 613-623

Tennekoon G, Zaruba M, Wolinsky J (1983) Topography of cerebroside sulfotransferase in Golgi-enriched vesicles from rat brain. J Cell Biol 97: 1107-1112

Trinchera M, Carrettoni D, Ghidoni R (1991a) A part of glucosylceramide formed from exogenous lactosylceramide is not degraded to ceramide but re-cycled and glycosylated in the Golgi apparatus. J Biol Chem 266: 9093-9099

Trinchera M, Fabbri M, Ghidoni R (1991b) Topography of glycosyltransferases involved in the initial glycosylations of gangliosides. J Biol Chem 266: 20907-20912

Trinchera M, Pirovano B, Ghidoni R (1990) Sub-Golgi distribution in rat liver of CMP-NeuAc GM3- and CMP-NeuAc:GT1bα2$\rightarrow$8 sialyltransferases and comparison with the distribution of the other glycosyltransferase activities involved in ganglioside biosynthesis. J Biol Chem 265: 18242-18247

van 't Hof W, Silvius J, Wieland F, van Meer G (1992) Epithelial sphingolipid sorting allows for extensive variation of the fatty acyl chain and the sphingosine backbone. Biochem J 283: 913-917

van Echten G, Iber H, Stotz H, Takatsuki A, Sandhoff K (1990) Uncoupling of ganglioside biosynthesis by Brefeldin A. Eur J Cell Biol 51: 135-139

van Echten G, Sandhoff K (1993) Ganglioside metabolism. Enzymology, topology, and regulation. J Biol Chem 268: 5341-5344

van Genderen IL, van Meer G (1993) Lipid sorting – measurement and interpretation. Biochem Soc Trans 21: 235-239

van Genderen IL, van Meer G, Slot JW, Geuze HJ, Voorhout WF (1991) Subcellular localization of Forssman glycolipid in epithelial MDCK cells by immuno-electronmicroscopy after freeze-substitution. J Cell Biol 115: 1009-1019

van Helvoort ALB, van 't Hof W, Ritsema T, Sandra A, van Meer G (1994) Conversion of diacylglycerol to phosphatidylcholine on the basolateral surface of epithelial (MDCK) cells. Evidence for the reverse action of the sphingomyelin synthase. J Biol Chem 269: 1763-1769

van Meer G (1989) Lipid traffic in animal cells. Annu Rev Cell Biol 5: 247-275

van Meer G (1993) Transport and sorting of membrane lipids. Curr Opin Cell Biol 5: 661-673

van Meer G, Burger KNJ (1992) Sphingolipid trafficking-sorted out? Trends Cell Biol 2: 332-337

van Meer G, Gumbiner B, Simons K (1986) The tight junction does not allow lipid molecules to diffuse from one epithelial cell to the next. Nature 322: 639-641

van Meer G, Stelzer EHK, Wijnaendts-van-Resandt RW, Simons K (1987) Sorting of sphingolipids in epithelial (Madin-Darby canine kidney) cells. J Cell Biol 105: 1623-1635

van Meer G, van 't Hof W (1993) Epithelial sphingolipid sorting is insensitive to reorganization of the Golgi by nocodazole, but is abolished by monensin in MDCK cells and by brefeldin A in Caco-2 cells. J Cell Sci 104: 833-842

Voelker DR (1993) The ATP-dependent translocation of phosphatidylserine to the mitochondria is a process that is restricted to the autologous organelle. J Biol Chem 268: 7069-7074

Vos JP, Lopes-Cardozo M, Gadella BM (1994) Metabolic and functional aspects of sulfogalactolipids. Biochim Biophys Acta 1211: 125-149

Warnock DE, Lutz MS, Blackburn WA, Young WWJ, Baenziger JU (1994) Transport of newly synthesized glucosylceramide to the plasma membrane by a non-Golgi pathway. Proc Natl Acad Sci USA 91: 2708-2712

Wattenberg BW (1990) Glycolipid and glycoprotein transport through the Golgi complex are similar biochemically and kinetically. Reconstitution of glycolipid transport in a cell free system. J Cell Biol 111: 421-428

Yahi N, Baghdiguian S, Moreau H, Fantini J (1992) Galactosyl ceramide (or a closely related molecule) is the receptor for human immunodeficiency virus type 1 on human colon epithelial HT29 cells. J Virol 66: 4848-4854

Young WWJ, Lutz MS, Blackburn WA (1992) Endogenous glycosphingolipids move to the cell surface at a rate consistent with bulk flow estimates. J Biol Chem 267: 12011-12015

Zeller CB, Marchase RB (1992) Gangliosides as modulators of cell function. Am J Physiol 262: C1341-C1355

Zurzolo C, van 't Hof W, van Meer G, Rodriguez-Boulan E (1994) VIP21/caveolin, glycosphingolipid clusters and the sorting of glycosylphosphatidylinositol-anchored proteins in epithelial cells. EMBO J 13: 42-53

NOTE TO THE ADDENDUM

The following contribution was received by the Editors after the complete manuscript had been sent to the Publisher.

Interleukin 1α and Tumor Necrosis Factor-α modulate the nuclear phosphoinositide signaling system

Sandra Marmiroli§, Alberto Bavelloni*, Irene Faenza*, Rashmi Sood*§, Spartaco Santi§, Stefania Cecchi§, Nadir M. Maraldi*§ and Andrea Ognibene*.

§Istituto di Citomorfologia Normale e Patologica, C.N.R.
*Laboratorio di Biologia Cellulare e Microscopia Elettronica, I.O.R.
*§via di Barbiano 1/10, 40136 Bologna, Italy

Existence of a nuclear inositol lipid cycle.

Growing interest has been currently focused on the role of phospholipids in signal transduction and cell growth. It is now well established that the breakdown products of several plasma membrane phospholipids act as signal molecules, i.e. as intracellular second messengers or as agonists that modulate cell functions. Despite the fact that phosphoinositides are a very minor fraction of membrane phospholipids, well known examples of phospholipid-derived signaling molecules include diacylglycerol and inositol trisphosphate, which are rapidly generated from PIP_2 by the action of a specific phospholipase C following stimulation of cell-surface receptors.

However, the concept of an actively regulated inositol lipid pool in the nucleus has gained ground only recently (Cocco et al. 1994, and references therein), and stems from the observation made by Smith and Wells in 1983 that rat liver nuclei possess PI kinase and PIP kinase.

Although at that time these enzymes were assumed to be associated with the nuclear envelope, the work of Cocco et al. on Friend erythroleukemia cells in 1987 quite unexpectedly showed that: membrane-free nuclei do possess both PI and PIP kinase and their substrates; some aspects of the metabolism of the PIs in Friend cell nuclei are affected by the state of differentiation of the cell (Cocco et al. 1987); it appears from these observations that PIP_2 is synthesised to a greater extent in nuclei of cells which have been induced to differentiate by treatment with different chemicals. However, the inclusion of exogenous PIP in the assay neutralised the differences in PIP_2 levels between undifferentiated and differentiated cells, ruling out the possibility of an increased PIP-kinase activity (Cocco et al. 1987). The

NATO ASI Series, Vol. H 92
Signalling Mechanisms – from Transcription Factors to Oxidative Stress
Edited by L. Packer, K. Wirtz

differentiation-related changes in PIs metabolism occur in the inner envelope or further inside the nucleus and are not the result of cytoplasmic contamination caused by the continuity between the ER (where PI is synthesised) and the outer nuclear envelope.
After this first body of evidence, and in spite of the careful morphological and biochemical analysis done to show the purity of the nuclear preparations, the existence of a nuclear lipid cycle autonomous from the "classical" one at the plasma membrane has been viewed sceptically by the scientific community, and the burning issue of the purity of the nuclear preparation needed to be unequivocally settled. The demonstration provided by Cocco et al. (1988) and by Divecha et al. (1990) that the observed changes in PIs metabolism were restricted to the nuclear compartment, while the cytoplasmic fraction was unaffected, proved conclusive. In addition, a direct comparison of PIP_2 synthesis in rat liver nuclei prepared with or without the membrane showed no differences in activity, suggesting that any PIs in the outer membrane do not participate directly in nuclear PI metabolism (Cocco et al. 1988).

Ultrastructural localisation of nuclear lipids.

Further insights to the precise nuclear localisation of the enzymes involved in the nuclear cycle and their substrates were obtained by Maraldi and co-workers. By means of either phospholipase-gold or immunogold electron microscopy, they were able to demonstrate that a significant part of the lipids remaining after removal of the membranes is associated with the interchromatin domains and is also retained by the inner nuclear matrix constituents (Maraldi et al. 1992a). More recently, the presence of a specific phospholipid involved in signal transduction, namely PIP_2, has been demonstrated in the nuclei of intact cells by ultrastructural immunocytochemical labelling (Mazzotti et al., in press). Moreover, the nuclear PIP_2 has been reported to undergo quantitative variations in the course of cell cycle -related events (Mazzotti et al., in press), similar to those reported previously for other nuclear lipids (Maraldi et al., 1992b; Maraldi et al., 1993).
Other key elements of the nuclear cycle, such as PI-PLC and PKC, have been found to reside within the nucleus and colocalize with their lipid substrate or cofactor, respectively.
Taken together, these findings confirm the actual presence within the nucleus of the main elements of the PI cycle, excluding any artifactual influence due to the extraction procedures. They also demonstrate that the system is not prevailingly associated with the nuclear envelope but with the nuclear matrix elements which modulate the structure of the chromatin and its functional activity.

Modulation of nuclear PLC by extracellular stimuli.

The demonstration that not only in cell-free systems but also in intact cells a nuclear lipid cycle exists and is responsive to extracellular stimuli provides a new framework in which to interpret previous findings (Cocco et al., 1988; Divecha et al., 1991).

Indeed, quiescent, serum-starved Swiss 3T3 cells treated with the mitogenic growth factor IGF-I showed a rapid decrease of ^{32}P PIP_2 labelling of isolated nuclei incubated with ^{32}P ATP. The cytoplasmic inositol lipid pool was totally unaffected by the treatment with IGF-I. Rather, it was largely responsive to bombesin which, on the contrary, was unable to evoke nuclear changes (Cocco et al., 1988).

A key experiment done in 1991 by Irvine and Divecha (who had set up a picomole-sensitive mass assay for the three inositides and for diacylglycerol), showed that the IGF I stimulation of 3T3 cells triggers a rapid and transient breakdown of PIP and PIP_2 and a parallel increase of DAG restricted to the nuclear compartment. Since in the same cells bombesin is able to trigger an identical response in the plasma membrane, the obvious conclusion suggested by these results is that, upon binding of IGF-I to the receptor, a signal is sent to the nucleus where it stimulates a nuclear phosphoinositide-specific phospholipase C that hydrolyses PIP_2 and produces IP_3 and DAG.

The existence of multiple PLC isoenzymes in mammalian tissues has been extensively demonstrated, by direct protein purification and molecular cloning (Rhee et al., 1989). The various PLC isoforms seem to be activated by different receptors and regulated by different mechanisms (Rhee 1992 and references therein). It became therefore necessary to investigate the subcellular distribution and regulation of the different isoforms in Swiss 3T3 cells.

Characterisation of nuclear PLC.

Following up the intriguing data from Irvine and Divecha, more recently Martelli et al. provided the evidence that the multiple isoforms of PLC are distributed differently within Swiss 3T3 cells: in particular, the γ isoform is predominantly cytoplasmatic while the β_1 isoform is predominantly nuclear; moreover, PLC δ is totally absent (Martelli et al., 1992). The nuclear localisation of PLC β has been subsequently confirmed in other cell types, such as PC 12 rat pheochromocytoma cells (Mazzoni et al., 1992) and rat liver nuclei (Divecha et al., 1993).

Moreover, the effect of IGF-I stimulation is largely neutralised (75%) by the antibody against the β form, while antibodies against other isoforms are totally ineffective (Martelli et al. 1992). In conclusion, there exists an autonomous phosphoinositide cycle in the nucleus of Swiss 3T3 cells that is capable of being modulated by an extra cellular stimulus. The key enzyme of this nuclear cycle, which is responsible for the modulation, is PLC β_1.
The necessity to know whether this nuclear response is unique to 3T3 or is common to other systems prompted us to undertake a study on human osteosarcoma SaOS-2 cells stimulated with Interleukin 1α (Maraldi et al., 1993; Marmiroli et al., 1994a); the data from these studies will be discussed extensively in the following paragraph.

Interleukin 1 and TNF stimulate nuclear PLC in SaOS-2 cells.

Human osteosarcoma SaOS-2 are bone-derived cells which posses high affinity receptor for the cytokine growth factor interleukin 1. Il-1 treated SaOS-2 cells participate in resorption of bone and cartilage and are a model to study bone resorption *in vitro* (Rodan et al., 1990).
Interleukin 1 mediates a wide range of biological responses, most likely because of the presence of IL-1 receptors on most cell types. Analysis of the sequence of the receptor cytoplasmatic domain provides very few clues as to what the post receptor events might be (Sims et al., 1988).
Hitherto, there has been much conflicting evidence concerning the earliest intracellular events following the binding of IL-1 to its receptor and the mechanisms by which this interaction is coupled to IL-1 regulation of gene transcription. It is very well known, indeed, that IL-1 activates several transcription factors, leading to modulation of gene expression. This action is particularly intriguing, not only because of its physiological implications but also because it cannot be explained by known signal transduction cascades.
In human osteosarcoma SaOS-2 cells, IL-1, alone (Marmiroli et al. 1994a) or in combination with TNF (Marmiroli et al. 1994b), is able to evoke a very rapid and transient activation of PLC. This response is confined to the nuclear compartment whereas the cytoplasmatic inositol lipid pool is not affected (Marmiroli et al. 1994a) . Although in the past many groups have failed to detect any PIP_2 breakdown in cells stimulated through the IL-1 receptor, our results do not directly contradict these earlier studies, as it is clear that the IL-1 induced PIP_2 hydrolysis and IP_3 production are to small to be measured in whole cell extracts.
Owing to the existence of several PLC isoforms, we have investigated their distribution in SaOS-2 cells by means of monoclonal antibodies whose specificity had been previously demonstrated (Mazzoni et al., 1992).

The western blotting analysis of the cytoplasmatic and nuclear fractions probed with specific MoAb against PLC β_1 and γ shows that while PLC γ is distributed in both compartments, PLC β_1 is detectable only in the nuclear compartment (Marmiroli et al. 1994a). These findings correlate well with the immunofluorescence analysis performed in cell monolayers probed with MoAb against PLC β_1 or γ (Maraldi et al., 1993). These experiments confirm that the distribution of these isoforms differ in that PLC γ is present in both cytoplasmatic and nuclear compartments whereas PLC β_1 is localised predominantly in the nucleus and it is only barely detectable in the cytoplasm. In order to investigate the fine ultrastructural localisation of PLC β_1 and γ, with the same antibodies we have probed *in situ* matrix preparations, in which the network of the inner nuclear matrix threads connecting the residual nucleolar remnants and the array of microfilaments belonging to the residual cytoskeleton are preserved side by side.

In the case of anti PLC β_1 the distribution of the gold particles was restricted to the inner nuclear matrix; PLC γ appeared mainly associated with the cytoskeletal filaments, and, only to a lesser extent, with the nuclear matrix (Maraldi et al. 1993). These findings clearly indicate that these proteins are not prevailingly associated with the nuclear envelope. Rather, a fraction of both PLC β and γ has been found associated with the nuclear matrix elements. Besides, the results on the subcellular localisation of PLC in SaOS-2, compared with previous results obtained in Swiss 3T3 cells (Zini et al., 1993) and PC 12 cells (Zini et al., 1994), clearly show that different cell types possess a different intracellular partitioning of the isoforms. This specificity could account for the broad range of responses that follows the hydrolysis of phosphoinositides.

Moreover, the increase of nuclear PLC activity in response to IL-1 was completely neutralised by the MoAb against the β_1-form (Marmiroli et al. 1994a), thus suggesting that the activation of this nuclear enzyme represents an early step in the IL-1 signaling pathway in SaOS-2 cells.

The observation that, after IL-1 or IL-1 plus TNF treatment, the nuclear PKC activity is rapidly increased (Ognibene et al., 1994) correlates well with the PIP_2 breakdown and the DAG and IP_3 formation caused by the stimulation of PLC. So far, 11 different PKC isoforms have been cloned in mammalian tissues (Dekker and Parker, 1994 and references therein). We have therefore investigated which isoforms are present in SaOS-2 cells and how they are distributed.

The western blotting analysis shows that all conventional PKCs are represented, while, between novel PKCs, only PKC ε is present (Marmiroli et al. 1994b). No differences in the intracellular partitioning of these isoforms have been detected. Unlike previously reported systems (Neri et al. 1994), the activity rise following IL-1 stimulation does not seem to be the result of a protein translocation. Rather, it is probably the activation of a constitutively nuclear enzyme.

Taken all together, these data strengthen the contention of the existence of a nuclear lipid signaling pathway and clearly indicate that the activation of a nuclear PLC isoform is an early step in the mechanism of action of IL-1 and TNF in SaOS-2 cells.

Future perspectives and concluding remarks

Having established the existence of an autonomous inositol lipid cycle in the nucleus of several cell systems, including the human osteosarcoma SaOS-2 cell system, we are presently exploring the ongoing enigma of the role of this nuclear pathway.
The notion that IL-1 and TNF could employ this nuclear pathway to effect signal transduction by their receptors, prompted us to investigate whether this pathway is important in ultimately modifying the production or the activity of a series of transcription factors, known to be modulated by IL-1/TNF, such as AP1 and NFkB.
To demonstrate this, the production of systems in which the expression of the key enzyme of the nuclear cycle, namely PLC β_1, can be switched on or off, will be very helpful.
Obviously, another major challenge for future studies is to establish the amino acid sequence of the putative nuclear PLC β_1 and its mechanism of regulation.
We can certainly expect more excitement as the impact of this nuclear pathway on the biology of signal transduction becomes clarified.

Acknowledgements. We are grateful to Aurelio Valmori for skilful technical assistance. This work has been supported by grants from "Ricerca Corrente 1994" Istituti Ortopedici Rizzoli.

REFERENCES

Bahk, Y.Y., Lee, Y.H., Lee, T.G., Seo, J., Ryu, S.H. and Suh, P.G. (1994) *J. Biol. Chem.* **269**, 8240-8245.

Cocco, L., Gilmour, R.S., Ognibene, A., Letcher, A.J., Manzoli, F.A. and Irvine R.F. (1987) *Biochem. J.* **241**, 765-770

Cocco, L., Martelli, A.M., Gilmour, R.S., Ognibene, A., Manzoli, F.A. and Irvine, R.F. (1988) *Biochem. Biophys. Res. Commun.* **154**, 1266-1272

Cocco, L., Martelli, A.M. and Gilmour, R.S. (1994) *Cell. Signall.* **5**, 481-485.

Divecha, N., Banfic, H. and Irvine, R.F. (1991) *EMBO J.* **10**, 3207-3214

Divecha, N., Rhee, S.G., Letcher, A.J. and Irvine R.F. (1993) *Biochem. J.* **289**, 617-620.

Dekker., L.V. and Parker, P.J. (1994) *TIBS* **19**, 73-77.

Maraldi, N.M., Zini, N., Squarzoni, S., Del Coco, R., Sabatelli, P. and Manzoli, F.A. (1992a) *J. Histochem. Cytochem.* **40**, 1383-1392.

Maraldi, N.M., Mazzotti, G., Capitani, S., Rizzoli, R., Zini, N., Squarzoni, S. and Manzoli, F.A. (1992b) *Advan. Enzyme Regul.* **32**, 73-90.

Maraldi, N.M., Zini, N., Santi, S., Bavelloni, A., Valmori, A., Marmiroli, S. and Ognibene, A. (1993) *Biol. Cell* **79**, 243-250.

Maraldi, N.M., Cocco, L., Capitani, S., Mazzotti, G., Barnabei, O. and Manzoli, F.A. (1994) *Advan. Enzyme Regul.* **34**, 129-143.

Marmiroli, S., Ognibene, A., Bavelloni, A., Cinti, C., Cocco, L. and Maraldi N.M. (1994a) *J. Biol. Chem.* **269**, 13-16.

Marmiroli, S., Faenza, I., Bavelloni, A., Santi, S., Cecchi, S., Squarzoni, S., Sood, R. and Ognibene, A. (1994b) *Molecular Mechanisms of Transcellular Signaling: 'From the Membrane to the Gene'* Abstract Book **40.**

Martelli, A.M., Gilmour, R.S., Bertagnolo, V., Neri, L.M., Manzoli, L. and Cocco, L. (1992) *Nature* **358**, 242-245

Mazzoni, M., Bertagnolo, V., Neri, L.M., Carini, C., Marchisio, M., Milani, D., Manzoli, F.A. and Capitani, S. (1992) *Biochem. Biophys. Res. Commun.* **187**, 114-120

Mazzotti, G., Zini, N., Rizzi, E., Rizzoli, R., Galanzi, A., Ognibene, A., Santi, S., Matteucci, A., Martelli, A.M. and Maraldi, M. (1995) *J. Histochem. Cytochem.*

Neri, L.M., Billi, A.M., Manzoli, L., Rubbini, S., Gilmour, R.S., Cocco, L. and Martelli A.M. (1994) *FEBS Lett.* **347**, 63-68.

Ognibene, A., Bavelloni, A., Faenza, I., Santi, S., Marmiroli, S. and Maraldi N.M. (1994) *Cell signalling: from membrane to nucleus.* Abstract Book **31**.

Rhee, S.G., Suh, P.H., Ryu, S.H. and Lee, K.Y. (1989) *Science* **244**, 546-550

Rodan, S.B., Wesolowski, G., Limjuco, G.A., Schimdt, J.A. and Rodan, G.A. (1990) *J. Immunol.* **145**, 1231-1237

Sims, J.E., March, C.J., Cossman, D., Widmer, M.B., MacDonald, H.R., McMahan, C.J., Grubin, C.E., Wignall, J.M., Jackson, J.L., Call, S.M., Friend, D., Alpert, A.R., Gillis, S., Urdal, D.L. and Dower, S.K. (1988) *Science* **241**, 585-589

Smith, C.D. and Wells, W.W. (1983) *J. Biol. Chem.* **258**, 9368-9373

Suh, P.G., Ryo, S.H., Won, C.C., Lee, K.Y. and Rhee, S.G. (1988) *J. Biol. Chem.* **263**, 14497-14504.

Zini, N., Martelli, A.M., Cocco, L., Manzoli, F.A. and Maraldi N.M. (1993) *Exp. Cell. Res.* 208, 257-269.

Zini, N., Mazzoni, M., Neri, L.M., Bavelloni, A., Marmiroli, S., Capitani, S. and Maraldi N.M. (1994) *J. Cell. Biol.* **65**, 206-213

INDEX

NATO ASI Series H

NATO ASI Series H

Vol. 21: **Cellular and Molecular Basis of Synaptic Transmission.** Edited by H. Zimmermann. 547 pages. 1988.

Vol. 22: **Neural Development and Regeneration. Cellular and Molecular Aspects.** Edited by A. Gorio, J. R. Perez-Polo, J. de Vellis, and B. Haber. 711 pages. 1988.

Vol. 23: **The Semiotics of Cellular Communication in the Immune System.** Edited by E.E. Sercarz, F. Celada, N.A. Mitchison, and T. Tada. 326 pages. 1988.

Vol. 24: **Bacteria, Complement and the Phagocytic Cell.** Edited by F. C. Cabello und C. Pruzzo. 372 pages. 1988.

Vol. 25: **Nicotinic Acetylcholine Receptors in the Nervous System.** Edited by F. Clementi, C. Gotti, and E. Sher. 424 pages. 1988.

Vol. 26: **Cell to Cell Signals in Mammalian Development.** Edited by S.W. de Laat, J.G. Bluemink, and C.L. Mummery. 322 pages. 1989.

Vol. 27: **Phytotoxins and Plant Pathogenesis.** Edited by A. Graniti, R. D. Durbin, and A. Ballio. 508 pages. 1989.

Vol. 28: **Vascular Wilt Diseases of Plants. Basic Studies and Control.** Edited by E. C. Tjamos and C. H. Beckman. 590 pages. 1989.

Vol. 29: **Receptors, Membrane Transport and Signal Transduction.** Edited by A. E. Evangelopoulos, J. P. Changeux, L. Packer, T. G. Sotiroudis, and K.W.A. Wirtz. 387 pages. 1989.

Vol. 30: **Effects of Mineral Dusts on Cells.** Edited by B.T. Mossman and R.O. Begin. 470 pages. 1989.

Vol. 31: **Neurobiology of the Inner Retina.** Edited by R. Weiler and N.N. Osborne. 529 pages. 1989.

Vol. 32: **Molecular Biology of Neuroreceptors and Ion Channels.** Edited by A. Maelicke. 675 pages. 1989.

Vol. 33: **Regulatory Mechanisms of Neuron to Vessel Communication in Brain.** Edited by F. Battaini, S. Govoni, M.S. Magnoni, and M. Trabucchi. 416 pages. 1989.

Vol. 34: **Vectors asTools for the Study of Normal and Abnormal Growth and Differentiation.** Edited by H. Lother, R. Dernick, and W. Ostertag. 477 pages. 1989.

Vol. 35: **Cell Separation in Plants: Physiology, Biochemistry and Molecular Biology.** Edited by D. J. Osborne and M. B. Jackson. 449 pages. 1989.

Vol. 36: **Signal Molecules in Plants and Plant-Microbe Interactions.** Edited by B.J.J. Lugtenberg. 425 pages. 1989.

Vol. 37: **Tin-Based Antitumour Drugs.** Edited by M. Gielen. 226 pages. 1990.

Vol. 38: **The Molecular Biology of Autoimmune Disease.** Edited by A.G. Demaine, J-P. Banga, and A.M. McGregor. 404 pages. 1990.

NATO ASI Series H

Vol. 39: Chemosensory Information Processing.
Edited by D. Schild. 403 pages. 1990.

Vol. 40: Dynamics and Biogenesis of Membranes.
Edited by J. A. F. Op den Kamp. 367 pages. 1990.

Vol. 41: Recognition and Response in Plant-Virus Interactions.
Edited by R. S. S. Fraser. 467 pages. 1990.

Vol. 42: Biomechanics of Active Movement and Deformation of Cells.
Edited by N. Akkas. 524 pages. 1990.

Vol. 43: Cellular and Molecular Biology of Myelination.
Edited by G. Jeserich, H. H. Althaus, and T. V. Waehneldt. 565 pages. 1990.

Vol. 44: Activation and Desensitization of Transducing Pathways.
Edited by T. M. Konijn, M. D. Houslay, and P. J. M. Van Haastert. 336 pages. 1990.

Vol. 45: Mechanism of Fertilization: Plants to Humans.
Edited by B. Dale. 710 pages. 1990.

Vol .46: Parallels in Cell to Cell Junctions in Plants and Animals.
Edited by A. W Robards, W. J . Lucas, J . D. Pitts, H . J . Jongsma, and D. C. Spray. 296 pages. 1990.

Vol. 47: Signal Perception and Transduction in Higher Plants.
Edited by R. Ranjeva and A. M. Boudet. 357 pages. 1990.

Vol. 48: Calcium Transport and Intracellular Calcium Homeostasis.
Edited by D. Pansu and F. Bronner. 456 pages. 1990.

Vol. 49: Post-Transcriptional Control of Gene Expression.
Edited by J. E. G. McCarthy and M. F. Tuite. 671 pages. 1990.

Vol. 50: Phytochrome Properties and Biological Action.
Edited by B. Thomas and C. B. Johnson. 337 pages. 1991.

Vol. 51: Cell to Cell Signals in Plants and Animals.
Edited by V. Neuhoff and J. Friend. 404 pages. 1991.

Vol. 52: Biological Signal Transduction.
Edited by E. M . Ross and K . W. A. Wirtz. 560 pages. 1991.

Vol. 53: Fungal Cell Wall and Immune Response.
Edited by J. P. Latge and D. Boucias. 472 pages. 1991.

Vol. 54: The Early Effects of Radiation on DNA.
Edited by E. M. Fielden and P. O'Neill. 448 pages. 1991.

Vol. 55: The Translational Apparatus of Photosynthetic Organelles.
Edited by R. Mache, E. Stutz, and A. R. Subramanian. 260 pages. 1991.

Vol. 56: Cellular Regulation by Protein Phosphorylation.
Edited by L. M. G. Heilmeyer, Jr. 520 pages. 1991.

NATO ASI Series H

Vol. 57: Molecular Techniques in Taxonomy.
Edited by G . M . Hewitt, A. W. B. Johnston, and J. P. W. Young .
420 pages. 1991.

Vol. 58: Neurocytochemical Methods.
Edited by A. Calas and D. Eugene. 352 pages. 1991.

Vol. 59: Molecular Evolution of the Major Histocompatibility Complex.
Edited by J. Klein and D. Klein. 522 pages. 1991.

Vol. 60: Intracellular Regulation of Ion Channels.
Edited by M. Morad and Z. Agus. 261 pages. 1992.

Vol. 61: Prader-Willi Syndrome and Other Chromosome 15q Deletion Disorders.
Edited by S. B. Cassidy. 277 pages. 1992.

Vol. 62: Endocytosis. From Cell Biology to Health, Disease and Therapie.
Edited by P. J. Courtoy. 547 pages. 1992.

Vol. 63: Dynamics of Membrane Assembly.
Edited by J. A. F. Op den Kamp. 402 pages. 1992.

Vol. 64: Mechanics of Swelling. From Clays to Living Cells and Tissues.
Edited by T. K. Karalis. 802 pages. 1992.

Vol. 65: Bacteriocins, Microcins and Lantibiotics.
Edited by R. James, C. Lazdunski, and F. Pattus. 530 pages. 1992.

Vol. 66: Theoretical and Experimental Insights into Immunology.
Edited by A. S. Perelson and G. Weisbuch. 497 pages. 1992.

Vol. 67: Flow Cytometry. New Developments.
Edited by A. Jacquemin-Sablon. 1993.

Vol. 68: Biomarkers. Research and Application in the Assessment of Environmental Health. Edited by D. B. Peakall and L. R. Shugart.
138 pages. 1993.

Vol. 69: Molecular Biology and its Application to Medical Mycology.
Edited by B. Maresca, G. S. Kobayashi, and H. Yamaguchi. 271 pages. 1993.

Vol. 70: Phospholipids and Signal Transmission.
Edited by R. Massarelli, L. A. Horrocks, J. N. Kanfer, and K. Löffelholz.
448 pages. 1993.

Vol. 71: Protein Synthesis and Targeting in Yeast.
Edited by A. J. P. Brown, M. F. Tuite, and J. E. G. McCarthy. 425 pages. 1993.

Vol. 72: Chromosome Segregation and Aneuploidy.
Edited by B. K. Vig. 425 pages. 1993.

Vol. 73: Human Apolipoprotein Mutants III. In Diagnosis and Treatment.
Edited by C. R. Sirtori, G. Franceschini, B. H. Brewer Jr. 302 pages. 1993.

NATO ASI Series H

Vol. 74: **Molecular Mechanisms of Membrane Traffic.**
Edited by D. J. Morré, K. E. Howell, and J. J. M. Bergeron. 429 pages. 1993.

Vol. 75: **Cancer Therapy. Differentiation, Immunomodulation and Angiogenesis.**
Edited by N. D'Alessandro, E. Mihich, L. Rausa, H. Tapiero, and T. R. Tritton. 299 pages. 1993.

Vol. 76: **Tyrosine Phosphorylation/Dephosphorylation and Downstream Signalling.**
Edited by L. M. G. Heilmeyer Jr. 388 pages. 1993.

Vol. 77: **Ataxia-Telangiectasia.** Edited by R. A. Gatti, R. B. Painter. 306 pages. 1993.

Vol. 78: **Toxoplasmosis.** Edited by J. E. Smith. 272 pages. 1993.

Vol. 79: **Cellular Mechanisms of Sensory Processing. The Somatosensory System.**
Edited by L. Urban. 514 pages. 1994.

Vol. 80: **Autoimmunity: Experimental Aspects.**
Edited by M. Zouali. 318 pages. 1994.

Vol. 81: **Plant Molecular Biology. Molecular Genetic Analysis of Plant Development and Metabolism.**
Edited by G. Coruzzi, P. Puigdomènech. 579 pages. 1994.

Vol. 82: **Biological Membranes: Structure, Biogenesis and Dynamics.**
Edited by Jos A. F. Op den Kamp. 367 pages. 1994.

Vol. 83: **Molecular Biology of Mitochondrial Transport Systems.**
Edited by M. Forte, M. Colombini. 420 pages. 1994.

Vol. 84: **Biomechanics of Active Movement and Division of Cells.**
Edited by N. Akkaş. 587 pages. 1994.

Vol. 85: **Cellular and Molecular Effects of Mineral and Synthetic Dusts and Fibres.**
Edited by J. M. G. Davis, M.-C. Jaurand. 448 pages. 1994.

Vol. 86: **Biochemical and Cellular Mechanisms of Stress Tolerance in Plants.**
Edited by J. H. Cherry. 616 pages. 1994.

Vol. 87: **NMR of Biological Macromolecules.**
Edited by C. I. Stassinopoulou. 616 pages. 1994.

Vol. 88: **Liver Carcinogenesis. The Molecular Pathways.**
Edited by G. G. Skouteris. 502 pages. 1994.

Vol. 89: **Molecular and Cellular Mechanisms of H^+ Transport.**
Edited by B. H. Hirst. 504 pages. 1994.

Vol. 90: **Molecular Aspects of Oxidative Drug Metabolizing Enzymes: Their Significance in Environmental Toxicology, Chemical Carcinogenesis and Health.**
Edited by E. Arınç, J. B. Schenkman, E. Hodgson. 623 pages. 1994.

NATO ASI Series H

Vol. 91: **Trafficking of Intracellular Membranes: From Molecular Sorting to Membrane Fusion.** Edited by M. C. Pedroso de Lima, N. Düzgüneş, D. Hoekstra. 371 pages. 1995.

Vol. 92: **Signalling Mechanisms – from Transcription Factors to Oxidative Stress.** Edited by L. Packer, K. Wirtz. 466 pages. 1995.

MIX
Papier aus verantwortungsvollen Quellen
Paper from responsible sources
FSC® C105338

If you have any concerns about our products,
you can contact us on
ProductSafety@springernature.com

In case Publisher is established outside the EU,
the EU authorized representative is:
Springer Nature Customer Service Center GmbH
Europaplatz 3, 69115 Heidelberg, Germany

Printed by Libri Plureos GmbH
in Hamburg, Germany